Electron Microscopy

Electron Microscopy
Principles and Techniques for Biologists

Second Edition

John J. Bozzola, M.S., Ph.D.
Center for Electron Microscopy
Southern Illinois University
Carbondale, Illinois

Lonnie D. Russell, M.S., Ph.D.
Laboratory of Structural Biology
School of Medicine
Southern Illinois University
Carbondale, Illinois

JONES AND BARTLETT PUBLISHERS
Sudbury, Massachusetts
BOSTON   TORONTO   LONDON   SINGAPORE

World Headquarters
Jones and Bartlett Publishers
40 Tall Pine Drive
Sudbury, MA 01776
978-443-5000
info@jbpub.com
www.jbpub.com

Jones and Bartlett Publishers Canada
P.O. Box 19020
Toronto, ON M5S 1X1
CANADA

Jones and Bartlett Publishers International
Barb House, Barb Mews
London W6 7PA
UK

Chief Executive Officer: Clayton Jones
Chief Operating Officer: Don Jones, Jr.
Publisher: Tom Walker
V. P., Sales and Marketing: Tom Manning
Marketing Director: Rich Pirozzi
Production Director: Anne Spencer
Production Editor: Rebecca S. Marks
Manufacturing Director: Therese Bräuer
Executive Editor: Brian L. McKean
Design: Carlisle Publishers Services
Editorial Production Services: Carlisle Publishers Services
Typesetting: Carlisle Communications, Ltd.
Cover Design: Anne Spencer
Printing and Binding: Quebecor Printing/Kingsport
Cover Printing: John Pow Company

Library of Congress Cataloging-in-Publication Data

Bozzola, John J.
    Electron microscopy : principles and techniques for biologists /
John J. Bozzola, Lonnie D. Russell. — [2nd ed.]
        p.   cm.
    Includes bibliographical references and index.
    ISBN 0-7637-0192-0 (hard)
    1. Electron microscopy.   I. Russell. Lonnie Dee.   II. Title.
QH212.E4B69   1998
570' .28 '25—DC21   98-40315
                                                    CIP

About the cover: Electron Microscopy, Second Edition, is the only book of its kind that covers all practical aspects of both scanning and transmission microscopy at an introductory level. The front cover illustrates a scanning electron microscope image of a moth's antenna. The section of antenna shown measures 2.5 mm in length. The specimen was air-dried, coated with 40 nm of palladium-gold, and examined at an accelerating voltage of 15kV, using a secondary electron detector. (Courtesy of S. J. Schmitt). The back cover image is an ultrathin section examined in the transmission electron microscope and shows densely packed flagellae of insect sperm (*Odontocerum albicore*). Each flagellum is about 0.4 µm in diameter. (Courtesy of R. Dallai).

Printed in the United States of America
02 01 00 99   10 9 8 7 6 5 4 3 2

To Our Parents

John, Sr.   *Lonnie, Sr.*

Angeline   *Jean*

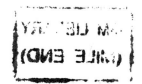

Brief Contents

Contents

Chapter 3

Specimen Preparation for Scanning Electron Microscopy   48

Chapter 4

Ultramicrotomy   72

Chapter 5

Specimen Staining and Contrast Methods for Transmission Electron Microscopy   120

Chapter 6

The Transmission Electron Microscope   148

Chapter 7

The Scanning Electron Microscope   202

Chapter 8

Production of the Electron Micrograph   240

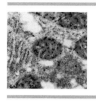

Chapter 9

Chapter 10

Chapter 11

Chapter 12

Miscellaneous Localization and Enhancement Techniques *310*

Chapter 13

Quantitative Electron Microscopy *320*

Chapter 14

Freeze Fracture Replication   342

Chapter 15

*Analytical Electron
Microscopy   368*

Chapter 16

*Intermediate and High Voltage
Microscopy   396*

Chapter 17

Tracers   406

Chapter 18

Image Processing and Image Analysis by Computer   414

Chapter 19

Interpretation of Micrographs   442

Chapter 20

*Survey of Biological
Ultrastructure   476*

Chapter 21

*Safety in the Electron
Microscope Laboratory*   *616*

Preface

THIS TEXTBOOK WAS WRITTEN out of a desire to teach electron microscopy to biologists in as simple a way as possible. It is an introductory textbook designed for those who are entering the field. In our own introductory electron microscopy courses, we have used several of the many excellent reference volumes available for electron microscopy. For beginners, however, the costs of several texts are prohibitive and the information overwhelming. Naturally, reference books tend to emphasize detail and theoretical aspects in place of the practical concerns that a beginner may have. Often, the various tissue preparation methodologies are given equal treatment, although there might be just a few methods that are used most of the time. It is our goal to provide a starting point for beginners and to direct the student toward any appropriate reference sources.

In writing this text we have emphasized mainstream methods in the hope that the forest will be seen and the trees ignored, at least for the time being. Our philosophy has been to introduce the topic and at the end of the chapter to provide references that deal more extensively with the topic. We have attempted to achieve a balance of theory and practical applications of electron microscopy. Our criteria for inclusion of a theory in the text was the down-the-road application of this theory which would either lead to an understanding of the major principles involved in the discipline or be a necessary prerequisite to perform biological electron microscopy. Thus, we may have deleted someone's favorite technique for the sake of an overall understanding of the topic.

It is generally a good principle that one should not enter into a venture, whether it be business or writing a textbook, with someone who has exactly the same skills that you do. We realized at the onset that our interests in electron microscopy are complementary. One of us is more interested in the theory, the techniques, and how the electron microscope can be used to achieve the desired result, while the other is interested primarily in the applications and the biological implications of the use of the electron microscope. From the start it was clear to both of us which chapters were our forté and how the material should be divided. In reading each other's chapters, we were able to offer a perspective from an orientation different than the writer. This helped to balance microscope theory with biological application.

We have not only described the workings of the electron microscope and the major techniques for biological specimen preparation but we have illustrated, in many instances, how these techniques have contributed to the biological sciences. Chapter 19, "Interpretation of Micrographs," serves as a tool for the novice to understand how various factors can influence what is seen in the micrograph. Chapter 20, "Survey of Biological Ultrastructure," serves as an introduction to a wide variety of animal and plant components for the novice, yet it emphasizes mammalian tissues where the bulk of biological electron microscopy is currently applied. Both chapters are profusely illustrated. In making the decision to write these chapters, we reasoned that knowledge of the microscope, microtome, and methods to prepare tissues was not sufficient to introduce a person to electron microscopy. Some experience in interpretation and examination of electron micrographs is also important. A novice can hardly begin to interpret a micrograph, whereas an investigator with considerable experience can glean a great deal from a micrograph. Individuals experienced in viewing micrographs and interpreting them take for granted the time spent in becoming familiar with electron micrographs.

It is the reader who will eventually determine if our philosophy in writing this text is a sound one. As students, the authors were of the opinion that everything that went into print must be faultless. As the years went by, this proved not to be a correct assumption. Authors have human frailties. It is in this vein that we, the authors, solicit your comments and suggestions on how to improve the text or to let us know what you find enjoyable about the text. We welcome exceptional micrographs for possible inclusion in subsequent editions of the text. If this book is instrumental in your career decisions or facilitates your research progress in electron microscopy, we would love to hear about it. That was our original intention.

Acknowledgments for the First Edition

We would like to express our thanks to the people who helped in so many ways in the production of this book. First of all, we thank our colleagues who shared with us their electron micrographs. Cindy Claybough, Karen Fiorino, Karen Schmitt, and Steve Mueller drew and helped improve upon most of the original line drawings and artwork. Dee Gates, Sushmita Ghosh, Cheri Kelly, Scott Pelok, Steve Schmitt, and Randall Tindall helped in the preparation of the illustrative electron micrographs and environmental photographs, and often served as models. John Richardson provided good advice on photographic considerations.

We would also like to acknowledge the many reviewers who critiqued the manuscript in its various stages of development. We benefited greatly from their long experience in teaching electron microscopy to both undergraduate and graduate students.

We are very grateful for the comments and suggestions of: Professor Richard F. E. Crang, University of Illinois at Urbana-Champaign; Professor William J. Dougherty, Medical University of Southern Carolina; Professor Laszlo Hanzely, Northern Illinois University; Professor Julian P. Heath, Baylor College of Medicine; Professor Harry T. Horner, Iowa State University; Dr. Morton D. Maser, Woods Hole Educational Associates; Dr. Judy A. Murphy, R. J. Lee Group; Professor Jerome J. Paulin, University of Georgia; Professor Lee D. Peachey, University of Pennsylvania; Professor David Prescott, University of Colorado; and Dr. Steve Schmitt, Southern Illinois University.

The editorial guidance offered by Joseph Burns, Paula Carroll and Judy Salvucci of Jones and Bartlett, and Joyce Jackson of Impressions was very much appreciated.

Finally, we thank our colleagues and supervisors in the Office of Research Development and Administration in the Graduate School, and the Physiology Department in the School of Medicine and College of Science for the freedom and understanding so essential to complete this project.

Acknowledgments for the Second Edition

Although it did not seem long ago that we completed the first edition of this textbook, we were very pleased when Jones and Bartlett requested a revision of the book for a second edition. Over the last six years we received valuable suggestions and micrographs for inclusion in this latest edition. We were particularly touched to hear from instructors and students who were using the book in various courses all over the world. Individually, we have spoken and lectured in numerous locations, often using selected chapters for short courses. It was flattering to both of us to hear comments about how pertinent and useful the book had been for others commencing in electron microscopy. We have incorporated nearly all of these suggestions in this second edition, and we hope that your input will be recognizable.

This revision has involved more people than did the first edition, and we can never adequately acknowledge the degree to which each person helped out. We would like to thank the reviewers of the second edition for a very thorough reading and scrutiny of the chapters. Judy Murphy of the electron microscopy program at San Joaquin Delta College in Stockton, California, spent many hours and many red ink pens on all of the chapters. Audrey Glauert, editor of the highly successful series, *Techniques in Electron Microscopy,* voluntarily critiqued most of the chapters and helped to improve all that she examined. Caroline Schooley (who kindly forwarded our book to Audrey Glauert) also added to the review. John Russ, author of *The Image Processing Handbook,* analyzed and improved our new chapter on digital imaging and image analysis. Many scientists and even some students contributed micrographs. Dee Gates, Steve Schmitt, and Randy Tindall again helped with the second edition and patiently endured the pressures to help us meet production deadlines.

The expert editing and layout by Kate Scheinman and Liz Ney of Carlisle Communications, Ltd. helped to improve the appearance and readability of this edition. Rebecca Marks and Brian McKean of Jones and Bartlett Publishers greatly expedited the revision and helped keep our efforts moving along in a very pleasant way.

Some of you may be aware of the disaster that occurred when our final manuscript (including nearly all of the original micrographs and artwork—some of which was hand drawn) was completely destroyed by a fire on an airplane en route to the publisher. We required over a year and a half to reconstruct the images and artwork, and we can never adequately thank all the contributors who kindly reprinted their figures. We hope that our Phoenix-like book, newly risen from the ashes, will be a useful guide for those of you embarking on the adventure we know as electron microscopy.

Electron Microscopy

Chapter 1

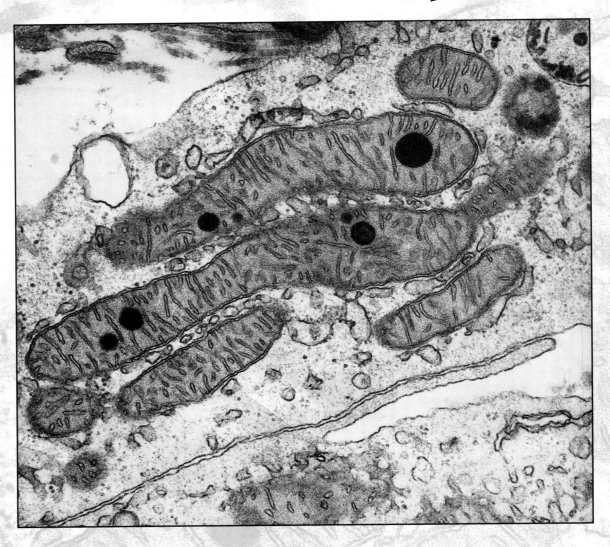

The Past, Present, and Future of Electron Microscopy

RARELY ARE NEW VISTAS OPENED to the human eye. The undersea world and perhaps air flight, outer space, or the moon have provided us with fascinating visual encyclopedias. The invention of the compound *light microscope* by the Janssens, manufacturers of eyeglasses, in 1590 opened the door to the microscopic world. Their microscope magnified objects up to 20 to 30 times their original size. In the next century, Antonie van Leeuwenhoek developed a simple (one lens) microscope, which represented a tremendous improvement in lens fabrication and permitted magnifications up to 300 times. By the beginning of the 20th century, objects could be magnified up to 1,000 times their original size, and particles that were only 0.2 μm apart could be distinguished from one another. Biological materials revealed a substructure as complex, variable, and dynamic as then could be imagined. In the 1930s another vista, as exciting as any of the rest, began to open: the submicroscopic world as viewed by the *electron microscope* (Figures 1.1, 1.2,

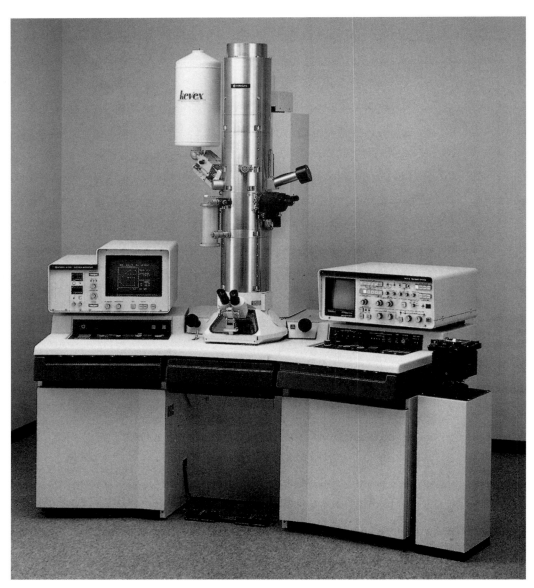

FIGURE 1.1 Modern electron microscope.
(Courtesy of Hitachi Scientific Instruments.)

FIGURE 1.2 This early electron microscope, known as the "Toronto Microscope," was operating in 1939 and was the first in North America.
(From a March 1940 issue of the New York Times, *courtesy Ladd Research Industries.)*

FIGURE 1.3 This electron microscope, built by W. Ladd, was used during World War II to examine rubber from German tires. This instrument helped the Allies build a tire that would compete with the much superior German tire. The instrument is now in the museum at the Armed Forces Institute of Pathology, Washington, D.C.
(Courtesy of Ladd Research Industries.)

and 1.3). The electron microscope took advantage of the much shorter wavelength of the electron. With the electron microscope, another thousandfold increase in magnification was made possible, accompanied by a parallel increase in resolution capability allowing biologists to both define and expand the world of light microscopy (Figure 1.4). Viruses, DNA, and many smaller organelles were visualized for the first time.

The electron microscope has profoundly influenced our understanding of tissue organization and especially the cell. It has given us the capabil-ity to visualize molecules (Figure 1.5) and even the atom.

Electron microscopy is defined as a specialized field of science that employs the electron micro-scope as a tool. Numerous techniques for tissue preparation and tissue analysis are included under the broad heading of electron microscopy. Although major contributions have been made utilizing the electron microscope, it is only one of many tools used to solve biological problems. *Electron microscopists* are those who use electron microscopes.

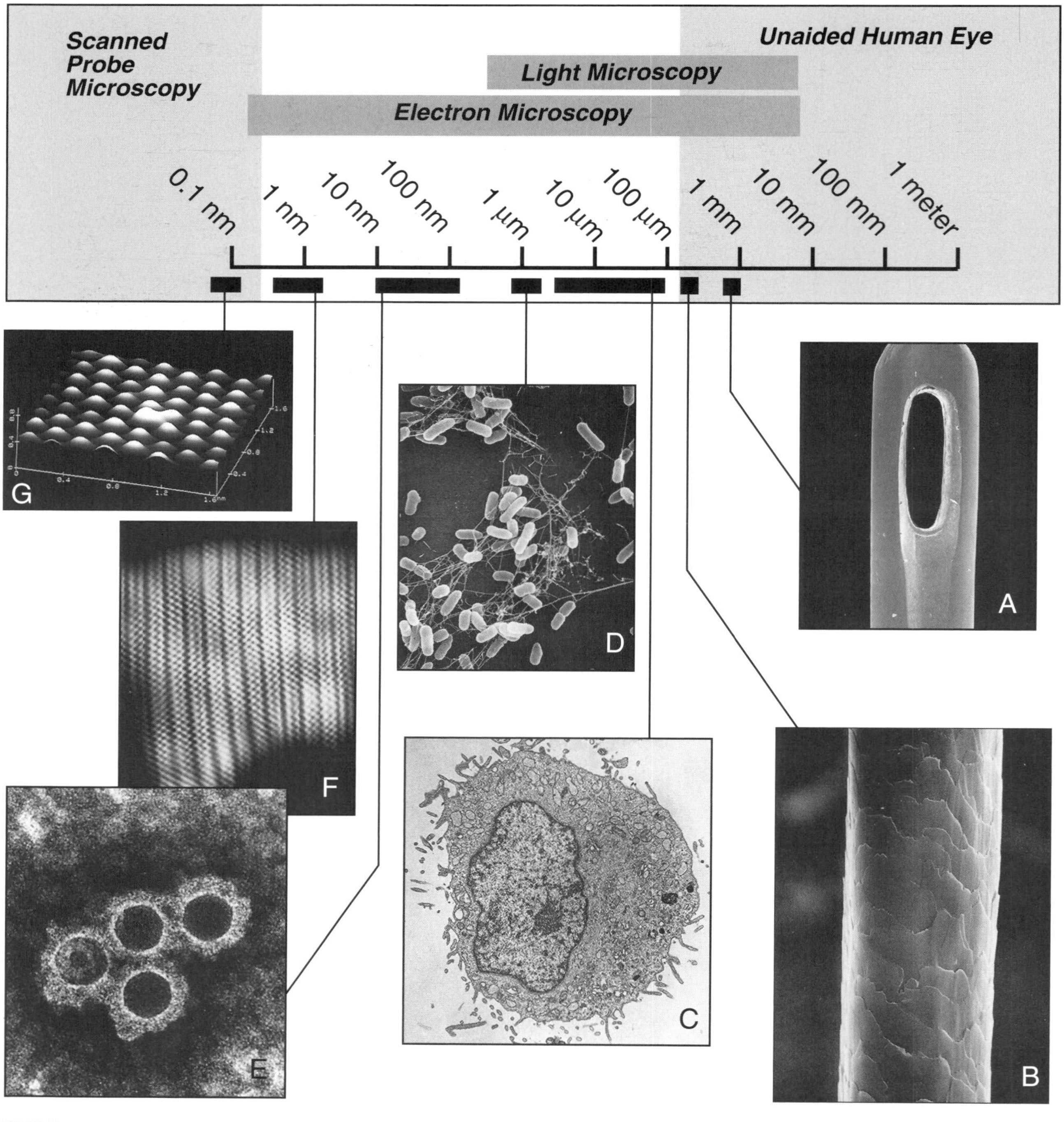

FIGURE 1.4 Range of resolution of various magnifying tools (*top*), the structures they are capable of resolving (*bottom, images*) and the size range of the structures (*middle, scale*). (A) Sewing needle viewed by scanning electron microscope (SEM), (B) human hair by SEM, (C) thin section of mammalian cell by transmission electron microscopy (TEM), (D) bacterial cells by SEM, (E) clump of four herpes viruses by TEM negative staining, (F) enzyme macromolecules by TEM negative followed by image processing. (*Courtesy of P. M. Hardwicke*) (G) scanned probe image of single atom defect (*center*) in graphite demonstrating the resolution of subtle atomic structures. (*Fig. 1.4G Courtesy of Digital Instruments, Inc.)*

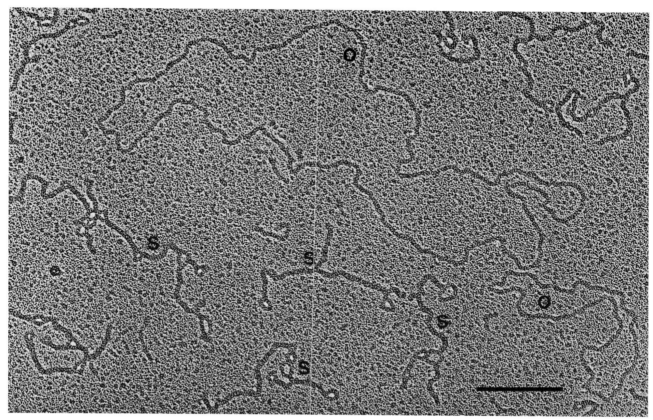

FIGURE 1.5 Micrograph of isolated DNA molecules that have been stained with uranyl acetate and shadowed with a thin coat of platinum. The micrograph shows open circular (O) and supercoiled (S) forms of DNA from plasmids. Bar = 0.25 μm.
(Courtesy of L. Coggins.)

Historical Perspective

It took 300 years to perfect the basic light microscope, but less than 40 years to refine the electron microscope. The world opened by electron microscopy has been an extremely exciting visual experience for biologists; however, it now is clear that the *descriptive age* of biological electron microscopy has passed. Beginning in the l940s and continuing well into the 1970s, descriptive biological electron microscopy thrived. Investigators were preoccupied with the discovery of the cell components. Discoveries were often so frequent and revolutionary that investigators continually questioned whether their findings accurately represented cell structure. Most findings proved to be real representations of cell structure. Some did not. But the period of questioning was essential in order to establish this new science as a credible one.

Ernst Ruska and Max Knoll are credited with developing the first electron microscope. Ruska's efforts were recognized in 1986 with the Nobel Prize in Physics. Shortly after the invention of the electron microscope, many individuals were occupied primarily with developing the instrument. Fascination with developing the instrument itself gradually gave way to using the instrument as a tool to answer biological questions, as biologists became increasingly interested in the applications of electron microscopes. A now classic paper by Porter and colleagues (1945) clearly demonstrated early on that the electron microscope could be used to study cells in detail.

The kinds of questions addressed by electron microscopists have matured. Initially, descriptive microscopy prevailed, but with time gave way to *experimental* and *analytical* approaches. The field of *cell biology,* heavily dependent on electron microscopy for its origins and early growth, has now matured into a multidisciplinary field that utilizes electron microscopy as one of several tools.

Many of the biologists who pioneered biological electron microscopy are still living. Some of these pioneers include Albert Claude, Don Fawcett, Earnest Fullam, Audrey Glauert, Hugh Huxley, Bob Horne, Charles Leblond, John Luft, George Palade, Daniel Pease, and Fritiof Sjostrand. Albert Claude and George Palade were awarded the 1974 Nobel Prize in Medicine for their accomplishments in cell biology employing electron microscopy. For accomplishments of a similar nature, Keith Porter received the Presidential Medal. It is still possible to hear firsthand about the excitement in the early days at hotbeds of electron microscopy such as the Rockefeller Institute in New York and the Cavendish Institute at Cambridge England.

Biological electron microscopists are, by and large, *anatomists* and *cell biologists.* However, individuals in many disciplines have come to use electron microscopy as a tool.

The techniques of basic biological electron microscopy were, at first, in the hands of a few investigators. Methods for preparing and sectioning tissue varied greatly among laboratories. Now, although minor variations exist, most techniques are standardized, freely available to everyone, and may be found in reference books. In spite of the numerous specialized texts on methodology, noth-

ing has replaced or is soon to replace practical, hands-on training in the laboratory with the guidance of knowledgeable, skilled personnel.

Electron microscope technologists have specialized training to carry out the day-to-day operations of the laboratory (Figure 1.6). In most instances, they have experience in biological aspects of electron microscopy, and many may have limited knowledge of the microscope's workings, which enables them to repair this equipment and its electrical components. Although technical personnel can handle more routine maintenance, most labs now have service contracts for outside specialists to perform periodic maintenance or make emergency service calls.

Electron microscopy in many universities and industrial settings has often become *centralized.* Centralization allows for many users of the same equipment and for a higher level of training of personnel as well as increases the breadth of the services that can be offered. The individual investigator then has time to concentrate on the biological problems rather than the technical problems of tissue preparation and equipment maintenance.

Electron microscopy has crossed disciplines to the degree that no single discipline can claim ownership of this tool. Anatomy, biochemistry, botany, cell biology, forensic medicine, microbiol-

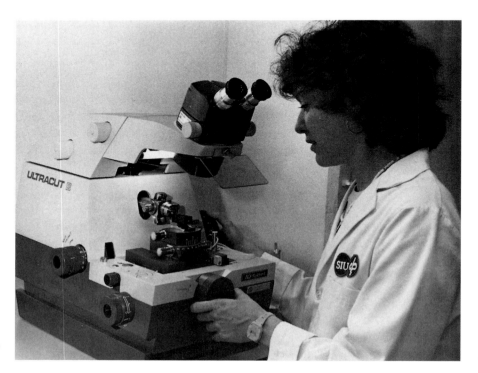

FIGURE 1.6 An electron microscope technician at work in a modern laboratory.

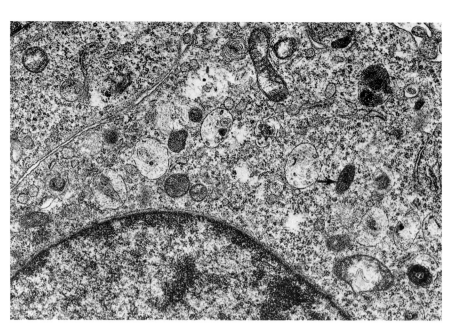

FIGURE 1.7 This micrograph is an example of the use of electron microscopy in diagnostic pathology. The arrow shows a type I melanosome in a human tumor cell, a structure that can only be resolved by electron microscopy. This melanosome is called a premelanosome because it has not produced melanin. If it had it would appear very dark. Thus, tumor cells at this immature, poorly differentiated state are more malignant and must be treated more aggressively. *(Micrograph courtesy of J. Harb.)*

ogy, pathology (Figure 1.7), physiology, and toxicology are biological and biomedical fields that rely heavily on the electron microscope. To some degree, virtually all biologically related journals publish electron micrographs. Some journals are devoted almost exclusively to papers on fine structure, while others emphasize technical advances in the field. The *International Federation of Societies of Electron Microscopy* and the *Microscopy Society of America (MSA)* meets yearly and serves as a format for the presentation of scientific papers and the introduction of commercial products. Many other countries have national microscopy organizations as well.

Development of the Electron Microscope

Two basic types of instruments are called electron microscopes (Figure 1.8). Both were invented at about the same time, but they have fundamentally different uses. The *transmission electron microscope* (TEM) projects electrons through a very thin slice of tissue (specimen) to produce a two-dimensional image on a phosphorescent screen. The brightness of a particular area of the image is proportional to the number of electrons that are transmitted through the specimen. The *scanning electron micro-*

scope (SEM) produces an image that gives the impression of three dimensions. This microscope uses a 2 to 3 nm spot of electrons that scans the surface of the specimen to generate secondary electrons from the specimen that are then detected by a sensor. The image is produced over time as the entire specimen is scanned. Figure 1.9A and B shows examples of SEM and TEM images of the same organism. A third, less used type of electron microscope, the *scanning transmission electron microscope* (STEM) has features of both the transmission and scanning electron microscopes. The STEM uses a scanning beam of electrons to penetrate thin specimens and determine the presence and distribution of the atomic elements in the specimen (see Chapter 15). Thus, its primary use is analytical.

Naturally, the development of the microscope preceded the development of techniques to be used with the microscope. The first electron microscopists, being engineers and physicists, were generally inexperienced in biological applications. In fact, the first microscope had no place for a specimen. Next, biological applications were sought, and the instrument was modified accordingly. However, even with capable instruments, major advances in biological applications did not happen immediately, but awaited the development of tissue preparation techniques, which were inadequate until the early 1950s.

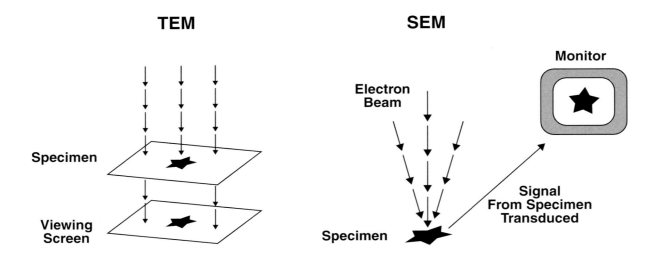

Two Basic Types of Microscopes

FIGURE 1.8   These are the basic differences between the transmission electron (TEM) and scanning electron (SEM) microscopes.

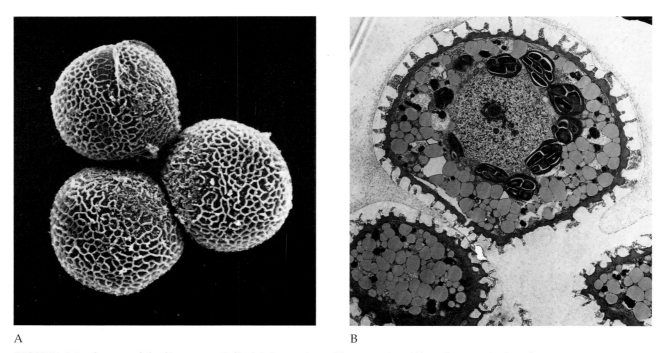

A   B

FIGURE 1.9   Spores of the liverwort, *Pallavicinia*, as viewed by scanning (A) and transmission electron microscopy (B). Note the features in the SEM that correlate the information revealed by the TEM. The SEM primarily reveals surface features while the TEM reveals internal details.
(Courtesy of K. Renzaglia.)

Historical Milestones in the Development of the Electron Microscope

1873	Abbe and Helmholtz independently showed that resolution depends on the wavelength of the energy source. This finding provided the theoretical promise of developing an electron microscope although, at this time, electrons had not been discovered.
1897	Thompson described negatively charged particles that were later termed electrons.
1924	De Broglie (1929 Nobel Laureate in Physics) demonstrated that electrons have properties of waves.
1926	Busch demonstrated that the path of electrons could be deflected by magnetic lenses in the same way that light could be defected by an optical lens.
1932	Knoll and Ruska developed the first electron microscope in Germany.
1937	Metropolitan Vickers was the first commercial enterprise to develop a prototype of an electron microscope.
1938	von Ardenne constructed the first scanning electron microscope.
1938–39	The Siemens Corporation introduced the first commercial transmission electron microscope.
1940–41	The RCA Corporation sold the first commercial transmission electron microscope in the United States. The subsequent generations of RCA microscopes proved invaluable to advances in microscopy in North America.
1941–63	Continued improvements were made in both resolution and convenient use of the transmission electron microscope, such that 0.2 to 0.3 nm resolutions were achieved. Improvements were made in the scanning electron microscope to achieve a resolution of about 10 nm.
1954	The Siemens Elmscope I electron microscope was introduced, an extremely popular and useful instrument that provided excellent transmission results in the hands of biologists.
1958	The Stereoscan, a scanning electron microscope readily used by biologists, was introduced by Cambridge Instruments.

1982	The scanning tunneling electron microscope was developed by H. Rohrer and G. Binnig (1986 Nobel Prize Laureates).
1990	Commercial transmission electron microscopes currently marketed: Hitachi (Japan), Topcon (Korea), JEOL (Japan), Philips (Holland), LEO (Germany). Commercial scanning electron microscopes currently marketed: AmRay (United States), Leica (United Kingdom), VG (United Kingdom), Hitachi (Japan), Topcon (Korea), JEOL (Japan), Philips (Holland), RJ Lee (United States), Leo (Germany).

Development of Preparative Techniques

About fifteen years after the development of the electron microscope by Knoll and Ruska, the first serious efforts were made to apply this technology to biological problems. Significant historical developments in tissue preparation techniques follow:

1934	Marton published the first electron micrograph of biological tissue. The image was inferior to conventional light microscope images (Figure 1.10).
1947–48	Claude popularized osmium fixation, a crude ultramicrotome and naphthaline embedding.
1949	Methacrylate was introduced as an embedding medium.
1950	Latta and Hartmann developed glass knives.
1952	Palade employed a buffering system to fix tissue in osmium tetroxide. Rapid progress in biological observation had its beginnings in the early 1950s.
1953	Porter and Blum introduced the first widely used microtome (Sorvall MT-1). Fernandez-Moran first used diamond knives to make ultrathin sections.
1956	Epoxy resin, first used by Glauert as an embedding medium, was refined by Luft. Luft also introduced potassium permanganate fixation. Palade and Siekevitz used the electron microscope to analyze cell fractions.
1957–63	Freeze fracture was developed.
1958	Watson introduced staining with heavy metals (lead and uranium).

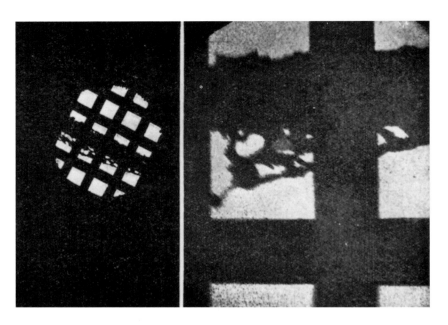

FIGURE 1.10 The first electron micrographs of a biological specimen were made by Dr. L. Marton in 1934. Plant tissue (*Drosera intermedia* of the sundew family) has been impregnated with osmium and is seen lying on a grid. The figure on the right is a magnified image of the one on the left. (*From Marton's original article in the Belgium Royal Academy of Science, 1934. Used with permission of the publisher.*)

1959   Horne introduced negative staining.
1961   Epon (Shell Oil Co.) was introduced as an epoxy embedding medium. Modern embedding media are refinements of Epon.
1963   Sabatini and coworkers introduced glutaraldehyde as a primary fixative.

Contributions to Biology and the Future of Electron Microscopy

Electron microscopy has contributed immensely to the field of biology. Just as we take for granted the basic descriptions that were important to the field of biology, it is easy to forget that virtually all organelles and cell inclusions were either discovered or resolved in finer detail using the electron microscope. Such descriptions have laid the foundation for experimental manipulations directed at unraveling cell function and understanding how cellular structure varies in normal, experimental, and diseased states. Electron microscopy is used not only to visualize biological materials, but also to analyze the chemical makeup and physical properties of biological materials (*analytical electron microscopy*).

The electron microscope has revolutionized our concept of cell structure and of how cells interact. Today's electron micrographs (Figure 1.11) are superior to the early images obtained. The techniques that apply to this instrument have multiplied and are undergoing refinement. Thousands of microscopes are used for biological applications in a variety of disciplines throughout the world.

As important as the development of the electron microscope itself are the various specialized techniques that are associated with electron microscopy and that have been developed to prepare and evaluate tissue. These include *autoradiography, freeze fracture, immunocytochemistry, electron diffraction, cryoelectron microscopy, elemental analysis, quantitative microscopy, high voltage microscopy,* all of which are covered in more detail in subsequent chapters. These techniques, along with experimental approaches, have largely replaced descriptive studies. Specialized techniques will undoubtedly prove worthwhile for a number of years to come.

The future of electron microscopy is bright. Computer-assisted imaging, image enhancement, and image storage are beyond the developmental stages and are now being viewed as essential accessories to the microscope. Automated quantitation, also in final developmental stages, will allow a more objective assessment of biological samples. It is a future goal to have the routine capability of imaging *living* systems at high resolution! Electron microscopy is expected to continue to meet the submicroscopic imaging needs of science and medicine.

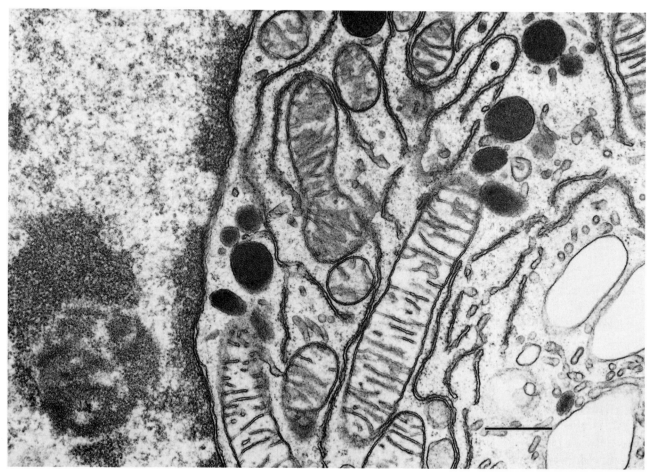

FIGURE 1.11   A recently taken electron micrograph showing the clarity of detail obtained by state-of-the-art microscopes and preparative techniques. A macrophage from mouse tissue. Bar = 0.5 µm.

Journals Devoted Primarily to Electron Microscopy

Micron (formerly *Electron Microscopy Reviews;* (Pergamon Press)

Journal of Electron Microscopy (Japanese)

Cell Vision: Journal of Analytical Morphology

Journal of Microscopy (English)

Biology of the Cell (formerly *Journal de Microscopie* or *Biologie Cellulaire;* French)

Journal of Ultrastructural Pathology

Microscopy Society of America

Scanning Microscopy

Ultramicroscopy

Selected Journals Publishing Electron Micrographs

Developmental Dynamics (formerly *American Journal of Anatomy*)

Anatomical Record

Cell and Tissue Research

Journal of Anatomy

Journal of Cell Biology

Journal of Histochemistry and Cytochemistry

Tissue and Cell

Microscopy and Microanalysis

Journal of Structural Biology (formerly *Journal of Ultrastructure and Molecular Structure Research* and *Journal of Ultrastructural Research*)

Ultrastructural Pathology

REFERENCES

Classic References

Aihara, K. and T. M. Mukherjee. 1993. Role of electron microscopy in global world health. *Microscopy Society of America Bulletin* 23 (4) (series of articles on the use of electron microscopy in pathology).

Brenner, S. and R. W. Horne. 1959. A negative staining method for high resolution electron microscopy of viruses. *Biochem Biophys Acta* 34:103–110.

Busch, H. 1926. Berechnung. der bahn von kathodenstrahlen in axialsymmetrischen electromagnitischen felde. *Ann Physik* 81:974–993.

De Broglie, L. 1924. Researches sur la theorie des quanta. Thesis, Paris: Masson and Cie. Also 1925. *Ann de Physique* 3:22–128.

Glauert, A. M., G. E. Rogers, and R. H. Glauert. 1956. A new embedding medium for electron microscopy. *Nature* 178: 803.

Knoll, M., and E. Ruska. 1932. Beitrag zur geometrischen elektronoptik. *Ann Physik* 12:607–61.

Latta, H., and J. F. Hartmann. 1950. Use of a glass edge in thin sectioning for electron microscopy. *Proc Soc Biol Med* 74:436–39.

Marton, L. 1934. La microscopie electronique des objects biologiques. *Bull Classe Sci Acad Roy Belg*, Series 5, 20:439–46.

Palade, G. E. 1952. Study of fixation for electron microscopy. *J Exptl Med* 95:285–98.

Palade, G. E., and P. Siekevitz. 1956. Liver microsomes: An integrated morphological and biochemical study. *J Biophys Biochem Cytol* 2:171–98.

Porter, K. R., and J. Blum. l953. A study of microtomy for electron microscopy. *Anat Rec* 117:685–710.

Porter, K. R., et al. 1945. A study of tissue culture cells by electron microscopy. *J Exptl Med* 81:233–46.

Sabatini, D. D., et al. 1963. Cytochemistry and electron microscopy. The preservation of cellular ultrastructure and enzyme activity by aldehyde fixation. *J Cell Biol* 17:19–57.

Watson, M. L. 1958. Staining of tissue sections for electron microscopy with heavy metals. II Application of solutions containing lead and barium. *J Biophys Biochem Cytol* 4:727–29.

Selected Historical References

Hawkes, P. W. 1985. The beginnings of electron microscopy. In: *Advances in electronics and electron physics*, Suppl 16, P. W. Hawkes, ed., Orlando: Academic Press.

Marton, L. l968. *Early history of the electron microscope.* San Francisco: San Francisco Press Inc., pp. 1–56.

Pfefferkorn, G. E. 1984. The early days of electron microscopy. *Scanning Electron Microscopy,* SEM Inc. 1:1–8.

Preven, D. R., and J. D. Gruhn. 1985. The development of electron microscopy. *Arch of Pathol Lab Med* 109:683–91.

Chapter 2

Specimen Preparation for Transmission Electron Microscopy

IN PREPARING SPECIMENS for transmission electron microscopy, virtually every step of the procedure affects the quality of the final electron micrographs. Therefore, it is important to process tissue according to prescribed methods and to understand what is happening to the sample in processing. Specimen processing should begin with careful planning and proceed with meticulous attention to detail. Since tissue processing represents considerable effort, it is better to execute it properly the first time rather than to repeat the entire procedure.

Tissue preparation for transmission electron microscopy may be divided into eight major steps: *primary fixation, washing, secondary fixation, dehydration, infiltration with transitional solvents, infiltration with resin, embedding, and curing.* (See Figure 2.1.)

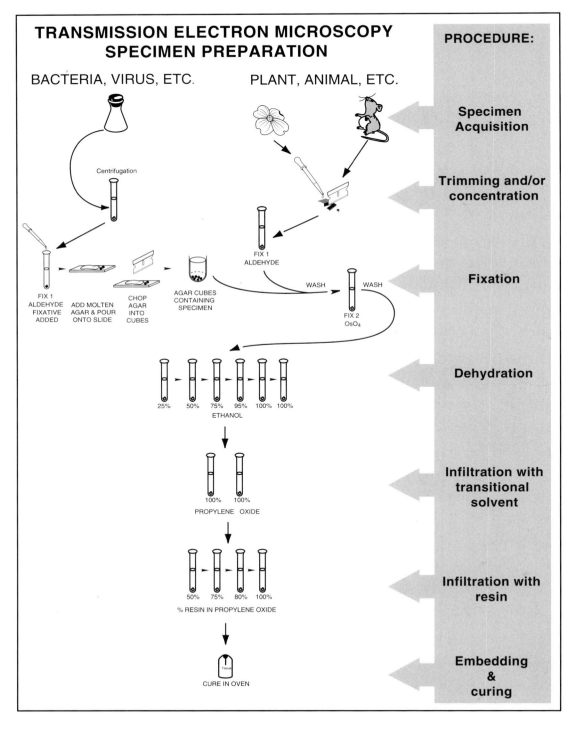

FIGURE 2.1
Steps in the preparation of biological specimens for transmission electron microscopy.

The process begins with living hydrated tissue and ends with tissue that is virtually water-free and preserved in a static state within a plastic resin matrix. The plastic resin mixture permeates the tissue, replacing all water within the cell and making the cell firm enough for sections to be cut.

Perhaps the least forgiving of all the steps is the tissue processing that occurs prior to sectioning. A poorly prepared specimen is useless to the investigator, whereas misadventures during the sectioning or staining process are somewhat more easily remedied. More sections can be cut and stained without a great amount of wasted time. Once tissue is processed in an acceptable manner and properly embedded in plastic, one may feel secure that the intended result should be obtainable.

There are many ways to prepare tissue for electron microscopy. This chapter will emphasize *mainstream* protocols. Less used variations on the general protocol are numerous and have been catalogued in reference books (Glauert, 1975; Hayat, 1981). As an overview of what follows, Table 2.1 shows a general tissue preparation protocol commonly used in laboratories where tissues are prepared for electron microscopy.

Fixation

Ideally, one purpose of fixation is to *preserve the structure of living tissue with no alteration from the living state*. This means that an ideal fixative must also halt potentially destructive autolytic processes at the time of fixation. Additionally, *fixation should protect tissues against disruption during embedding and sectioning and subsequent exposure to the electron beam*. Fixatives always fall short of these ideals; there are always some artifacts induced by the fixation process (see Chapter 19). Often one selects a particular fixation protocol for its ability to preserve one ultrastructural feature over another. For example, membrane structure may be desired. In some instances, artifacts or poor fixation of particular structures are tolerated at the expense of other beneficial features.

Historically, osmium tetroxide, which had limited use in light microscopy, was the first fixative used for electron microscopy (Claude, 1948). Tissues were either immersed in the fixative or exposed to osmium tetroxide vapors. Osmium tetroxide was later used in a physiological buffering system (Palade, 1952). Subsequently, potassium permanganate was

TABLE 2.1 General Tissue Preparation Scheme for Electron Microscopy

Activity	Chemical	Time Involved*
Primary Fixation	tissue is fixed with 2–4% glutaraldehyde in buffer	1–2 hr
Washing	buffer (three changes at 4°C, one of which may be overnight)	1–12 hr
Secondary Fixation	osmium tetroxide (1–2%: usually buffered)	1–2 hr
Dehydration	30% ethanol**	5 min
	50% ethanol	5–15 min
	70% ethanol	5–15 min
	95% ethanol (2 changes)	5–15 min
	absolute ethanol (2 changes)	20 min ea
Transitional Solvent	propylene oxide (3 changes)	10 min ea
Infiltration of Resin	propylene oxide: resin mixtures; gradually increasing concentration of resin	overnight–3 d
Embedding	pure resin mixture	2–4 hr
Curing (at 60°C)		1–3 d

*The specified times do not include the time involved in preparation of chemicals.
**Some recommend omitting the 30% and 50% ethanol steps.

introduced as a fixative (Luft, 1956). It was not until 1963 that the standard glutaraldehyde-osmium tetroxide protocol, used in most laboratories today, was developed (Sabatini et al., 1963). Primary fixation with glutaraldehyde was followed by secondary exposure (postfixation) to osmium tetroxide.

Most fixation protocols developed subsequently are modifications of the two-step basic procedure described above. A primary fixative was developed that combined glutaraldehyde and a low concentration of formaldehyde (up to 4%), which allowed more rapid initial fixation of the tissue because formaldehyde penetrates tissue more readily than glutaraldehyde does (Karnovsky, 1965). A secondary fixative using an osmium tetroxide solution which is reduced with ferrocyanide was introduced to enhance preservation of membranes and glycogen (Karnovsky, 1971). In the following years, numerous fixative variations based on the glutaraldehyde-osmium tetroxide protocol were developed for particular cells or tissues.

One major historical development in fixation methodology should be noted: the use of an intact living animal's vascular system to introduce the fixative into the tissues. The introduction of vascular or organ *perfusion* fixation (Palay, 1962) allowed rapid exposure of fixatives to tissue elements and more closely achieved the goals set for a fixative.

The Mechanism of Chemical Fixation for Electron Microscopy

Glutaraldehyde

The structure of glutaraldehyde (MW 100.12) is shown as follows:

$$\underset{H}{\overset{O}{\diagdown}}C - CH_2 - CH_2 - CH_2 - C\underset{O}{\overset{H}{\diagup}}$$

Glutaraldehyde is a five-carbon compound containing terminal aldehyde groups. Glutaraldehyde's attribute as a fixative is in its ability to cross-link protein by virtue of the terminal aldehyde groups that make the molecule bifunctional. Specifically, the aldehyde groups react with ϵ-amino groups of lysine in adjacent proteins, thereby cross-linking them. The general reaction which takes place between protein and glutaraldehyde as illustrated:

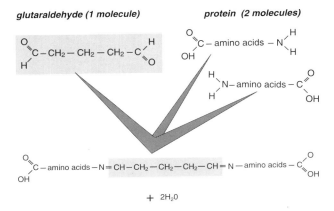

glutaraldehyde (1 molecule)   *protein (2 molecules)*

$$+ \ 2H_2O$$

In the above reaction, it is not necessarily assumed that both ends of the glutaraldehyde molecule react with the protein. Some glutaraldehyde terminal groups may remain unreacted.

Other possible types of glutaraldehyde and protein reactions have been proposed. In many instances, the preceding simplified reaction may turn out to be more complex and involve amino acids other than lysine. Glutaraldehyde will react to some degree with lipids, carbohydrates, and nucleic acids; thus its specificity within the cell may not be limited to protein. Regardless of the nature of the reactions taking place, the overall cross-linking of components within the cell is usually widespread.

Because protein is a universal constituent of cells, present in the cytosol and membrane constituents and within cytoskeletal elements, one could envision that glutaraldehyde could cross-link soluble proteins to each other and link these to fixed cytoskeletal and membranous constituents throughout the cell. Thus, the constituents of the cell would be unified into a single interlocking structure or meshwork held together by a multitude of glutaraldehyde molecules. If some constituents were not cross-linked, they would nevertheless be trapped in the meshwork if they are too large to escape. In the process of cross-linking, glutaraldehyde will also denature protein. Low concentrations of glutaraldehyde have less denaturing effects than higher concentrations. The extent of protein denaturation has a profound effect on many localization protocols for most proteins (see Chapters 9 and 10).

The rate of penetration of glutaraldehyde into tissues is very slow. The deeper the penetration of glutaraldehyde into a tissue, the slower it will advance from that point on. Much depends on the nature of the tissue being used; glutaraldehyde penetrates more slowly into compact tissue with multiple membrane layers to penetrate than it does into tissue that has large fluid spaces. Glutaraldehyde can be expected to penetrate at a rate of less than 1 mm per hour if compact tissue is immersed into the glutaraldehyde solution. Therefore, to obtain rapid immersion fixation, the immersed tissue should be under 1 mm in thickness in at least one dimension.

Pale-colored tissue fixed with glutaraldehyde will acquire a yellow tint with time. The change is extremely gradual and difficult to notice unless one compares the color of unfixed tissue with the fixed tissue. A gradual and progressive change in firmness of the tissue usually accompanies glutaraldehyde fixation. This is especially well demonstrated when tissues are perfused with glutaraldehyde.

CAUTION: Handle glutaraldehyde carefully. Its ability to fix tissues is not restricted to the tissues being prepared for electron microscopy. Exposure to skin will cause the skin to turn yellow and harden. Usually, the exposed (yellowed) skin will be sloughed off in a few days. Avoid prolonged breathing of glutaraldehyde and contact with the eyes. Repeated exposure to glutaraldehyde may cause contact dermatitis. Glutaraldehyde should be used under a well-ventilated hood and disposed of in a proper manner.

Osmium Tetroxide

Osmium tetroxide, often erroneously called "osmium" in the electron microscopy vernacular, is a compound with a molecular weight of 254.2. The osmium tetroxide molecule is symmetrical and contains four double-bonded oxygen molecules.

Osmium tetroxide works as a secondary fixative of tissue structure by reacting primarily with lipid moieties. It is widely believed that the unsaturated bonds of fatty acids are oxidized by osmium tetroxide. This reduced heavy metal adds density and contrast to the biological tissue.

Osmium tetroxide has a number of stable oxidation states and is soluble in polar and nonpolar solvents. Numerous other types of reactions of osmium tetroxide have been described that explain the stabilization of many cell components by osmium. Furthermore, the molecular weight of osmium tetroxide is sufficiently high to be effective in scattering electrons. Thus, osmium tetroxide is also an important stain. With time, tissues exposed to osmium stain intensely black as viewed with the naked eye. Osmium tetroxide also acts as a *mordant* (a substance capable of combining with stain or dye at a later time). For eletron microscopy, osmium tetroxide is a mordant for enhancing lead staining (see Chapter 5).

The penetration rate of osmium tetroxide is often slower than glutaraldehyde. Osmium tetroxide generally will not penetrate compact tissues of more than 0.5 mm in one hour. Very little additional penetration of tissue occurs after one hour. Specimen cubes that are much larger often show a core of unblackened tissue that is virtually useless for electron microscopy (Figure 2.2). The concentrations of both glutaraldehyde and osmium tetroxide are diluted within the specimen as these fixatives penetrate tissues. Usually, the quality of fixation at the center of a large tissue block cannot be expected to be as good as the peripheral area of the tissue. Furthermore, peripheral tissues may be over fixed. Osmium tetroxide is a nonpolar (uncharged) compound. However, it penetrates both charged and uncharged surfaces and is soluble in both polar and nonpolar tissue components.

The length of time tissue is fixed in osmium tetroxide has an effect on the appearance of the tissue. Too short exposure to osmium tetroxide results in under fixation. Prolonged exposure results in extraction of tissue components, especially later during the dehydration process. Exposures of tissues to osmium tetroxide should be generally limited to under 1.5 hours; however, notable exceptions exist (impervious seeds, botanicals, and some bacteria, for example, need longer exposure times).

FIGURE 2.2 Drawing of a tissue section incompletely penetrated by osmium tetroxide.

├─────2mm─────┤

CAUTION: Be extremely careful in handling osmium tetroxide, whether it is in crystalline or solution form. The fixative properties of osmium can be seen not only on tissue samples prepared for electron microscopy, but also on the living tissues of the individuals performing the fixation. Osmium tetroxide vapors are dangerous to the conjunctivum and the cornea of the eye and to the respiratory and alimentary membranes. Osmium tetroxide should always be used under a properly ventilated hood. Osmium tetroxide is moderately volatile. Osmium tetroxide vapors smell similar to chlorine bleach, a smell not readily forgotten in the event of accidental exposure to the substance. Avoid direct contact by wearing impervious plastic or rubber gloves.

Used or excess osmium tetroxide should be properly sealed and disposed of according to current safety standards. Osmium tetroxide should not be stored in plastic containers. Spills can be cleaned up with absorbent material (wood shavings or cat litter will do). Osmium tetroxide can be reduced by corn oil to a less harmful substance and then given to a properly licensed agency for final disposal.

Selection of a Fixative and a Buffer

Fixative

In selecting a fixative, it is important to determine what has worked in the past for the particular tissue under consideration with the particular conditions employed. Literature is generally available for the particular tissue to be prepared, and it should be consulted prior to making a final decision on which fixative to use. Usually, recent literature is the most helpful because the methods of fixation have improved continually over the years, and the standards by which fixation quality is judged have become higher with time. Current literature sources will often refer the reader to other sources for the detailed methodology.

It is important to determine what tissue features are to be emphasized in the study. If ribosomes are to be examined, then a fixative should be used that emphasizes ribosomes in the tissues, such as glutaraldehyde followed by osmium tetroxide. In such an instance, permanganates or osmium-ferrocyanide mixtures should be avoided because ribosomes are extracted or do not appear prominent with these protocols.

Buffers

If a buffering system is not used with the primary fixative, the pH of the tissue is drastically lowered during the fixation procedure. As a consequence, numerous artifacts may be produced. Buffering systems that maintain physiologic pH (e.g., 7.2 to 7.4 for most mammalian tissues and 6.8 for bacteria and plant material) result in fewer artifacts. Again, the literature should be consulted since the physiological pH for particular biological systems may vary from 5.5 to 7.5 or even higher.

There are several choices of buffering conditions available. Appropriate buffering systems for a particular biological tissue have usually been worked out, and the information is readily available in the literature. Common buffering systems in use are phosphate, cacodylate, and tris-maleate. Organic buffers such as PIPES and HEPES are often used in tissue culture work and may sometimes be used in the primary fixative. Phosphate buffer is often the buffer of choice since it is nontoxic and physiologically compatible with cells. A simplified guide to preparing some buffers is given in Table 2.2. When using Table 2.2, prepare solutions A and B and mix according to the formula given. In the formula, **X** refers to the amount of solution in the pH adjustment table (Table 2.3) that will yield a desired pH. Please note that chemicals used to make certain buffers come in hydrated forms so that the amounts needed will vary from those given in Table 2.3.

Just as important as the selection of the buffer is the *osmolarity* of the buffering system. The osmolarity of the buffer, as it contributes to the osmolarity of the overall fixative, is important since osmolarity changes induce shrinkage or swelling of tissue due to osmotic effects. Artifactual changes in cell size should be avoided or minimized. Literature sources are the best indication of the appropriate osmolarity for a tissue under consideration. The older literature shows that investigators paid meticulous attention to developing a buffering system with a physiologic osmolarity. Physiologic osmolarity (320 milliosmoles for most mammalian extracellular fluids) was at one time considered an ideal goal when preparing a fixative.

The total osmolarity of the fixative includes both the osmolarity of the buffer and that of any added substance(s) (such as NaCl) and that of the fixative (a 3% solution of glutaraldehyde alone is 300 milliosmoles). As it turns out, slightly hyperosmolar solutions often give the best empirical fixation results. Some tissues, however, appear to show little difference in preservation in buffering systems when molarities ranging from 0.05 to 0.2 are used, while others are very sensitive to buffer osmolarity. The osmolarity of the buffer and added substances (such as salts and sugars) are very important factors in assessing overall fixative osmolarity.

TABLE 2.2 Preparation of Common Buffers Utilized in Electron Microscopy

Buffer	pH Range	Stock Solutions		Formula
		A	B	
Cacodylate	5.0–7.4	0.2 M sodium cacodylate (42.8 g Na $[CH_3]_2AsO_2 \cdot 3\ H_2O$) per liter of distilled H_2O	0.2 N HCl	25 ml A + **X** ml B made. Bring volume up to 100 ml with distilled H_2O.
Phosphate	5.7–8.0	0.2 M monobasic sodium phosphate (27.6 g $NaH_2PO_4 \cdot H_2O$ per liter of distilled water)	0.2 M dibasic sodium phosphate (53.65 g $Na_2HPO_4 \cdot 7\ H_2O$ or 71.7 g $Na_2HPO_4 \cdot 12\ H_2O$ per liter of distilled water)	**X** ml A + (100 − **X**) ml B. Bring volume up to 200 ml with distilled water for 0.1 M buffer.
Tris-maleate	5.2–8.6	0.2 M tris acid maleate (24.2 g tris-[hydroxymethyl] amino-methane + 23.2 g maleic acid or 19.6 g maleic anhydride per liter)	0.2 N NaOH	25 ml A + **X** ml B made up to 100 ml

TABLE 2.3 pH Adjustment (see **X** under "Formula" in Table 2.2)

pH	Cacodylate	Phosphate	Tris-maleate
5.0	23.5		
5.2	22.5		3.5
5.4	21.5		5.4
5.6	19.6		7.8
5.8	17.4	92.0	10.3
6.0	14.8	87.7	13.0
6.2	11.9	81.5	15.8
6.4	9.2	73.5	18.5
6.6	6.7	62.5	21.3
6.8	4.7	51.0	22.5
7.0	3.3	39.0	24.0
7.2	2.1	28.0	25.5
7.4	1.4	19.0	27.0
7.6		13.0	29.0
7.8		8.5	31.8
8.0		5.3	34.5
8.2			37.5
8.4			40.5
8.6			43.3

The fixatives (glutaraldehyde, formaldehyde and osmium tetroxide) are thought to alter the permeability of the cell membrane to some degree. The osmotic effect of the fixative on the cell appears not to have a large impact on the cell. It may be that the rapid penetration of the buffer and its contents is what effect cells initially. *The selection of one particular buffering system over another is often based on the results obtained after examination of the tissue.*

The osmolarity of the fixative can be adjusted by increasing the osmolarity of the buffer or by adding substance(s) to the buffer such as sucrose, glucose, or sodium chloride. Calcium or magnesium chloride is often added to the buffer to preserve certain features within the tissue (such as membranes and DNA).

CAUTION: Some common buffers used in electron microscopy are extremely toxic and carcinogenic. For example, cacodylate buffer contains arsenic. S-collidine buffer, besides having a disagreeable odor, is also extremely toxic because of its pyrimidine components. Veronol-acetate buffer has barbiturates. Exercise extreme care when preparing a buffer. Avoid breathing the buffer salt powders and avoid contact of the buffer constituents with the skin. Use gloves and a fume hood and a safe method of disposal. Do not allow buffers to dry in glassware. They become airborne and are accidentally inhaled.

Obtaining and Preparing Buffered Glutaraldehyde Fixative

Assuming a standard glutaraldehyde prefixation and osmium tetroxide postfixation combination is to be used, the preparation of a solution of 3% buffered glutaraldehyde and another of 1% osmium tetroxide will be described.

Electron microscopy suppliers sell glutaraldehyde in two grades. The one most commonly used is *electron microscopy grade* solutions of glutaraldehyde, which vary in glutaraldehyde content from 8 to 70% and are usually packed in sealed ampoules (Figure 2.3) or screw cap bottles. Usually, 100 ml of electron microscope grade solutions are the maximum quantities available in a single container. The cost for a 10 ml ampoule of 50% electron microscopy grade glutaraldehyde in 1998 was about $7 and a 100 ml bottle was about $32. The use of electron microscope grade of glutaraldehyde is reasonably cost efficient if tissues are to be immersed in glutaraldehyde. If perfusion fixation of large animals is to be undertaken, *biological grade glutaraldehyde* is available in large volumes at a considerably lower price (1 gal or 3.8 L of a 50% solution costs about $50). Biological grade glutaraldehyde is less pure, but is considered adequate for most perfusion fixation purposes.

Any glutaraldehyde that is not sealed in an ampoule under dry nitrogen will polymerize with time and produce a whitish haze when added to the buffer. If this occurs, it should be discarded. Glutaraldehyde not in ampoules should be stored in a refrigerator and used within a month of receipt from the supplier. Storage of glutaraldehyde at room temperature will lead to its rapid polymerization. Glutaraldehyde may be frozen, if desired, to extend its shelf life.

Buffer, made according to Table 2.2, is checked for the appropriate pH. The buffer should be made just before use. Using 50% glutaraldehyde is recommended because the final concentration of glutaraldehyde is readily calculated. The amount added to make 100 ml of fixative is twice the glutaraldehyde percentage needed (e.g., 6 ml of a 50% glutaraldehyde per 94 ml buffer will produce a 3% glutaraldehyde solution). After the solution is prepared, it is brought to the appropriate temperature where it is ready for use.

Obtaining and Preparing Osmium Fixative

Osmium can be purchased in ampoules, which are sealed under dry nitrogen, in either crystalline form

FIGURE 2.3 Sealed ampoules of glutaraldehyde, 4% osmium tetroxide, and crystalline osmium tetroxide (*left to right*).

or in aqueous solutions of about 4% (Figure 2.3). Properly sealed osmium tetroxide is stable indefinitely in either form. The cost of 1 gm of osmium tetroxide (in 1998) was approximately $32, and 20 ml of a 4% solution was approximately $40.

From the crystalline form, one may prepare a 4% solution of osmium tetroxide in a container capped with a Teflon-lined screw cap and sealed with Parafilm. Such a solution is stable for many months. Osmium tetroxide dissolves slowly in distilled water at room temperature. It can be dissolved rapidly in distilled water warmed on a steam bath followed by agitation. Use of a magnetic stirring bar also will facilitate its entry into solution. It is not necessary to refrigerate osmium solutions, but if osmium is refrigerated, vapors often escape and darken the cooler. To retard vaporization, the sealed end of the bottle should be wrapped in waxed film such as Parafilm and the entire container wrapped in foil and placed in a Ziplock bag. Another way to store osmium tetroxide is to place it in a sealed glass container that is then placed in a second sealed glass container.

A stock solution of osmium tetroxide (2–4%) is prepared in distilled water. Prior to use it is diluted in buffer to a final concentration of 1 to 2%. The use of buffer with osmium is questionable since the tissues have already been fixed in glutaraldehyde and tonicity and pH issues are no longer important.

Immersion and Perfusion Fixation

One can either immerse tissue into fixative or use the blood-vascular system of organ perfusion to fix tissue. During perfusion, the fixative is distributed to the tissues through the blood vessels. From there, fix-

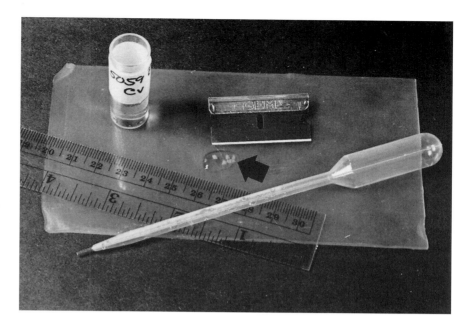

FIGURE 2.4 Setup appropriate for dicing tissue to be prepared for electron microscopy. Tissue (*arrow*) is placed on a sheet of dental wax and diced with a sharp razor blade. A ruler has been placed nearby to show the tissue block size, which should not exceed 1 mm after dicing.

atives will penetrate into the tissue spaces and to the various tissues. In some instances there may be no accessible vascular system through which to perfuse tissues. Biological specimens may be so small as to render them impossible to perfuse. In such instances the entire organism may be fixed by immersion.

Immersion

It is important to bear in mind the slow penetration rate of glutaraldehyde. If one elects the immersion technique, then the size of compact tissues must be *less* than 1 mm thick for glutaraldehyde to penetrate prior to the time that autolytic changes will take place. Usually, a sharp, clean razor blade is used to dice tissue into small cubes or slices. The tissue is cut on dental wax (Figure 2.4) or waxed paper with sawing motions of the razor blade so as not to compress the tissue too severely. Fixative is flooded over the tissue prior to the dicing procedure. At no time is the tissue allowed to dry! After the tissue is cut, the pieces are immersed in glutaraldehyde in a small vial (Figure 2.5). The volume of fixative should exceed the tissue volume by at least 5-fold and preferably 10-fold.

The minimum fixation period is usually an hour, but the fixation process has been extended for many hours, days, or even weeks. There is a controversy over how long material can stay in the primary fixative. Some believe that it is important to fix in glutaraldehyde for a few hours and to proceed immediately to secondary fixation with osmium tetroxide. Their argument is that osmium tetroxide is a fixative as well, and time should not elapse

FIGURE 2.5 Vial suitable for tissue processing for electron microscopy. Also shown is the vial cap and a dime for size comparison.

needlessly in which tissues may undergo artifactual changes. Others believe that the specimens may stay in glutaraldehyde for prolonged periods without artifactual change. If there is any doubt, the tissue should be processed from the glutaraldehyde step into wash buffer without delay and then into osmium tetroxide as soon as the glutaraldehyde fixation is deemed optimal.

Fixation of pellets produced by centrifugation often presents a penetration problem when glutaraldehyde is used as a primary fixative. If the pellet is not loosened from the centrifuge tube or has been spun down to the bottom of a pointed centrifuge tube, glutaraldehyde will have poor access to the tissue. Essentially, access of glutaraldehyde to pelleted tissue is from one side of the pellet since the pellet is tightly adherent to the centrifuge tube. Pellets must be in the order of 0.5 mm or less in thickness for adequate penetration of glutaraldehyde. If they can be loosened from the centrifuge tube prior to glutaraldehyde fixation without resuspension of the tissue, their exposure to glutaraldehyde and subsequent chemicals is enhanced. Pelleted material may be fixed in suspension and pelleted at a later time. However, fixed suspended material often fails to form a solid pellet.

Perfusion Fixation

In most tissues, cells are only a few μm from their vascular supply. When possible, perfusion of fixative using the vascular system as a delivery route is usually deemed to be the optimal route for tissue fixation. Most perfusions can be accomplished *without handling the tissue* and without the need to cut fresh tissue, a manipulation that results in excess pressure on the fragile tissue and distortion of its components.

One must be knowledgeable of the circulatory system of the species under consideration. The most common perfusion scheme employs whole-body perfusion. In a mammal, an apparatus such as shown in Figure 2.6 is used to introduce solutions into the left ventricle of the heart or into a major blood vessel of an appropriately anesthetized animal. The perfusion bottles are placed at about 4 feet above the animal being perfused. The general plan for vascular perfusion is to introduce fixative into the arterial system and to allow its egress from a cut made in the venous system. An outlet for blood and perfusion solutions (perfusate) is made in an area such as the right atrium or ventricle. For example, a needle is inserted into the left ventricle and solution is allowed to flow from the left ventricle to the aorta where it is distributed to most of the capillary beds

of the body. It returns to the heart via the venous system and exits the heart through a cut made in the right atrium or ventricle.

The solution initially introduced into the heart is usually a physiological solution such as saline (mammals) or Ringers (nonmammalian vertebrates), sometimes with an anticoagulant. It is perfused at room temperature to avoid changes in blood vessels that are related to temperature. Blood is cleared from the vascular system in this manner. The organ will usually blanch when it is cleared of blood. Exposure of glutaraldehyde to blood without prior clearing of the system will cause coagulation of blood and result in the failure of the perfusion. Most perfusion systems have a valving system (Figure 2.6) that allows switch over, at an appropriate time, from saline to glutaraldehyde. The exposure of the anesthetized animal to glutaraldehyde may cause involuntary twitching of the animal's muscles. The properly anesthetized animal is not conscious of pain during this procedure. In fact death comes while the animal is anesthetized and almost immediately upon introduction of glutaraldehyde.

Perfusion is usually conducted for about 20 to 45 minutes. During this period, the animal's organs harden and take on a yellowish tint. After perfusion, the desired organ is removed and diced into small (under 1 mm) pieces with a sharp razor blade. While being cut, the tissue should be immersed in fixative on a piece of dental wax. It is common to place the small pieces of tissue in a vial containing glutaraldehyde fixative for an additional hour or two to insure that fixation is adequate (postperfusion immersion).

Perfusion methodology may not necessarily involve whole body fixation. It may be partial body fixation or simply organ perfusion. Often one can gain access to the vessels leading to an organ and introduce a perfusion needle into the vessel (preferably an artery). Egress of fixative is by cutting the final drainage vein from the organ. The literature is the best source of information on what has been successful historically in terms of perfusion for a particular tissue. In many cases, blood will not clear from the organ without addition of vasodilators or anticoagulants such as procaine hydrochloride or heparin, respectively.

Fixation Conditions

Many specific details become important in the fixation of various tissues. For almost every tissue, there are special needs that must be taken into account in order to obtain optimal fixation. These cannot be

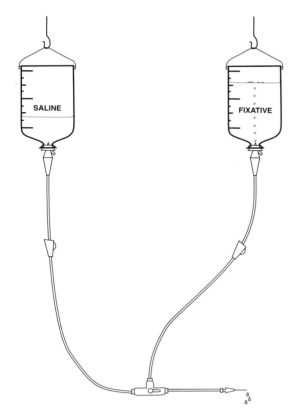

FIGURE 2.6 Drawing of gravity-fed perfusion apparatus. One of the two bottles contains saline to clear the vascular system and the other contains fixative. A valve allows only one of the solutions to flow at a time. A needle at the end of the tubing is used to tap into the animal's vascular system.
(From Histological and Histopathological Evaluation of the Testis *used with permission of Cache River Press.)*

dealt with effectively within the space limitations of this chapter. For specific information about a variety of protocols that meet specific fixation needs, see Hayat's reference volume (1981), which is devoted entirely to the theoretical bases and practical methods of fixation for electron microscopy.

Primary fixation and secondary fixation of tissues in osmium tetroxide is often performed at 4°C. With respect to primary fixation, tissue autolysis and structural changes are minimized at this temperature. However, some subcellular elements such as microtubules are cold sensitive. Thus, immersed tissue may be fixed prior to artifactual changes, although it is well-recognized that osmium tetroxide penetration at 4°C is slower than at room temperature. Secondary fixation in the cold (4°C) is sometimes advocated for osmium tetroxide. Empirically, the results may be better or unchanged for certain tissues that are fixed at room temperature as compared with fixation in the cold.

During primary fixation, the tissue should be gently and periodically agitated (for example every 10 minutes). After agitation, tissue surfaces which are up against the container vial will be dislodged and exposed to the fixative.

Less osmium tetroxide than glutaraldehyde is needed to fix tissues. The fluid level of osmium tetroxide in the vial should be two to three times the height of the tissue.

Some investigators will utilize the same buffering system during both the primary and secondary fixation protocols. There may be little or no difference in the results obtained by using either a buffering system or distilled water as a vehicle for osmium tetroxide. Caution must be exercised when using some organic buffers for postfixation since they will react with osmium tetroxide.

Microwave-Assisted Specimen Preparation

The use of microwave ovens in the preparation of specimens for both transmission and scanning electron microscopy has increased significantly over the past several years and has become a routine procedure in many research pathology laboratories (Chen et al., 1983; Leong, 1991 and 1993; Login and Dvorak, 1994). The major reasons are due to a decrease in the fixation time from hours or minutes to seconds, enhanced structural preservation, better retention of lipids and soluble proteins, better preservation of antigenic reactivity, shorter polymerization times for embedding plastics (Giammara, 1993), and accelerated staining times for sectioned specimens.

Microwaves are a form of nonionizing radiation (see Table 6.1) that interact with dipolar molecules (such as water and the polar side chains of proteins) and cause a 180° back and forth rotation of the molecules at a rate of over 2 billion cycles per second. As molecular agitation increases, the diffusion rate of all molecules in solution, including fixatives, increases. In addition, organic reactions are enhanced by microwave irradiation. Heat is also generated as a result of molecular friction, but it is not the reason for the rapid fixation because heat alone does not provide adequate ultrastructural preservation. Important parameters for microwave fixation are: (a) expose the specimens to microwaves for no more than 30 to 60 seconds, (b) keep temperatures below 50 to 55°C, (c) maintain a small specimen and container size, (d) determine the proper location in the microwave to place the specimen,

FIGURE 2.7   Research-grade microwave oven. *(Courtesy of Electron Microscopy Sciences.)*

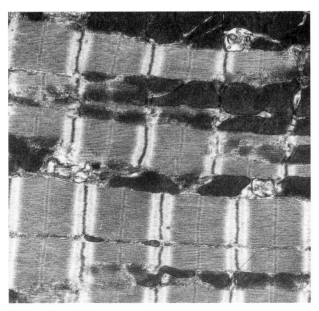

FIGURE 2.8   Electron micrograph of mouse heart fixed by ultrafast microwave fixation. The contractile elements of the muscle fibers, as well as the mitochondria and sarcoplasmic reticulum, are very well preserved. *(Reprinted with permission from G. R. Login, et al., in* Microwave Processing of Materials II. *W. B. Snyder, Jr., W. H. Sutton, M. F. Iskander, D. L. Johnson, Eds., Materials Research Society, Pittsburgh, 1991, vol. 189, pp. 329.) Magnification A ×16,740.*

(e) include a container of water to absorb excess microwave energy (Login and Dvorak, 1994).

Although it is possible to use the higher powered (650W and above) conventional microwave ovens for preparing specimens, industrial or research grade units (Figure 2.7) offer many advantages: higher power with less specimen heating, high volume ventilation ports to exhaust toxic fumes, better control of temperature and power levels, and approximation of energy focal points where specimens should be located. With any microwave oven, it is essential to calibrate the oven to accurately locate the energy focal points and to determine power settings and times appropriate for the specimen. This is a time-consuming task requiring attention to many details, but well worth the effort because considerable time will be saved after establishing a routine procedure. The best starting point is to follow the steps described by Login and Dvorak (1994) who developed a set of exercises that logically guide one through the calibration of any microwave oven. We have included a typical microwave protocol for illustrative purposes. It is unlikely, however, that one will be able to follow this specific protocol without first calibrating the microwave oven. See Figures 2.8 and 2.9 for examples of specimens prepared by microwave irradiation. See also the reference book by Kok and Boon (1992).

After tissues have been fixed for the appropriate time (10–60 seconds) in a primary fixative (glutaraldehyde/formaldehyde), the sample vial is removed and immediately shaken to equalize the temperature. The specimen is rinsed in several changes of buffer followed by postfixation in 1 to 2% osmium tetroxide

for no more than 60 seconds in the microwave. Specimens may then be dehydrated in the appropriate liquid (alcohol, acetone) and placed in various mixtures of dehydrant and epoxy embedding resin for periods of 10 to 15 minutes.

CAUTIONS: It is safest not to place alcohols or volatile reagents greater than 50% inside the oven (Login and Dvorak, 1994). This is a major consideration in poorly ventilated microwave ovens in which potentially explosive fumes will build up. It is critical to have a cooled water load in the microwave to maintain temperatures below 55°C.

Following several changes in embedding resin, the specimens are transferred into flat, silicon embedding molds situated in calibrated locations within the oven and the resin is polymerized. Typical polymerization conditions may be at 50% power in a 700 W microwave for 15 minutes. Alternatively, since it is difficult to maintain temperatures in flat embedding molds, one may place the specimens into BEEM capsules which are then tightly capped and submerged in a plastic container (Tupperware™, for example) of water. While under water, the capsules are microwaved for 15 minutes at 100% power with a temperature restriction

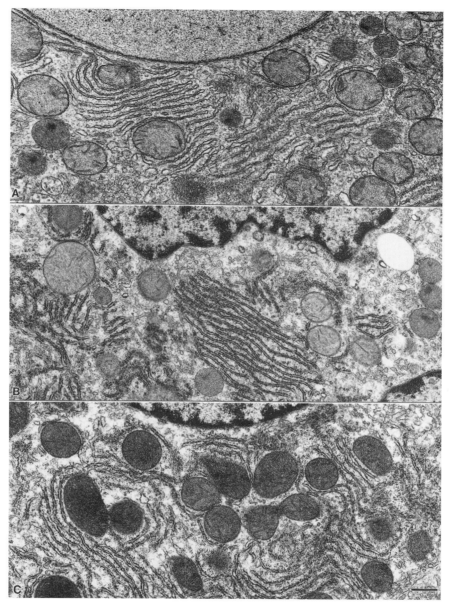

FIGURE 2.9 Transmission electron micrographs of rat liver fixed by three fast-microwave methods (Biorad H2500 microwave processor, 600 watts). In A–C, stacks of rough endoplasmic reticulum are present and are not dilated. Mitochondria, peroxisomes, and lipid bodies are well-preserved by all microwave methods. In A, nuclear chromatin is finely dispersed. In B–C, the nuclei show chromatin condensation along well-preserved nuclear membranes. (A) Fast, primary microwave-chemical fixation in 2% 0.2 M S-collidine buffered osmium tetroxide for 6 seconds, irradiated from ~20 to 43°C, (B) sequential fast, primary microwave-chemical fixation in a mixed aldehyde fixative (2% paraformaldehyde, 2.5% glutaraldehyde, 0.025% calcium chloride, in 0.1 M sodium cacodylate buffer, pH 7.4) for 7 seconds, irradiated from ~20 to 55° C, followed by fast, microwave-chemical postfixation in osmium for 7 seconds, irradiated from ~20 to 42° C, or (C) fast, primary microwave-chemical fixation in a mixed aldehyde fixative followed by standard postfixation in osmium for 2 hours at room temperature. Specimens were not exposed to additional osmium and were stored in 0.1 M sodium cacodylate buffer, pH 7.4 at 4° C for 2 weeks (A) or 24 hours (B, C), prior to processing the central 1 mm^3 of the 5 mm^3 tissue blocks. Marker bar=1.0 µm.
(Reprinted with permission of G. Login and J Histochem Cytochem.)

of 95°C, followed by 60 minutes without temperature restriction. The water level must be maintained so that the capsules are always submerged. Some investigators bypass microwave polymerization and use a conventional oven set at 95°C for 5 hours.

Freezing Method for Specimen Preparation

Although it is possible to chemically fix specimens in seconds and even milliseconds (Login et al., 1991) using microwaves, ultrarapid freezing is thought to provide superior ultrastructural preservation because it will freeze all molecules in place in a more natural manner. No combination of chemical fixatives is able to accomplish this; hence freeze-fixation is currently the only way to fix diffusible molecules and ions in place. Unfortunately, except under special situations in which the specimen is kept frozen until further processed or viewed in the frozen state, freeze-fixation is reversed when specimens thaw and molecules are free to shift about.

For living specimens to be properly preserved by freezing, this must occur in an ultrarapid manner at a rate of 10^4 to 10^5°C per second to avoid the formation of damaging ice crystals within the cells. At this rate, water molecules are unable to organize into ice crystals and the molecules instead form amorphous or vitreous ice. The goal of ultrarapid freezing, therefore, is to achieve vitrification of cellular water. A major hurdle to achieving vitrification has to do with the size of specimens that are frozen. As the depth of tissue increases, the rate at which heat can be removed is greatly diminished so that good preservation is achievable to a depth of only 200 to 500 μm using the high pressure freezing method (Studer et al., 1989). Most of the presently available procedures for ultrarapid freezing use various methods. For a review of the most commonly used cryo methods, see the review article by Gilkey and Staehelin (1986). We have summarized some of these methods in the following discussion.

Plunge freezing is the simplest and most widely used method and is achieved by plunging the specimen into a liquid cryogen. This has been quite successful with extremely thin specimens such as aqueous suspensions of virus particles or subcellular components suspended as a thin film (100 nm or less) over the mesh of an electron microscope grid that is plunged into ethane. The vitrified specimens may then be viewed in a special holder designed to maintain the frozen con-

dition inside of the transmission electron microscope. Some spectacular micrographs have been published using this technique (Adrian et al., 1984). It is also possible to plunge freeze larger specimens, however, the depth of vitrification is restricted to 1 to 2 μm. It is possible to use a liquid nitrogen slush as a cryogen. This is prepared by placing a styrofoam container of liquid nitrogen into a vacuum jar and evacuating slightly using only a rotary pump. Eventually, the liquid nitrogen becomes slushy and may be used to plunge freeze specimens after being removed from the vacuum.

Slam freezing, also termed cold metal block freezing, achieves ultrarapid freezing rates by rapidly contacting (e.g., slamming) the specimen onto the polished surface of a copper or silver block that has been chilled to −196 to 269°C with liquid nitrogen or liquid helium. This may be accomplished using modified cryo pliers, precooled silver hammers, or, most efficiently, using commercially available devices that gravity-drop a rod containing the specimen mounted on the end onto the chilled metal. Vitrification is achieved to a depth of 10 to 15 μm.

Propane jet freezing is achieved by sandwiching a 200 to 500 μm thick specimen (suspension or slice) between two metal plates that are then clamped into a device that directs jets of liquid propane chilled with liquid nitrogen against both sides of the specimen plates. The rapidly flowing jet of propane achieves vitrification of the specimen to a total depth of approximately 40 μm.

High-pressure freezing is similar to propane jet freezing except that the specimen is pressurized to 2100 atmospheres to enhance freezing. High pressure lowers the freezing point of water as well as the rate of ice crystal formation. After the specimen is frozen by the cryogen at high pressures, the specimen sandwiched between two metal plates is dropped into liquid nitrogen. To retrieve the sample, the two metal plates are separated under liquid nitrogen and the specimen further processed as desired. Currently, high-pressure freezing achieves good freezing to a depth of 500 μm.

After specimens have been ultrarapidly frozen, several options exist. In the case of plunge freezing, they may be viewed directly in the TEM. Other specimens may be viewed directly in the SEM using special cryo stages (Figure 7.31) or they may be further processed using cryoultramicrotomy (Chapter 4), freeze fracturing/etching (Chapter 14) or they may be processed using freeze-substitution.

In freeze-substitution (see review by Steinbrecht and Muller, 1987), the frozen specimens are maintained at a low temperature of $-80°C$ in an organic solvent such as acetone, ethanol, or methanol containing such chemical fixatives as an aldehyde or osmium tetroxide. Over a period of days or weeks, the frozen water in the cell is substituted by the organic solvent and some of the frozen molecules react with the fixatives. A gradual dehydration and fixation take place that chemically stabilizes structures and molecules that might normally be distorted or displaced. After a rinse in solvent to remove unreacted fixatives, the specimen is gradually infiltrated with a low viscosity resin such as Spurr's epoxy resin or Lowicryl acrylic resin (Armbruster et al., 1982) and polymerized at the low temperature. After polymerization, the plastic capsule can be treated as a conventionally processed specimen and sectioned and stained for viewing in the TEM. Examples of specimens that were fixed using high-pressure freezing followed by freeze-substitution are shown in Figure 2.10. In most instances, similar results may be obtained using freeze-substitution alone.

Popular Fixation Protocols Other than Glutaraldehyde-Osmium Tetroxide

Glutaraldehyde followed by osmium tetroxide is considered the standard fixation protocol, capable of stabilizing the maximum number of different cell components. Most specialized fixation techniques are modifications of the glutaraldehyde-osmium protocol. A few are described.

Karnovsky's Fixative

This popular fixative utilizes a relatively low percentage of formaldehyde in the primary fixative. The use of formaldehyde with glutaraldehyde achieves a more rapid overall penetration of fixative (Karnovsky, 1971) than with glutaraldehyde alone. It is theorized that the formaldehyde temporarily stabilizes structures that are later more permanently stabilized by glutaraldehyde. Formaldehyde penetrates about five times faster than glutaraldehyde. Because of the addition of formaldehyde (and calcium chloride) to the glutaraldehyde fixative, the osmolarity of the total fixative is about 2110 milliosmoles, far in excess of what would be considered physiological osmolarity in mammalian systems. Since the effect of fixative osmolarity on tissue structure is far less than that of the buffer and its added constituents, there seems to be negligible effect of the added formaldehyde with its high osmolarity on tissue structure.

Karnovsky's fixative is often used when penetration of fixatives is considered to be poor or when immersed tissue is of a larger size than usual. The formaldehyde-glutaraldehyde method has been used extensively for nervous tissue. The quality of tissues fixed with Karnovsky's method is similar to a good fixation with glutaraldehyde alone.

Formaldehyde (CH_2O) is a gas, but may be obtained in liquid form in 37 to 40% solutions called *formalin*. When mixing fixatives, formalin is usually not employed because it contains numerous impurities such as methanol. Instead, powdered, polymerized formaldehyde or *paraformaldehyde* is used to prepare formalin. Since paraformaldehyde goes into solution with great difficulty at room temperature, the distilled water in which it will be dissolved is first heated to 60 to 80°C under a ventilated fume hood. A small amount of sodium hydroxide or preferably potassium hydroxide is added to the heated solution and stirred until the solution clears.

The following protocol should be followed when preparing 100 ml Karnovsky's fixative:

1. Dissolve 4 gm paraformaldehyde powder in 50 ml distilled water by heating to 60 to 70°C under a hood. Add two to four drops of 1N KOH, stirring until the solution clears.
2. After cooling add 10 ml of 50% glutaraldehyde.
3. Bring to 100 ml with 0.2 M sodium cacodylate or phosphate buffer (at the proper pH).
4. If cacodylate buffer is used, then 50 mg calcium chloride ($CaCl_2$) is added slowly to the solution.
5. Tissues are perfusion fixed routinely or immersion fixed for 2 to 5 hours. Standard washing and postfixation with osmium tetroxide and dehydration are employed.

Karnovsky's original fixative, containing 5% glutaraldehyde and 4% formaldehyde, is often modified to yield fixatives with lower concentrations of both formaldehyde and glutaraldehyde (e.g., 2% paraformaldehyde and 2.5% glutaraldehyde).

Osmium-Reduced Ferrocyanide

This method is a slight modification of the standard glutaraldehyde-osmium tetroxide protocol (Karnovsky, 1971; Russell, 1978). It is especially suited for demonstration of membranes and glyco-

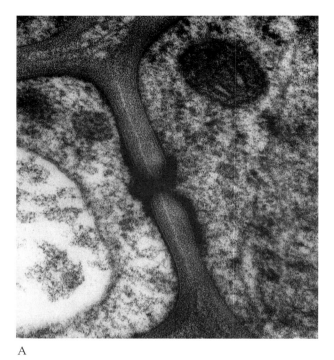

A

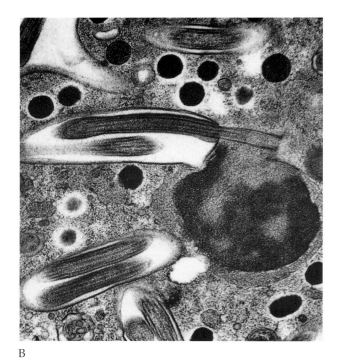

B

FIGURE 2.10 Examples of specimens fixed using high-pressure freezing followed by freeze-substitution. Samples were frozen using high-pressure freezing and transferred into the −80°C substitution fluid containing 2% osmium tetroxide and 0.05% uranyl acetate in acetone for 3 to 4 days. Specimens were gradually warmed to ambient over a period of 7 hours and rinsed in acetone several times. Samples were embedded in Epon-Araldite over a period of 32 or more hours and hardened at 60°C for 48 hours. Sections were stained in aqueous uranyl acetate followed by Reynold's lead citrate. (A) Longitudinal section through portion of a simple septation in the cell wall (CW) of the plant pathogenic fungus *Exobasidium*. (B) Gametogenesis in the foraminiferan *Trilocululina oblonga* showing flagella (F) associated with each gametic nucleus.
(C) Longitudinal section through the root tip of the plant *Arabidopsis*.
(Courtesy of B. Richardson for A and C; S. Goldstein and B. Richardson for B.)

C

gen. The drawbacks are that ribosomes are poorly preserved and the cytoplasmic matrix appears less dense than when observed with the traditional glutaraldehyde tissue preparation regimen. However, the added prominence of membranous elements within the cytoplasm gives an overall pleasing appearance to the tissue. Many of the micrographs used in this text are from tissues prepared using this method (see Chapter 20, Figures 20.1, 20.6, and 20.73).

Fixation with glutaraldehyde is according to standard methods published in the literature. The tissues are washed thoroughly in buffer overnight (three or four changes) and fixed in a freshly prepared mixture consisting of 1% osmium tetroxide and 1.25% potassium ferrocyanide (Russell, 1978). It is important to use ferrocyanide and *not* ferricyanide. To prepare the final solution, equal volumes of 2% osmium tetroxide and 2.5% aqueous potassium ferrocyanide are combined. The final solution appears brown. The tissue may not blacken quite as nicely as seen with osmium tetroxide, but may turn a dark brown. It will later blacken during the standard alcohol dehydration steps.

Potassium Permanganate

Historically, the development of the permanganate fixatives (Luft, 1956) preceded the standard glutaraldehyde-osmium tetroxide fixation protocol (Sabatini et al., 1963). Although now used only occasionally, permanganates are useful to fix certain plant specimens and yeast cells.

Permanganate-fixed tissues, especially plant tissues, are noted for their preservation of membranous elements. The remainder of the tissue has a pleasing appearance, but upon close examination appears washed out with few nonmembranous elements visible or prominent. Permanganate is known to penetrate tissues more rapidly than glutaraldehyde. The mechanism of fixation is not known, although several theories are postulated (Hayat, 1981) to explain its reactivity with tissue components. An example of permanganate fixation is shown in Chapter 20, Figure 20.173.

Many protocols using permanganate are available; however, the most common is to use a buffered solution of 3% potassium permanganate ($KMnO_4$) for 2 hours in the cold (4°C). The specimens are then rinsed in several changes of buffer overnight and dehydrated in a standard protocol. Postfixation in osmium tetroxide is not used with potassium permanganate fixed tissues.

Fixative Additives

There are several substances that may be added to fixatives to enhance fixation of particular structures. For example, tannic acid added to the glutaraldehyde fixative will enhance membranes (Kalina and Pease, 1977), microtubules (Burton et al., 1975), and microfilaments (Goldman, et al., 1979; Maupin and Pollard, 1983). Apparently tannic acid acts as a mor

dant to enhance later staining with heavy metals. The techniques to enhance particular features of the cell are numerous and are detailed in reference text books (Glauert, 1975; Hayat, 1981).

Fixation of Plant Tissues

In general, fixation of plant tissues is more difficult than animal tissues. Although a glutaraldehyde-osmium tetroxide combination is generally the most effective fixative for plant tissues, there are several minor modifications in this protocol that have their basis in differences in animal and plant specimens. There are also several reasons why plant tissues do not fix as well as animal tissues. Often plants are covered by waxy substances or cuticles that are hydrophobic and impede fixative penetration. The low protein content of plants renders glutaraldehyde less effective at cross-linking protein. The cell wall of plants presents a partial boundary to penetration of glutaraldehyde. The vacuole of plants rapidly dilutes the fixative concentration, necessitating the use of fixative of higher concentrations. A pH of 7.0 usually is used during fixation. Generally, when fixing plant tissues, it is not as important to begin fixation immediately after collecting the specimen (e.g., a leaf) since autolytic changes take place more slowly.

Preparation of plant cells may vary somewhat from the traditional methods employed for mammalian cells. For instance, while similar types of buffers may be employed (phosphates being the most common), the pH tends to be somewhat lower, at pH 7.0 to 7.2. Fixation times are usually longer (from 2 to 4 hours, on average), and dehydration and resin infiltration protocols are also lengthened. In some impenetrable specimens such as seeds, it is necessary to remove or abrade the seed coat in order to permit the reagents to enter. Some investigators may even resort to fixation times in glutaraldehyde and osmium tetroxide from 12 to 18 hours or more. This may seem excessive and even questionable; however, in dormant specimens and exceptionally dense tissues the protracted fixation regimes may be necessary. Low viscosity embedding resins such as Spurr's epoxy resin or LR White acrylic resins are recommended (see section "Embedding" later in the chapter).

HINT: A good resin formulation consists of taking one part of complete Spurr's resin and mixing it with one part of complete epoxy 812. The mixture yields a resin with good penetration and excellent contrast and stainability properties.

A commonly used protocol for plants follows.

Primary Fixation	immersion of small pieces of tissues in 2–4% phosphate buffered glutaraldehyde or formaldehyde/ glutaraldehyde mixtures	2–4 hr
Washing	buffer rinses (3 ea for 30 min) and possibly 1 overnight rinse	2–18 hr
Secondary Fixation	1–4% buffered osmium tetroxide	2–4 hr
Dehydration Series	25% ethanol	20 min
	50% ethanol	20 min
	75% ethanol	20 min to overnight
	95% ethanol	30 min
	100% (2–3 changes)	30 min ea
Transitional Solvent	propylene oxide (3 changes)	20 min ea
Resin Infiltration	2 parts propylene oxide : 1 part Spurr's	1–2 hr
	1 part propylene oxide : 1 part Spurr's	1–4 hr
	1 part propylene oxide : 1 part Spurr's	1–4 hr
	complete Spurr's resin	1–4 hr to overnight
Embedding	complete Spurr's resin	1–4 hr
Curing	60°C	48 hr

Washing

After primary fixation with glutaraldehyde, the tissue is usually washed in the same buffer vehicle used in the glutaraldehyde fixation step. Washing is extremely important because it eliminates any free unreacted glutaraldehyde that remains within the tissue. Aldehydes remaining from the primary fixation will be oxidized by osmium tetroxide. Some protocols call for one or two 10-minute washes of the tissue in buffer. Unreacted glutaraldehyde will diffuse as slowly outward from the tissue as inward so that a minimum of a few hours of washing with at least three changes of buffer is recommended. (Residual glutaraldehyde or partially polymerized aldehyde may generate a "peppery" background upon combination with osmium tetroxide.) Several rinses with at least one overnight wash in buffer will eliminate most of the unreacted glutaraldehyde. It is desirable to include $CaCl_2$ (about 5 mM) to prevent extraction and to enhance preservation of nucleic acids. Since small quantities of buffer are not costly, it is advisable to fill the vial to the top with buffer to extract as much glutaraldehyde from the tissue as possible. Washing is usually performed at 4°C overnight.

Dehydration

Dehydration is the process of replacing the water in cells with a fluid that acts as a solvent between the aqueous environment of the cell and the hydrophobic embedding media. Water is a highly polar molecule that is, by far, the major component of virtually all cells. Common dehydrating agents are ethanol or acetone. Ethanol is the most widely used dehydration agent. The general philosophy of the dehydration step is to replace water within the tissue gradually by using a graded series of dehydration agents. Usually 30 or 50% ethanol or acetone is the first solvent tissue is exposed to after secondary fixation followed by 70%, 85%, 95%, and absolute ethanol or acetone. As one reaches higher concentrations of the dehydrating agent, the time that tissue is exposed to the dehydration agent and the number of changes of the dehydration agent are increased in order to eliminate the small amount of water remaining in the tissue. Ethanol is preferred to acetone as a dehydrating agent because anhydrous acetone absorbs water from the atmosphere and is a more powerful extractor of lipids within the cell. Molecular sieves can be used to desiccate alcohol for ready use in dehydration protocols.

CAUTION: The fine grit from the ceramic molecular sieves may attach to the specimens and become embedded with them, damaging a glass or diamond knife at the time of sectioning.

The hygroscopic nature of dehydrants presents a problem because of their absorption of water from the air. Absolute ethanol or acetone left open for a short time will absorb water to the degree that it will not be capable of eliminating all of the water from the tissue. It is extremely important to keep dehydrants sealed tightly and not open them too often. Bottles of dehydrants (100%) that have been opened and resealed and have been around the laboratory for months must be suspected of having absorbed water. They can be used to make the lower percentage (e.g., 50%) dehydrants used in the initial steps of dehydration. To avoid water absorption during the later phases of dehydration, it is important to use sealed vials. When replacing one dehydrant with another, it is necessary to do so as rapidly as possible to avoid drying of the tissue.

Acetone, ethanol, and other dehydration agents tend to extract lipids from the tissue. Lipid droplets may appear nonhomogeneous or may have lost their electron density due to dehydration steps. Osmium tetroxide used during tissue fixation will help stabilize unsaturated lipids during the dehydration process.

Use of Transitional Solvents

Replacement of the dehydration solution by another intermediary solvent that is highly miscible with the plastic embedding medium is usually necessary to interface with the embedding media. The standard solvent used is *propylene oxide,* a highly volatile and potentially carcinogenic liquid. It will also further dehydrate the tissue. Like absolute alcohol or ethanol, it is highly hygroscopic, and the same precautions mentioned previously for these substances should be taken with propylene oxide. Usually more than one change of propylene oxide is necessary to replace the alcohol. Although most embedding media are directly miscible with dehydrating agents, most protocols employ a transitional solvent between the dehydrant and the resin to speed up the infiltration process. Another transitional solvent is acetone.

Infiltration of Resin

Infiltration is the process by which dehydrants or transition fluids are gradually replaced by resin monomers, most of which are very viscous like *pancake syrups.* Epoxy mixtures are introduced gradually into the tissue block after dehydration. The solvent, propylene oxide, is mixed with the epoxy and placed into vials with the tissue. Gradually, the epoxy-solvent ratio is increased until pure epoxy is used. The pure resin specimens are transferred into molds or capsules containing the resin and are final-

ly placed into an oven where the epoxy components polymerize to form a solid. A typical infiltration schedule follows.

1 part propylene oxide : 1 part epoxy	15 min–1 hour
1 part propylene oxide : 2 parts epoxy	15 min–2 hours
1 part propylene oxide : 4 parts epoxy	15 min–2 hours
pure epoxy	overnight

Place tissue and pure epoxy in embedding mold and place in oven.

During infiltration by the epoxy, the vials of tissue are gently agitated on a turntable that positions the tissue vials at about 30° from vertical and slowly rotates the tissues (Figure 2.11). Tissue vials should be closed so that moisture from the air does not enter them.

Some tissues may be processed up to pure liquid resin and stored for weeks in the freezer. Under these conditions, the resin will polymerize very slowly. It is often wise to place some of the samples in epoxy and store them in the freezer. If there are problems with embedding, such as holes in the tissue due to poor

FIGURE 2.11 Rotary mixer used to infiltrate tissues. Vials containing tissue are situated in holes of the mixer.

infiltration, then the tissue in the freezer may be processed backward (usually to propylene oxide), reinfiltrated, and then embedded again.

Embedding

The liquid embedding media must harden to form a solid matrix that thoroughly permeates the tissue. In common practice, this is accomplished by using epoxy monomers that harden with time and under certain curing conditions. Thus the tissue, which at one time was primarily hydrated, is solid and stable after embedding.

Epon Embedding

In the majority of instances, an epoxy embedding media is used. In the early 1960s, an epoxy embedding media, marketed by Shell Oil Co. under the brand name of Epon (or Epon 812), was introduced as an embedding medium by A. Glauert and subsequently modified by J. Luft (Luft, 1961). For many years, it was extremely popular and, therefore, used extensively. Epon, as such, is no longer available but has been replaced by similar types of epoxy resins. Although Epon is no longer in use, much of the embedding for electron microscopy used this embedment. Thus for historical purposes it is worthy of note.

The components of different epoxies are generally the same. Epon and its successors are made by immediately thoroughly mixing the resin, the hardeners, and the accelerator. Using Epon embedding media as an example, the components are as follows:

Resin	Epon 812 or substitute
Hardeners	DDSA (dodecenyl succinic anhydride)
	NMA (nadic methyl anhydride)
Accelerator	DMP-10, DMP-30, BDMA (benzyldimethylamine)

The hardness of the Epon mixture is controlled by the relative amounts of DDSA and NMA, the former producing softer blocks and the latter harder blocks.

Measuring Embedding Media

Because the ingredients of resin mixtures are usually highly viscous and difficult to clean from volumetric cylinders, it is not easy to mix them based on volume. The most convenient and accurate way to measure embedding media is to weigh the compo-

nents on a top-loading balance. First, the container is tared (or zeroed), and the first ingredient is added to reach the weight called for in the instructions. The second ingredient is added to reach the cumulative weight total of the first and second ingredients. Most embedding media can be hardened into materials of various degrees of firmness by varying the proportions of the ingredients. See the manufacturer's recommendations to obtain the necessary hardness.

CAUTION: All of the components of embedding media have the potential to cause skin irritation and rash. Use gloves when handling them and work in a well-ventilated room. Fumes must be avoided since epoxys are allergenic and are suspected of being carcinogens.

Mixing Embedding Media

It is recommended that the resin and the hardener be thoroughly mixed in a disposable plastic beaker using a resin mixer, a glass rod, or a tongue depressor. The accelerator is then added and mixed thoroughly with the above. Vigorous mixing for at least 7 minutes will assure that polymerization will be adequate. Mixing that is too vigorous will incorporate numerous air bubbles into the plastic that will interfere with embedding. It is possible to remove the bubbles by placing the resin mix under vacuum. Resin is not easily cleaned from the container or from the stirring rod, so the use of disposable materials is suggested. Contaminated utensils and partially empty containers are placed at 60°C for 48 hours to polymerize the resin prior to disposal.

Other Embedments and Their Use

There are many types of embedding media. For the most part, they are referred to by their brand names, although in some instances there may be little difference between one brand and another. Some embedding media are chemically unique and have special uses, whereas others are replacements for Epon and show minor difference from the Epon formula. The most common epoxy embedments in use today are *Araldite*, the epoxy replacements for Epon (see above) and *Spurr's*. Spurr's, formulated by Dr. A. R. Spurr, is an epoxy embedding medium of low viscosity. *Lowicryl* is a low viscosity acrylic embedding medium that can be used at low or high temperatures for immunocytochemistry. Denaturation of proteins is minimized at low temperatures, and the embedding medium will accept a small percentage of water remaining in the tissue. The use of propylene oxide thus may be avoided. LR White is a low viscosity acrylic embedding medium that can also be used for cytochemistry (see below). *Methacrylate (GMA or HPMA)* is a water-miscible embedding medium that is also often used for cytochemical studies, although it has often been used routinely for a variety of purposes. Water-soluble embedding media such as glycol methacrylates do not necessitate complete dehydration prior to embedding. They, like LR White described above, cause minimal denaturation of protein during tissue processing.

Suppliers provide specific instructions for mixing embedding media and curing of the embedment with the purchase of the embedding kits. As examples, recommended methods are provided here for four of the more common embedding media: EMbed 812, Araldite 502, Spurr's, and LR White.

Epoxy Resin 812

The standard replacement for Epon is a resin with the number 812, which is made by various manufacturers. Several brand names (Poly/Bed, EMbed, etc.) have been used to further characterize the mixture.

To make 812 embedding media:

TABLE 2.4 Epon Embedding Media (Glauert, 1991)

	Soft	Medium	Hard
812 Resin	20 ml (24 g)	20 ml (24 g)	20 ml (24 g)
DDSA	22 ml (22 g)	16 ml (16 g)	9 ml (9 g)
NMA	5 ml (6 g)	8 ml (10 g)	12 ml (15 g)
BDMA	1.4 ml (1.5 g)	1.3 ml (1.5 g)	1.2 ml (1.4 g)

The 812 resin can be prepared by measuring the ingredients either by volume or by weight as shown in Table 2.4. By varying the components, it is possible to prepare resin mixtures that will polymerize into plastics of various hardness. The more NMA, the harder the plastic; the more DDSA, the softer the final plastic. The more commonly used formulation is the medium mixture, which is suitable for most specimens. Harder specimens, such as plants, will require the hard mixture. Soft mixtures are useful with softer specimens and are easier to cut with a glass knife. For extremely thin sections, as may be cut with a diamond knife, the resin mixture should be formulated to give a rather firm block.

For infiltration and embedding schedule for epoxy mixture 812, see the schedule in Table 2.1.

Araldite

Araldite embedding media is an epoxy mixture characterized by uniform hardening and little shrinkage in the process. Sections show good stability under the electron beam. Two popular formulations of Araldite are available (CY-212 and 502). Araldite can be combined with one of several other epoxy mixtures to derive the benefits of both (Table 2.5).

To make Araldite 502:

TABLE 2.5 Araldite 502 Embedding Medium (Glauert, 1991)

Araldite 502	19.0 ml	(22.0 g)
DDSA	21.0 ml	(21.0 g)
Dibutyl phthalate (optional)	0.6 ml	(0.6 g)
BDMA	1.2 ml	(1.3 g)

The Araldite and DDSA can be mixed first and stored at 4°C in a tightly sealed container in the refrigerator for up to six months. After warming the closed container to prevent moisture condensation, the BDMA is added, and the ingredients are mixed thoroughly just before use.

Infiltration and embedding schedule for Araldite 502:

propylene oxide (2 changes)	15 min ea
1 propylene oxide: 1 resin mixture	1 hr
1 propylene oxide: 4 resin mixture	3–6 hr
pure resin	1 hr
pure resin mixture for polymerization	overnight 35°C
	next day 45°C
	next day 60°C

Satisfactory results can be obtained by overnight polymerization of resin mixture at 60°C.

Spurr's

The lowered viscosity of Spurr's (1969) embedding medium facilitates its penetration into a variety of tissues that are otherwise difficult to penetrate. Moreover, its routine use for tissues of all hardnesses is frequently advocated. The resin, vinylcyclohexane dioxide (VCD), is a di-epoxide that yields highly cross-linked polymers. The relative amount of flexibilizer (diglycidyl ether of polypropyleneglycol or DER 736) in the preparation regulates hardness, although a separate hardener (nonenyl succinic anhydride or NSA) is often used. The accelerator (dimethytlaminoethanol or DMAE) governs the speed of the reaction. See Table 2.6 to make Spurr's.

TABLE 2.6   Spurr's Embedding Media (Spurr, 1969)

Component	Firm Mixture	Soft Mixture	Rapid Cure
VCD Resin	10.0 g	10.0 g	10.0 g
DER 786 Flexibilizer	6.0	7.0	6.0
NSA Hardener	26.0	26.0	26.0
DMAE Accelerator	0.4	0.4	1.0
Cure Time (70°C)	8 hr	8 hr	3 hr

The medium is prepared by weighing components on a top-loading balance in a disposable plastic beaker. The accelerator should be added last after thoroughly mixing all of the other components. Then a final, thorough mixing should be undertaken. Embedding media is freshly prepared for each use or stored at approximately −20°C. The medium is compatible with ethanol or acetone dehydrants and with propylene oxide. The resin is very sensitive to traces of moisture (yields very brittle blocks), so anhydrous conditions must be scrupulously maintained.

CAUTION: Spurr's resin components, especially VCD, are quite toxic and are carcinogenic.

Guidelines for Successful Epoxy Resins

Glauert (1991) has summarized the proper formulations and handling of the Epon and Araldite epoxy resins. Her recommendations are as follows:

1. Store the basic components (resins, hardeners and catalysts) at room temperature rather than in the refrigerator or freezer.
2. In order to reduce viscosity and facilitate mixing, the components may be warmed up to 60°C. After combining the warmed components and stirring for 5 to 10 minutes, the mix is maintained at room temperature. The complete resin mixture should be made up freshly for each embedding rather than storing mixtures in the freezer for more than several days.
3. Although it is possible to speed up polymerization of the resin at elevated temperatures, more consistent blocks will be obtained if one does not exceed 60°C.

4. The recommended accelerator for the Epon substitutes is the less viscous reagent, BDMA, used at 3% final concentration rather than DMP-30. Since moisture may adversely affect the accelerators, they should be stored under anhydrous conditions and discarded after the bottle has been open for 6 months.
5. Use a verified formula for the preparation of the resins since several errors exist in the literature and in some reference books. Formulation based on volume is preferable to gravimetric procedures. The use of the anhydride:epoxide ratio corrections are not necessary.

Finally, prudence dictates that one should run a trial formulation and polymerization on any new or questionable batch of components rather than risking the loss of valuable specimens.

Combining Different Resins

One may take advantage of the desirable properties of different resins by combining the completed mixtures in various proportions. This is especially true for the epoxy resins. For instance, equal volumes of the complete resin mixes for epoxy 812 and Spurr's low viscosity medium gives an embedding medium with good penetration and reasonable contrast. Likewise, Mollenhauer (1964) showed that a mix consisting of 100 ml of the complete epoxy 812 combined with 60 ml of the Araldite 502 mix gave a plastic especially suited for plant materials that was easier to section than either epoxy used alone.

LR White

LR White is an acrylic resin often used for enzyme and immunohistochemistry and cytochemistry. Its low viscosity allows rapid embedding, and its hydrophilic nature allows the penetration of water molecules that contain substrates or antibodies that may be brought into apposition with the sections. The fixation protocol is governed by the need to preserve enzymes and antigens for subsequent localization (see Chapters 9 and 10). Ethanol dehydration is preferred. Infiltration times in resin as short as 3 hours are possible, although overnight infiltration is suggested. Gelatin capsules are filled with the resin and sealed to keep oxygen from interfering with the polymerization process. The embedded tissue blocks are cured at 60°C with little to no temperature fluctuation for one day. A rapid tissue processing protocol for LR White is given later in this chapter and a method to embed tissue culture cells has been developed (Hirst, 1991).

Low Pressure Removal of Solvents and Air Bubbles

This optional step employs the use of a sealed container connected to a vacuum pump to remove infiltration solvents and air bubbles. Small amounts of infiltration material remaining within tissues are volatilized by this procedure. A large number of bubbles appearing at the surface indicate significant amounts of infiltration agent remaining in the tissue. In addition, small air bubbles within the embedment and clinging to the tissue will enlarge and surface under vacuum conditions. Some investigators advocate warming of the resin prior to mixing to minimize bubble formation.

To remove bubbles, pump the chamber containing open vials down to about 0.1 Pa. Watch through the chamber window for bubble formation and stop the pump when bubble formation ceases.

Curing of the Embedment

The final step in embedding is curing the tissue blocks. In the case of acrylics, curing involves linear polymerization of the mixture. With epoxy resins, a chain of polymerized resin is formed that crosslinks to other chains of resin to form an integrated meshwork that totally permeates the tissue. For most resins, increased temperature accelerates the rate of hardening. Ultraviolet light can be used in combination with catalysts to polymerize resins. Lowacryl resin may be hardened at low temperatures using ultraviolet light. From 12 hours to 3 days are necessary to harden most embedments. If capsules can be indented with a thumbnail then they are not fully hardened.

Embedding Containers

The resin is placed in polyethylene capsules, and the tissue is placed into the resin to gradually sink to the bottom. The capsules are a size that will fit most microtome chucks. The most common capsule goes by the brand name of BEEM (Better Equipment for Electron Microscopy; Figure 2.12A). These polyethylene containers come in various configurations. Most have pointed tips to form a pyramid-shaped protrusion allowing the specimen to protrude from the "block," as the hardened epoxy mass is termed. The pointed tip facilitates trimming the block after the epoxy has hardened (see Chapter 4). Special holders are manufactured to hold some BEEM capsules (Figure 2.12B). The holders can support about 25 capsules each and are convenient for moving capsules into an oven for hardening.

Silicone rubber embedding molds, although incompatible with certain resins, are available in various shapes (Figure 2.12B) to allow more accurate orientation of a specimen than in a BEEM capsule container. Tissues may be oriented to some extent in BEEM capsules by a process known as flat embedding (Figure 2.13). A razor blade is used to cut off the pointed end of a BEEM capsule. The cap to the BEEM capsule, which is rarely used anyway, is detached and the specimen is placed in the cap and oriented as desired. A small amount of epoxy mixture is added. The remaining cylinder is fit snugly into the cap and filled with epoxy mixture. One disadvantage of this procedure is that after hardening, large amounts of epoxy must be removed from the sides of the specimen to select the specimen for sectioning. A hand-held rotary grinding device is useful in removing large amounts of excess hardened epoxy (Figure 2.14).

CAUTION: Dust from a rotary grinder can be dangerous. Use a fume hood when grinding specimens.

Embedding Labels

There must be some means to identify the specimens at a later time. In a large experiment, for example, one may embed 10 specimens of each type and have 40 or more treatment groups. These capsules may be easily confused with one another unless they are individually marked. If a BEEM capsule is used, then a narrow strip of paper may be inserted inside the top of the capsule (Figure 2.15). It should have a pencil or typewritten identification code.

CAUTION: Some carbon typewriter ribbons or inks from pens may dissolve upon contact with the epoxy mixture.

The code should designate the experiment (usually a number), the treatment (preferably a letter), and the block number (usually a number). It is time consuming to make labels by hand; however labels can be printed relatively easily on a computer and cut out by hand (Figure 2.15). Computers can be used to make and print hundreds of labels in minutes.

Removal of Tissue Supports

Both BEEM and gelatin capsules may be removed by using a razor blade to cut them free from the

FIGURE 2.12A Two types of BEEM capsules.

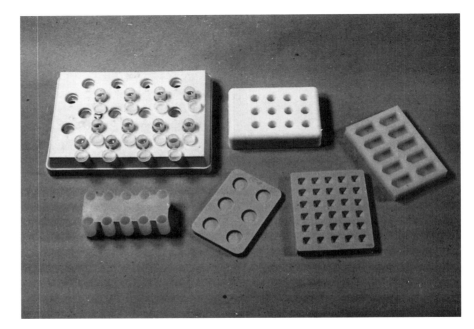

FIGURE 2.12B BEEM capsule holders containing BEEM capsules are shown at the top and bottom (*left*). Several embedding molds of various types are shown (*right*).

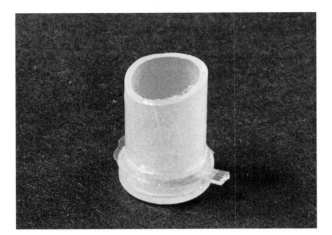

FIGURE 2.13 Flat embedding using BEEM capsules. The pointed tip of the BEEM capsule is cut off from the capsule, and the BEEM capsule cap is removed and place over the cut end. The cap forms a flat surface for flat embedding of tissues.

FIGURE 2.14 Rotary grinding device for removing excess epoxy around a tissue block.

hardened block, but this is a dangerous procedure that often results in cut fingers. BEEM capsule removers are an effective tool to accomplish the same purpose, and this equipment has a long life span (Figure 2.16). They are a worthwhile investment for anyone who uses BEEM capsules regularly. Exercise caution when using BEEM capsule removers with soft blocks, since they may distort the plastic and the tissue.

Tissues can be removed readily from silicone rubber mold holders, since the rubber is flexible and easily peeled away from the embedment.

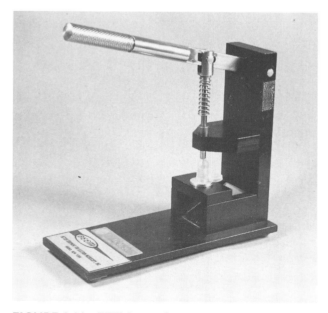

FIGURE 2.15 Computer-generated embedding labels (*left*) and embedding label placed on a BEEM capsule (*right*).

Embedding Cell Fractions

It was recognized early in its development that electron microscopy could be used to analyze *cell fractions* (Palade and Siekevitz 1956; Novikoff, 1956; de Duve and Beaufay, 1981) or to examine minute organisms that could be pelleted in a test tube. Elaborate fractionation schemes allowed the purification of plasma membranes and intracellular membranous components, and electron microscopy was often used to check the purity of such fractions or to perform cytochemical localization in pelleted material. Pelleted material is treated differently than other tissue during tissue preparation. The pellet should be formed in a rounded, but not pointed, centrifuge tube (Figure 2.17). The pellet may be sufficiently tight to allow use of a flexible scraper to remove the unfixed pellet from the side of the tube, and to be able to chop the pellet into small fragments while, at the same time, avoiding resuspension of the material. If so, pellets can be processed routinely by gently adding and withdrawing agents during the tissue preparative scheme. An example of pelleted plasma membranes is shown in Figure 2.18.

It is likely that the fractions will be too fragile to remove from the tube. The pellet should be fixed in the centrifuge tube. The pellet thickness must be under 0.5 mm for glutaraldehyde and osmium tetroxide to penetrate rapidly. The pellet thickness is controlled by the amount of material placed in the centrifuge tube initially. Fluids are added gently by running them down the side of the tube. Most tissues that appear pale initially

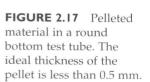

FIGURE 2.16 BEEM capsule remover.

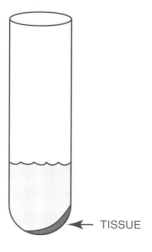

FIGURE 2.17 Pelleted material in a round bottom test tube. The ideal thickness of the pellet is less than 0.5 mm.

← TISSUE

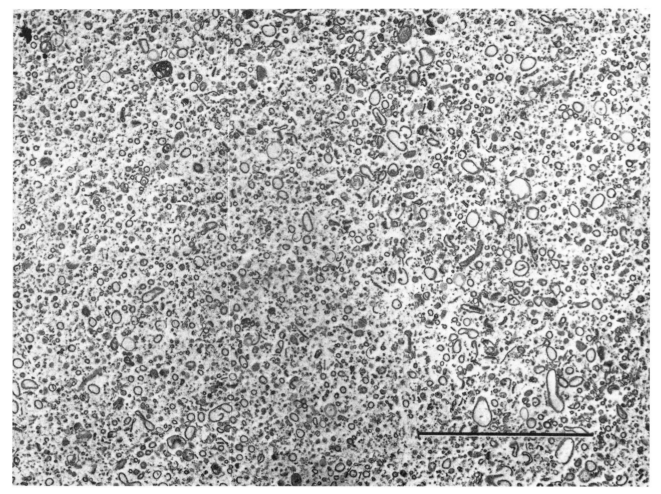

FIGURE 2.18   A section through a pellet of plasma membranes. The membranes were fragmented from the surface of cells, and in the process they formed the small vesicles seen in this figure. Vesicles were pelleted and processed for electron microscopy in the pelleted form. Bar = 0.75 μm.

will yellow slightly under the influence of glu-taraldehyde. When the surface of the tissue impacted against the tube has yellowed, one can be assured that glutaraldehyde has permeated the tissue.

Another method of dealing with cell pellets or cell suspensions is to fix the cell suspension in glu-taraldehyde, centrifuge the cells, and then suspend the centrifuged cells in warm (45°C) 2% agar or agarose in buffer. Upon cooling to room tempera-ture, the agarose may be cut into cubes with a razor blade and the cubes processed as usual (wash, osmicate, dehydrate, etc.).

Osmium tetroxide, an even slower penetrant than glutaraldehyde, is less likely to penetrate the pellet than glutaraldehyde. Often glutaraldehyde fixation imparts sufficient stability to the pelleted material to allow one to dislodge the pellet from the side of a test tube and process it intact. If the pellet is thin enough (under 0.5 mm), one can visualize the osmium tetroxide penetration by observing the darkening of the material on the side where the material impacts the container. The pellet may be able to be dislodged with a flexible scraper at this time. If not, it should be processed in its container through its dehydration, infiltration, transitional solvents, embedding, and curing steps. The con-tainer can usually be removed by a special method. Some microcentrifuge tubes and other tubes will fit snugly into microtome chucks.

Embedding Tissue Culture Cells

It is often desirable to process tissue culture cells in their original container if the container is not affected by any of the materials during tissue processing. Polystyrene, the most common type of plastic used for culture dishes and bottles, is dissolved by such tissue processing fluids as acetone or propylene oxide. Polypropylene containers are resistant to these chemicals. Fluids must be added gently to the containers during the processing steps in order to avoid disturbing the fragile monolayers. During the dehydration steps, the plastic covers or caps should be placed on the containers to prevent absorption of water by the alcohols.

It is also possible to grow cells on light microscope slides or coverslips that have been previously sterilized by autoclaving or by submersion in 70% ethanol for 20 minutes. After growing the cells on the microscope slide or coverglass, the substrate containing the cells is gently rinsed in physiologically balanced buffer and fixed by immersion in an aldehyde fixative for 15 to 30 minutes. Following several rinses in isotonic buffer, the cell/substrate assembly is placed into an osmium fixative for 30 minutes, rinsed in distilled water several times, and dehydrated in an ethanol series of progressively increasing concentrations up to absolute ethanol. After several quick dips into propylene oxide (taking care not to allow the cells to dry out), the cells are covered with several drops of a 1:1 mixture of propylene oxide and embedding medium for 5 minutes. Care must be taken to cover the cells completely with this mixture and not to allow the cells to become dry. This procedure must be conducted in a fume hood with caution since the mixture is extremely flammable and toxic. Next, one fills a standard BEEM capsule with the plastic embedding medium and, after draining the monolayer of the 1:1 mixture, the capsule is inverted on top of the area containing the cells. The slide or coverslip containing the inverted BEEM capsule is then placed into a 60°C oven and hardened for 24 hours. The BEEM capsule at 60°C (containing the partially hardened resin) may be snapped off of the glass substrate and the specimen block placed back into the oven for another 24 hours. Usually, the capsule snaps away from the glass substrate cleanly, and no glass shards are removed with the embedded cells. If the capsule does not release smoothly, one may gently pry the capsule away by slipping a single-edged razor blade between the slide and the capsule. If problems are still encountered, some researchers claim that placing the slide onto a small block of dry ice or cooling with liquid nitrogen and then prying with the razor blade will facilitate the release. One must be aware, however, that sometimes the two may not be separated cleanly by any method, and glass chips (that may damage a diamond knife) will be incorporated into the plastic.

A container is available commercially that permits one to culture cells directly onto the surface of a specially treated microscope slide (Figure 2.19). The sterile chambers consist of a slide to which a plastic chamber with a removable lid is attached. Organisms that have been grown onto the glass slide may be fixed and embedded *in situ* by gently pouring all of the fixation and embedding reagents into the chamber. Since the plastic chamber will be dissolved with propylene oxide or acetone, one should substitute absolute ethanol during infiltration with the embedding medium. Ethanol is miscible with the Epon substitutes. After hardening of the resin, the thin layer of plastic containing the cells is stripped from the glass slide and sectioned. For more details, see the article by Bozzola and Shechmeister (1973) who also describe a method for processing the same tissues for viewing by light microscopy, as well as SEM and TEM, to obtain a correlative picture. Alternative methods for conducting correlative microscopy are given in the reference book by M. A. Hayat (1987).

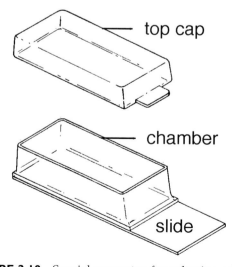

FIGURE 2.19 Special apparatus for culturing cells on a microscope slide. All fixation and embedding reagents are gently poured into the chamber and, after hardening, the plastic layer containing the specimen is removed and sectioned.

Suitable Containers for Tissue Processing

A variety of clean containers can be used for tissue processing. Usually, glass vials about 10 to 30 mm in diameter with tight snap-on caps are the most convenient (Figure 2.5). It is important to economize on chemicals such as osmium tetroxide, so vials that are too large should be avoided. Labels must be securely fastened to the vials, and permanent markers should be used on the labels. It is wise to label the vial and *then* apply a tape over the label to protect it from solvents.

The vials and/or their lids should be compatible with the chemicals used to process tissues, otherwise they may dissolve in the process. For instance, propylene oxide will dissolve many types of culture dishes or plastic test tubes. Polystyrene tubes or culture dishes are dissolved by propylene oxide and acetone, whereas polypropylene or glass containers are safe to use. If there is any doubt, solvents should be placed in the container to test their suitability. The same is true for the vial caps or liners for vial caps.

Rapid Tissue Processing Protocols

Typically, processing tissue for ultrastructural examination takes a week or more. In many instances, it is necessary to process tissues rapidly for ultrastructural examination. For example, pathologists commonly process biopsy specimens rapidly to obtain a speedy diagnosis. Several protocols have been developed that shorten the times necessary for the various steps described previously. One method is cited below although others are available (Miller, 1982).

A rapid method for processing *epoxy-embedded tissues* has been provided by Bencosme and Tsutsumi (1970) and Johannessen (1973). The entire process takes about 5 hours. The steps are listed as follows:

1. Fix tissue in fixative of choice for 30 minutes. The dimensions of the tissue block should not exceed 0.5 mm.
2. Rinse in 3 changes of 0.1 M sodium cacodylate buffer for 5 minutes each.
3. Postfix in 1% osmium in collidine-HCl buffer for 20 minutes.

4. Rinse extensively and stain *en bloc* with 2% aqueous uranyl acetate for 20 minutes.
5. Dehydrate in ethanol:
   - 50%   3 minutes
   - 70%   3 minutes
   - 95%   3 minutes
   - 100%   5 minutes
   (3 changes each)
6. Add 2 changes of propylene oxide for 5 minutes each.
7. Infiltration:
   - 1:1 mixture of epoxy and propylene oxide for 15 minutes
   - 3:1 mixture of epoxy and propylene oxide for 20 minutes
   - pure epoxy for 10 minutes
8. Embed in fresh epoxy and place in oven at 75°C for 45 minutes. Subsequently, transfer capsules to a 95°C oven for 45 minutes. Allow capsules to cool at room temperature before sectioning.

NOTE: Steps 1–7 should be carried out with the tissue vials in a rotator to allow even and rapid penetration of chemicals.

A rapid method for processing LR White embedded tissues has been developed by David Leaffer. The method allows sectioning in as little as 5 hours.

1. Fix tissue in 4% glutaraldehyde in 0.1 M cacodylate buffer for 1 to 2 hours. Tissue blocks under 1 mm across need less fixation time (e.g., 15–30 min).
2. Wash in cacodylate buffer for about 2 minutes.
3. Postfix in 1% osmium tetroxide for 1 hour.
4. Stain *en bloc* with uranyl acetate for 1 hour.
5. Dehydrate in ethanol:
   - 30%   3 minutes
   - 50%   3 minutes
   - 80%   3 minutes
   - 100%   5 minutes
   (2 changes each)
6. Infiltration and embedding: LR White, 20 minutes (2 changes each)
7. Embed tissues in LR White and place in a 75 to 85°C oven for 30 minutes.

Automatic Tissue Processors

Automatic tissue processors are available for electron microscopy. Such devices are expensive and are justified only when the volume of tissue pro-

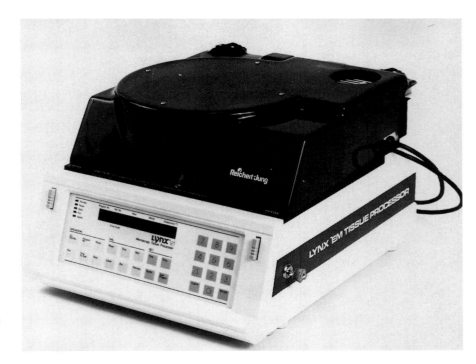

FIGURE 2.20 Automatic tissue processor. *(Courtesy of Leica.)*

cessing is great. Computerized programs allow fixation, dehydration, use of transitional solvents, infiltration, and resin infiltration to be carried out in a fully automated fashion (Figure 2.20).

Tissue Volume Changes During Specimen Preparation

Each chemical utilized in tissue processing has an effect on tissue volume. This problem cannot be avoided, but one should try to minimize volume changes as much as possible. For example, buffers are used with a range of osmolarities that maintain tissue volume. When performing careful morphometric studies (see Chapter 13), one should be aware of tissue volume changes.

The cell interior is maintained at a higher osmotic pressure than its exterior. The use of slightly hypertonic fixatives, as compared to the external environment of the cell, is advisable to prevent swelling of the cell. As a general rule, cells shrink slightly after glutaraldehyde fixation, but expand beyond their unfixed size in osmium tetroxide fixation. Subsequently, dehydration and embedding cause tissues to shrink by 10% or more.

Judging Adequate Specimen Preparation

It is easier to describe artifacts in specimen preparation than it is to describe normal features of specimens. Chapter 19 deals with many of the common artifacts of specimen preparation from fixation onwards in the tissue preparation protocol. Experience is the best guide to judging adequate specimen preparation. Perhaps the simplest way to describe the well-prepared specimen is to say that it is aesthetically pleasing to the eye. The old adage "What looks good is good" often applies, although not always, since some tissue artifacts are also very pleasing to the eye. Good looking tissues are the combined product of many steps in the preparative scheme. Once one is used to viewing tissues that are adequately prepared, it is easier to judge them in the future.

Tissues that are well fixed show intact membranes. There is an evenness in the density of the cytoplasm that indicates no areas of clumping of cytoplasmic matrix. The cytoplasm should have a distinctly increased density as compared with the extracellular space or nontissue spaces. The smooth and rough endoplasmic reticulum are tubular and saccular, respectively. Completely round vesicles of endoplasmic reticulum are usually distortions of the natural form of the endoplasmic reticulum. The ribosomes are expected to be attached to the rough endoplasmic reticulum in tissues known to have plentiful bound endoplasmic reticulum. Mitochondria have a finely granular matrix and are not swollen, nor is their matrix rarefied (clear of any tissue constituent or cytoplasmic matrix). The membranes of the nuclear envelope are evenly spaced except at pore regions. Membranes and cellular constituents should appear prominent, especially if the fixation protocol was selected to highlight specific components.

Tissue should be easily sectioned. If holes are present in tissue blocks, then either air bubbles were embedded with the tissue, or dehydration or infiltration were inadequate. Water remaining in the specimen during dehydration will not allow most embedding media to infiltrate the specimen adequately. By examining the surface of the sectioned block with a dissecting scope under reflected light, it can be determined if holes remain in the embedment or if they are caused by the sectioning knife.

Tissue blocks may not harden sufficiently during the curing process, posing a sectioning problem. This usually is the result of improper mixing of the ingredients of the embedding mixture. Either ingredients were added in the wrong proportions or sufficient time was not spent in mixing them. In addition, defective or insufficient accelerators/catalysts may prevent proper polymerization. Water may be present in certain anhydrous plastic monomers and will interfere with proper polymerization.

REFERENCES

Adrian, M., J. Dubochet, J. Lepault, and A. W. McDowall. 1984. Cryo-electron microscopy of viruses. *Nature* 308:32–36.

Armbruster, B., E. Carlemalm, R. Chiovetti, R. M. Garavito, J. Hobot, E. Kellenberger, and W. Villiger. 1982. Specimen preparation for electron microscopy using low temperature embedding resins. *J Microscopy* 126:77–85.

Bencosme, S. A., and V. Tsutsumi. 1970. A fast method for processing biological material for electron microscopy. *Lab Invest* 23:447–51.

Bozzola, J. J., and I. L. Shechmeister. 1973. *In situ* multiple sampling of attached bacteria for scanning and transmission electron microscopy. *Stain Technol* 48:317–25.

Burton, P. R., R. E. Hinkley, and G. B. Pierson. 1975. Tannic acid-stained microtubules with 12, 13, and 15 protofilaments. *J Cell Biol* 65:227–33.

Chen, E. C., et al. 1983. A fine structural study of microwave fixation of tissues. *Cell Biol Int Rep* 7:135.

Claude, A. 1948. Studies on cells: Morphology, chemical constitution of and distribution of biochemical functions. *Harvey Lecture* 43:1921–64.

de Duve, C., and H. Beaufay. 1981. A short history of tissue fractionation. *J Cell Biol* 91:293s–99s.

Demaree, R. S., R. T. Giberson, and R. L. Smith. 1995. Routine microwave polymerization of resins for transmission electron microscopy. *Scanning* 17:V25–V26.

Giammara, B. L. 1993. Microwave embedment for light and electron microscopy using epoxy resins, LR White, and other polymers. *Scanning* 15:82–87.

Gilkey, J. C., and A. Staehelin. 1986. Advances in ultrarapid freezing for the preservation of cellular ultrastructure. *J Electr Microsc Tech* 3:177–210.

Glauert, A. M. and P. R. Lewis. 1998. *Biological specimen preparation for transmission electron microscopy*. Volume 17: Practical Methods in Electron Microscopy. A. M. Glauert, ed. Portland Press, London and Miami.

Glauert, A. M. 1991. Epoxy resins: an update on their selection and use. *Microsc and Analysis* (September): 15–20.

Goldman, R. E., B. Choojnacki, and M-J. Yerna. 1979. Ultrastructure of microfilament bundles in baby hamster kidney (BHK-21) cells: The use of tannic acid. *J Cell Biol* 80:759–66.

Hayat, M. A. 1981. *Fixation for electron microscopy.* New York: Academic Press.

Hayat, M. A. 1987. *Correlative microscopy in biology: Instrumentation and methods.* New York: Academic Press.

Hirst, E. M. A. 1991. An easy method for orientated embedding tissue culture cell monolayers in LR White resin for postembedding immunocytochemistry. *J Electr Micros Tech* 17:456–58.

Johannessen, J. V. 1973. Rapid processing of kidney biopsies for electron microscopy. *Kidney Internat* 3:46–52.

Kalina, M., and D. Pease. 1977. The preservation of ultrastructure in saturated phosphatidyl cholines by tannic acid in model systems and type II pneumocytes. *J Cell Biol* 74:726–41.

Karnovsky, M. J. 1965. A formaldehyde-glutaraldehyde fixative of high osmolarity for use in electron microscopy. *J Cell Biol* 27:137A.

Karnovsky, M. J. 1971. Use of ferrocyanide-reduced osmium tetroxide in electron microscopy. In *Proc 14th Ann Meet Amer Soc Cell Biol*, p. 146. Abstract 284.

Kok, L. P., and M. E. Boon. 1992. Microwave cookbook for microscopists, art and science of visualization, 3d ed. Coulomb Press, Leiden, The Netherlands. ISBN 90–71421–20–1.

Leong, A. S-Y., M. E. Daymon, and J. Milios. 1985. Microwave irradiation as a form of fixation for light and electron microscopy. *J Path* 146:313–21.

Leong, A. S-Y., and Grove, D. W. 1990. Microwave techniques for tissue fixation, processing and staining. *EMSA Bull* 20:61–65.

Leong, A. S-Y. 1991. Microwave fixation and rapid processing in a large throughput histopathology laboratory. *Pathol* 23:271–73.

Leong, A. S-Y. 1993. Microwave for diagnostic laboratories. *Scanning* 15:88–98.

Login, G. R., and A. M. Dvorak. 1985. Microwave fixation for electron microscopy. *Amer J Path* 120:230–45.

Login, G. R., S. Kissell, B. K. Dwyer, and A. M. Dvorak. 1991. A novel microwave device designed to preserve cell structure in milliseconds. In W. B. Synder Jr., W. H. Sutton, M. F. Iskander, and D. L. Johnson (eds.), *Microwave processing of materials II.* Materials Research Society, Pittsburgh, 189:329–46.

Login, G. R., and A. M. Dvorak. 1994. *The microwave tool book.* Beth Israel Corp. ISBN 0–9642675–0–0.

Login, G. R., and A. M. Dvorak. 1994. Methods of microwave fixation for microscopy. A review of research and clinical applications: 1970–1992. In W. Graumann, Z. Lojda, A. G. E. Pearse, and T. H. Schiebler (eds.), *Prog Histochem Cytochem* 27/4, Gustav Fischer Verlag, New York, 127 pp.

Luft, J. H. 1956. Permanganate: a new fixative for electron microscopy. *J Biophys Biochem Cytol* 2:799–802.

Luft, J. H. 1961. Improvements in epoxy embedding materials. *J Biophys Biochem Cytol* 9:409–14.

Mascorro, J. A., and G. S. Kirby. 1991. Viscosity characteristics and hardening rates for EMbed 812 and LX 112 with alternative anhydride and catalyst choices. *Proc Electr Microsc Soc Amer* 49:292–93.

Maupin, P., and T. D. Pollard. 1983. Improved preservation and staining of HeLa cell actin filaments, clathrin-coated membranes, and other cytoplasmic structures by tannic acid-glutaraldehyde-saponin fixation. *J Cell Biol* 96:51–62.

Miller, L. A. 1982. Practical rapid embedding procedure for transmission electron microscopy. *Lab Med* 13:752–56.

Novikoff, A. B. 1956. Preservation of the fine structure of isolated liver cell particulates with polyvinylpyrollidone-sucrose. *J Biophys Biochem Cytol* 11:65–66.

Palade, G. E. 1952. A study of fixation for electron microscopy. *J Exptl Med* 95:285–98.

Palade, G. E., and P. Siekevitz. 1956. Pancreatic microsomes: An integrated morphological and biochemical study. *J Biophys Biochem Cytol* 2:171–200.

Robards, A. W. and A. J. Wilson (Principal Eds). 1993. Basic biological preparation techniques for TEM. In: *Procedures in Electron Microscopy.* 5:1.1 to 5:9.10. John Wiley & Sons, Ltd (UK). ISBN: 0–471–92853–4.

Russell, L. D., and S. Burguet. 1978. Ultrastructure of Leydig cells as revealed by secondary tissue treatment with a ferrocyanide:osmium mixture. *Tissue and Cell* 9:99–112.

Sabatini, D. D., K. Bensch, and R. J. Barrnett. 1963. Cytochemistry and electron microscopy. The preservation of cellular ultrastructure and enzymatic activity by aldehyde fixation. *J Cell Biol* 17:19–58.

Steinbrecht, R. A., and M. Muller. 1987. Freeze-substitution and freeze-drying. In *Cryotechniques in Biological Electron Microscopy.* Steinbrecht, R. A. and Zierold, K. (eds.), Springer-Verlag, New York, pp. 149–172.

Studer, D., M. Michel, and M. Muller. 1989. High pressure freezing comes of age. *Scanning Electr Microsc Supp* 3:253–69.

van de Kant, H. J. G., D. G. de Rooij, and M. E. Boon. 1990. Microwave stabilization versus chemical fixation. A morphometric study in glycomethacrylate- and paraffin-embedded tissue. *Histochem J* 22:335–40.

Chapter 3

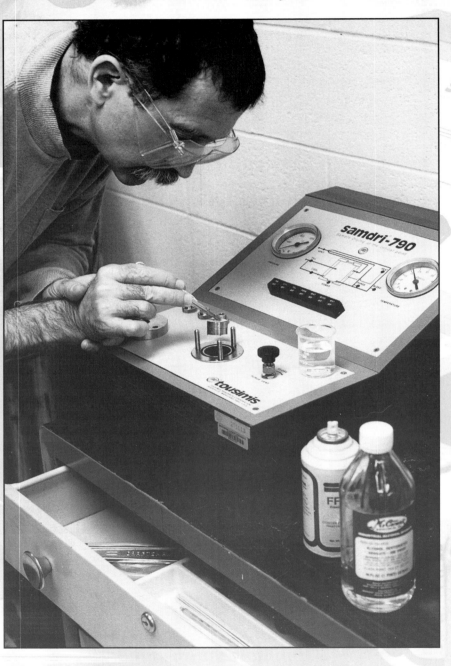

Specimen Preparation for Scanning Electron Microscopy

MANY OF THE STEPS involved in preparing biological specimens for scanning electron microscopy (SEM) are similar to the initial steps used in preparing samples for transmission electron microscopy (TEM). As presented in Figure 3.1, a generalized sequence for preparing specimens for SEM involves rinsing surfaces to remove debris, stabilizing in an aldehyde fixative followed by osmium tetroxide, rinsing in distilled water, dehydrating, mounting the sample on a metal specimen stub, and coating the specimen with a thin, electrically conductive layer. There are, however, several significant differences in preparing specimens for SEM as compared to TEM.

Surface Cleaning

Since the surface of the specimen usually is the region of interest in SEM, one must first clean the surface of materials that might otherwise obscure these features. Materials such as mucus, secretions, red blood cells, bacteria, broken cell debris, silt, dust, and detritus must be removed *prior* to fixation or this material may be chemically fixed to the specimen surface and be impossible to remove later. The extent to which one cleans the specimen surface depends on the nature of the specimen, the chemical makeup of the surface, and the environment from which the specimen is removed. Under ideal situations, the specimen is quickly rinsed in an appropriately buffered solution of the appropriate pH, temperature, and osmotic strength so as to mimic the natural milieu. One can then proceed to fix and further process the specimen. On the other hand, it may be necessary to clean the surface of obfuscating coatings as described in the following box.

How to Remove Surface Coatings

Even seemingly stubborn material can sometimes be removed from specimen surfaces by repeated gentle rinsing or flushing of the surface with a mild buffer. This may be accomplished by using a pipette, plastic squeeze bottle, or a dental irrigation device to direct gentle pulses of a rinse solution over the surface. When studying the structure of the endothelial cells lining the blood vessels in vascularized tissue, one should first perfuse the animal with physiologically balanced solutions to remove red blood cells and serum proteins and then perfuse the animal with the fixative. Otherwise, the cellular and proteinaceous components of the blood serum will be fixed onto the surface of the endothelium.

On difficult-to-remove coatings, one may incorporate various enzymes (mucinases, for example), mild detergents, or surfactants in the rinse solution prior to fixation. With hardy samples that are dry, it may be possible to use a camel's hair brush and mild bursts of microscopically clean, compressed air to dislodge debris. In all of these instances, one must treat all surfaces as gently as possible to avoid introducing artifactual changes. Always consult the published literature to see how others have cleaned specimen surfaces.

Buffers and Fixatives

The same criteria for the selection of buffers and fixatives as described in Chapter 2 apply for SEM. Although some researchers feel that buffers for use in SEM should be slightly hypotonic in contrast to the buffers used in TEM, this option is by no means clearly substantiated. When in doubt, rely on published SEM protocols until experience proves otherwise. If such protocols are not available, the same fixatives used in TEM studies will, in most cases, be suitable for SEM. These protocols use the standard, non-coagulating aldehyde-followed-by-osmium-tetroxide fixation regimes that have been used satisfactorily in TEM for many years. Under certain circumstances, it may be possible to employ some types of coagulating fixatives (Craf's, Bouin's, FAA) that are used in light microscopy.

As detailed in the previous chapter on specimen preparation for TEM, one must consider pH, buffer type, tonicity, concentration of fixative, temperature, and time in fixative. Specimen size may be a lesser concern in SEM fixation, since one is generally interested in viewing the outermost structures, which contact the fixative first and are fixed immediately. Nonetheless, one should strive to maintain a size of no more than 1 mm in one of the dimensions. This will ensure not only that underlying tissues are adequately fixed to support the surface features, but that the dehydration steps will be more efficient. A listing of some commonly used fixatives is given in Table 3.1. Microwaving can also be used to accelerate fixation and specimen preparation steps for scanning electron microscopy (Kok and Boon, 1992; Riches and Chew, 1984).

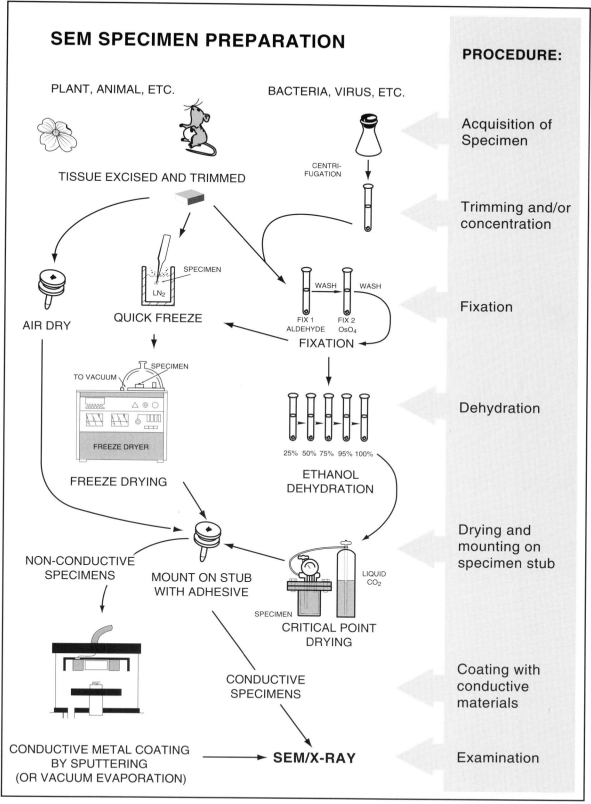

FIGURE 3.1 Schematic showing sequence of events for processing biological specimens for SEM. *(Modified from original figure provided by Judy Murphy.)*

TABLE 3.1   Fixatives Commonly Used in SEM

Specimen	Fixative	Buffer System	Reference
Procaryotes	• glutaraldehyde • osmium tetroxide • FAA (10% formalin, 85% ethanol, 5% glacial acetic acid)	cacodylate, phosphate veronal-acetate none	Watson et al. 1984
Fungi	• glutaraldehyde/OsO$_4$ followed by OsO$_4$ • OsO$_4$ vapors • glutaraldehyde followed by aqueous uranyl acetate	cacodylate, phosphate none cacodylate	Watson et al. 1984
Aquatic Organisms (protozoa, sponges, metazoa)	• glutaraldehyde/formaldehyde • Parducz (6 parts of 2% aqueous OsO$_4$ plus 1 part saturated aqueous HgCl$_2$—freshly prepared) • glutaraldehyde followed by OsO$_4$	cacodylate none phosphate, cacodylate sea or pond water	Maugel et al. 1980
Higher Plants	• glutaraldehyde followed by OsO$_4$ • FAA alone or followed by OsO$_4$ in buffer • formaldehyde followed by freeze drying • osmium vapors	phosphate buffer phosphate none none	Falk, 1980
Animals	• glutaraldehyde or glutaraldehyde/formaldehyde followed by OsO$_4$ • OsO$_4$ • glutaraldehyde/formaldehyde • FAA	cacodylate or phosphate cacodylate or phosphate various none	Nowell and Pawley, 1980

As can be seen in Table 3.1, various combinations of fixative and buffers are possible. In addition, it is possible to add various chemicals (tannic acid, lysine, ferrocyanide), or salts (MgCl$_2$, CaCl$_2$) to the fixative to effect a better preservation. More detail may be obtained by consulting the references by Murphy and Roomans (1984), Postek, et al. (1980) and Revel, et al. (1983) as well as the current scientific literature.

Working with Individual Cells or Macromolecules

During fixation for TEM, one normally works with multicellular organisms or tissues that may be handled as dissected cubes (described in the previous chapter). Small specimens (tissue culture or bacterial cells, viruses, individual molecules such as actin, etc.) pose the same problems they do in preparing them for TEM. A convenient way to deal with cells as small as 0.4 μm is to trap the cells on microfilters. In this procedure, a reusable cartridge is loaded with a filter of appropriately sized pores (the smallest is 0.2 μm), and the suspension of cells is gently passed through a syringe to settle on the filter (Figure 3.2). Alternatively, suction may be applied to draw liquid through the filter. In both cases, excessive force must not be applied or the cells will be damaged. After a brief rinse in the proper solution, the fixative is passed over the cells and left in contact with the monolayer for 15 to 60 minutes. After rinsing, the cells may be dehydrated by passing alcohols through the filter. The filter then is removed and dried prior to mounting, coating, and viewing in the SEM. To enhance conductivity, some microporous filters are constructed of silver.

Another method for dealing with cells or macromolecules is to use a substrate (such as a slide or coverglass) that has been coated with either poly-L-lysine (Mazia, 1975) or Alcian blue. The microscope slides are prepared for coating by first cleaning them using alcohol or a commercial acid cleaner. The cleaned slides are then coated by flooding them for 10 minutes with an aqueous solution containing 1 mg/ml of the reagent. After rinsing in distilled water, the slides may be used immediately or stored after drying. These reagents are effective adhesives for cells since they impart a positive charge to the negatively charged glass surfaces. Cells have a net negative charge and are therefore attracted to the now positive surface of the substrate. A fluid (culture medium or buffer) containing the specimen may subsequently be placed on the slide and the specimen allowed to settle by gravity onto the substrate. After rinsing and fixation, the adherent specimen is then processed (dehydrated, critical point dried, coated with metal, etc.) while attached to the substrate.

Rinsing and Dehydration

After fixation, tissue is rinsed in distilled water several times to remove buffer salts. The specimen is then either freeze-dried or critical point dried (see next section). In rare instances, it may be possible to air-dry samples after fixation and rinsing, but *most biological tissues will collapse, flatten, or shrink under these conditions.* For instance, it is likely that some

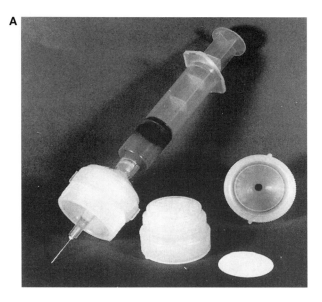

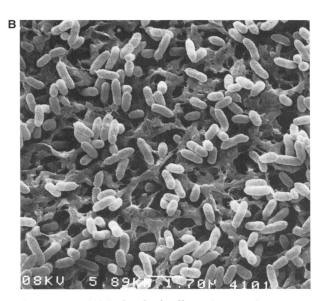

FIGURE 3.2 (A) Individual cells or tiny specimens suspended in a buffer system may be deposited onto microporous filters by passage through a filtering device. Fixatives are then passed through the syringe and over the cells, followed by ethanolic dehydration. The filter holder is then opened, and the filter is removed, dried, and mounted onto a specimen stub. After coating for conductivity, the filter surface is then examined by SEM. (B) Bacterial cells trapped on a microporous membrane filter and subsequently processed for SEM. Marker bar = 1.7 μm. *(Courtesy of R. de la Parra, Millipore Corp.)*

specimens (such as embryonic tissues) would shrink to half of their normal volume when air-dried. A major problem inherent in air-drying involves the passage of the receding air/water interface through the specimen. During air-drying, the *surface tension* forces associated with this interface are approximately 2,000 PSI (Anderson, 1951). Most biological samples will be flattened by such forces. To prevent this collapse, the passage of the interface through the tissues must be avoided. This may be accomplished using freeze-drying, critical point drying, or other vacuum drying procedures as described in the following sections.

Specimen Drying Techniques

Critical Point Drying

Most soft biological specimens are dried using this procedure since it is fast and gives reliable results in the majority of cases. In a commonly followed protocol, specimens are dehydrated in an ethanol series (30, 50, 75, 95, 100%) and transferred into a dehydrant-filled, cooled vessel termed a "bomb" (e.g., a vessel for compressed gases). After sealing the vessel, the dehydrant (usually ethanol) is displaced with a pressurized *transitional fluid* such as liquid carbon dioxide or Freon. Following several changes of transitional fluid to ensure complete displacement of the dehydrant, the bomb is completely sealed and the temperature of the transitional liquid is raised slowly by the application of heat. Pressure inside the vessel begins to rise in response to the heat. Eventually, the transitional fluid will reach a *critical point*, or a particular temperature/pressure combination specific for the transitional fluid, at which point a transition occurs and the density of the liquid phase equals that of the vapor phase. Dif-

ferent transitional fluids have different combinations of temperature and pressure to achieve their respective critical points (Table 3.2). Water is unsuitable for use as a transitional fluid since it has a critical temperature of 374°C and a critical pressure of 3,184 psi (i.e., biological specimens would be destroyed under these conditions).

This technique is successful because, at the critical point, the specimen is totally immersed in a dense vapor phase devoid of the damaging liquid/air interface one wishes to avoid. After achieving the critical point, the heat is maintained at the critical temperature (to prevent condensation of the vapor back to a liquid), and the vapor is slowly released from the chamber until the vessel is at atmospheric pressure. The dried specimen may then be mounted on a specimen stub, coated with a metal for conductivity (see Chapter 5) and viewed in the SEM. Figure 3.3 is a diagram of a typical critical point drying apparatus; Figure 3.4 is a photograph of a commercially available unit.

When placing small chunks of tissue in the critical point dryer, one may wish to use special holders to minimize damage and to maintain the identities of the specimens. Such holders are readily fashioned from polyethylene BEEM embedding capsules (see Chapter 2) by perforating the plastic with a hot dissecting needle (Figure 3.5A). For smaller specimens that may be lost through such perforations, one may fashion another type of holder by first trimming off the sealed end of the capsule to form an open cylinder. BEEM capsule covers are then bored out to remove most of the plastic so that only a circular ring remains. This ring is used to hold a tightly woven nylon mesh in place by slipping it over the mesh on both ends of the cylinder (Figure 3.5B). Slips of paper containing specimen information written in pencil (NOT INK!) are placed inside the chambers for identification purposes. Specimens may be transferred into the cham-

TABLE 3.2   Dehydrants and Transitional Fluids Used in Critical Point Drying

Dehydrant	Transitional Fluid	Critical Temp °C	Critical Pressure PSI
Ethanol, Amyl Acetate	Liquid CO_2	31.1	1,073
Acetone	Freon 116	19.7	432
Ethanol	Freon 23	25.9	701
Ethanol/Freon	Freon 13	28.9	561

The investigator can choose from several different transitional fluids based on the critical point desired. Liquid carbon dioxide is more commonly used (and less destructive to the ozone layer of our planet) than are the Freons (trademark of DuPont de Nemours and Company).

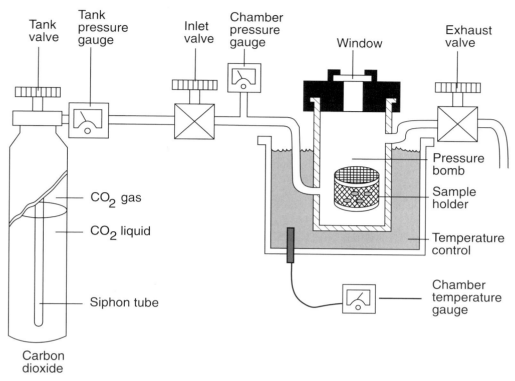

FIGURE 3.3 Diagram of critical point drying apparatus. The temperature of the pressure bomb may be regulated by a water bath or an electric heating element.

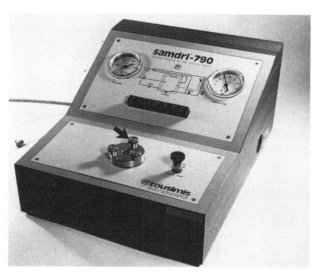

FIGURE 3.4 Photograph of a semiautomatic critical point drying apparatus. Specimens are placed into the recessed specimen chamber and the lid is sealed using the three thumb screws (*arrow*). Processing of the specimen is carried out by pressing a series of buttons (*above the specimen chamber*). The large dial (*upper left*) indicates the pressure inside the specimen chamber while the dial (*upper right*) indicates the temperature.

bers prior to or after dehydration. Care must be taken not to allow the specimen to air-dry, however. Commercial specimen holders are also available from original equipment manufacturers and from many EM supply houses.

For other types of specimens that may be attached to glass microscope slides, coverglasses or on micropore filters, one may fashion special holders out of plastic syringe barrels, plastic tubing, or even metal washers. These special holders will prevent the specimens from shifting in the critical point dryer and possibly damaging the specimen. Figure 3.6 shows some of these devices.

The critical point drying procedure is not without its problems. Some types of tissues may undergo significant shrinkage, ranging from 10% to 15% for nerve tissues to 60% for embryonic tissues (Boyde and Maconnachie, 1981). In addition, the transitional fluids are solvents that may extract steroids, carotenoids, porphyrins and actin. Additional information about critical point drying may be found in the articles by Boyde (1978) and Cohen (1977, 1979).

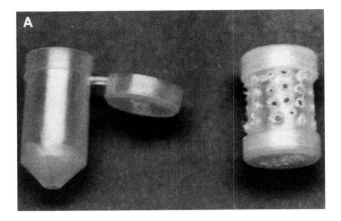

FIGURE 3.5 Devices for holding small specimens. (A) Unmodified (*left*) and modified (*right*) BEEM capsule used for embedding TEM specimens. Holes have been made in the side, and both ends have been capped off after insertion of specimen. (B) BEEM capsule with ends modified to accommodate either a screen mesh or micropore filter.

FIGURE 3.6 Commercially produced stainless steel holders for securing specimens during the critical point drying process.

Safety Precautions with Critical Point Drying

Dangerous pressures build up inside the bomb, and a few instances of the vessel cover rupturing have been reported. Modern instruments have protective discs that burst prior to the development of such pressures inside the chamber. Windows covering the chamber are of specially tested quartz or glass and should be checked for nicks or cracks before each use. It is recommended that the critical point dryer be placed inside of a fume hood or behind a shatter resistant window (plexiglass, for instance). One is cautioned not to look directly down into the pressurized chamber. Instead, a metal mirror (not glass) should be used to examine the chamber contents. The mirror should be mounted on a movable support that will fall away from the user in the event of a window failure. A well-ventilated room is necessary since the gases used in this process will displace oxygen and may asphyxiate or sicken unwary users.

Freeze-Drying

A second method for avoiding the water/air interface during specimen drying is to freeze the specimen rapidly while still in the aqueous phase. The frozen specimen is transferred into a special chamber designed to maintain the temperature lower than $-80°C$, and the apparatus is rapidly evacuated into the 10^{-1} Pascal (Pa) range or higher. While frozen, the solid ice gradually undergoes sublimation to the gaseous phase and is either absorbed by desiccants placed into the chamber or removed by the vacuum system. Depending on the size of the specimen, temperatures used, and the vacuum level achieved, this process may take from several hours to several days to complete.

Although it is possible to freeze unfixed specimens and then to freeze-dry them, the freeze-drying process usually begins after the specimen has been fixed and rinsed extensively to remove traces of fixatives. Sometimes the specimen is infused with a cryoprotectant such as sucrose, glucose, glycerol, ethanol, DMSO, or dextran to reduce ice crystal formation and damage. Unfortunately, most of these materials will not be sublimed away and will remain behind obscuring surface features. Ethanol is probably the most useful cryoprotectant since it is removed by evacuation. Experimentation is in order to determine whether or not a cryoprotectant is needed when attempting freeze-drying for the first time.

Freeze-Drying Procedure

In a typical procedure, possibly after infusion with cryoprotectant, fresh or fixed, ethanol dehydrated tissue is plunged into a liquid nitrogen chilled fluid such as isopentane, liquid Freon, supercooled liquid nitrogen, or liquid propane. (*DANGER: Propane is explosive!*) These fluids, termed *quenchants*, are used instead of plunging the specimen directly into liquid nitrogen, which tends to boil vigorously when specimens are placed in it. Without a quenchant, the gaseous phase generated by the vigorous boiling of liquid nitrogen insulates the specimen and slows the rate of freezing so that damaging ice crystallization occurs.

NOTE: Extremely rapid freezing rates are important in order to avoid ice crystal formation. When freezing rates are faster than 140 K per second, water is converted into an amorphous solid, avoiding crystal formation. For more information, see the review article by Gilkey and Staehelin (1986).

Quenchants wet the specimen surfaces but do not boil vigorously, so that the heat will be rapidly removed from the specimen to effect a rapid freezing. Unfortunately, the freezing rate is rapid only to a depth of about 15 to 10 μm, so that underlying tissues will undergo extensive ice damage. Because the surface features are probably the ones being investigated, this may not be of consequence.

After the specimens have been frozen in the quenchant, they may be stored in liquid nitrogen indefinitely, or they may be transferred to the cooled stage of the freeze-dryer. The stage may be chilled with liquid nitrogen or cooled using an electronic Peltier cooling system. Figure 3.7 shows a typical freeze-drying apparatus used in SEM specimen preparation.

After the specimen has been transferred onto the chilled stage rapidly (to avoid condensation of moisture), the apparatus is covered, evacuated to about 10^{-2} Pa, and maintained under vacuum/cold conditions as long as necessary for drying. After drying, the specimen stage is gradually warmed to room temperature, dry air is admitted to the chamber, and the specimen may be mounted onto the specimen stub for coating and viewing in the SEM. If the dried specimen must be stored for any period of time prior to coating and viewing, it must be stored in a desiccator.

A modification of the conventional method of freeze-drying was described by deHarven and colleagues in 1977. In this procedure human leukocytes attached to a poly-l-lysine coated coverglass were fixed in 2.5% glutaraldehyde in culture medium for 30 minutes at room temperature and dehydrated in an ethanol series. After standing in absolute ethanol for several minutes, the coverslips were plunged into an ethanol bath that had been chilled to $-85°C$. The specimens were left in the syrupy thick ethanol for several minutes and then transferred to the precooled ($-75°C$) stage of a freeze-dryer. The chamber was evacuated to 0.1 Pa for 20 minutes and the cooling stopped until the specimen warmed up to $0°C$. The temperature of the stage was then raised to 30 to $35°C$ and the dried specimen removed, coated, and examined by SEM.

NOTE: It is possible to fashion a simple freeze-dryer by cooling a large block of copper or brass in liquid nitrogen, placing the quenchant-frozen specimen into a recessed area on the block, covering the recess with another recessed block, and placing the chilled device inside a vacuum evaporator. The block will remain cold inside of the vacuum for several hours, probably long enough for small specimens to freeze-dry. Covering the specimen is essential in this device in order to prevent condensation of water or vacuum pump oil onto the specimen surface.

FIGURE 3.7 Freeze-drying apparatus showing specimen chamber where frozen hydrated specimens may be placed (arrow). The specimen is maintained at $-80°C$ while under vacuum until all solidified water has been removed by sublimation.

One advantage of freeze-drying is that much less shrinkage occurs with most specimens compared to critical point drying. Disadvantages include the problems associated with rapid freezing and transfer into the apparatus, the time involved in drying the tissues (sometimes days versus hours in the critical point drying procedure), and problems with nonevaporable deposits left behind on surfaces. Boyde (1978) has discussed the pros and cons of the freeze-drying technique.

Air-Drying Procedure

Under certain conditions it may be possible to air-dry biological specimens. Usually air drying of specimens that were in water, acetone or alcohols results in excessive shrinkage, cracking and collapse of fine structures as cilia or filapodia. However, some solvents such as ethanol and hexamethyldisilazane may be used to air-dry certain types of tissues.

> The bacterial preparation shown in Figure 3.2 was prepared by fixing the cells in glutaraldehyde followed by osmium tetroxide. After a brief rinse in distilled water, the membrane filter was dehydrated in an ethanol series (35, 45, 55, 65, 70, 85, 95 and 100% for 5 minutes each) followed by immersion in hexamethyldisilazane for 5 minutes. The filters were then allowed to air-dry at room temperature for 30 minutes prior to mounting on an SEM stub, coating with Pd/Au, and examination in the SEM.

Formerly, fluorocarbon compound, trade named Peldri II (Ted Pella Company), was used with good results on a wide variety of biological specimens (Kennedy et al., 1989). Unfortunately, it is no longer available.

Specimen Fracturing Procedures

If one wishes to expose the interior of biological specimens, it is possible to fracture the specimen using either a cryofracturing or a dry fracturing procedure as described in the following sections.

Cryofracturing

One method of exposing the interior of cells for viewing by SEM is to freeze the tissue in liquid nitrogen and fracture the frozen specimen while it is at liquid nitrogen temperatures. After the sample has been fractured, it is dehydrated, mounted onto a specimen stub, coated with a metal for conductivity, and examined in the SEM.

If one is working with unfixed tissues, it is desirable to cryoprotect them (10% glycerine or 10% to 25% dimethyl sulfoxide in buffer) prior to freezing in order to avoid ice crystal damage. Small pieces of tissue are essential in order to permit as rapid cooling rates as possible. Otherwise, even cryoprotected specimens may be damaged by ice crystal formation. (See Chapters 2 and 14 for information on rapid freezing and cryoprotection methodologies.)

One reason for freezing unfixed rather than fixed specimens is to permit the washing out of soluble proteins, salts, and other solutes when the fractured cells are thawed. These intracellular materials might otherwise prevent viewing deeply into the cell interiors, or they may obscure such ultrastructural details as the fine filaments that make up the cytoskeleton. A second reason for working with unfixed specimens may involve the use of cytochemical or immunocytochemical probes on the fractured and thawed cells (i.e., some enzymes or antigens may be denatured by fixation).

Since the use of chemically unfixed materials in the cryofracturing technique may lead to the introduction of a number of artifacts (ice crystals, migration of unstabilized cellular components, etc), one should first attempt the procedure using fixed specimens. If these prove unsuitable, then unfixed materials may be used. A good review of the cryofracturing procedures may be found in the articles by Tanaka (1981) and Haggis (1982). Tanaka's procedure is summarized below and illustrated in Figure 3.8.

> ## Cryofracturing Procedure of Tanaka (1989)
>
> 1. Excise small pieces of tissue (1 × 1 × 5 mm or less) and immediately place the tissue into 1% osmium tetroxide in a buffer of the appropriate type, pH, and molarity for the tissue being studied. Generally this is accomplished at 4°C for 60 to 90 minutes.
>
> 2. Rinse the tissue in isotonic buffer solution (three changes, 15 minutes each), and cryoprotect the fixed tissues by passage through 25 and 50% dimethyl sulfoxide (DMSO) aqueous solutions for 30 minutes each.
>   *CAUTION: DMSO is immediately absorbed through the skin and carries with it any toxic substances (i.e., arsenic salts of cacodylate buffers)*

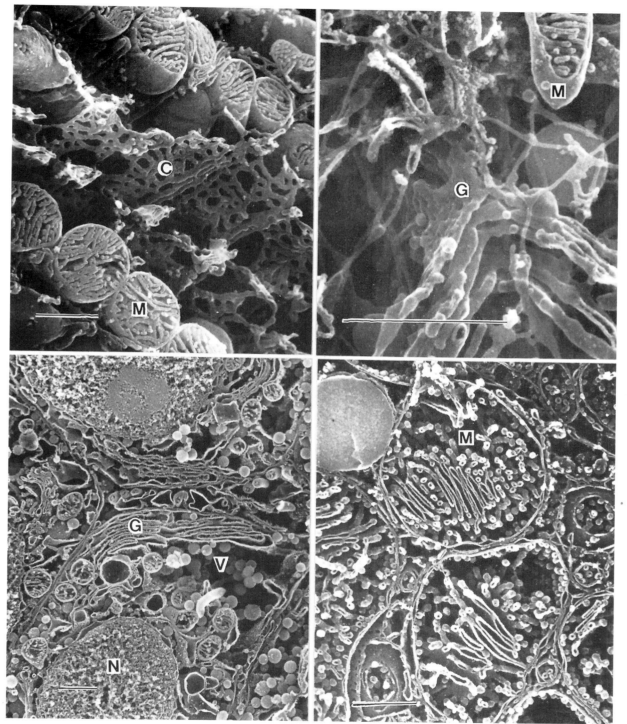

FIGURE 3.8 Examples of cryofractured tissues prepared by the technique of Tanaka (1981). Note the three-dimensional structural detail present: netlike nature of the smooth endoplasmic reticulum (C) surrounding the mitochondria (M). Golgi (G) bodies are present in some cells and nuclei (N) and vesicles (V) are readily visible. Marker bars = 1 μm.
(Courtesy of K. Tanaka.)

directly into the bloodstream. Wear rubber gloves and avoid splashing the DMSO solutions onto unprotected areas.

3. Rapidly freeze and adhere the tissue onto the surface of a metal plate by chilling with liquid nitrogen. Crack the frozen tissue block using a liquid nitrogen chilled scalpel and a hammer. This should be accomplished as rapidly as possible to prevent the formation of frost on the specimen surface. (To avoid frosting, it is best to fracture the specimen while it is submerged in liquid nitrogen.)

4. Immediately place the cracked, frozen tissues into 50% DMSO (at room temperature) to thaw.

5. Rinse several times in buffer solution to remove the DMSO, and place the tissues into 1% osmium tetroxide for 1 to 2 hours followed by 0.1% osmium tetroxide solution in the appropriate buffer. Keep in this solution for 24 to 72 hours at 20°C. (This step is used to remove cytoplasmic materials that might otherwise obscure ultrastructural details.)

6. Transfer the tissues into a freshly filtered 2% tannic acid solution for 2 hours followed by a 1% osmium tetroxide solution for 1 hour, followed by a buffered solution of 2% tannic acid for 12 hours, and then 1% osmium tetroxide for 1 to 2 hours. (This procedure builds up an electrically conductive coating of osmium.)

7. Rinse the tissue in distilled water (three changes, 15 minutes each), and dehydrate the specimens in an ethanol series.

8. Critical point dry the dehydrated tissues as described earlier in this chapter.

9. Carefully mount the specimens onto a specimen stub as outlined later in this chapter.

10. Coat the specimens with a heavy metal such as platinum (see "Methods for Metal and Carbon Evaporation" in Chapter 5) with a thickness of approximately 20 nm.

11. Observe in a high resolution SEM.

Dry Fracturing

Specimens that are not delicate may be fractured after fixation, dehydration and critical point drying. Several methods may be used to fracture the dry specimen. If the tissue block was attached to a stub, the specimen may be fractured with a sharp scalpel or razor blade. In some tissues, double-stick tape may be pressed gently onto the surface of the specimen and pulled to strip away the

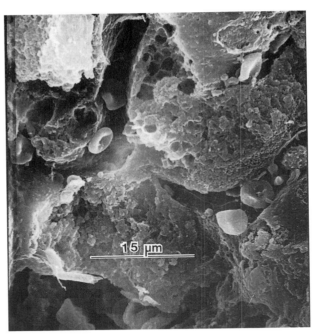

FIGURE 3.9 Specimens may be embedded in paraffin and the hardened paraffin fractured following the procedure of Shennawy et al. (1983) to expose internal features. Mouse liver showing red blood cells inside of lacunar spaces. Magnification bar = 15.0 μm. The excessive brightness shown in some areas of the micrograph is due to the phenomenon termed "charging," resulting from the buildup of high-voltage static charge that repels excessive amounts of electrons into the secondary electron detector.
(Courtesy of M. Doran and S. Pelok.)

uppermost layers. A fine needle may be used to dissect away tissue and expose the area to be studied.

Obviously this method may result in crushing or deforming delicate specimens, so it may be more desirable to embed the specimen in a supporting matrix such as paraffin prior to fracturing. In a procedure described by Shennawy et al. (1983), tissues are fixed, dehydrated, and embedded in paraffin using standard protocols employed by light microscope histologists. The paraffin block containing the specimen is trimmed to the level of the specimen, cooled, and the paraffin block is broken by snapping it in half. The two halves are deparaffinized using xylene, and the tissue blocks are placed into absolute ethanol, critical point dried, and examined in the SEM after mounting and coating with a heavy metal. Good results have been obtained using a variety of mammalian tissues (Figure 3.9).

Replication Procedures

If direct observation of a biological surface is not possible, a replica of the surface may be examined in the SEM. In instances where sacrifice of the live specimen is not feasible, or if one wishes to sequentially study changes on an identical specimen surface, or if the specimen undergoes unacceptable alteration when using more conventional procedures, then replication may be in order. Other reasons for using replicas are if a unique or rare specimen is too large to fit into the SEM specimen chamber, if the specimen is too precious to subdivide, if it consists of an interior surface (airways, vasculature, etc), or if the specimen is damaged by the vacuum or the electron beam. The three basic types of replicas are negative surface replicas, positive surface replicas, and corrosion casts.

Negative Surface Replication Using Cellulose Acetate Film

This simple but limited technique involves impressing an acetone-softened sheet of cellulose acetate film onto a surface and allowing the acetate to harden after the acetone evaporates. Since acetone is dropped onto the surface to be replicated, this procedure can be conducted only on hard, non-living surfaces such as bone, chitinous exoskeletons of

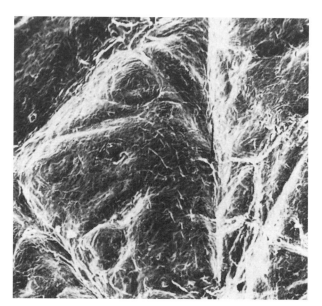

FIGURE 3.10A Cellulose acetate negative replication procedure of human skin surface.
(Courtesy of O. Crankshaw and SEM, Inc.)

insects, dried plant materials, etc. The specimen should have relatively low topography (since the procedure does not replicate high areas particularly well) and it should obviously not be damaged by the acetone. The technique has low resolution capabilities and is useful in magnifications up to several thousand times only (Figure 3.10A).

Cellulose Acetate Negative Replication Procedure

1. Cut a piece of cellulose acetate sheeting into a size suitable for the specimen (1" square, for example).
2. Apply several drops of acetone to the dry surface of the specimen and immediately press the cellulose acetate film onto the wet surface. Hold it in position under the thumb for about 1 minute. Be careful not to slip the tape along the surface during the drying procedure or the replica will be damaged. The specimen surface should not be excessively moistened with acetone or the film will become wet on the side against the thumb.
3. Gently peel away the acetate film, and keep this first replica for study purposes. Generally, the first peel is used to clean particulate debris from the specimen surface and may be discarded. On the other hand, some materials (pollen, bacteria, fibers, etc.) that were extracted from the specimen surface during replication may be of interest to the investigator. This type of replica is called an extraction replica.
4. A second replica is made of the now cleaned surface, and the acetate film is allowed to dry for 10 minutes in a clean environment. Keep track of which side was in contact with the specimen surface.
5. Mount the replica on a specimen stub with the replicated surface facing up. Care must be taken if glues or cements are used, since they will soften and distort the replica. It is best to use double-stick carbon tape rather than a glue that is dissolved in an organic solvent.
6. Coat the mounted replica with a conducting metal by sputter coating or thermal evaporation (see Chapter 5).

Negative/Positive Replication Using Silicone, Resin

The use of silicone based compounds for surface replication is more desirable than cellulose acetate because living specimens with high topography may be examined with better resolutions (Figure 3.10B). The technique is more complicated, however,

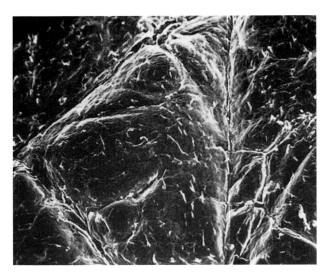

FIGURE 3.10B   Silicone negative impression material of human skin surface.
(Courtesy of O. Crankshaw and SEM, Inc.)

and requires several types of ingredients—some of which may be toxic (i.e., epoxy resins). With this procedure, one first makes a negative replica using the silicone impression material and then fills this negative impression with a plastic monomer (Spurr's epoxy resin, for example) to make a positive replica. The replication materials (silicones, epoxy resins, etc.) may be obtained from most EM or dental supply houses.

Silicone, Resin Positive/Negative Replication Procedure

1. Place the specimen to be replicated in a stable position. If animals or plants are used, it is important that movement of the specimen be minimal during replication.
2. Mix the silicone (polyvinylsiloxane or Xantropren Blue may be used) according to manufacturer's directions.
3. Clean the surface to be replicated by swabbing it with lintless cloths soaked in buffer solutions and followed by 70% ethanol, if possible, to remove loose debris and oils.
4. Apply the silicone impression material to the surface. This should be done as quickly as possible since the silicone may harden rapidly. The material should be spread thinly and evenly over the surface using compressed air to spread a thin layer across the surface. A thin layer is preferred initially since it will conform better than an overly thick one.
5. Wait the recommended amount of time for the material to harden.
6. Apply a thicker layer of silicone material to support the initial thin layer and allow it to harden as well.
7. Gently peel away the silicone replica from the specimen surface.
8. Examine the replica under a dissecting microscope for the presence of air bubbles or other irregularities and repeat the procedure if necessary.
9. If one wishes to examine the negative replica, it should be mounted on a specimen stub and coated for electrical conductivity.
   To examine a positive replica, it is first necessary to prepare a suitable low viscosity plastic resin such as Spurr's epoxy resin, Quetol, or one of the acrylics (see Chapter 2 for epoxy resin formulations).
10. Pour the plastic resin into the silicone negative impression carefully to avoid trapping air bubbles. Should bubbles be noticed, they may be dislodged using a small wooden probe, or the silicone and resin assemblage may be placed into a chamber under gentle vacuum to remove them.
11. Harden the plastic resin by placing the silicone and resin assemblage into an oven at 60°C overnight.
12. After cooling, gently peel the silicone away from the hardened positive plastic impression.
13. Mount the positive plastic replica on a specimen stub, coat with metal for conductivity, and examine in the SEM.

Corrosion Casting of Animal Vasculatures

One can make an extensive series of impressions of the endothelial lining of blood vessels of a recently killed animal by first perfusing the animal with a physiological fluid to flush away blood (see Chapter 2, "Perfusion fixation"), followed by the injection of a methyl methacrylate resin mixture (Hossler et al., 1986; Olson, 1985). Upon polymerization of the casting compound, the tissue is dissolved in strong caustic solutions so that only the impression of the vessels remains (Figure 3.11). This technique has proven valuable to render the fine structure of animal vasculatures.

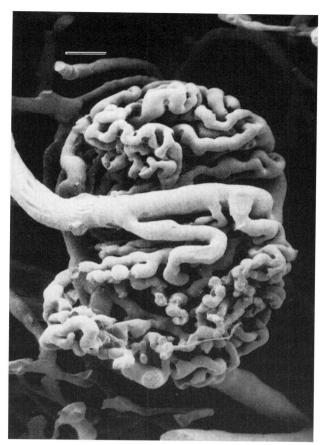

FIGURE 3.11 Corrosion cast of pig kidney glomerulus revealing formation of efferent arteriole within the glomerulus. Marker bar = 25 μm.
(Courtesy of B. J. Moore and K. R. Holmes.)

Corrosion Cast Replication Procedure

1. Kill the animal by injection and perfuse it with a physiologically balanced salt solution as described in Chapter 2.

2. After the blood vessels are totally clear and tissues have blanched, perfuse the vasculature with freshly prepared methyl methacrylate solution using a flow rate of 2 ml/min until the casting material totally fills the vessels.

3. Allow the animal to remain undisturbed until polymerization of the casting solution. Follow manufacturer's directions for the exact timing and temperatures to use for the polymerization.

4. Place the tissues in a corrosion solution composed of 20% NaOH or KOH to dissolve the organic material. This may take 12 to 24 hours,

depending upon the mass of tissue to be dissolved. The process may be accelerated by elevating the temperature of the solution to 50°C.

5. Rinse the vascular cast in distilled water. Extensive rinsing is necessary, and ultrasonication of the casts in distilled water is usually necessary to remove stubborn debris.

6. Dehydrate the cast by placing it into 95% ethanol and allowing it to air- or critical point dry.

7. Cut the cast into manageable sections, mount onto a specimen stub, coat with metal for electrical conductivity, and examine in the SEM.

A number of artifacts may be introduced during the replication process, so that familiarity with normally processed tissues is essential before one embarks on a study employing this method. It is obvious that examination of the actual specimen rather than a replica is more desirable; however, when this is not possible, replication does offer an alternative. For more information, see the references by Crankshaw (1984), Hodde and Nowell (1980), Hossler, et al. (1986), Olson (1985), Pameijer (1978), Pfefferkorn and Boyde (1974) and Scott (1982).

Specimen Mounting

After specimens have been fixed, dehydrated and dried using an appropriate protocol they may be attached to a metallic *stub* (usually made of aluminum) and then coated with a metal prior to insertion into the SEM. Most biological specimens must be coated with conducting metals or carbon (Chapter 5) to prevent the buildup of high voltage static charges that will degrade the quality of the SEM image.

Although specimens are sometimes glued directly to the stub using an adhesive, some specimens may first be attached to number of different substrates (glass microscope slides, coverglasses, cleaved mica sheets, microporous filters, polished metals such as aluminum, stainless steel, beryllium, etc.) and then attached to the stub.

An important first step in deciding which stub and glue to use is to determine the type of signal to be collected. In the standard viewing mode, secondary electrons are collected so that aluminum stubs and metallic coatings are permissible. On the other hand, if X-ray analysis or backscattered signals are to be used, then carbon stubs with non-metallic, carbon-based glues and evaporated carbon coatings of the specimen are in order. Whenever possible, the surface of the stub or substrate should be as smooth and free of structure as possible to prevent confusing backgrounds. The stubs must be cleaned in organic solvents to remove oils used during the manufacturing process and handled with gloves or forceps to prevent the transfer of body oils onto the specimen and specimen holders.

As shown in Figure 3.12, stubs may be purchased or modified to fulfill different mounting requirements. One must always ascertain whether or not such stubs will fit into the SEM chamber and be manipulable without bumping into or damaging sensitive components (polepieces, wires, detectors) in the specimen chamber.

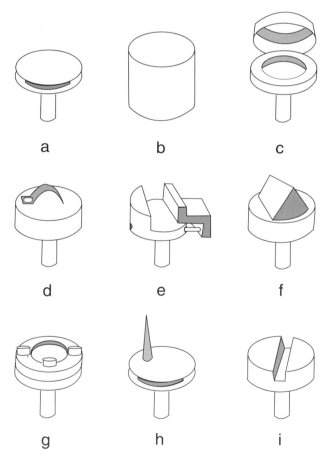

a **b** **c**

d **e** **f**

g **h** **i**

FIGURE 3.12 Specimen stubs used for mounting biological specimens for viewing in the SEM. The standard types are shown in a and b, while others are modifications to hold or pin down various types of specimens.
(Courtesy of Judy Murphy and SEM, Inc. Redrawn with permission.)

Criteria for Selecting Specimen Adhesives

Not many commercially available glues are adequate for use in SEM, since they do not fulfill some of the following important criteria. Suitable glues must:

* not damage specimens
* not contaminate the microscope
* provide good tackiness and still be easy to apply
* be resistant to the electron beam, heat, and vacuum
* provide a smooth, nonconfusing background.

For larger specimens, as well as for attaching substrates onto stubs, a number of glues and sticky tapes have been evaluated. Of the countless glues that have been tested, polyvinyl chloride (trade named "Microstik"), alpha-cyanoacrylate (Super Glue, Eastman 910), and cellulose nitrate (Duco china cement) combined with an equal volume of silver conductive paint or carbon paint appear to be the most useful. Care must be taken not to use too much glue, since it will extend the pumpdown time of the vacuum system and may wick into the specimen and damage it. In order to ensure that the glues are thoroughly dried it is advisable to place them into a 45 to 50°C oven for several hours prior to insertion in the SEM.

Adhesive tapes, especially carbon conductive tape with adhesive on both sides, are particularly useful for holding specimens and substrates. Other useful double-stick tapes include Scotch Double-Coated Tapes #665 and 666 and Scotch Adhesive Tape #463. In general, tapes are less desirable than glues since they tend to emit gases and break down under high beam currents.

A convenient way to quickly attach some specimens is to use transfer tabs available from some photo supply stores and EM supply houses (Figure 3.13). The tabs bear a dab of adhesive that may be transferred onto the stub by light pressure. Specimens may then be carefully placed onto the adhesive and pressed lightly, if possible, to ensure a good bond. If delicate specimens are being attached to the stub, then light bursts of compressed, clean air may be used to press the specimen into the adhesive.

Most biological specimens that have undergone drying are extremely brittle and prone to damage unless handled extremely carefully. Among the tools that may be used to pick up and mount specimens onto substrates or stubs are fine pointed jeweler's forceps, wooden applicators or toothpicks, dissecting needles, eyelash probes (a single eyelash mounted on end of toothpick), and

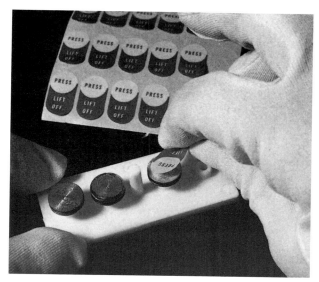

FIGURE 3.13 Adhesive transfer tabs are used to deposit a small amount of adhesive onto SEM stubs. The adhesive is adequate to hold most small specimens on the stub.

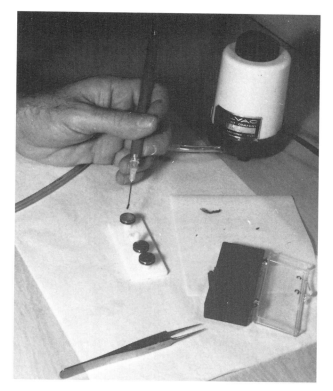

FIGURE 3.14 Vacuum needle used to pick up and deposit small delicate specimens onto SEM stubs.

micropipettes. A handy device for picking up specimens is the vacuum needle shown in Figure 3.14. A small aquarium pump is used to generate a slight vacuum through the hollow needle that is placed onto the specimen. Finger pressure over a hole in the handle causes an increase in the suction so that the specimen is held on the tip of the needle.

No pressure should be exerted on the specimen, but it should be lifted onto the stub or substrate (previously coated with the proper glue or double stick tape) using the appropriate tool. Once the solvents have thoroughly dried, the specimen stubs may be labeled on the underside with a permanent marker pen, stored in a dust-free desiccator, or coated for conductivity and viewed in the SEM. For more details on specimen mounting methods, see the review article by Murphy (1982). In addition, the excellent collection of articles edited by Murphy and Roomans (1984) covers all of the steps necessary for fixation, drying, mounting and coating of specimens for SEM.

Specimen Coating for Conductivity

In the standard protocols, after specimens have been mounted onto the specimen stub, they are coated with a thin layer (approximately 20 to 30 nm) of a conductive metal such as gold, platinum or gold/palladium alloy. Metal coatings usually are applied by sputter coating (see next section) or by thermal evaporation as described in Chapter 5. The metal coatings prevent the buildup of high-voltage charges on the specimen by conducting the charge to ground (i.e., the coating is contiguous with the aluminum specimen stub that is connected to the grounded specimen stage of the SEM). In addition, the metal coatings serve as excellent sources of secondary electrons as well as helping to conduct away potentially damaging heat.

Sputter Coating Procedure

The most commonly used method for coating SEM specimens with a thin layer of a metal is the *sputter coating* procedure. Several different designs of instrumentation are possible and have been reviewed by Echlin (1978, 1981). Probably the most commonly used system is the direct current sputtering device similar to the one diagrammed in Figure 3.15. Details of the sputtering chamber and the principle involved in the sputtering process are illustrated in Figure 3.16.

In the sputtering process, after placing the metal stub containing the attached specimen into the specimen chamber, the chamber is evacuated to approximately 0.1 Pa using a rotary vacuum pump.

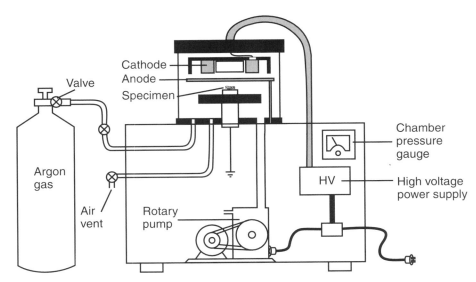

FIGURE 3.15 Diagram of parts of a commonly used sputter coater. A rotary pump is used to evacuate the specimen chamber to remove atmospheric gases and to permit the introduction of argon gas into the chamber. The application of a high voltage to the target causes the ionization of the argon into Ar^+ molecules that strike the cathode (target) and eject metal atoms that then coat the specimen.
(Redrawn from Postek et al. 1980.)

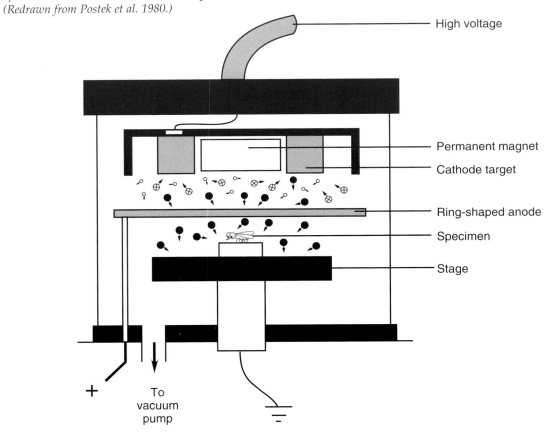

FIGURE 3.16 Detailed diagram of specimen chamber of a sputter coater. The metal cathode target is struck by ionized Ar^+ molecules (⊕) to cause the ejection of atoms of the metal target (●) that eventually coat the specimen. A permanent magnet is placed inside of the target to deflect potentially damaging electrons (○) away from the specimen and toward the anode ring. To further protect the specimen from excessive heating, the stage may be cooled with water or by electronic means.
(Redrawn from Postek et al. 1980.)

The purpose of evacuation is to remove water and oxygen molecules that might damage the surface of the specimen.

The pumping procedure may be somewhat extended if this is the first time the chamber is being evacuated or if the specimen has any residual moisture present due to adsorbed water vapor or organic molecules from the solvents used to glue the specimen onto the stub. It is best to minimize these molecules by placing the mounted specimen into a drying oven overnight. The temperature of the oven should be as high as possible—but not so high as to damage the specimen. It is very important not to pump the specimen chamber for long periods of time since this will lead to a backstreaming of vacuum pump oil onto the specimen surface (see the discussion of problems associated with vacuum systems in Chapter 6). Normally, if one cannot achieve the desired vacuum in a properly sealed system within 15 to 20 minutes, then either the specimen or the specimen chamber has adsorbed water or organic molecules. Drying the specimen or cleaning the chamber are in order.

It is important to reemphasize that one must remove traces of moisture and oxygen from the specimen chamber, since these molecules will be broken down in the high voltage fields inside the chamber into highly reactive molecules that will oxidize and damage the specimen surface. For this reason, one evacuates the chamber and replaces the atmosphere present with an inert gas such as argon that will not react with the specimen.

After the specimen chamber has achieved the desired vacuum level, an inert gas such as argon is slowly introduced into the chamber. The flow of argon is adjusted so that the vacuum can be maintained at approximately 6 to 7 Pa. A negatively charged, high voltage (1 to 3 kV) field is applied to the area of the chamber termed the *target* and the argon gas molecules are ionized into Ar^+ and electrons. The negatively charged target, composed of a heavy metal such as gold or palladium/gold, is struck with such force by the Ar^+ molecules that some of the metal atoms of the target are ejected. These atoms are bounced about by the various ions (Ar^+ and electrons) present in the chamber and eventually strike the specimen surface to gradually build up a metallic coating. The fact that the metal particles are knocked about in random paths (rather than in straight lines) is important, since they will strike the specimen at various angles and thereby more uniformly coat the irregularly shaped specimen surfaces. (This is in contrast to the evaporative method of coating specimens with heavy metals described in Chapter 5, where the metal atoms travel in straight lines and are unable to reach areas not in a direct line with the source of the metal.) The sputtering process is continued until a proper thickness of metal coating has accumulated on the specimen surface. The thickness needed depends upon the topography of the specimen: specimens with low topographies need less coating than those with higher topographies. Generally a coating of 15 to 40 nm is adequate for most specimens.

Estimation of Coating Thickness

Some sputter coater systems (see Figure 3.17) have a digital thickness monitor consisting of a crystal of quartz that oscillates at a certain frequency. As metal coating builds up on the crystal that is placed close to the specimens, the oscillation frequency is altered in proportion to the thickness of the coating. A simpler method may involve placing a white piece of paper or a glass coverglass into the chamber and observing the darkening of the material until the desired shade of gray or proper reflectance color is obtained. In any case, for accuracy of measurement, one should establish a series of standards for comparison to recently coated specimens. The thickness of the standards is determined by including a piece of plastic (such as epoxy or methacrylate) in the chamber during the coating process and then cutting an ultrathin section across the coating and measuring the thickness in the TEM. See Chapter 4 for details of ultramicrotomy procedures.

Potential Problems Associated with Sputter Coating

1. **Thermal Damage** may be caused by radiation from the target as well as by the electrons striking the grounded specimen. This may be minimized by (a) deflecting the electrons away from the specimen using a permanent magnet centered in the target, (b) using an anode ring placed above the specimens to attract the electrons, (c) cooling the specimen stage, or (d) pulsing the high voltage—rather than applying the voltage continuously—to prevent heat buildup.

2. **Surface Contamination** may occur due to backstreaming of oil from the rotary pump. This may be minimized by (a) not pumping for extended periods of time on the sealed specimen chamber, (b) maintaining a positive flow of argon gas over the specimen surface and into the vacuum pump, or (c) installing filters or

traps in the rotary pump line to the specimen chamber.

3. **Surface Etching** will occur if water vapor, oxygen, or carbon dioxide molecules remain inside of the specimen chamber and are ionized in the high voltage field. One must use an inert gas such as argon rather than normal air to generate the ions that strike the target. Water vapor may be detected by noting the color of the glowing plasma generated during the sputtering process. If the color is blue rather than lavender, one must suspect contamination by water vapor and take the appropriate measures (e.g., dry the specimen or clean the specimen chamber). Sometimes it is possible to clean a specimen surface by etching it deliberately. Normally, this is done by reversing the polarity of the instrument so that the specimen is made highly negative and the argon ions strike and erode the specimen rather than the target.

Noncoating Techniques

It is also possible to render biological specimens electrically conductive by methods other than coating by sputtering or thermal evaporation. One limitation of the standard methods of sputtering and thermal evaporation (in contrast to specialized procedures of high resolution sputtering and electron beam evaporation) is the diminished resolution imposed on the specimen surface due to the use of thick coatings of 20 nm or more. In addition, these coatings may not be continuous (leading to localized charging), the specimens may be damaged by heating during the coating process, and the process requires specialized equipment.

Several alternatives for rendering biological specimens electrically conductive have been developed over the years. All of these methods involve the deposition of thin layers of the conductive metals onto the specimen by reducing the metal from aqueous salt solutions of the metal. Since the thickness of the coating layers is in the range of 4 to 20 nm, resolutions of the same order may be expected. These procedures have been reviewed by Murphy (1978, 1980) and Murakami et al. (1983). An example of one method is given in the following box.

FIGURE 3.17 Photograph of a commercially available sputter coater. The argon tank is to the right of the sputter coater. This unit is equipped with quartz crystal monitor to give a digital readout of the thickness of metal coating deposited on the specimens.

Tannic Acid/Osmium Noncoating Technique (Takahashi, 1979)

1. If immersion fixation is used, start at step 4. If perfusion fixation is used, the tissues are initially perfused in an appropriately buffered salt solution to clear the blood from the vessels (see "Perfusion fixation" in Chapter 2). Following this, the tissue is perfused at 4°C for 15 minutes with a freshly prepared mixture of 1.5% glutaraldehyde and 0.5% tannic acid in a buffer system appropriate for the tissues being studied. (*Note: If a phosphate buffer is used, the final solution should be filtered immediately before use.*)

2. Perfuse 10 minutes with a freshly prepared mixture of 1.5% glutaraldehyde and 1% tannic acid in the appropriate buffer.

3. Cut tissues into small pieces of 1 mm^3 and rinse in buffer for 1 hour.

4. Immerse tissue into 2% osmium tetroxide in buffer for 2 hours at 4°C.

5. Rinse in buffer three times (5 minutes each).

6. Immerse in a freshly prepared mixture of 8% glutaraldehyde and 2% tannic acid for 12 hours. (Note: This solution should be changed three times over the 12 hour period. Cacodylate buffer is recommended since phosphate buffers may precipitate during this time period.)

7. Rinse in buffer three times (5 minutes each).

8. Immerse for 2 hours in a 2% solution of osmium tetroxide in cacodylate buffer.

9. Rinse in buffer three times (5 minutes each).

10. Repeat steps 6 through 9.

11. Dehydrate tissue, critical point dry, mount on stub using conductive paint, and examine in SEM.

Specimen Storage

It may be necessary to store the fixed and dried specimens either before or after mounting onto the SEM stub. In either case, certain precautions must be observed to ensure that artifactual changes are not introduced during this period. The specimens are extremely fragile and brittle and may be damaged by contact even with a soft object such as a camel's hair brush. It is best to handle the specimen in an area that is not to be studied in the SEM. Normally, one would store unmounted specimens in small cardboard or plastic boxes (pillboxes) or even petri dishes lined with filter paper or small pieces of aluminum foil (to facilitate picking up the specimens). Specimens that have been mounted onto stubs may be stored in commercially available boxes designed to hold the stubs securely, or one may easily make storage holders by drilling holes in plastic to accommodate the stubs (Figure 3.18).

If plastic containers are used, one must be aware that static charges will often cause unmounted specimens to become displaced, often sticking onto the covering lids. If several different types of specimens are placed into the same container, they may become mixed up. Glass or cardboard containers are less likely to develop such static charges. All containers must be dry, clean and dust-free to prevent contamination of the specimen surfaces.

The specimens must be stored in a dry environment to prevent rehydration of the specimen. Most laboratories store specimens in a glass or plastic desiccator containing a drying agent of some sort. One should be careful to check that the agent is not exhausted (most have color indicators) and that it is sequestered away from the specimens since most drying agents tend to be powdery on their surface. One may cover the drying agent with lintless lens tissues to prevent the transfer of powdered desiccant onto the specimen. The dust-free container must be tightly sealed and kept closed when specimens are not being transferred. Often silicone greases are used to seal the jars, so take care not to contaminate the specimen with these lubricants. Finally, one should avoid storing specimens in the desiccator if they are outgassing solvents (amyl acetate, glues, etc), since the desiccant and specimen will adsorb these fumes and may be damaged. Often one may detect such fumes by sniffing the jar upon opening. If fumes are detected, another jar should be used or the desiccant replaced.

The dried specimens are quite hygroscopic and will absorb water from the environment. This leads to the swelling and shrinking of the specimen as it is moved into or out of the desiccator. Such volume changes may cause fine fracturing of the specimen surface coatings, and delicate fea-

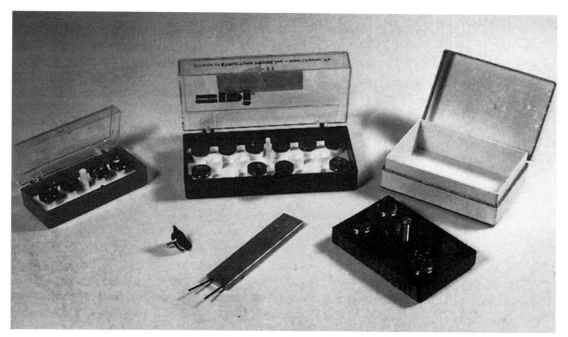

FIGURE 3.18 Various types of containers for storing specimen stubs. Some are commercially available while others may be easily fabricated.

tures may be damaged. It is best to keep the specimen in the sealed container until mounting and viewing is necessary.

REFERENCES

Anderson, T. F. 1951. Techniques for preservation of three-dimensional structure in preparing specimens for the electron microscope. *Trans NY Acad Sci* 13:130–34.

Becker, R. P. and O. Johari eds. 1979. *Cell surface labeling.* Scanning Electron Microscopy, Inc. (AMF O'Hare, Il) 344 pp.

Boyde, A. 1978. Pros and cons of critical point drying and freeze drying for SEM. *Scan Electr Microsc* II:303–14.

Boyde, A., and E. Maconnachie. 1981. Morphological correlations with dimensional change during SEM specimen preparation. *Scan Electr Microsc* IV:27–34.

Cohen, A. L. 1977. *A critical look at critical point drying—theory, practice and artifacts.* SEM/1977/I. IIT Research Institute, Chicago, IL., 525–36.

Cohen, A. L. 1979. Critical point drying—principles and procedures. *Scan Electr Microsc* II:303–23.

Cohen, A. L., and M. Shaykh. 1973. *Fixation and dehydration in the preservation of surface structure in critical point drying of plant material.* SEM/1973. IIT Research Institute, Chicago, IL., 371–78.

Crankshaw, O. S. 1984. Instruction of replica techniques for scanning electron microscopy. *Scan Electr Microsc* IV: 1731–7.

DeHarven, E., N. Lampen, and D. Pia. 1977. Alternatives to critical point drying. *Scan Electr Microsc* I: 519–24.

Echlin, P. 1978. Coating techniques for scanning electron microscopy and X-ray microanalysis. *Scan Electr Microsc* I:109–32.

———. 1981. Recent advances in specimen coating techniques. *Scan Electr Microsc* I: 79–90.

Falk, R. H. 1980. Preparation of plant tissues for SEM. *Scan Electr Microsc* II:79–87.

Gilkey, J. C., and L. A. Staehelin. 1986. Advances in ultrarapid freezing for the preservation of cellular ultrastructure. *J Electr Microsc Tech* 3:177–210.

Haggis, G. H. 1982. Contribution of scanning electron microscopy to viewing internal cell structure. *Scan Electr Microsc* II:751–63.

Helinski, E. H., G. H. Bootsma, R. J. McGroarty, G. M. Ovak, E. deHarven, and J. L. Pauly. 1990. Scanning electron microscopic study of immunogold-labeled human leukocytes. *J Electr Microsc Tech* 14:298–306.

Hodde, K. C. and J. A. Nowell. 1980. SEM of micro-corrosion casts. *Scan Electr Microsc* II:89–106.

Hossler, F. E., Douglas, J. E., and Douglas, L. E. 1986. Anatomy and morphometry of myocardial capillaries studied with vascular corrosion casting and scanning electron microscopy: a method for rat heart. *Scan Electr Microsc* IV:1469–75.

Kennedy, J. R., R. W. Williams and J. P. Gray. 1989. Use of Peldri II (a fluorocarbon solid at room temperature) as an alternative to critical point drying for biological tissues. *J Electr Microsc Tech* 11:117–25.

Kok, L. P., and M. E. Boon. 1992. Microwave cookbook for microscopists, art and science of visualization, 3d ed. Coulomb Press, Leiden, The Netherlands. ISBN 90–71421–20–1.

Maugel, T. K., D. B. Bonar, W. J. Creegan, and E. B. Small. 1980. Specimen preparation techniques for aquatic organisms. *Scan Electr Microsc* II:57–77.

Mazia, D., G. Schatten and W. Sale. 1975. Adhesion of cells to surfaces coated with polylysine. *J. Cell Biol* 66:198–200.

Murakami, T., N. Iida, T. Taguchi, O. Ohtani, A. Kikuta, A. Ohtsuka and T. Itoshima. 1983. Conductive staining of biological specimens for scanning electron microscopy with special reference to ligand osmium impregnation. *Scan Electr Microsc* I:235–46.

Murphy, J. A. 1978. Non-coating techniques to render biological specimens conductive. *Scan Electr Microsc* II:175–93.

———. 1980. Non-coating techniques to render biological specimens conductive/1980 update. *Scan Electr Microsc* I:209–20.

———. 1982. Considerations, materials and procedures for specimen mounting prior to scanning electron microscopic examination. *Scan Electr Microsc* II:657–96.

Murphy, J. A., and G. M. Roomans, eds. 1984. *Preparation of biological specimens for scanning electron microscopy.* Scanning Electron Microscopy, Inc., AMF O'Hare, Il., 344 pp.

Nowell, J. A. and J. B. Pawley. 1980. Preparation of experimental animal tissue for SEM. *Scan Electr Microsc* II:1–19.

Olson, K. R. 1985. Preparation of fish tissues for electron microscopy. *J Electr Microsc Tech* 2:217–28.

Pameijer, C. H. 1978. Replica techniques for scanning electron microscopy, a review. *Scan Electr Microsc* II:831–36.

Pfefferkorn, G. and A. Boyd. 1974. Review of replica techniques for scanning electron microscopy. *Scan Electr Microsc* I:75–82.

Postek, M. T., K. S. Howard, A. Johnson and K. L. McMichael 1980. *Scanning electron microscopy: A student's handbook.* Ladd Research Industries, Inc., Burlington, Vt., 305 pp.

Rebhun, L. J. 1972. Freeze-substitution and freeze-drying. IN *Principles and techniques of scanning electron microscopy,* Vol.3, chapter 5, M. A. Hayat, ed., Van Nostrand Reinhold Co., N.Y.

Revel, J-P., T. Barnard, G. H. Haggis and S. A. Bhatt, eds. 1983. *The science of biological specimen preparation for microscopy and microanalysis.* Scanning Electron Microscopy, Inc., AMF O'Hare, Il., 245 pp.

Riches, D. J., and E. C. Chew. 1984. The use of microwaves for fixation in electron microscopy. In *Proceedings of the 3rd Asia-Pacific Conference on electron microscopy,* M. F. Brown, ed., Hentexco Trading Co., Hong Kong, 257–60.

Scott, E. C. 1982. Replica production for scanning electron microscopy: a test of materials suitable for use in field settings. *J Microsc* 125:337–41.

Shennawy, I. E., D. J. Gee and S. R. Aparicio. 1983. A new technique for visualization of internal structure by SEM. *J Microsc* 132:243–46.

Takahashi, G. 1979. Conductive staining method. *Cell* 11:114–23.

Tanaka, K. 1989. High resolution electron microscopy of the cell. *Biol of the Cell* 65:89–98.

Watson, L. P, A. E. McKee and B. R. Merrell. 1984. Preparation of microbiological specimens for scanning electron microscopy. *Scan Electr Microsc* II:45–56.

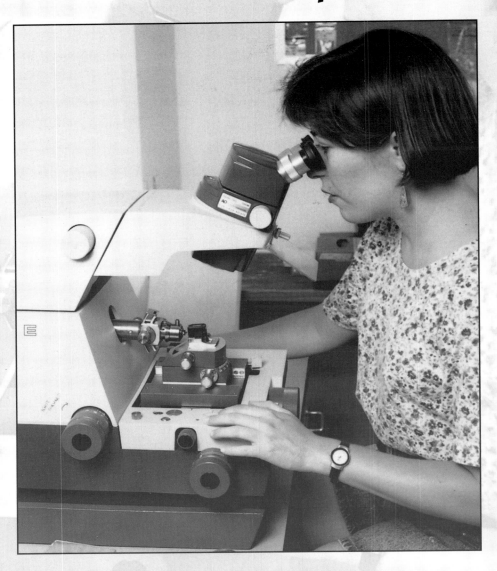

Chapter 4

Ultramicrotomy

ULTRAMICROTOMY is a procedure for cutting specimens into extremely thin slices, or *sections,* for viewing in the transmission electron microscope (TEM). Sections must be very thin because the 50 to 125 kV electrons of the standard electron microscope cannot pass through biological material much thicker than 150 nm. In fact, for best resolution, sections should be from 30 to 60 nm. This is roughly equivalent to splitting a 0.1 mm thick human hair into 2,000 slices along its long axis, or cutting a single red blood cell into 100 slices. Ultramicrotomy is a demonstrably demanding technique that requires much practice and patience from the beginning microtomist, who must pay attention to such details as specimen preparation and embedding, preparation of knives and specimen support grids, and cleanliness of all reagents and utensils. A thorough working knowledge of all equipment involved in the ultramicrotomy process is essential.

When ultramicrotomists refer to "thin" sections, they mean sections from 50 to 100 nm in thickness, which are suitable for viewing in the transmission electron microscope. Semithin or "thick" sections in the 0.5 to 2 μm range, are about 10 to 20 times thicker than the thin sections. These thick sections, or *survey sections* as they are often called, are viewed in the light microscope to determine if the right area of the specimen is in position for thin sectioning. It is very common practice to view a thick section in the light microscope first before proceeding with ultramicrotomy or thin sectioning.

To cut thin sections from relatively soft biological specimens, it is first necessary to infiltrate suitably fixed and dehydrated specimens with a liquid plastic that is then hardened or polymerized as described in Chapter 2. The plastic serves as a very hard matrix that provides support to the tissue as the knife passes through the specimen. Another method of hardening the tissue for ultramicrotomy is rapid freezing followed by cutting at temperatures of −80 to −140°C. Known as *cryoultramicrotomy,* this technique will be discussed at the end of the chapter.

The specialized instrument used for cutting sections is called an *ultramicrotome* (see Figure 4.1). Ultramicrotomes advance the specimen in precise, repeatable steps by using either a mechanical or thermal advancement mechanism. In the standard procedure, after the sections have been cut and are floating on the surface of water contained in the trough of the knife, they are picked up on a copper screen mesh, or grid, and stained for contrast using salts of a heavy metal prior to viewing in the transmission electron microscope.

The sequence of steps in preparing thin sections for examination involves: *trimming or shaping the specimen block, preparing ultramicrotome knives and specimen support grids, cutting sections in the ultramicrotome, picking up the sections onto the specimen grid, and staining them to enhance contrast.* These steps are illustrated in Figure 4.2 and in Chapter 5.

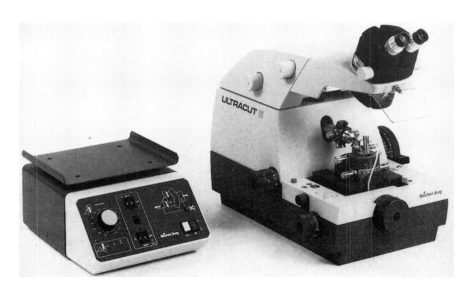

FIGURE 4.1 Ultramicrotome for cutting thin sections of plastic-embedded specimens. *(Courtesy of Leica.)*

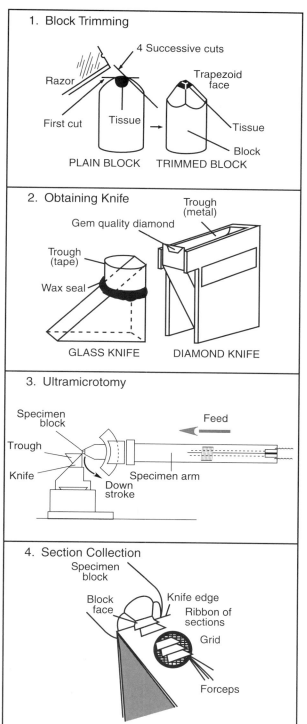

1. Block Trimming

4 Successive cuts

Razor

First cut   Tissue

Trapezoid face

Tissue

Block

PLAIN BLOCK   TRIMMED BLOCK

Specimen blocks are trimmed to expose the specimen and to form a trapezoidal cutting face. Usually, a single-edged razor blade is used for this step, which is done by hand while observing the process under a dissecting or stereomicroscope. Automatic trimmers are also available.

2. Obtaining Knife

Trough (metal)

Gem quality diamond

Trough (tape)

Wax seal

GLASS KNIFE   DIAMOND KNIFE

The knife used to cut sections may be made from a special grade of plate glass just prior to use. It is necessary to attach a trough to the glass knife to hold the water onto which the sections will be cut. Alternatively, a diamond knife may be used that has a large metal trough to hold the water.

3. Ultramicrotomy

Specimen block

Trough

Knife

Down stroke

Feed

Specimen arm

The trimmed specimen block is placed into the ultramicrotome and a thick (1-2 μm) section is usually cut and examined in the light microscope to verify that the correct specimen area is in position. Subsequently, ultrathin sections (60-80 nm) may be cut using a mechanically or thermally advanced specimen arm.

4. Section Collection

Specimen block

Block face

Knife edge

Ribbon of sections

Grid

Forceps

Sections that have been cut onto the water surface in the trough are picked up using a mesh or specimen grid usually made of copper. The sections adhere to the grid upon removal from the water and are dried in a dust-free environment. The sections may also be stained using special heavy metal salts such as lead citrate and uranyl acetate.

FIGURE 4.2   Illustration and brief explanation of steps involved in ultramicrotomy process. *(Redrawn with permission of J. Murphy.)*

Shaping the Specimen Block

To prepare for ultramicrotomy, it is necessary to shape the plastic specimen block into a small cutting face in order to minimize the stresses imposed on the cutting edge of the knife as the specimen is passed over it. This is usually done in two stages. The specimen block is first *rough trimmed* to exclude excess plastic matrix and to expose the surface of the specimen. A thick section is made from the rough-trimmed block and examined in the light microscope. After locating the areas of interest in the thick section and on the block face, further fine trimming is done to remove unwanted areas of the specimen. The usual goal in fine trimming is to produce a truncated pyramid with its sides sloped at 45 to 60 degrees, as shown in Figure 4.3. A pyramidal shape is most often used because it is a more stable structure than, for instance, one with non-sloped sides. The pyramid should be no larger than 0.5 to 1.0 mm on either of the two parallel sides. Especially when using a glass knife, it is important to reduce these dimensions as much as possible to prevent vibrations generated as the specimen contacts the knife.

Rough Trimming by Hand

Although several instruments are available to shape the block in a precise manner, they are quite expensive (approaching the cost of an ultramicrotome) and may not be justified especially since precision trimming can be done with most ultramicrotomes or even with a single-edged razor blade as follows.

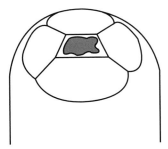

FIGURE 4.3 Specimen block that has been rough trimmed to give general shape to the block. After examination of a thick section taken from the block, the block will be fine trimmed to a considerably smaller shape containing only specimen (*shown in black*).

Shaping the Block Using Razor Blades

1. Clamp the specimen block into an appropriate holder or *chuck*, and place the chuck into a stable holder under a dissecting or stereo microscope. Some ultramicrotomes have adapters so that one can use the ultramicrotome optics and accessories for this step.
2. Cut into the specimen by making a series of slices with a new, single-edged razor blade that has been cleaned with xylene or acetone to remove oils. The slices should be made parallel to the tabletop and should stop just before the desired area in the specimen is reached (Figure 4.4A). If osmium was used as a fixative, the specimen will be black and readily seen as the trimming takes place. It may be possible even to see the desired area on the flat top surface of the blackened specimen by adjusting the lighting and looking for the reflectance on the cut block surface under high magnification in the stereo microscope. The gross structure of the tissues can be seen at the block face due to differential reflections of the various tissue components as the bright light is reflected from the block face surface (Figure 4.5). If it is not possible to select the desired area in this manner, then it will be necessary to cut 0.5 to 2.0 μm thick survey sections as described in the following section. This is the most common way to determine the tissue orientation. The razor cuts should be as thin as possible to prevent removal of large chunks from the specimen block by haphazard fracturing rather than cutting.
3. Begin making a series of very thin razor cuts to a depth of 2 to 3 mm and at a 60 degree angle relative to the desk top along one side of the block (Figure 4.4B). Detach the slices from the block by carefully making one horizontal cut.
4. Rotate the specimen block 180 degrees and begin a second series of 60 degree cuts until the side of the specimen is reached (Figure 4.4C). These cuts will generate a narrow specimen platform on the top of the pyramid. The platform should be wide enough so that no important areas of the specimen are cut off, but not so wide as to cause sectioning difficulties. If in doubt about what to trim, do not trim away any areas until a survey section has been examined in the light microscope.
5. Rotate the specimen block and make a third and fourth series of side cuts at a 60 degree angle to generate a trapezoid-shaped pyramid top (Figure 4.4D, E).

Viewed from Side   **Viewed from Above**

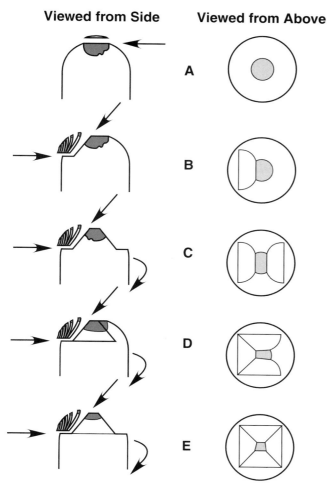

A

B

C

D

E

FIGURE 4.4 Steps in trimming a plastic specimen block. Biological specimen is seen as a darker area in the plastic. Straight arrows indicate razor cuts made in the plastic, while curved arrows indicate turns to be made in the orientation of the block.
(Redrawn with permission from Ultramicrotomy *by N. Reid, 1975.)*

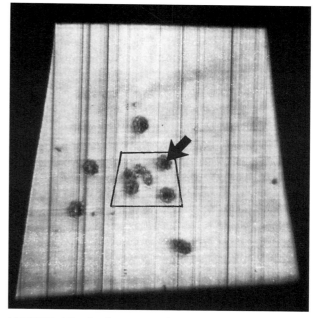

FIGURE 4.5 Surface view of rough-trimmed specimen block. Gross structure of individual plant cells (*arrow*) within tissue may be seen in the reflected light. Striations running through the block face are knife marks caused by nicks in the glass knife edge. This block face is approximately 5 mm across at the top. The superimposed trapezoidal shape indicates the shape that the fine trimmed block will assume after excess plastic has been removed using a razor blade.

The roughly trimmed block (Figure 4.5) may be many times larger than the finished block used for thin sectioning; however, it will be fine trimmed once the orientation of the tissue has been determined and important areas located by light microscopy. If the researcher is fortunate enough to be working with homogeneous samples (e.g., clones of cell cultures or cell fractions) and has no need for orientation of sections or selecting areas and retrimming, then the block can be fine trimmed as described in a subsequent section.

Rough Trimming by Machine

Most experienced microtomists prefer to rough trim specimens by hand using razor blades, since it is possible to complete the process in about 2 to 10 minutes. It also is possible to shape the block accurately on the ultramicrotome following the steps for hand trimming (described previously) except that the block is mounted in the ultramicrotome and shaped using a glass knife. Two of the angled sides of the block are produced by shifting the ultramicrotome knife 30 to 40 degrees relative to the initial flat face and approaching the block from the side using the coarse advance mechanism of the ultramicrotome. After the two sloped sides are formed, the specimen chuck is rotated in the specimen arm and the angled knife is used again to remove the third side and, following a third specimen block rotation, the fourth side (Figure 4.4E). A variety of faces can be made by varying the rotation of the specimen chuck in the arm of the microtome.

Special specimen trimmers may also be used in some research laboratories. One such trimmer (unfortunately, no longer available) is the LKB Pyramitome (Figure 4.6A), which resembles an ultramicrotome except that the Pyramitome can cut only thick sections in the micrometer range. The Pyramitome shapes blocks much more rapidly than the standard ultramicrotome because it has special rotating specimen and knife holders to orient the specimen and knives precisely. In addition, it is possible to cut thick sections on the Pyramitome, evaluate them in the light microscope, and then use a special accessory to superimpose the specimen image in the microscope slide over the block face. This permits very accurate fine trimming of the block face in the Pyramitome.

Another specially designed instrument is the Leica Trimmer (Figure 4.6B), which precisely shapes the block face with a rapidly spinning routerlike tip. Inventive microtomists also have used high speed grinders (available at hardware stores) to remove large portions of plastic rapidly and roughly shape the block (Chapter 2). However, use all of these units cautiously since enough heat may be generated during the process to distort or even melt the embedding plastic or, worse yet, to grind away the specimen inadvertently.

CAUTION: The fine dust produced during the grinding process is probably injurious to one's health, so be careful not to inhale the potentially carcinogenic plastic.

The Mesa Trimming Procedure

In 1966, DeBruijn and McGee-Russell described a special type of trimming that is especially conservative of specimen. This procedure, which can be accomplished only by using machine trimming, is called the *mesa* trim (so named because the shaped block resembles geological features called mesas). After forming the initial flat cut at the top of the block to expose and locate the specimen (Figure 4.7A), the glass knife is moved to the right of the desired specimen area, and a series of cuts are made parallel to the specimen surface to a depth of approximately 15 to 25 μm (Figure 4.7B). Following this, the specimen is rotated 90 degrees, and sections are cut again parallel to the surface to the same depth as the first series of sections (Figure 4.7C). The specimen chuck is now rotated two more times through 90 degrees, and the third and fourth cuts are made on either side of the desired area and to the same depth as the previous two cuts (Figure 4.7D, E). The end product is a small, raised, boxlike mesa with vertical sides cut by the glass knife (Figure 4.7F). After the raised mesa of the specimen has been sectioned down, it is possible to trim another mesa in the same or another area. The advantage of this technique is that a minimum of specimen is removed during the trimming process, making it possible to trim and examine many other areas of the specimen block.

FIGURE 4.6A Specimen trimmer for shaping specimen blocks. The LKB Pyramitome uses glass knives to shape the specimen block. The specimen is advanced several micrometers at each cut. The operator adjusts the orientation of the knives and the specimen to obtain precisely shaped blocks. In addition, the instrument may be used to cut micrometer-thick sections for evaluation by light microscopy.

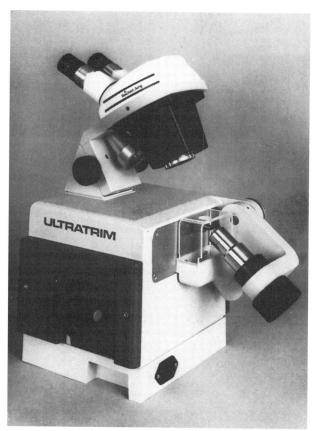

FIGURE 4.6B The Leica trimmer may be used for shaping the plastic specimen block by means of a rapidly spinning routerlike tip. The specimen is positioned in a clamping chuck, and the specimen is moved across the router to trim the block to the desired shape. (*Courtesy of Leica.*)

Thick Sectioning

If orientation or a preview of the specimen is desired, as is normally the case, the roughly trimmed block should be mounted into the ultramicrotome and 0.5 to 2.0 μm sections should be cut. Most microtomists use glass knives to cut such thick sections, which are picked up using the needle from a syringe (Figure 4.8), an eyelash probe, wire loops, or a wooden spatula made by sharpening an applicator stick to a fine point. Special types of diamond knives may also be used and sections are retrieved from the water trough using a wire loop. The sections are then transferred onto a tiny droplet of distilled water on an alcohol- or acid-cleaned microscope slide by inverting the transfer device so that the section side contacts the water droplet, which draws the section onto the slide. The slide is transferred to a hot plate (70 to 90°C) until the droplet evaporates. After the slide has cooled, a drop of staining solution (Millipore-filtered 1% toluidine

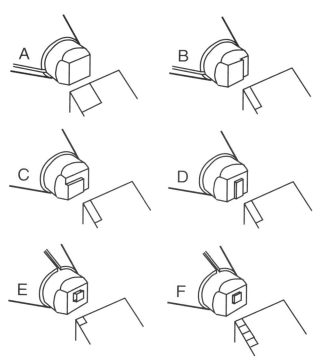

FIGURE 4.7 Steps in trimming a specimen block using the "mesa" technique.
(*Redrawn with permission from* Ultramicrotomy *by N. Reid, 1975.*)

FIGURE 4.8 Retrieval of thick section from cutting edge of glass knife using the beveled needle from a tuberculin syringe. The trimmed block, viewed from above, is indicated by an arrow. The section is located just under the block and the syringe point is seen coming into the picture in the lower right.

blue dissolved in 1% aqueous sodium borate solution) is placed over the dried section and reheated on the hot plate until the edges of the stain begin to dry and turn an iridescent green color. The slide is then gently rinsed in distilled water from a wash bottle and dried by heating it on the hot plate prior to examination in a light microscope. Do not direct the wash water directly on the sections or they may be removed from the slide. Figure 4.9 (small inset) shows a section 1 to 2 μm thick as viewed in a light microscope; the larger print is of the same specimen after ultramicrotomy as viewed in the TEM. In this case, the thick section on the glass slide was covered with epoxy resin (i.e., reembedded in epoxy), and after removal from the slide, the reembedded *section* was again sectioned in an ultramicrotome to generate the thin section shown.

It may be desirable to collect numerous plastic sections for viewing. If this is the case, one can collect long *ribbons* of connected sections using a scoop readily fashioned from laboratory plastic tubing (Figure 4.10A). Using such scoops, it is possible to mount 200 to 400 sections on a single glass slide (Figure 4.10B). If such collections become routine, it is recommended that an older diamond knife or a knife constructed of industrial diamond be used, since the attached water trough is large enough to accommodate the scoop easily, and a much wider variety of specimens may be cut with the diamond.

Fine Trimming

After the plastic survey sections have been examined by light microscopy and the area of interest located (see Figure 4.5), it is often necessary to retrim the plastic block to exclude areas that are not of interest and to optimize the size of the block face for thin sectioning. This must be done carefully under the stereomicroscope until a cutting block of reasonable size is obtained. These final cuts must be made with a sharp, clean razor blade so that the block sides parallel to the knife edge are quite smooth. It is very important that the two sides parallel to the knife edge be exactly parallel in order to produce straight ribbons of sections. If these two sides deviate from the parallel, curved ribbons will be produced that will be difficult to handle and fit onto a grid (Figure 4.11A). In such cases, one will be able to cut only individual sections as shown in Figure 4.11B. A properly trimmed block will permit the microtomist to cut long ribbons of serial sections (Figure 4.12) that are much easier to maintain in the order cut.

Although the trapezoid-shaped block face is most commonly used, other shapes may be more appropriate in certain situations. For example, with elongated specimens such as fibers or rootlets, a block face that accommodates the shape of the specimen is more appropriate (Figure 4.13). In this

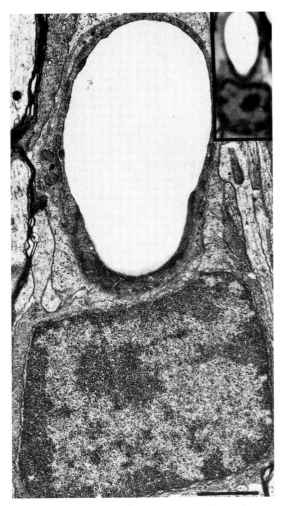

FIGURE 4.9   (*Inset*) Thick, 1 to 2 μm section of microglial nerve cell viewed in the light microscope. The large micrograph was obtained after the thick section was reembedded in epoxy plastic, and resectioned in an ultramicrotome to generate the ultrathin section shown. The large clear space is an empty capillary. The electron micrograph is magnified 5 times more than the light micrograph. The image of the thick section is a mirror image of the electron micrograph. Marker bar = 1 μm. (*Courtesy of J. A. Paterson.*)

case, one side may be 3 to 4 times longer than the adjacent side. These types of blocks are best cut in the final thin sectioning process by orienting the shorter side parallel to the knife edge and keeping the shorter dimension under 0.5 mm if possible. It is also possible to trim a block to a square face, the obvious advantage being that one can approach the knife from any of the four equivalent sides. Specimens that are difficult to section can be trimmed to a very small triangular face. In this case, one approaches the knife edge with the point of the tri-

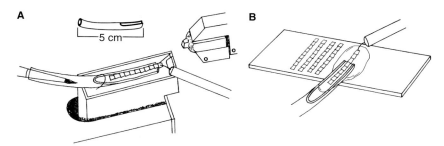

FIGURE 4.10 (A) Device fashioned from small bore laboratory tubing for picking up long ribbons of sections from the water trough of a diamond knife. (B) Transfer of long ribbons of continuous sections onto a glass slide. After drying onto the slide, the thick sections can be stained and examined in the light microscope. Ultrathin sections may be subsequently cut from the block face.
(Courtesy of S. M. Royer and of The Williams and Wilkins Co.)

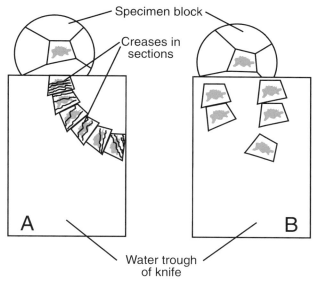

FIGURE 4.11 (A) Curved ribbons of sections floating in the knife trough of an ultramicrotome knife. Problems were caused by a lack of parallel sides on top and bottom edges of trimmed block face. (B) It still may be possible to section improperly trimmed blocks by cutting individual sections rather than attempting to allow a ribbon to form.

FIGURE 4.12 Properly trimmed specimen blocks give rise to continuous ribbons of sections. These photomicrographs were taken through the stereomicroscope of the ultramicrotome and represent the view that the microtomist would hope to see during the sectioning procedure.

angle rather than one of the three flat sides (Figure 4.14). Although ribbons will not be produced with some of the block shapes described, individual sections of high quality can be organized into groups with an *eyelash probe* (fashioned by cementing an eyelash onto the end of an applicator stick) and the grouped sections picked up on a grid.

HINT: Due to the potential shortage of eyelashes, it is possible to purchase such hairs or to use the fine hair from a dalmatian dog. The inventive microtomist may be able to come up with even more alternatives!

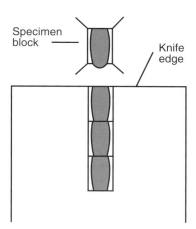

FIGURE 4.13 Plastic specimen block trimmed to accommodate an elongated specimen.

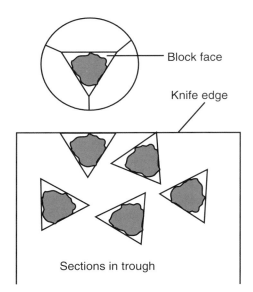

Block face

Knife edge

Sections in trough

FIGURE 4.14 Plastic specimen block trimmed in a triangular block face. This shape is used for specimens that may be otherwise difficult to section.

Types of Ultramicrotome Knives

Very early in the development of the ultramicrotomy technology, it was determined that metal knives made by meticulously honing standard histology knives or razor blades were too dull to cut ultrathin sections. Such soft and fragile knives rapidly lost their cutting edges when the thin sections required for transmission electron microscopy were attempted. In 1950, Latta and Hartmann discovered that the edge of broken plate glass could be used to cut satisfactory sections. With minor changes, glass is still used extensively in ultramicrotomy for trimming specimen blocks as well as for cutting ultrathin sections. However, certain types of hard specimens such as bone, plants, and thick-walled spores are difficult to cut even with a good glass knife because the edge dulls too quickly. Fernandez-Moran first indicated that a gem-quality diamond could be used to fabricate a knife for use in the microtomy process. Diamond was a logical material to use, since it is the hardest material known (Table 4.1).

It is also possible to use industrial diamonds for ultramicrotome knives. However, industrial diamonds, in contrast to the near flawless crystal of the gem diamond, contain more imperfections in the crystalline lattice, which result in microscopic flaws in the cutting edge. Although such diamonds are

suitable for cutting 0.5 to 1 μm survey sections, they are not recommended for high-quality ultrathin sectioning purposes because knife marks visible only in the electron microscope will result. Industrial-diamond knives are frequently referred to as "histo" knives because the thick sections are more suitable for histological studies conducted at the light microscope level. If the researcher intends to cut orientation or semithick sections on a routine basis, histo knives are recommended because they are superior to glass and come with a large water trough attached.

CAUTION: Although gem diamonds are very hard materials, they may be damaged if sections thicker than 1 μm are cut. This is especially true if the specimen itself is hard. Always follow the recommendations of the manufacturer in this matter.

Glass Knives

It has long been theorized that, because glass is a supercooled liquid, such knives should be prepared the same day they are used since their sharpness may be diminished by slight molecular flow at the edge. It is now generally accepted that glass knives can be prepared several days in advance without any noticeable effect on their sharpness as long as they are kept in a dry, dust-free environment such as a glass-enclosed desiccator.

Originally, microtomists would purchase ¼" plate glass from a local glazier and, using a com-

TABLE 4.1 MOHS Hardness Scale

Hard	
10	Diamond, carborundum
9	Chromium, sapphire, hard steel
8	Topaz, zirconia
7	Flint, osmium, quartz, hard glass
6	Feldspar, pumice, nickel
5	Asbestos, hard iron, soft glass, bone
4	Platinum, soft iron, marble, brass
3	Boric acid, copper, gold
2	Aluminum, rock salt, calcium
1	Asphalt, potassium
Soft	

Source: Handbook of Chemistry and Physics, *59th ed. West Palm Beach, Fl: (CRC Press, 1978–1979) eds. R. C. Weast and M. J. Astle, F-24.*

mon scoring wheel glass cutter, score and break the glass into 1″ wide strips using specially modified glazier's pliers (Figure 4.15). It is still possible to produce high-quality glass knives with such tools, but this technique is both time consuming and difficult for beginning microtomists to master. As a result, several devices are available for routinely producing high-quality knives. The most common instrument is the Leica Knifemaker (Figure 4.16). The manufacturer also supplies high-quality glass strips for use in the instrument, thereby eliminating another variable.

During the manufacturing process, glass can be formed into plates either by floating the molten glass on molten tin to produce a very flat, uniform surface or by passing the molten glass through a series of rollers. Leica glass is currently produced by the latter

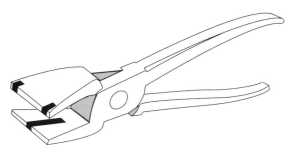

FIGURE 4.15 Modified glazier's pliers for breaking plate glass. Black areas in jaw are strips of electrical tape used to create raised points to press against the glass that is placed into the pliers.
(Courtesy of Leica.)

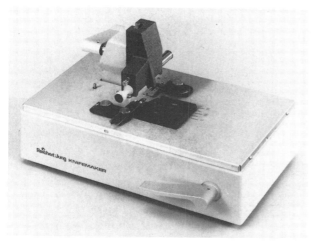

FIGURE 4.16 A commonly used knifemaker for precisely breaking glass strips into ultramicrotome knives.
(Courtesy of Leica.)

procedure so that the stresses are oriented along the rolling direction. As a consequence, the commercially produced glass strips are scored parallel to these stresses. The unpolished glass is quite uniform in thickness, very hard, and has consistent breaking properties. Such glass may be used in the knifemaker or it may be broken into knives using glazier's tools.

Preparing Glass Knives with a Commercial Knifemaker

The Leica Knifemaker is a precision instrument for securing, scoring, and breaking 1″ glass strips first into squares and then into diagonal knives. If one follows the manufacturer's directions and makes all critical adjustments, this instrument will permit microtomists to produce good quality knives on a regular basis. Problems usually arise when inexperienced individuals attempt to readjust the settings on the knifemaker or mishandle the glass during the cleaning or alignment process. Glass knives are made on this instrument as follows:

1. Wash the glass strips with a standard dishwashing detergent and a lint-free cotton cloth. The strips will be extremely slippery and sharp, so care is in order. Should the glass slip from one's grip, do not attempt to catch it but step away and allow it to fall.

2. Rinse the glass with tap water and dry the strip by wiping gently with paper towels or allow it to dry in a dust-free environment. Again, take care not to cut oneself as the towel is passed over the sharp corners. The glass strips can be wrapped individually in dry paper towels and taped shut for long-term storage.

3. Using cotton or nylon gloves to prevent contaminating the glass, pass a proper length of glass strip into the knifemaker with the factory score line down. Score and break the strip into squares.

4. Place the squares in the machine with the glass oriented as shown in Figure 4.17, and score a diagonal following the manufacturer's directions. Be sure to engage the rubber damping cushion, which absorbs the shock of the break and greatly extends the useful length of the cutting edge. Slowly rotate the breaking knob or lever until a solid-sounding break is heard.

In addition, it has been demonstrated that placing 1 to 2 drops of distilled water along the diagonal score line of the glass just prior to breaking greatly improves the quality and length

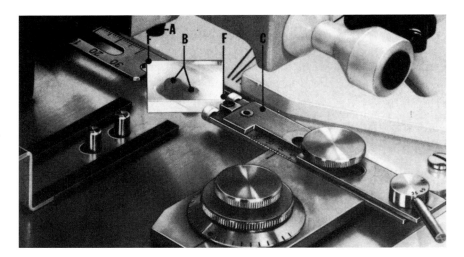

FIGURE 4.17 Glass squares being placed into knifemaker for diagonal scoring. Two studs (A) in clamping head firmly press down against studs (B) in base to effect the break. Sliding guides or holders (C) hold the square in place so that the score can be made between the two F points.
(Courtesy of Leica.)

of the cutting edge (Slabe, Rasmussen, and Tandler, 1990).

5. Remove the two halves, evaluate them carefully under a stereomicroscope, and mount a water trough on the knife as described in "Evaluation of Glass Knives."

Manually Crafted Knives

If one does not own a knifemaker, it is possible to make knives from the commercial glass strips by scoring the strips with a cutting wheel purchased from a hardware store. Glass strips are quite economical and will result in much higher quality knives than can be obtained using glass purchased from a glazier. One principle to follow is the "balanced break" concept in which one applies equal pressure on each side of the score by breaking the glass strip into equal halves in the following manner.

Breaking Glass Knives by Hand

1–2. Same as in section dealing with knifemaker.
3. While wearing cotton or nylon gloves, place the glass strip with the factory score line down on a piece of graph paper ruled in 1″ divisions. Score the strip into 1″ squares using the hand-scoring wheel and a ruler. The score should not extend to the very edges, but should stop within 0.5 to 1 mm. A diamond scorer should not be used since it often gouges out rather than scores a uniform line on the glass.
4. Using modified glazier's pliers, place the single raised point of the jaws under the score that falls in the middle of the glass strip. Increase pressure on the pliers until the glass breaks in half (Figure

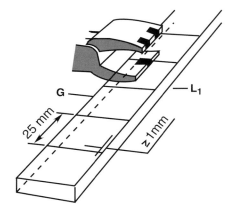

FIGURE 4.18 Breaking of glass strip into 25 mm squares using glazier's pliers. G indicates the smooth, nonscored portion of the strip where the good part of the knife edge will begin. L_1 indicates the serrated edge generated by factory scoring.
(Courtesy of Leica.)

4.18). Continue breaking the remaining portions of the glass strip into equivalent halves until all strips are broken into 1″ squares.

5. Examine one of the squares and orient it so that the factory score marks are down. It is easy to tell the factory scored side since this ¼″ face will be quite flat compared to the faces broken by hand. Make a diagonal score running to within 1 mm of the corner adjacent to the one good edge (Figure 4.19). Although the ideal knife score should run from corner to corner to give a true 45 degree angle, this is not possible using the hand-scoring method. Instead, the diagonal should be directed slightly toward the good edge indicated in the figure.

6. Grasp the square with the glazier's pliers so that the raised point in the jaw is under the diagonal score. The tip of the pliers should form a right angle relative to the score line. Increase pressure until the glass is broken in two (Figure 4.19). The knives can be stored in a dust-free area until ready for use.

Evaluation of Glass Knives. Carefully examine the two knives produced by bisecting the glass square. Each half should be nearly identical when placed side by side. Pay particular attention to the heel or shelf at the base of each knife (see Figure 4.23B). Knives with good edges usually exhibit very shallow heels, typically between 0.5 and 1 mm. Generally, one of the knives will be of high quality for ultramicrotomy: the one that broke onto the smooth, good edge of the commercial glass strip. The other knife, or *counter piece* as it is sometimes called, will have a much shorter usable cutting edge and can be used for rough trimming or for cutting thick sections. Both knives should be evaluated as follows.

Secure the knife in the microtome or under a stereo microscope with a movable focussed light source. Most modern microtomes provide excellent sources of lighting for evaluation of the knife edge. Move the light source and knife to obtain a very narrow, bright strip of reflected light running along the very edge of the knife. Some patience is required since the goal is to obtain a thin sliver of light along as much of the edge as possible. Imperfections, nicks or discontinuities in the edge will show up as dark specks, whiskers, or rough and serrated areas. As you look down on the edge, note the stress line that usually starts in the *left-hand* corner and arcs down toward the heel of the knife (Figures 4.20 and 4.21). Generally, the best part of the knife begins a short distance from the left corner and may extend from ⅓ to ½ way across the knife. Only careful examination at the highest magnifications will reveal these desirable areas. The actual angle of glass knives is usually greater by 10 or more degrees than the angle scored (45 degrees) since the fracture of the knife does not follow the score line as it exits the glass square.

A number of knives can be made and stored for several days, but take care not to contact the fragile edge with anything, including fingers. If grease from fingers contacts the area of the water trough, it will be contaminated and it may leak due to a poor seal. Special boxes can be constructed or purchased to store the knives in a dry and dust-free environment (Figure 4.22).

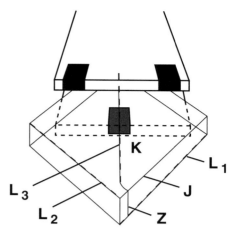

FIGURE 4.19 Breaking diagonally scored 25 mm glass strips with a glazier's pliers. J is the smooth edge of the glass strip. Best part of the knife edge (Z) will begin at intersection with J. L_1 is factory score mark. L_2 is score used to break 25 mm squares. L_3 is diagonal score used to make a good knife (K). The half of the glass knife on the left is termed the counter piece and may also be usable for sectioning purposes. Notice how L_3 does not extend to the corner and how the break veers away from the corner. This effectively enlarges the final angle of the knife edge over the angle that was scored onto the glass strip. *(Courtesy of Leica.)*

Attachment of Water Trough or Boat

After making the knife and prior to cutting either thick or ultrathin sections, a trough or boat to contain the water used for floating sections is attached to the knife by a variety of means. The most commonly used boat can be fashioned from electrician's, masking, or silver audiovisual slide tape by wrapping a small length around the sloped part of the knife and using paraffin or nail polish to secure a watertight seal at the base (Figure 4.23A). These knives should dry for at least 30 minutes prior to use or solvent may dissolve into the flotation fluid and create contaminants that adhere to the sections. Since some of the adhesives and solvents used in formulating the tapes may contaminate the trough, electron microscope suppliers or experienced microtomists should be consulted about the best brands of tapes to use.

More elaborate plastic or metal troughs can be purchased from microtome or knifemaker manufacturers (Figure 4.23A). Such troughs fit the glass tightly to make a good seal and can be rapidly attached to the glass. The major advantage, however, is the rather large area provided for maneuver-

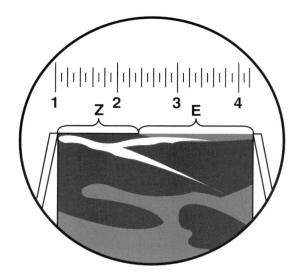

FIGURE 4.20A   Drawing representing top view of glass knife showing stress line curving down toward base of knife. The stress line begins in upper left and marks the point of intersection of J and Z referred to in Figure 4.19. Z is good part of knife edge. E is part of edge containing imperfection or nicks. *(Courtesy of Leica.)*

ing various tools to collect many sections. Although they are considered disposable, troughs can be cleaned of wax in a solvent such as xylene and reused many times.

CAUTION: Use xylene in a fume hood since it is toxic and flammable.

A trough for containing only one or two sections can be rapidly prepared by placing a droplet of molten paraffin 4 to 5 mm down from the knife edge. The hydrophobic properties of the cooled wax will repel a water droplet placed above it and provide a small area for floating the sections. Although it is not recommended for quality work, since sections are difficult to retrieve, it is adequate for quickly cutting a survey section or two.

Diamond Knives

Diamond knives are delicate instruments usually costing several thousand dollars each. Consequently, only experienced microtomists are entrusted with their use. The natural gemstones used are usually pale yellow, of regular crystal structure, and of the greatest possible purity. Initially, the large stones are cleaved into smaller segments, which are then ground into small slabs on a turntable using

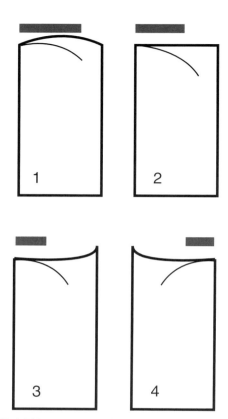

FIGURE 4.20B   Diagram showing four possible cutting edges in glass knives. (1) shows a cutting edge that is bowed up. Such knives are generally sharp, durable and have a long usable edge. (2) shows a commonly encountered knife with a generally straight edge and slightly shorter cutting edge. (3 and 4) show knives having projections or horns to the right or left. Areas near the horns are always full of imperfections; however, some parts of the edge may be usable. The dark bar shows areas of the edge that may be used for general cutting purposes. Always evaluate the glass knife edge prior to cutting sections.

FIGURE 4.21   View of cutting edge of glass knife showing imperfections in the right half of the cutting edge.

FIGURE 4.22 Plastic box for storing glass knives prior to use. The cover is shown in the raised position with eight glass knives inside of the box.

FIGURE 4.23A (*Left*) One method of attaching a tape trough to a glass knife involves sealing the bottom using either wax or fingernail polish (*arrow*). (*Right*) A second type of commercially available plastic or metal trough that has been attached to the glass knife using wax or nail polish for a watertight seal (*arrow*).

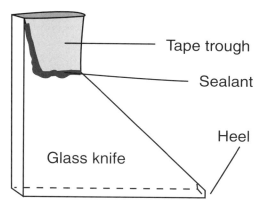

FIGURE 4.23B Drawing of a glass knife showing trough, trough sealant, and heel or shelf at base of knife. The best glass knives have a very shallow heel, less than 1 mm in height.

diamond dust. The orientation of the crystalline lattice is maintained to within 2 degrees of a predetermined value during the grinding. The grinding process often reduces the stones by 50% of their original weight. The diamond is mounted into a soft metal shaft (Woods metal) and final polishing to a sharp edge is conducted under conditions that manufacturers consider proprietary. The shaft containing the final edge is then mounted in a metal trough or boat and cemented, usually with an epoxy plastic. Typical diamond knives are shown in Figure 4.24.

The final angle of the knives may vary from 40 to 60 degrees. Generally, the smaller-angled knives are used for sectioning softer biological specimens and are capable of cutting thinner sections since the knives are sharper. The larger-angled knives, on the other hand, have sturdier edges and are more suitable for cutting harder specimens such as bone, tooth, or even metals. Such knives are not as sharp and generally will not cut the very thinnest of sections. A reasonable compromise is a 45 to 50 degree angled knife. Such knives are sufficiently sturdy to last many years and are capable of cutting sections on the order of 50 to 60 nm. Diamond knives will cut a greater variety of specimens and can section specimens with varying hardness and containing hard

inclusions such as crystals or asbestos fibers. The very best that a diamond knife can achieve is to cut extremely thin, 30 to 60 nm, sections from small block faces and 100 nm sections from block faces that are several millimeters wide.

The lifetime of a diamond knife may vary from a single use to several decades depending on the hardness of the specimen as well as how carefully one handles the instrument. The cutting edge is extremely thin, probably only several molecular layers, and susceptible to damage. Any forward or backward pressure, even a touch with a fingertip or filter paper, will damage the edge, so great care must be exercised in cleaning the knife.

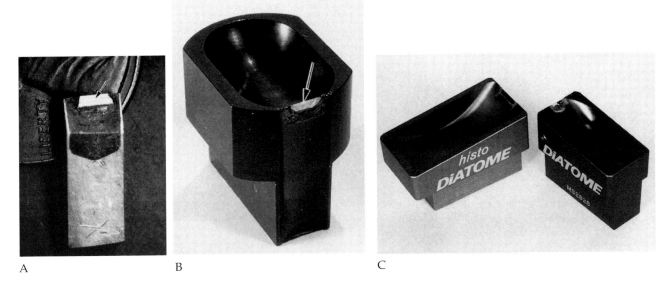

A   B   C

FIGURE 4.24 (A) Gem-grade diamond (*arrow*) mounted in Woods metal holder and polished to give a 3 mm long cutting edge. This unit is mounted into the large metal trough shown in B and C. (B) Photograph of a diamond knife used in ultramicrotomy. Arrow indicates cutting edge. (C) Photograph of a "histo" diamond knife (*left*) used for cutting thick plastic sections in the range of 1 μm. Compare the size of the cutting edge and trough to the diamond knife used for ultramicrotomy (*right*). Industrial-grade diamonds are used in the histo knife, whereas gem-grade diamonds are used in the standard ultramicrotome knife.

REMEMBER: Sections thicker than 1 μm should never be cut using a gem-grade diamond knife since the cutting edge may be damaged.

Cleaning Diamond Knives

After sectioning with a diamond knife, any adherent sections are removed by wiping with an eyelash probe and rinsing the edge with distilled water from a squirt bottle. Debris lodged on the edge itself can be removed by gently passing a *cleaning stick* soaked in water, a sharpened shaft of Styrofoam, or pith-wood (usually supplied with the knife) along the cutting edge (Figure 4.25). Using a toothpick or orange-wood stick may harm the edge, so check with the knife manufacturer for the proper types of cleaning sticks to use with a specific knife. No cleaning solutions (acids, bases, harsh detergents, or strong organic solvents such as acetone) should be used, and strict adherence to the manufacturer's recommendations is urged.

Occasionally, the cutting edge becomes difficult to wet with the trough liquid. Overnight submersion of the knife in 1% Tween 20 or mild detergent in distilled water usually solves this problem. If the edge is still dif-

ficult to wet, passing a cleaning stick over the edge may help. Wetting the cleaning stick with 50% ethanol may help to remove any stubborn debris. Usually, problems of edge wetability can be traced to allowing debris or sections to dry onto the edge. Therefore, should one leave the knife in the microtome for any length of time, it is advisable to overfill the boat and not allow the level to go below the knife edge. Once the edge has been cleaned, the knife is put back into its special holder while still wet. It is dangerous to use canned or compressed gas to dry the edge since small particles or even organic contaminants such as oil may be impacted onto the edge of the knife.

Diamond knives eventually dull through extended use; however, they can be resharpened usually at 50 to 70% of the cost of a new knife. Some manufacturers guarantee a certain number of resharpenings before replacement is necessary. If damage to the edge is too extensive, perhaps due to accidental contact, it may not be possible to resharpen the knife and replacement will be needed. To maintain the integrity of the knife edge, it is highly recommended that one knife be assigned to one person whenever possible, rather than several investigators sharing the knife.

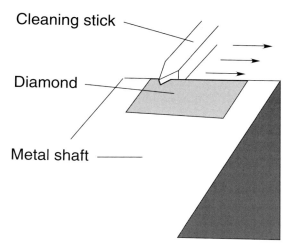

FIGURE 4.25 Method of passing a cleaning stick along the cutting edge of a diamond knife. Some manufacturers supply the styrofoam rods that can be used.

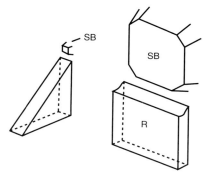

FIGURE 4.26 Comparison of RALPH knife (R) to standard glass knife. Note relative sizes of specimen blocks (SB).

Histo Knives

Histo diamond knives are used to cut sections in the 0.2 to 2 μm thickness range from block faces larger than 1 mm. Histo knives are constructed from industrial diamonds and manufactured by a process different than that used for gem-grade knives. They cost approximately one-third that of the higher quality knives. Such 5 to 6 mm knives, while not able to cut ultrathin sections, are cared for in exactly the same way as knives made from natural diamonds.

A special type of glass knife, approximately twice as wide as the standard 6.4 mm knife, can be constructed using a thicker grade of glass available from various EM suppliers and the Leica company. When such knives are manufactured as described previously, the cutting edge is curved, making it difficult to cut large block faces. Fortunately, it is possible to make *RALPH* knives (named after the late Dr. Paul Ralph) with theoretically unlimited lengths and very straight edges. With the RALPH knife, the cutting edge runs along the broad, 1″ flat side of the glass strip rather than along the narrow 6.4 mm thick side (Figure 4.26). The RALPH knife is used most often for cutting paraffin or softer acrylic and epoxy plastic embedded specimens for light microscopy. It will last probably as long as a metal knife. In cutting harder epoxy plastics, however, the knife usually dulls after 20 to 50 cuts. This type of knife is finding frequent use in light microscopy since the edge is much sharper and more durable than the standard metal knife. As with standard glass knives, the RALPH knives can be made by hand or with specially designed instruments. The very best that a RALPH knife can achieve is to cut 10 × 10 mm block faces in the 0.2 to 1.0 μm range.

Trough Fluids

The troughs of all the knives described must be filled with a fluid in order to support the sections as they are cut. Double-distilled water is the most commonly used type of fluid onto which the sections are floated. Occasionally, if the knife edge is difficult to wet using plain distilled water, one can add a drop of Tween 20 or PhotoFlo (Kodak) detergent to each 100 ml of distilled water in order to improve the wetting properties. If sectioning is in progress when wetability problems are encountered, and one does not wish to replace the fluid in the knife, one may dip a dissecting needle or pin into the detergent and touch the tip of the needle to the trough fluid to transfer the proper amount of detergent to the trough. Some researchers may use a trough fluid composed of 1 to 10% ethanol or acetone in distilled water to improve the wetability properties; however, concentrations any higher than 10% are to be avoided since they may damage the epoxy cements used to seal certain diamond knife shanks into the aluminum trough. Always consult the directions included with the diamond knife for precautions on the use of trough fluids. It is necessary to micropore filter all floatation fluids in order to remove bacteria or other particulates that may be present in distilled water that has been standing for several days.

HINT: Micropore filters should be changed every several weeks since bacteria will grow on and decompose the filter and give rise to filter debris that will contaminate the sections. In addition, one should pass through (and discard) the first 10 to 15 ml of water from the filter to remove any humectants used in the manufacture of the filters.

Grids

A specimen grid is the electron microscope analog of the glass slide used in light microscopy. As the name implies, grids are screens or fine, mesh supports upon which sections and liquid suspensions may be deposited for transport and viewing in the electron microscope. The standard size grid is 3.05 mm in diameter. Most grids have one side that is more brilliant than the other, so microscopists often refer to putting specimens on the dull side of the grid.

Grids were originally made by weaving fine copper wires into a gauze that was then flattened by rollers to remove the undulating surfaces inherent in the weaving process. Discs were then produced with a device similar to a paper punch. At the present time, grids are manufactured using an electrolytic process in which metals are deposited onto templates to produce an extremely fine surface free of undulations, burrs, and irregular open spaces.

A wide variety of shapes and configurations is currently available (Figure 4.27) using metals such as gold, platinum, nickel, stainless steel, or rhodium. In fact, just about any nonmagnetic metal can be used to manufacture grids. Nonmagnetic metals are specified since magnetizable metals (such as iron) would interfere with the images formed by the electromagnetic lenses in the microscope. Stainless steel and nickel often exhibit traces of magnetism so that the grids must be demagnetized prior to use (see Chapter 5, Figure 5.7).

Metals are normally used in the fabrication of grids because they will conduct away heat and electrostatic charges in the specimen resulting from bombardment by the electron beam. *Copper* is still

| Mesh | Percentage Open Area | |
	Regular	Thin Bar
50	80	—
75	72	—
100	64	—
150	49	—
200	46	84
300	29	77
400	23	70
1000	—	55

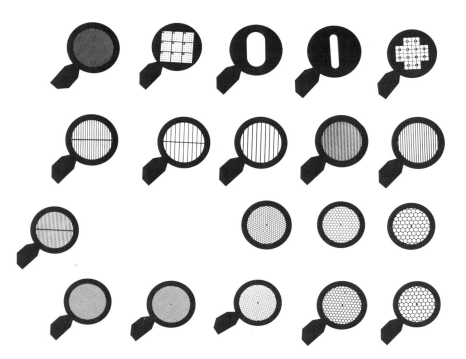

FIGURE 4.27 Specimen grids are available in a variety of mesh designs. The percentage of open area in normal and thin-bar grids is shown in the accompanying table.
(Courtesy of Ted Pella, Inc.)

considered standard for most purposes. However, the more inert metals may be needed if sections are to be subjected to strong oxidants or acids used in certain cytochemical procedures. If analytical studies are to be conducted on the specimen, then grids constructed of carbon or nylon may be used in place of metals that might interfere with the analysis.

The most popular mesh size is 200 (which means 200 bars per inch), but mesh sizes ranging from 50 to 1,000 are available. Larger meshes provide more unobstructed viewing areas, while the smaller mesh sizes provide better support when viewing extremely thin sections or negatively stained specimens. The thickness of the bars forming the mesh will affect the percentage of viewing area. "Thin-bar" grids are available that change the open area from 46 to 84% in 200 mesh grids. Besides cross-hatched mesh, grids are also available with hexagon-shaped mesh. Although the percentage of open area is the same as in the cross-hatched grids, it is felt that the sections are better supported by the hexagons.

In applications requiring a totally unobstructed view of the sections, *single hole* or *slot grids* may be used (Figure 4.28). In these cases, support for the specimen must be provided by a thin membrane of plastic or carbon that bridges the hole or slot. *Support films,* as they are called, are relatively electron transparent.

All grids should be cleaned prior to use by either sonication or swirling in acetone followed by ethanol. After allowing the grids to settle to the bottom of a small beaker and decanting the liquid, the beaker is inverted over a filter paper and the invert-

FIGURE 4.28 Single hole and slot grids are used whenever large, unobstructed views of specimens are desired. The large slot is 1.5 mm wide. In addition, "finder-grids" may be used which have coordinates along the sides of the gridwork that are useful for relocating areas that were examined previously.

Handling Grids

Grids are extremely fragile, thin, and difficult to handle with anything but fine-pointed jeweler's forceps. Picking up grids from a very flat surface such as glass or plastic may present a problem, so most grids are placed onto filter or lens papers and retrieved from these surfaces. Often, it is easier to pick up sections if the grid is bent slightly at the very edge. This is done by grasping the grid at the rim and raising the forceps as the grid is kept flat against a hard flat surface such as a glass slide or plastic petri dish. Usually the bend forms a 30 to 40 degree angle. Locking forceps can be made by slipping a small O-ring over the handles of the forceps. By slipping the ring up and down the handle, the points may be opened or locked in the closed position. Grids that have been coated with a plastic film must be very carefully bent to avoid breaking or tearing the film. Pure carbon films are extremely fragile and will probably break if one attempts to bend standard grids containing these types of films. Instead, special grids with tabs or handles (see Figure 4.27) should be used with these support films so that the bend takes place at the handle rather than at the grid edge. Prior to viewing the specimen in the TEM, the tabs are removed with a razor blade or a special detabber or fingernail clipper. Carbon films are best prepared on grids made of nickel since they are more rigid and less likely to be damaged in handling.

When grids are wetted during various procedures, care must be taken to remove the fluid remaining between the points of the forceps before releasing the grid onto filter paper. Otherwise, the specimen grid will be drawn in between the points, often damaging the specimen. Before releasing the grid, wick the fluid from between the points using a small wedge of filter paper (Figure 4.29). Special locking forceps are commercially available, as are anticapillarity forceps that prevent the trapping of water between the forceps points. Several different types of forceps are shown in Figure 4.30.

Grids may be stored in petri dishes, specimen side *up*, with filter paper liners for short periods of time. Glass dishes are preferable to plastic, since static electricity may cause the grids to jump onto the lid resulting in a loss of orientation. For long-term, safer storage, special grid storage boxes are recommended. Such boxes consist of 20 to 100 numbered slots so that the precise location of all specimens can be recorded until the grids are needed for viewing. Several types of grid storage boxes are shown in Figure 4.31.

ed beaker (with the grids stuck to its bottom) and filter paper are placed into a 60 degree oven until dry. The cleaned grids usually fall onto the filter paper when the solvent has evaporated. The grids are separated and stored under dust-free conditions. Individual grids can also be cleaned by dipping the grid (held in a tweezer) 5 to 10 times into 4% nitric acid. After several rinses in distilled water, the grids can be used immediately to pick up sections.

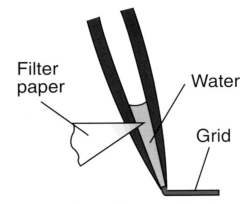

FIGURE 4.29 Water can be removed from between points of forceps by using a piece of filter paper. For actual photograph, see Figure 5.6.

A

FIGURE 4.30 (A) Examples of two different types of fine-pointed forceps used to handle electron microscope grids. A grid may be picked up and secured in the forceps by sliding a small rubber O-ring down the shaft toward the points. (B) Close-up photograph of specially designed forceps that lock upon release of finger pressure, thereby securing the grid. One arm of the forceps has a slight bend at the tip to prevent the accumulation of water. These are therefore termed "anti-capillarity forceps."

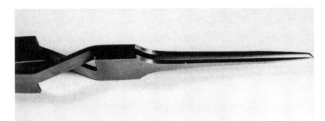

B

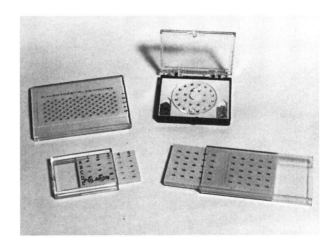

FIGURE 4.31 Different types of commercially available storage boxes for keeping track of specimen grids.

Support Films for Grids

Whenever possible, grids without any supporting films should be used. Most sections will withstand the rigors of handling and viewing in the electron microscope without damage; however, very thin sections as well as some of the acrylic plastics are unstable and prone to damage unless some additional support is provided. Usually 20 to 40 nm thin films of plastic and/or carbon are placed onto the grids to provide this extra support. An ideal supporting film should be thin and electron transparent as well as strong.

Plastic Films

Collodion

This plastic is also known as Celloidin, pyroxylin, or cellulose nitrate. Collodion is one of the first plastics developed and was originally used as the clear base for motion picture films. Composed of cellulose nitrate, made by dissolving cellulose in nitric acid, Collodion is extremely flammable when stored as dry chips or wools. It is used for making support films since it provides adequate support and is relatively easy to make. Collodion is not as stable under the electron beam as are Butvar and Formvar plastics, so it is necessary to use thick 40 to 50 nm films for equivalent support. Although one can purchase the material as

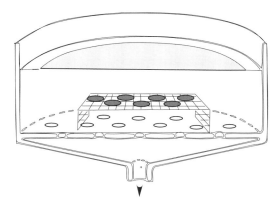

FIGURE 4.32 Buchner funnel containing platform onto which grids will be placed for coating. The funnel is filled with distilled water and then the platform and grids are arranged under the water prior to floating the appropriate film on the water surface.
(Courtesy of L. W. Coggins and IRL Press Limited.)

solid plastic chips, it is more convenient (and safer) to purchase purified solutions from commercial suppliers.

Preparing Collodion-Coated Grids

1. Make or obtain a 2 to 3% solution of Collodion in n-amyl acetate. This should be stored in a dark bottle (ultraviolet light breaks down and weakens Collodion) and the bottle placed in a desiccator such as a resealable plastic bag with silica gel to prevent the absorption of moisture.

2. Construct a platform out of stainless steel screen and place it on the bottom of a clean glass petri dish or Buchner funnel that has been equipped with a clamped piece of tubing. The platform should be about 2″ × 3″ and be raised by about ¼″ to ½″ (Figure 4.32).

3. Pour clean distilled water into the container to cover the platform by at least ½″. It is important that the water be free of lint and microorganisms.

4. Place the grids, dull side up, on the top of the platform. The submerged grids must not overlap, but should be placed close together.

5. With a clean, dry, glass Pasteur pipette, take up a small volume of the Collodion solution. With the pipette held 1 cm above the water surface, release one drop of the solution onto the water surface. The solution should rapidly spread out to form a uniformly silver layer of plastic after the solvent has evaporated. This first film, which is used to clean the water surface of lint, is removed and discarded using a forceps.

6. A second film is made in the same manner and allowed to dry for 1 to 2 minutes. Examine the film using a fluorescent light held directly over it. One should see an intact, continuous silvery sheet free of lint, tears, and irregular patches of color. Do not proceed until a film that fulfills these criteria is obtained.

7. Lower the water level by draining the Buchner funnel or by aspirating water from the petri dish. Position the Collodion membrane over the grids using the tip of the fine forceps as the water is lowered. When the water has been totally drained, remove the platform and place it in a 40 to 50°C dust-free oven for drying. (An alternate method is to lift the screen containing the grids up under the Collodion film using a forceps.)

Individual coated grids may be removed and stored until needed, or a light coating of carbon may be evaporated onto the films to further stabilize them (see Chapter 5). It is important to evaluate the films in the TEM prior to use to ascertain that they are sturdy and free of holes. Films that break or fall apart in the microscope indicate that the Collodion content must be increased, while holes indicate the presence of water in the solution. If water is present, the solution should be discarded. The correct thickness of film, indicated by a silver/gold interference color, is important since an overly thick film will degrade both contrast and resolution, while a thin film will drift or break in the microscope. As is the case with all plastic films, the coated grids will be good for several weeks, but should be as freshly prepared as possible for optimum results.

Butvar/Formvar

Butvar (polyvinyl butyral) and Formvar (polyvinyl formal) plastics are much stronger films than Collodion and can, therefore, be much thinner (25 to 30 nm) in practice. Unfortunately, since drops of the plastic solution do not readily spread over a water surface, these films cannot be conveniently prepared using the drop technique used with Collodion. Instead, a thin layer of the plastic film is stripped from a glass slide onto a water surface.

Preparing Butvar- or Formvar- Coated Grids

1. Clean a standard microscope slide in 95% ethanol by wiping with laboratory tissues. Various brands of slides should be evaluated in this procedure until a suitable one is found. In order to enhance separation of the plastic film, some microscopists place a tiny droplet of dishwashing detergent onto the slide and use a lintless cloth to rub away most of the detergent leaving an extremely thin molecular layer.
2. Dip the clean, dry slide into a 0.3 to 0.5% solution of Butvar or Formvar in chloroform (or ethylene dichloride) in a tall glass cylinder. Withdraw and hold the slide in the vapor phase above the liquid for 30 seconds or so. Completely withdraw the slide and allow it to drain vertically on a filter paper. This must be done in a dry, dust-free environment.
3. Place the coated slide on a filter or lens paper and score around the periphery of the slide (3 to 4 mm in from the edge) using a razor blade or fine needle. The object is to break the continuity of the film to facilitate its removal, but not to deeply scratch the glass slide.

4. Slowly lower the slide held at a shallow angle into the dish of water described for making Collodion films so that the film is detached from the slide (Figure 4.33).
5. Lower the water level so that the film is deposited on the grids.

A major problem with this technique is the failure of the plastic film to detach from the glass slide. Numerous methods have been suggested to enhance detachment. Breathing on the coated slide to cause condensation of water vapor immediately before immersing it into the water seems to help. Coating the cleaned, dry slides with the wetting agent Victawet (E. F. Fullam, Inc.) in a vacuum evaporator sometimes helps. To accomplish this, a piece of Victawet about the size of a grain of rice is placed in the wire basket of a vacuum evaporator and slowly heated under the high vacuum. The vaporized Victawet will condense onto the slides. The coated slides are then wiped with a lintless filter paper to remove excess agent just prior to dipping into the plastic solution. It is important to remove the excess Victawet or films perforated with holes will be produced.

An alternate method for coating the slides involves placing the dull side of the grids directly onto the floating plastic film and touching them lightly with the forceps to enhance adherence. The grids are picked up by bringing a clean microscope slide down onto the film and then reversing the slide under water so that the side of the slide with the filmed grids emerges from the water first. This method requires some practice. An easier method involves rolling the floating film and grids up onto the outside surface of a clean glass cylinder or beaker by contacting the cylinder to the edge of the film and rolling the cylinder so that the grids and overlying film go onto it. The plastic films may also be picked up onto Parafilm, filter paper, or stainless steel mesh. After the plastic film has dried, the grids may be carefully removed from the substrate so that the films remain on the grids. It is best to evaluate the filmed grids under a fluorescent light and a stereomicroscope to detect tears, holes, or overlapped areas of plastic film.

Perforated or Holey Films

The *holey film* (Figure 4.34) has many uses in electron microscopy. Not only is it used to support certain types of particulate specimens for high resolution work, but it also makes an excellent standard for checking astigmatism, resolution, and stability of the microscope. These films can be prepared in several different ways.

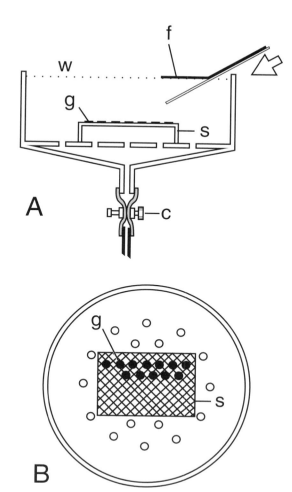

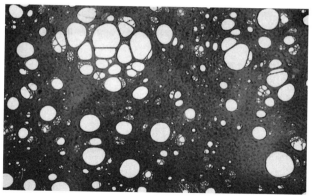

FIGURE 4.34 Electron micrograph of perforated or holey plastic film that is often used to evaluate the performance of the transmission electron microscope. Such films may also be used to support tiny specimens (crystals, for example) that have been suspended across the holes.

FIGURE 4.33 (A) Detachment of Formvar or Butvar film (f) from glass slide and onto surface of water (w) in Buchner funnel. Grids (g) are placed under water onto the wire mesh shelf (s). The water is drained from the apparatus by loosening the clamp (c) so that the film is gently lowered onto the grids on the platform. The platform is removed and allowed to dry before removing the coated grids. (B) Top view of Buchner funnel apparatus illustrated in Figure 4.33A, here showing the placement of the grids (g) on the stainless steel screening (s). The open circles represent the drain holes in the support base of the Buchner funnel. *(Courtesy of L. W. Coggins and IRL Press Limited.)*

Preparation of Holey Plastic Films

Bayer and Anderson (1963) described a rather easy method to prepare holey films by dipping a slide into a 0.3 to 0.4% solution of Formvar in ethylene dichloride. Hold the slide in the vapor phase above the liquid for 1 to 2 minutes allowing a very thin layer to form by drainage of the solution to the base of the slide. The slide is rapidly withdrawn from the vapor and immediately breathed upon to condense moisture into the film before it dries in the atmosphere. *IMPORTANT: The film must be thin and the moisture must be condensed into the film before the solvent evaporates. Failure to make satisfactory perforated films is usually due to problems with one or both of these points.* The film is then scored with a razor blade, floated onto a water surface, and grids applied as described in the preceding section.

A second method, described by Harris in 1962, involves suspending a small volume of glycerine in the Formvar solution by vigorous shaking. This generates tiny droplets in the Formvar since the two phases are not miscible. The size of the droplets is determined by the amount of glycerine added. To generate holes of 25, 15, 7 or 5 μm, respectively, 1 ml of glycerine is added to one of the following volumes of the Formvar solution: 8, 16, 32, or 120 ml, respectively.

Dip a slide into the vigorously shaken Formvar-plus-glycerine suspension immediately after shaking, and withdraw it to drain in the vapor phase of the ethylene dichloride in the tall cylinder. Remove the slide and, after 10 to 15 minutes of drying at room temperature, breathe on the slide (or expose it to the steam rising over boiling water) to help loosen the film and float it onto the water surface. After the grids have been picked up, the dry film may be coated with carbon as described in the sections that follow. The very best perforated films will be made by dissolving away the Formvar backing as described in the section entitled "Pure Carbon Films."

Carbon-Coated Plastic Films

It is highly desirable to evaporate carbon (or graphite) onto the plastic films to strengthen them further, since layers of carbon as thin as 2 to 5 nm impart great strength to the plastic. In fact, carbon may be used sometimes to strengthen sections on uncoated grids. Carbon coating may be conducted as follows.

Preparation of Carbon-Coated Plastic Films

1. Place the plastic-coated grids, or grids containing sections, into a vacuum evaporator on a clean glass surface with the plastic film side up and evacuate the chamber to 1×10^{-3} Pa.
2. Slowly heat the carbon rod or braid until it first turns red, and maintain this condition for 10 to 15 seconds to outgas any adsorbed contaminants. Rapidly increase the current to the carbon until evaporation takes place. It is important that the temperature be raised *rapidly* in bursts lasting 5 seconds or so rather than bringing up the current slowly. This will avoid heating and damaging the fragile plastic films. The grids should be placed at least 6" to 8" away from the electrodes for the same reason. The thickness of the carbon may be estimated by placing a white index card with a raised object such as a screw on the card near the specimen. As the carbon is evaporated, the card darkens except in the area shaded by the screw. Hence, one can readily compare uncoated areas to carbonized areas to more readily gauge the thickness. A light tan color, compared to the white shaded area, will give a coating of 4 to 7 nm in thickness. Details of carbon evaporation are given in Chapter 5.
3. After the electrodes have cooled for 3 to 5 min, air is admitted to the vacuum system and the grids are removed.

Pure Carbon Films

Carbon films without any plastic substrates are more desirable supports because they are considerably thinner, less electron dense, and much stronger than plastic alone. Unfortunately, since they are difficult to make, pure carbon films are not used as often as the combination films, except when high stability and resolution are required.

Methods of Preparing Pure Carbon Films

Method 1. Place individual Formvar/carbon-coated grids, prepared by the method described previously, onto a stainless steel platform submerged in chloroform. The filmed sides should be facing up. The apparatus is identical to the one described for preparing Collodion-coated films (see Figure 4.33) except that it is filled with chloroform rather than water. After several minutes, the Formvar bridging the open spaces on the grid will be dissolved leaving only the carbon. The Formvar sandwiched between the carbon and copper of the grid should remain undissolved and serve as a cement to hold the carbon in place. The length of time (several minutes) for the dissolution of the plastic must be determined by trial and error since too long of an exposure will dissolve all of the Formvar and the carbon will detach. After the chloroform is drained, the stainless steel mesh is placed in a 60 degree oven to permit the chloroform to evaporate. If one uses Collodion/carbon grids in this technique, amyl acetate solvent must be used to dissolve the Collodion. *DANGER: This procedure must be conducted in a fume hood since chloroform is a potential carcinogen and an anesthetic.*

Method 2. Another method of preparing carbon supports is coating a Victawet-treated slide (see section on Butvar/Formvar film making) with a layer of carbon in a vacuum evaporator. Instead of using a glass slide, it is also possible to make a substrate for the carbon by delaminating a 1×3 inch mica sheet into two thin layers by slipping a clean, single-edged razor blade between the layers and prying them apart. This should be done in a fume hood since mica will shower thousands of fine particles into the air during this process. *CAUTION: Do not breathe the mica particles; they are potentially dangerous.* Transfer the slide or freshly cleaved mica sheet into a vacuum evaporator and coat with a light tan coating of carbon.

Prepare adhesive-coated grids by first placing clean grids (dull side up) on a filter paper and then placing one drop of "grid glue" on top of each of the grids. *Grid glue* is prepared by dissolving the adhesive from 2" of transparent tape (Scotch Brand) in 10 ml of ethylene dichloride, discarding the tape, and keeping the solution.

Place the adhesive-coated grids, dull side up, on top of the submerged stainless steel rack described in "Preparing Collodion-Coated Grids." Slowly lower the carbon-coated slide or mica sheet under the water so that the carbon layer will float free on the water surface. Lower the water level until the carbon comes to rest on top of the grids. Dry the rack in a 60 degree oven overnight.

If one is unsuccessful with this latter method, score the carbonized slide or mica sheet into 3 × 3 mm squares using a clean razor blade, and float off the tiny carbon squares onto a water surface. Lift up each square individually onto an adhesive-coated grid. Blot the excess water by touching the edge of the grid to a filter paper and placing the grids in an oven to dry.

The Ultramicrotome and the Sectioning Process

Development of the Ultramicrotome

The development of the modern ultramicrotome was the major breakthrough that made possible the routine sectioning of tissues for subsequent viewing of cellular ultrastructure in the transmission electron microscope. Prior to the advent of such instruments, only naturally thin specimens such as viruses, bacteria, and edges of whole cells could be examined. Internal cytological details were only surmised based on light microscopy combined with cellular fractionation and biochemical analysis of the fractions. *The ultramicrotome was one of the most important ancillary instruments developed in the field of biological transmission electron microscopy.*

As early as 1939, Von Ardenne attempted to cut sections using a modified histological microtome that cut wedge-shaped sections using metal knives. The thinnest region of the wedge was examined hoping to find sufficiently thin areas. Several years later, improvements by O'Brien and McKinley as well as by Fullam and Gessler led to the development of high-speed or "cyclone knife" microtomes in which the specimen was clamped in a fixed holder and the metal knife spun at 12,500 rpm by an electric motor. As the specimen was advanced into the spinning blade, by means of a fine screw, 0.1 μm or thicker sections were cut and rapidly dispersed into the air in all directions, making collection of sections rather haphazard. Investigators tried even more rapidly spinning blades until it became clear that such high speeds actually hindered their efforts.

A significant advance in the development of the modern ultramicrotome appeared in 1953,

when Keith Porter and J. Blum reported the development of two different microtomes and sectioning technologies, the principles of which are used even today. The first was a *thermal advance* ultramicrotome. The plastic-embedded specimen was mounted on the end of a horizontal metal bar and passed over a glass knife. On the return stroke, the specimen avoided bumping the back of the knife by passing to one side. Thermal advancement of the specimen was achieved by heating the metal bar in which the specimen was mounted with a gooseneck reading lamp. Sjostrand (1953) subsequently reported a refinement of this ultramicrotome, which underwent extensive refinement and gave rise to the LKB (now Leica) instrument line of thermal advancement microtomes. In the most modern versions, the specimen arm is advanced by a heating coil while being passed over the knife. The specimen avoids contact with the back of the knife during the vertical return stroke since the knife is retracted 15 μm by a powerful electromagnet. When the specimen reaches the top of its return, the magnet is shut off so that the knife can spring back into position as the specimen is allowed to drop by gravity. The speed of the fall of the specimen arm is controlled by a coil motor so that rates of 0.1 to several mm per second may be achieved. A diagram of the basic features of the microtome are shown in Figure 4.35A.

The second "improved" instrument, the *Porter-Blum MT1,* used a *mechanical advancement* mechanism in which a screw thread advanced a pivot arm that acted as a fulcrum or lever to press the specimen arm forward (Figure 4.35B, C). Since the forward motion of the screw was reduced 1:200 times by the fulcrum, the specimen could be advanced in increments as low as 25 nm. In 1962, the *Sorvall MT2* instrument was introduced, which featured a motorized or hand-driven specimen arm, more precise control of section thickness, and retraction of the specimen arm on the return stroke in order to avoid the back of the knife. Several models later, and following a major redesign of the instrument, an improved microprocessor-controlled version, the *MT-X,* was manufactured by RMC, Inc. (Figure 4.36).

Basic Features of All Ultramicrotomes

Although certain designs and features may vary, all modern ultramicrotomes consist of several basic components that permit: (a) fine advancement of

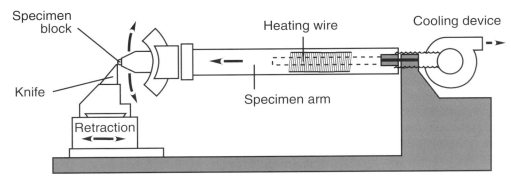

FIGURE 4.35A   Diagram showing thermal advancement mechanism used in the older LKB (Leica) ultramicrotome. A heating wire wound around the specimen arm is heated to cause the arm to expand. Since the rear of the arm is anchored to a solid base, the expansion of the arm is in the direction of the knife. A motor moves the arm up and down at fixed intervals. The rate at which the arm drops, however, is adjustable by the operator (e.g., the cutting speed). Thickness of the sections is determined by the amount of heating of the specimen arm since the interval between cuts is fixed. In order to avoid bumping the specimen on the back of the knife, the knife is retracted several μm by means of a powerful electromagnet.
(Courtesy of Leica.)

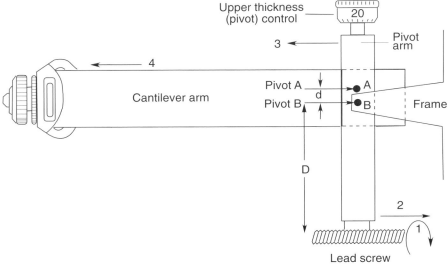

FIGURE 4.35B   Diagram showing mechanical advancement mechanism used in the Porter-Blum MT1 ultramicrotome. A fine threaded screw was turned a small amount during each cutting cycle (*arrow 1*). The turn of the screw moved the pivot arm backwards in the direction indicated (*arrow 2*). Since the pivot arm was fixed at a pivot point on the frame of the microtome, the backward motion of the pivot arm resulted in a slight amount of forward movement of the cantilever arm (*arrows 3 and 4*). By adjustment of the upper thickness control, the pivot point was moved either up or down. For instance, if the pivot point was moved to point B, a thicker section would be cut than if the pivot point were at point A. These principles are shown in the simplified diagram shown in Figure 4.35C.
(Courtesy of RMC, Inc.)

the specimen by mechanical or thermal means; (b) precise orientation of the specimen and knife; (c) coarse mechanical advancement of the knife for approaching specimen and thick sectioning; (d) control of cutting speed; (e) knife avoidance on return stroke; and (f) magnification and illumination of the sectioning process.

The Sectioning Process

The cutting of ultrathin sections is a demanding technique for beginning microtomists. It is important to become thoroughly familiar with the features and details of operation of the particular ultramicrotome being used before attempting to cut

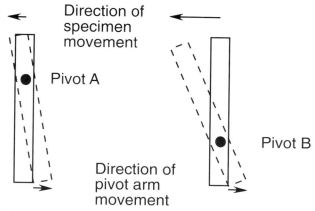

Direction of specimen movement

Pivot A

Direction of pivot arm movement

Pivot B

FIGURE 4.35C Conceptual diagram showing the effect of moving the pivot point of the pivot arm from point A (*left*) to point B (*right*). Since the cantilever arm is moved forward by larger increments at point B, thicker sections will be cut compared to point A.

sections. Patience, concentration, and a certain degree of manual dexterity are required of all microtomists. It is essential to have adequately embedded and trimmed specimens as well as high quality knives and specimen support grids. The environment surrounding the microtome, as well as all tools and reagents, must be immaculate. Detailed records of the protocols followed must be kept in order to repeat the experiment. One would proceed in the following manner.

Specimen Insertion

Once a specimen block has been trimmed using one of the methods described in the previous sections, the chuck bearing the block is placed into the arm of the ultramicrotome and the chuck oriented as closely as possible to the arrangement described in the "Specimen Orientation" section later in the chapter. Although the orientation will be only approximate at this point, it will greatly facilitate subsequent readjustment and decrease the possibility of damaging the knife.

Water Level Adjustment

A knife is mounted in the holder and slightly overfilled with double-distilled water that has been microfiltered using 0.45 μm filters. The edge of the knife should be wetted at this point. If not, use a fine, mascara-free eyelash probe (made by gluing an acetone-cleaned eyelash to the end of an applicator stick) to brush the water onto the edge. While looking down on the water surface, lower the water level using a small syringe until a silver to silver-gray reflection is obtained with the fluorescent light of the microtome (Figure 4.37). It may be necessary to move the light to obtain this reflection. The level should not be so low that the knife edge becomes dry. Never let sections dry down onto the knife edge: they are difficult to remove once dried. If wetability of the knife edge is a problem, then 10% ethanol or acetone may be used in place of distilled water; however, the level must be watched because these fluids evaporate rapidly.

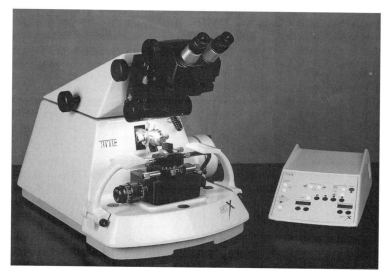

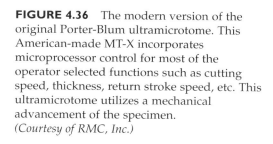

FIGURE 4.36 The modern version of the original Porter-Blum ultramicrotome. This American-made MT-X incorporates microprocessor control for most of the operator selected functions such as cutting speed, thickness, return stroke speed, etc. This ultramicrotome utilizes a mechanical advancement of the specimen. (*Courtesy of RMC, Inc.*)

A

FIGURE 4.37A The knife is overfilled with microfiltered distilled water to ensure that the knife edge is wetted along its entire length. A diamond knife is shown clamped into the knife holder of the microtome.

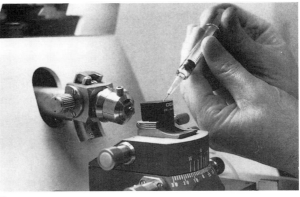

B

FIGURE 4.37B The water level is lowered by withdrawing water using a small syringe until a silver-gray reflection of the fluorescent light used to illuminate the boat is seen on the water surface. The knife edge must remain wetted.

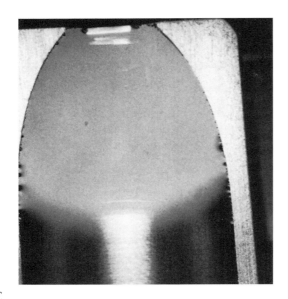

C

FIGURE 4.37C Microtomist's view through the stereomicroscope of the microtome showing the silver-gray reflection seen when the water level in the knife is properly adjusted.

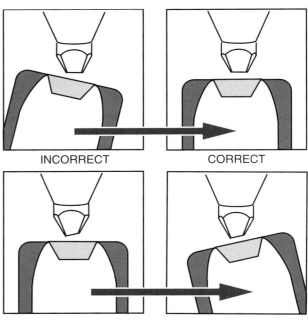

D

FIGURE 4.37D One must orient the knife parallel to the top surface of the specimen block. In the top-left panel, it is necessary to rotate the knife counterclockwise to orient the cutting edge parallel to the specimen surface (*top-right*). In the bottom-left panel, the surface of the specimen block was trimmed at an angle so it is necessary to rerotate the knife counterclockwise to make it parallel to the specimen block (*bottom-right*). All drawings show the microtomist's view of the top of the knife and specimen block.

Cutting Range Adjustment

If one observes the movement of the specimen arm in an ultramicrotome during the cutting process, it is apparent that there are several zones along the path where certain events take place. Shortly after the arm begins its downward journey from its topmost position, it slows down for a distance of 5 to 15 mm. This is the so-called *cutting range* where the specimen is passed over the knife edge and the section is cut. It is important that the knife height be adjusted so that it is well within this range, otherwise, the specimen movement will be too rapid and unsatisfactory sections or damage to the knife will result. On most ultramicrotomes, the location as well as the length of the cutting range along the cutting stroke can be adjusted by the operator.

After passing through the cutting range, the speed of the specimen arm picks up and it avoids the knife edge on its return to the top of its travel as discussed previously in "Development of the Ultramicrotome." By the time it returns to the topmost point of its travel, the specimen has been advanced the appropriate distance in nanometers established by the microtomist using the specimen thickness setting on the ultramicrotome.

Clearance Angle Adjustment

The top of the knife in its holder is inclined slightly toward the specimen to prevent the specimen block from rubbing the back of the knife during the cutting stroke. A 4 degree angle is used as a starting point, but some experimentation may be necessary as outlined in the section "Methodical Sectioning" (see Figure 4.38). Diamond knives usually come with a clearance angle recommended by the manufacturer.

Knife Advancement

The knife edge must be adjusted so that it is parallel to the specimen cutting face (Figure 4.37D). Mechanically advance the knife as close as possible to the specimen block face without actually contacting it. This will require the use of both coarse and fine knife movement controls. This is done by looking directly down upon the block and moving the knife stage forward while observing the closing gap between the block and the back of the knife. On some ultramicrotomes, when the microtome light source is adjusted properly, a shadow of the knife is projected onto the block face when viewed through the stereomicroscope. On such instruments, the gap between the knife and the block is easily visualized by the diminishing shadow as the knife is moved forward (Figure 4.39). The knife should also be

FIGURE 4.38 Knife holder with diamond knife clamped into position. Note the clearance angle adjustment scale (*arrow*) that is used to measure the angle of the knife inclination relative to the specimen.

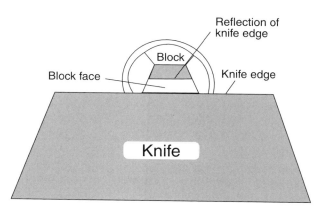

FIGURE 4.39 Advancement of the knife toward the specimen block can be evaluated in some microtomes by examining the reflection of the knife in the block face of the specimen. It is necessary to adjust the light properly to obtain this view. By studying the reflection, one can not only determine the relative distance of the knife from the specimen, but also use the reflection to align the knife edge parallel to the specimen block face.

moved laterally to bring a good cutting edge of the knife opposite the block face.

Specimen Orientation

The block is oriented in the arm of the ultramicrotome so that the edge that first comes in contact with the knife is parallel to the knife edge and horizontal to the tabletop (Figure 4.40). It is also important to adjust the slant of the specimen cutting face so that it moves

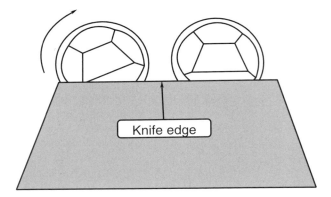

FIGURE 4.40 Orientation of the specimen block so that the leading edge of the block is parallel to the knife edge is accomplished by rotating the specimen block holder while it is in the ultramicrotome arm. This may be done by using the stereomicroscope of the ultramicrotome to look directly into the block face. Great care must be taken at this step so that the knife is not damaged by moving the block face into the knife edge.

parallel to the cutting plane (Figure 4.41). This ensures that the knife will cut a full section rather than a chip from the specimen surface. The best way to achieve this is to establish a reflection of the knife edge in the block face and to observe the gap between the actual edge and its reflection in the block face. If the gap remains constant as the specimen moves past the knife edge during a downstroke of the microtome arm, then the vertical adjustment is correct. However, if the distance changes, it will be necessary to change the tilt of the specimen in the vertical plane. This is difficult for beginning microtomists; however, if one has just faced and trimmed the block in the microtome, it will be properly oriented. If not, some trial adjustment may be necessary before a full section is cut.

Two major mistakes that may be made during orientation of the specimen include ramming the block face into the knife and not retracting the knife after making adjustments which results in cutting an extremely thick section.

IMPORTANT: The knife should always be moved back away from the block during the specimen orientation procedure to avoid damaging the knife edge.

HINT: Try to keep the vertical orientation in the arc adjustment of the specimen chuck in the microtome (Figure 4.41) always in the same position so that specimen chucks can be removed and returned to exactly the same position each time.

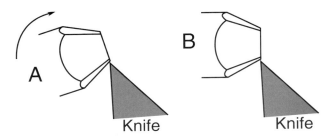

FIGURE 4.41 Adjustment of the slant of the specimen cutting face. In order to cut full sections, it is necessary to move the specimen block from the position shown in A to the position shown in B. All ultramicrotomes have mechanical adjustments for achieving this reorientation of the specimen block.

Knife Contact

Using the knife fine-advancement controls, move the knife as close as possible to the block face. This is done under high magnification with good illumination while advancing the knife in increments of 1 μm or less. After an initial thick section is made, the specimen fine-advancement mechanism is activated. If it is essential to collect the very first section, the specimen fine advancement mechanism must be activated several micrometers before initial contact with the block face. This will require patience as the distance is progressively narrowed in steps of 50 to 90 nm.

Ultrathin Sectioning

After adjusting the thickness control to an appropriate setting (usually 80 to 90 nm, initially), ultrathin sections are cut (Figure 4.42) and collected in the trough of the knife. The goal is to produce a uniformly thin section at each pass of the specimen over the knife. One usually starts on the thicker side and adjusts the thickness setting downward until the desired section thickness is obtained. Actual thickness of the sections is determined not so much by the machine settings, but is based on the interference colors generated by shining a fluorescent light onto a water surface to establish a silvery background reflection over the water. An estimate of thickness is given by these colors. Interference colors are generated when white light reflected from the water surface passes through the plastic sections and is refracted so that the emerging light waves interfere with those coming from the water surface.

Most microtomes have an *automatic mode* to move the specimen through its path over the knife and into its return knife-avoidance regime. In older microtomes that lack the automatic feature (e.g.,

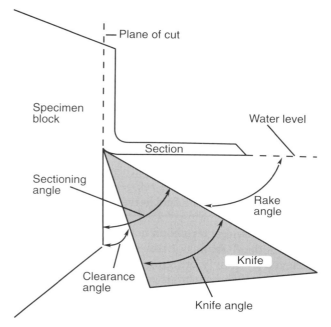

FIGURE 4.42 The various angles formed by the knife and specimen block must be proper in order to achieve good sections. Compression of the section occurs as it is cut by the knife and floats onto the water surface. The compression can be relieved by slightly heating the sections as described in the text. Thickness of the sections is estimated based on the interference colors on the silvery water surface (see Figure 4.37C). Colorless to gray sections = 30–60 nm, silver-gray to silver = 50–70 nm, gold = 70–90 nm, purple = 100–190 nm, and blue = 200+ nm in thickness.

Sorvall MT1), handwheels are turned to move the specimen arm through its path. Microtomes with more precise control of the cutting speed use a motor-driven belt to move the arm. Some models (LKB, for example) control the speed of gravitational drop with a motor connected to the specimen arm by a cord. The *Huxley Microtome* employs an oil-filled piston to control the rate of drop of the specimen arm by adjusting the size of the hole through which the oil exits the piston.

If the *cutting speed* is adjustable, an initial setting would be 0.5 to 1 mm/sec for diamond knives and 2 to 3 mm/sec for glass knives. If no problems are encountered, sufficient numbers of sections are collected in the knife trough. An eyelash probe may be used to move the sections in the trough by gently touching the probe to the section (Figure 4.43). Take care not to press too hard against the section, since it has a strong tendency to adhere to the eyelash probe. Should this happen, the section should be removed with a filter paper (not the fingers because body oils will contaminate the probe and trough).

Section Retrieval

The sections should be moved into clusters of 2 to 4 sections with an eyelash probe. Generally, this number of sections will fit conveniently onto the specimen grid (which has been prepared as described earlier in this chapter). The objective will be to place the sections in the center 2/3 of the grid, since the speci-

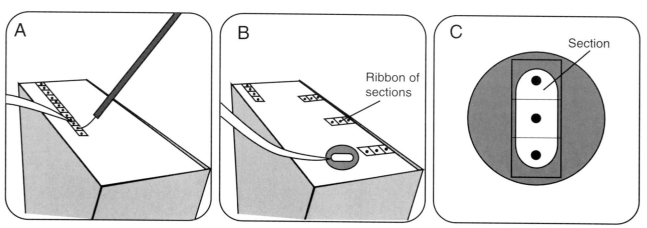

FIGURE 4.43 Sections can be oriented in the water trough by using an eyelash probe to lightly touch the section and move it as desired. (A) Long ribbons of sections can be broken into smaller ribbons using two probes. (B) Ribbons or individual sections can be moved into groups that will conveniently fit onto a single grid. (C) Slot grid showing three sections along the slotted opening. The area that will be viewable in the TEM labeled "section." *(Courtesy of Leica.)*

It may be necessary to collect a complete series of continuous or *serial sections* as they come off the microtome. In this instance, all sections must be collected or accounted for and grids containing grid bars should be avoided, since they would obstruct a complete view of the cellular details. Usually, when serial sections are taken, it is desirable to trace one or more structures in all of the sections, necessitating the use of single hole or slot grids with a supporting film. Although quite a demanding undertaking, it is possible for the experienced microtomist to collect a complete series of such sections. It will be necessary to use slot or single hole grids with a supporting membrane and to keep careful records of each series of sections as they are made. The specimen block should be trimmed to a shape such as a trapezoid so that the order of sections in a ribbon can be readily ascertained.

A good diamond knife and automatic ultramicrotome with a mechanical advance is desirable if all of the serial sections must be the same thickness. A mechanical advance is preferred, since it is often necessary to stop microtomy, pick up sections or make other adjustments, and then resume sectioning with a minimum disturbance to the order and continuity of the sections. Thermally advanced units will continue to advance during this time period, so that a much thicker section than desired may be cut upon resumption of sectioning. Nonetheless, it is possible to do serial sectioning on thermally advanced ultramicrotomes by collecting groups of sections in various areas of the trough and retrieving them when the ultramicrotome advance arm is being cooled. Since the arm will have retracted during the collection period, it will be necessary to advance the knife several micrometers in order to resume sectioning. This should be done carefully using micrometer and finer mechanical advancements to avoid cutting thicker than desired sections.

Collecting Serial Sections onto Single Hole or Slot Grids

Rather than attempting to collect the sections directly onto a coated slot grid, it is best to collect them as follows. Grasp an uncoated slot grid in a forceps so that the dull side of the grid is down, and bring the slot or hole directly down on the sections so that they are corralled by the grid (Figure 4.46A). Upon lifting the grid straight up, surface tension will carry the sections in the water that fills the void in the grid. The grid is then placed, sections still on top, onto a plastic film stretched over a 3.5 to 5 mm hole drilled out of a platform made of aluminum stock (Figure 4.46B).

HINTS: (I) To enhance adherence of the grid to the Formvar film, dip the clean grids into a separate 0.5% Formvar solution with a forceps and immediately place the grids on a filter paper until thoroughly dry. (2) The film coated platforms are prepared by first floating the plastic onto a water surface and bringing the platform up from under the film. (3) Care must be taken not to puncture the thin film on the platform with the forceps or edge of the grid. (4) A number of holes can be drilled out of the aluminum so that 25 or more grids can be placed onto one such platform. (5) It is possible either to purchase the ready-made platforms (Structure Probe, Inc.) or to purchase punched stainless steel sheeting (Small Parts, Inc.) that may be cut into the desired sizes of platforms.

The platform is kept warmed to 60°C most of the time (except possibly during the transfer of sections) in order to facilitate spreading of the sections and to prevent wrinkling. After the grids have dried completely, the flat head of a nail or other instrument, about the same size as the grid, is used to push the grids down onto a filter paper placed under the platform (Figure 4.46C). The grids may then be stained and examined in the transmission electron microscope as described in the next two chapters.

Factors Affecting Sectioning

A number of conditions may affect the quality of sections produced. It is unusual when sectioning problems can be attributed to instrument failure, since most microtomes are reliable, sturdy devices that normally last many years before service is needed.

Location of the ultramicrotome is an important consideration since it is more sensitive to vibrations than the modern TEM. The room must be clean and dust-free, quiet and free of distractions, of uniform temperature, without drafts, and have good lighting (preferably dimmable). Most ultramicrotomes have built-in mechanisms for neutralizing minor vibrations; however, it may be necessary to place the ultramicrotome on a special vibration-damping table or possibly to relocate the microtome in another room if vibrations become excessive.

The *dehydration and embedding* steps may cause problems if the specimen is not adequately dehydrated or if the plastic has not totally infiltrated the cells. Poorly mixed or only *partially polymerized plastic* is a problem sometimes encountered by beginning investigators. Care must be taken when weigh-

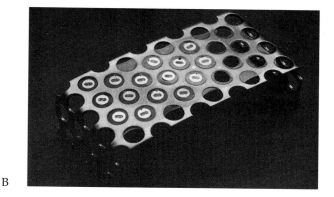

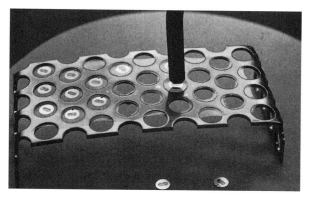

FIGURE 4.46 (A) Short ribbons of serial sections may be picked up by placing a grid directly down onto the sections so that they are corralled by the grid. Upon lifting up the grid, the sections will be retained in the thin film of water that remains in the slot. It is then possible to stretch the sections by passing the grid close to a heated wire as shown in Figure 4.44B. (B) The slot grid containing the stretched sections is placed onto a plastic film that extends over the holes in an aluminum platform. The sections remain corralled by the slot grid and will become attached to the plastic film upon evaporation of the water. (C) When completely dried, the slot grids containing the sections on the plastic film are removed by pushing down with a blunt surfaced tool as shown.

ing and mixing the resin monomers. The *firmness of plastic* should match as nearly as possible the hardness of the specimen to minimize the knife's skipping as it passes through areas of different hardness. Normally, one uses harder plastic formulations for diamond knives, while glass knives are better able to section softer blocks.

Proper *trimming* of the block is extremely important to generate parallel, smooth sides that yield straight ribbons of sections rather than individual sections that scatter in the trough and are hard to manipulate and collect. The block face should be as small as possible and consist mostly of specimen. Larger block faces are more difficult to section, even on a diamond knife, so try to maintain the block face at 0.5 to 1.0 mm. When razor blades are used to trim the block, be very careful not to leave small chips of razor blade steel embedded in the block. These will destroy both glass and diamond knives.

The knife must be sharp, securely locked in place in the holder, and free of any defects on the cutting edge. Always check the condition of the knife edge before commencing sectioning. Even diamond knife edges may contain dirt or sections that could affect the quality of sections produced. If multiple users share a diamond knife, it is best to maintain a logbook to keep track of problems with the knife.

Although a 4 degree *clearance angle* will be a good sectioning angle most of the time, it may be necessary to vary this parameter. For example, when sections are being dragged over the knife edge rather than being severed from the block completely or if *chatter* is encountered, then the clearance angle should be adjusted.

Chatter will be seen as closely spaced parallel lines running perpendicular to the direction of cut when the sections are viewed in the electron microscope. If one is able to see the striations in sections as they float in the boat or by using a light microscope, then they probably are caused by *low-frequency vibrations* under 1 KHz and are most likely due to vibrations in parts of the ultramicrotome or in the building itself. The latter type of problem may be difficult to solve without purchasing or building vibration damping equipment or even relocating the ultramicrotome. Fortunately, the striations caused by vibrations may be so widely spaced that they may not present a problem for many areas of the specimen, especially at higher magnifications.

Detecting and Preventing Vibrations

It is possible to confirm the presence of low-frequency vibrations in several ways. The easiest method is to place a petri dish in the problematic area and fill it to the brim with water. Observe the surface of the water for movement. This must be done in a draft-free environment (underneath a large inverted beaker or bell jar) so as not to confuse air currents with vibrations. An even more sensitive method uses a small amount of mercury in a petri dish. Shine a stationary laser (as from a laser pointer pen attached to a ring stand) onto the mercury pool and observe the reflected laser spot on the wall or ceiling. Vibrations are immediately evidenced by the movement of the laser spot. *CAUTION: mercury is a toxic compound. Do not contact the liquid with bare hands. Use pipettes instead. Keep the liquid contained in a plastic vessel and report any spills to pollution control authorities. Do not use mercury unless you are familiar with the proper precautions.*

NOTE: An inexpensive vibration-damping table can be constructed by mounting a heavy platform of slate on top of handballs that have been secured on top of a sturdy table.

Chatter caused by *high-frequency vibrations* that can be seen only in the transmission electron microscope are nearly always due to embedding problems or wrong sectioning parameters. When chatter extends over the entire section, then the clearance angle, knife angle, and cutting speed must be varied as described in the next section of this chapter. If only small areas inside the cells are showing chatter, it may not be possible to readily correct the problem.

The *cutting speed,* or rate at which the specimen is passed over the knife edge, is another important parameter affecting the quality of sections. Normally, one uses 1 mm/sec with diamond knives, while 2 mm/sec seems better suited to glass knives. However, the hardness of the specimen, angle and sharpness of the knife, as well as thickness of the section

all should be considered when selecting this speed. Although it is thought that harder specimen blocks should be cut at slower speeds (1 mm/sec or less), some specimens such as bone and tooth may section better at faster speeds (5 to 10 mm/sec). Faster speeds will cause greater compression of the sections so that it may be necessary to apply heat or organic vapors such as xylene or chloroform over the surface of the sections so they can return to nearly normal size. To avoid organic fumes, use an electrically heated wire loop (Figure 4.44B). Such loops can be purchased from several EM supply houses. Sections are stretched by placing the grid containing the sections on the droplet of water between the wire loop and observing the stretching process under a dissecting microscope. Do not overheat the sections and do not allow the water to completely dry out. Once stretched, the moist grid is placed onto a filter paper to dry.

Methodical Sectioning

Manufacturers of diamond knives have often predetermined such parameters as knife angle, optimum clearance angle, and cutting speed. However, they may need to be determined empirically if problems occur when using glass, and sometimes even diamond, knives. It is necessary to use a methodical approach such as outlined in the flow diagram devised by the LKB (Leica) company as follows:

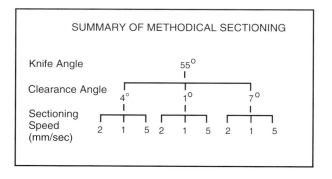

Start with a knife of a known angle and try cutting the specimen using a 4-degree clearance angle and cutting speeds of 2, 1, and 5 mm/sec. If high-quality sections are not produced, the clearance angle is changed to 1 degree and the same three speeds again attempted. Should this also prove unproductive, then a 7 degree clearance angle is used with the three sectioning speeds. If sections are still not obtained, make another glass knife with a different knife angle and repeat the process. Try knife angles of 55, 60, and 65 degrees in this test.

A Guide to Sectioning Problems and Causes

Failure to cut any sections: (a) microtome at end of fine advance, (b) dull knife, (c) specimen block too soft, (d) knife or block not secure, (e) negative clearance angle, (f) wet block face, (g) vibrations, (h) temperature fluctuations nearby.

Thickness variations from one entire section to another: (a) dull knife, (b) feed set too low, (c) bumping of instrument, (d) microtome problem, (e) drafts and temperature variations, (f) loose component—knife, specimen, mechanical advancement lock, (g) block too large or soft, (h) wrong knife angle or cutting speed.

Thickness variations within same section: (a) inhomogeneous specimen—areas of hard/soft, (b) large areas of plain plastic around specimen, (c) low frequency vibrations, (d) dull portion of knife.

Wrinkled sections: (a) block face too large, (b) soft block, (c) dirty or dull knife, (d) clearance angle too great, (e) water level too low, (f) cutting speed too fast, (g) knife loose or angle wrong.

Compressed sections: (a) block too soft, (b) inadequate expansion—try organic vapors or heat, (c) cutting speed too fast.

Chatter: (a) high-frequency vibrations during sectioning process—try different cutting speed, knife, and clearance angles, (b) block too tall with small base, (c) dull knife or soft block, (d) block or knife slightly loose, (e) knife too high in holder.

Specimen block lifts section on return stroke: (a) water level too high, (b) block face dirty, wet, or hydrophilic in nature due to nature of specimen, such as sperm cells, or that of the embedding plastic, (c) clearance angle too small, (d) dirty knife or wet back, (e) static electricity on block face.

Block face gets wet: (a) see a–e, above, (b) block face too large, (c) cutting speed too slow.

Sections are dragged over knife edge: (a) cutting speed too slow, (b) knife sitting too low in holder, (c) water level too high, (d) clearance or knife angle too low.

Sections have holes: (a) bubbles in resin, (b) incomplete infiltration, (c) hard objects in specimen.

Specimen drops out of block: (a) poor infiltration, (b) block too soft.

Sections have striations perpendicular to knife: (a) nick in knife edge, (b) dirt on knife edge, (c) knife damaged by hard material embedded in block.

Sections do not form ribbons: (a) block edges not parallel to edge of knife or each other, (b) water level wrong, (c) cutting speed too slow, (d) static electricity on block.

Ribbon of sections is curved: (a) leading and trailing edges of block not parallel, (b) compression on one side of section.

Sections hard to see: (a) illumination angle wrong, (b) water level wrong, (c) viewing angle wrong.

Sections hard to move in boat: (a) flotation fluid is contaminated and should be changed (replace with clean fluid containing trace of detergent or touch needle dipped into detergent into the water trough of knife).

Sections stick to eyelash probe: (a) dirty eyelash or roughened end of eyelash, (b) bearing down on section too much with eyelash.

Sections move away from grid: (a) dirty grid—clean in acid/alcohol or flame naked grids in alcohol lamp.

Cryoultramicrotomy

Cryoultramicrotomy is a procedure for cutting ultrathin frozen sections of biological specimens. Cryoultramicrotomy appears to offer several advantages over conventional ultramicrotomy techniques. One major advantage was originally thought to be *speed,* since it should be possible to freeze and section a specimen in 1 to 2 hours. This would offer the pathologist excellent diagnostic capabilities similar to those currently obtained using cryostat sectioning of surgical biopsies. To date, however, such uses for cryoultramicrotomy have not been realized on a routine basis probably because methods to obtain conventionally embedded and sectioned specimens in 4 to 5 hours have been developed (see Chapter 2).

A second and more practical reason for using cryoultramicrotomy is either for the localization of antigens using immunocytochemical procedures (Chapter 9) or for the elemental analysis of diffusible substances or ions by X-ray analysis (Chapter 15). In the latter case, it appears that cryoultramicrotomy offers probably the only way that one can localize diffusible substances.

For immunocytochemical procedures, it is best to first try a conventional approach by fixing in aldehyde fixatives (1% glutaraldehyde/2% formaldehyde) followed by rapid dehydrating in ethanol and embedding in an acrylic resin (Lowicryl or L.R. White resin) that is polymerized at relatively low temperatures (−20 to 40°C, respectively). The plastic-embedded specimens can be sectioned on a conventional ultramicrotome and stained using the immunocytochemical procedures outlined in Chapter 9. If this should fail after repeated attempts, then one may consider resorting to cryoultramicrotomy. Realize, however, that this may involve a considerable investment of time and finances (the cryokit attachment for most ultramicrotomes can cost as much as an ultramicrotome alone) without any assurances of success. A conservative approach would be to contact a researcher currently using the technique and to collaborate on the research.

The technique for cryoultramicrotomy was first reported in the scientific literature in 1952 by Fernandez-Moran; however, specifics of the procedure were not made available in that publication. In 1964, W. Bernhard and M. T. Nancy described a procedure that gave good ultrastructural details of biological tissues. In this procedure, the specimen was fixed in 2.5% glutaraldehyde and, after rinsing in buffer, transferred into a 10 to 20% solution of gelatin to provide a supporting matrix. Following a dehydration in 30% glycerol, the specimen was placed onto a metal specimen carrier and frozen in liquid nitrogen. After a quick trimming of the frozen specimen block with a chilled razor blade, sections were cut either into a trough containing 50% DMSO (as an antifreeze agent for the trough fluid) or onto a dry knife at −50°C. If a dry knife was used, the dry frozen sections were transferred onto a Formvar-coated grid and flattened onto the grid using slight pressure. This work was accomplished by placing a conventional ultramicrotome inside of a cryostat or cabinet that had been chilled from −50 to −70°C. While such an arrangement affords a stable temperature control, it is not possible to achieve temperatures lower than −70°C in a cryostat. This is an important point, since ice crystal damage occurs if the temperature rises above −70°C. The use of a trough fluid for the collection of cryosections was eventually abandoned due to the thickening of antifreeze agents at such low temperatures.

In 1972, many significant contributions to the field of cryoultramicrotomy were published independently by Christensen, Dollhopf, and Persson. These researchers reported the development of kits or attachments for the Porter-Blum, Reichert, and LKB ultramicrotomes, respectively. Instead of plac-ing the conventional ultramicrotome inside of a cryostat, an insulated chamber was fashioned around the specimen and knife. Since this area was cooled with liquid nitrogen, it was possible to regulate the temperature down to −170°C. Some of the kits even permitted independent control of the temperature of the knife and the specimen. Once these cryokits were made available commercially, researchers were able to refine the specimen preparation techniques, and many technical papers were published during the mid-1970s and 1980s. Examples of these attachments installed on two ultramicrotomes are shown in Figure 4.47. A diagram illustrating the design of the cryo-chamber of the RMC instrument is shown in Figure 4.48. Out of these research and development efforts, two major procedures for obtaining ultrathin frozen sections on dry knives have come to be used most often.

Dry Retrieval Method

In his 1971 article, A. K. Christensen described not only the development of an apparatus for cutting ultrathin frozen sections, but he also detailed the first readily usable procedure for cutting and collecting ultrathin frozen sections. Photographs of the first commercially available cryounit for the Porter-Blum MT2 ultramicrotome are shown in Figure 4.49. The Christensen procedure was used to produce dry cryosections of unfixed tissues. The reason for cryosectioning unfixed specimens and picking them up under dry conditions (as opposed to the sucrose droplet retrieval method of Tokuyasu, which is described subsequently) is to prevent the loss of diffusible substances. See Christensen's original paper for more details. The Christensen procedure is illustrated in Figure 4.50.

Dry Knife, Wet Retrieval Method

This method was originally described by Tokuyasu in 1973 and is the predominant method for producing cryosections for use in cytochemical and immunocytochemical procedures. The chief advantages are that the technique is relatively simple, reproducible, and yields high-quality sections with good preservation of enzymatic and immunological reactivities. The steps for this procedure are as follows:

1. The tissues are cut into appropriately sized pieces (or centrifuged into a compact pellet if individual cells are used). This is usually done rapidly at temperatures of 4 to 5°C in order to minimize ultrastructural changes.
2. If a fixative is desired, the tissues are placed into the chilled fixative for an appropriate amount of

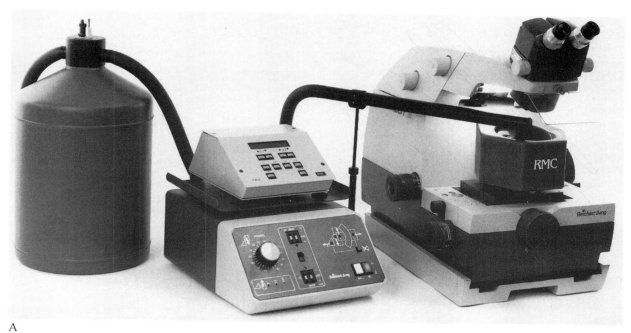

A

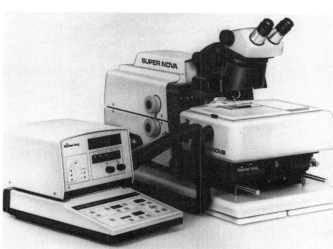

B

FIGURE 4.47 Cryoultramicrotomes for cutting ultrathin frozen sections. All ultramicrotome manufacturers have accessories for cryoultramicrotomy. (A) A universal cryoattachment that fits all commonly used brands of ultramicrotomes is now available. The unit, sold by RMC, Inc., is shown installed on a Reichert-Jung ultramicrotome. Reichert-Jung and RMC also sell different units designed specifically for their own brands of ultramicrotome.
(Courtesy of RMC, Inc.)

(B) The LKB Cryo Nova ultramicrotome set up for cutting ultrathin frozen sections.
(Courtesy of Leica.)

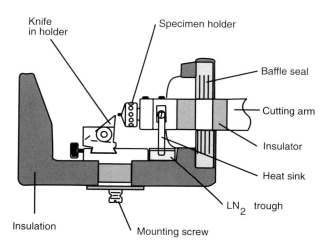

FIGURE 4.48 Diagram of interior of cryoultramicrotome for cutting ultrathin frozen sections.
(Courtesy of RMC, Inc.)

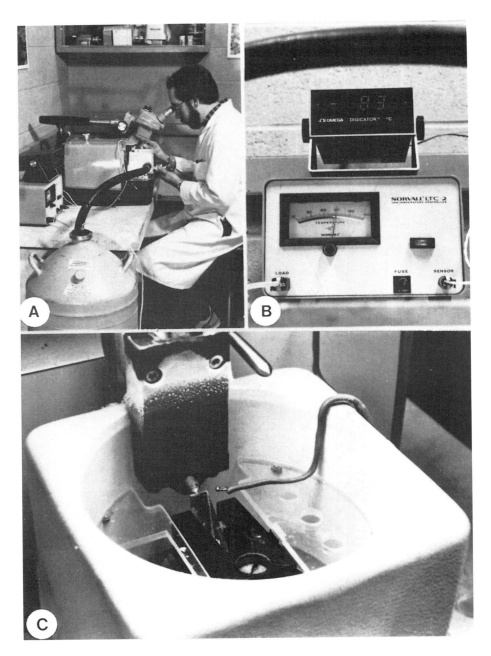

FIGURE 4.49 The first commercially available version of the Christensen cryoultramicrotomy kit for use on the Porter-Blum MT2B ultramicrotome. (A) Overall view of the ultramicrotome with attached cryochamber (*white area in center of top left photograph*). The large tank in the foreground contains liquid nitrogen that is converted to cold gas and transported to the cryochamber for cooling. (B) Low temperature controller for setting and maintaining the temperature of the cryochamber. The digital thermometer on the top of the controller is for sensing the actual temperature inside the cryochamber. (C) Close-up view of the cryochamber. The wire extending down into the chamber and approaching the knife edge is the temperature sensor for the digital thermometer. Inside the cryochamber one can see the glass knife and a specimen being cut by the knife. Note the formation of frost on the top part of the specimen arm. The constant flowing of cold, dry nitrogen gas out of the cryochamber prevents frost from forming inside the chamber. Plexiglass shelves surround the knife and permit the storage of chilled grids and other tools (eyelash probes, rods for flattening sections, etc.).
(Courtesy of A. K. Christensen and Wiley-Liss Publishers.)

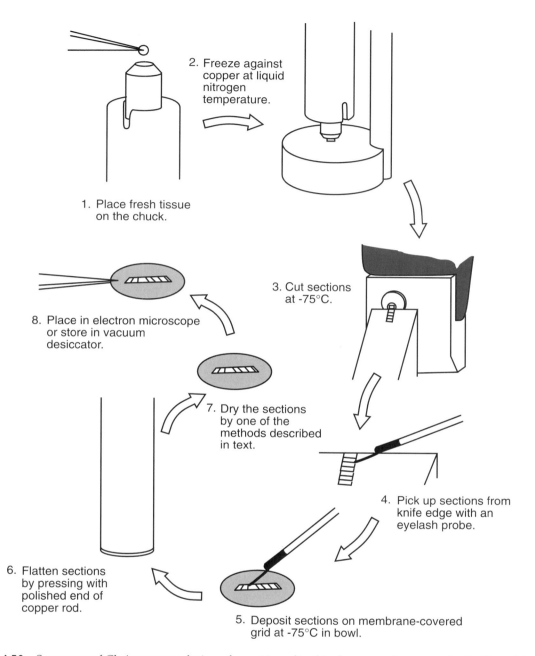

2. Freeze against copper at liquid nitrogen temperature.

1. Place fresh tissue on the chuck.

3. Cut sections at -75°C.

8. Place in electron microscope or store in vacuum desiccator.

7. Dry the sections by one of the methods described in text.

4. Pick up sections from knife edge with an eyelash probe.

6. Flatten sections by pressing with polished end of copper rod.

5. Deposit sections on membrane-covered grid at -75°C in bowl.

FIGURE 4.50 Summary of Christensen technique for cutting ultrathin frozen sections on a dry knife and transferring the dry sections onto a coated grid. The sections could then be freeze-dried by leaving them in the dry atmosphere of the cryochamber.
(Redrawn with permission of A. K. Christensen.)

time. Normally, one uses only an aldehyde fixative such as 1 to 2% glutaraldehyde or a mixture of 0.5 to 2% glutaraldehyde and 1 to 4% formaldehyde for 15 to 30 minutes. Other aldehyde fixative combinations and times may be more appropriate, depending on the investigation.

3. Rinse in buffer to remove unreacted fixative. The type of buffer, as well as its tonicity, will depend upon the tissue. In addition, it is recommended that 0.1 M glycine be incorporated into the buffer in order to inactivate any residual aldehyde groups that may be associated with tissue components. This will help to minimize nonspecific staining.

4. Place the specimen into 2.1 to 2.3 M sucrose (or 1.8 M sucrose in 15 to 20% polyvinyl pyrollidone) in phosphate buffered saline (0.1 M phosphate buffer containing 0.9% NaCl) for 15 to 60 minutes. The time for infusion of the sucrose into the tissue depends on the size of the tissue.

5. Mount the specimen pieces onto metal holders (usually copper stubs or pins). It is particularly useful to shape the specimen pieces at this time (rather than after they have been frozen) so that they have a wide base in contact with the copper pin and so that they have a pointed tip. This is similar to trimming plastic specimen blocks into the pyramid shape.

6. Quick-freeze the mounted specimen by contacting it to a chilled metal surface or by plunging the copper holder into liquid nitrogen. If plunge freezing is used, a quenchant such as isopentane or Freon or even nitrogen slush is needed to enhance the freezing rate. The specimen must be moved or shaken while submerged in the coolant in order to achieve rapid freezing. The goal is to *vitrify* the remaining water inside the cells (i.e., to freeze the water so rapidly that it forms solid amorphous water rather than ice crystals). The specimens can be stored indefinitely in liquid nitrogen.

7. Transfer the frozen specimen (while still submerged in liquid nitrogen) into the cryochamber of the ultramicrotome and proceed to cut ultrathin frozen sections (Figure 4.51A) as follows.

8. Advance the chilled knife carefully toward the specimen until it nearly contacts it. Use the fine mechanical advancement of the microtome until the first section is cut. Take care not to cut a section thicker than 1 μm; otherwise the block

face may be shattered and damaged. In addition, only the uppermost 5 μm or so of tissue will have undergone vitrification, and ice damage may become more pronounced as one proceeds deeper into the tissue block.

NOTES: Some ultramicrotomes permit one to adjust the temperature of the knife independently of the specimen. It may be necessary to experiment with various knife temperatures until a proper combination is obtained. A good starting point is to set the knife at the same temperature as the specimen. Either glass or diamond knives may be used. Glass knives appear to last longer at the colder temperatures, so diamond knives may not be needed. As in conventional sectioning, the glass knife must be of high quality as determined by careful examination just prior to use. Modern knife makers can produce such high quality knives for cryoultramicrotomy. Plain knives, without attached troughs, are used since sections will be cut onto a dry knife.

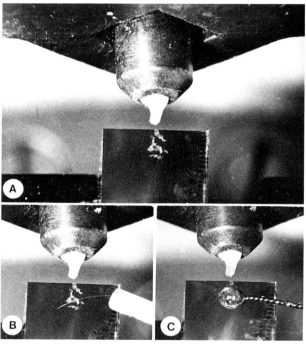

FIGURE 4.51 (A) Ultrathin frozen sections shown collecting on a glass knife. The specimen is the large white object in the center of the photograph. (B) Manipulation of the frozen sections is accomplished by a chilled eyelash probe. (C) The sections are picked up on the underside of a drop of 2.3 M sucrose suspended in the platinum wire loop. *(Courtesy of A. K. Christensen, T. E. Komorowski, and Wiley-Liss Publishers.)*

9. Begin cutting ultrathin sections using the automatic advance mechanism of the ultramicrotome. A cutting speed of 1 mm/sec, or less, is a good starting point. A fast return cycle is recommended in order to minimize thermal changes to the specimen block.

10. As the dry sections are cut, they will collect along the knife edge. These sections may have a clear to slightly translucent appearance and may exhibit interference colors similar to plastic sections floating on a water surface. These colors, however, cannot be used to estimate the thickness. Use an eyelash probe to move the sections down the knife so that they will not be in the way of the next section. Although one may occasionally obtain ribbons of sections, normally, individual sections are formed. Continue cutting sections as long as possible. If static electricity becomes a problem, a static eliminator (gun or radioactive strip) can be employed.

11. Move the sections into groups that will fit conveniently onto a grid using a chilled eyelash probe (Figure 4.51B). Pick up the sections using a 1 to 2 mm wide platinum loop that has been dipped into 2.3 M sucrose in phosphate buffered saline so that a droplet of the sucrose solution remains in the loop (Figure 4.51C). Approach the group of sections with the droplet and contact the sections to the underside of the droplet. Take care to allow the droplet to cool but not to freeze in the cryochamber, otherwise the sections may not attach to the droplet. Do not contact the knife with the droplet or it may freeze onto the knife.

12. Withdraw the droplet containing the sections outside of the cryochamber. The droplet containing the sections will probably have frozen by this time. Allow several seconds for the droplet to thaw—as evidenced by a transition from milky white to clear. Touch the underside of the droplet (containing the sections) onto a Formvar-and carbon-coated grid (Figure 4.52) so that the sections will be transferred onto the grid.

13. Immediately pick up the grid (absolutely avoid drying of the sections), and float the grid (sections down) onto a droplet of phosphate buffered saline containing 1% bovine serum albumin or 2% gelatin.

14. Proceed to stain the sections using the appropriate cytochemical or immunocytochemical procedure (see following examples).

FIGURE 4.52 Transfer of the thawed, cryoultramicrotomed sections onto Formvar-coated grids is accomplished by touching the underside of the sucrose droplet (*section side*) to the grid. *(Courtesy of A. K. Christensen, T. E. Komorowski, and Wiley-Liss Publishers.)*

If cytochemical procedures are not needed, then the sections can be stained by floating the grids on four changes of distilled water (to remove the phosphate buffer) followed by flotation on 0.5% aqueous uranyl acetate as described by Tokuyasu in his 1989 paper. After a quick rinse in distilled water, the grid can be placed onto a filter paper to dry. For an example of results obtainable using this procedure see Figure 4.53.

Example of the Use of Frozen Sections in an Immunocytochemical Localization Procedure

Bastholm, L., M. H. Nielsen, and L.-I. Larsson. 1987. Simultaneous demonstration of two antigens in ultra-thin cyrosections by a novel application of an immunogold staining method using primary antibodies from the same species. *Histochemistry* 87:229–231.

Procedures. In this study two different antigens, human pituitary growth hormone and synthetic human ACTH, were localized using two different rabbit antisera against each of the hormones. After cryosectioning mouse pituitary glands using the Tokuyasu procedure, the sections were reacted with the rabbit antiserum against human growth hormone followed by gold-labeled goat anti-rabbit antibodies. The gold-labeled antibodies were used at a high con-

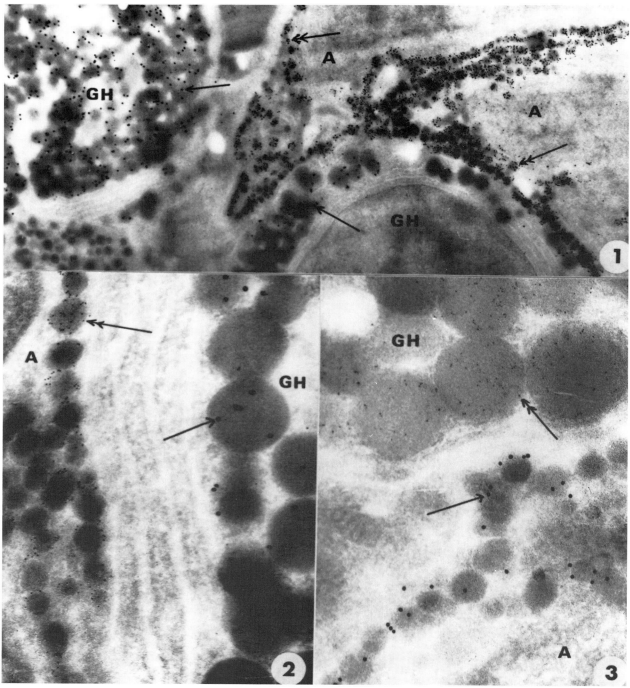

FIGURE 4.53   Ultrathin cryosection of mouse pituitary showing the localization of growth hormone (GH) in the micrographs. ACTH hormone (A) is indicated. Note the difference in sizes of the gold particles (particles associated with ACTH(A) are smaller than GH particles). The probes used were gold-labeled goat anti-rabbit antiserums.
(Courtesy of L. Bastholm, M. H. Nielsen, L.-I. Larsson, and Springer-Verlag Publishers.)

centration so that all of the rabbit antibodies were totally saturated. After treating the sections with paraformaldehyde fumes to stabilize the rabbit and goat antibodies, the sections were treated with a glycine-containing buffer to inactivate residual aldehyde groups, and reacted with the rabbit antibody against ACTH. Following a staining with gold-labeled goat anti-rabbit antibodies (in this case the gold particles were of a different size than was the first gold label), the sections were rinsed and ultimately stained in uranyl acetate.

Results. The researchers were able to discriminate against the two hormones since two differently sized gold probes were used (see Figure 4.53). Since a rabbit antiserum was used to localize two different antigens, it was very important that the first set of rabbit antibodies be totally saturated and stabilized (hence the use of excess gold anti-rabbit antibodies followed by formaldehyde linking of the antigen-antibody complex) before the second gold labeled anti-rabbit stain was used. The electron micrographs shown in Figure 4.53 reveal that the two hormones are localized in the granules of different cells. One cell type is responsible for the production of the growth hormone (GH, on the micrograph), while a different cell was responsible for the production of the ACTH (A, on the micrograph). Note that the gold particles are of sufficiently different sizes (15 nm versus 5 nm) that the different antigenic sites are clearly distinguishable. The researchers found that either size of gold probe could be used with either antigen and that the order of the application of the different sizes did not matter in this instance.

For further reading on immunocytochemical applications, see Chapter 9 and some of the selected references at the end of this chapter. The combination of cryoultramicrotomy and the use of double labeling represents a challenging yet powerful technique for the localization of multiple antigenic sites. Before embarking on cryoultramicrotomy, however, one should first attempt the localization using conventionally embedded specimens as described in Chapters 2 and 9.

REFERENCES

Ultramicrotomy

Abad, A. 1988. A study of wrinkling on single-hole coated grids using TEM and SEM. *J. Electron Microsc Tech* 8:217–22.

Bayer, M. E., and T. F. Anderson. 1963. The preparation of holey films for electron microscopy. *Experientia* 19:1–3.

Coggins, L. W. 1987. Preparation of nucleic acids for electron microscopy. In *Electron microscopy in molecular biology: A practical approach,* J. Sommerville and U. Scheer, eds. IRL Press, Oxford, England. Part of The Practical Approach Series.

DeBruijn, W. C., and S. M. McGee-Russell. 1966. Bridging a gap in pathology and histology. *J. Royal Microscopical Society* 85:77–90.

Fernandez-Moran, H. 1953. A diamond knife for ultrathin sectioning. *Exptl Cell Res* 5:255–56.

Fullam, E. F., and A. E. Gessler. 1946. A high speed microtome for the electron microscope. *Rev Scient Instrum* 17:23–31.

Harris, W. J. 1962. Holey films for electron microscopy. *Nature* 196:499–500.

Latta, H., and J. F. Hartmann. 1950. Use of a glass edge in thin sectioning for electron microscopy. *Proc Soc Exp Biol Med* 74:436–39.

Lindner, M., and P. Richards. 1978. Long-edged glass knives ('Ralph Knives'): Their use and the prospects for histology. *Science Tools* 25:61–67.

O'Brien, H. C., and G. M. McKinley. 1943. New microtome and sectioning method for electron microscopy. *Science* 98:455–56.

Porter, K. R., and J. Blum. 1953. A study in microtomy for electron microscopy. *Anat Rec* 117:685–709.

Royer, S. M. 1988. A simple method for collecting and mounting ribboned serial sections of epoxy embedded specimens. *Stain Technol* 63:23–26.

Slabe, T. J., S. T. Rasmussen, and B. Tandler. 1990. A simple method for improving glass knives. *J Electron Microsc Tech* 15:316–17.

Von Ardenne, M. 1939. Die Keilschnittmethode, ein Weg zur Herstellung von Microtomschnitten mit weniger als 10^{-3} mm. Starke fur electronenmikroskopische Zwecke Z. Wissensch. *Mikroskopie* 56:8–15.

Cryoultramicrotomy

Appleton, T. C. 1974. A cryostat approach to ultrathin 'dry' frozen sections for electron microscopy: A morphological and x-ray analytical study. *J Microsc* 100:49–74.

Bastholm, L., M. H. Nielsen and L.-I. Larsson. 1987. Simultaneous demonstration of two antigens in ultrathin cryosections by a novel application of an immunogold staining method using primary antibodies from the same species. *Histochemistry* 87:229–31.

Bernhard, W., and M. T. Nancy. 1964. Coupes a congelation ultrafines de tissu inclus dan la gelatine. *J Microscopie* 3:579–88.

Christensen, A. K. 1971. Frozen thin sections of fresh tissue for electron microscopy, with a description of pancreas and liver. *J Cell Biol* 51:772–804.

Christensen, A. K., and T. E. Komorowski. 1985. The preparation of ultrathin frozen sections for immunocytochemistry at the electron microscope level. *J Electron Microsc Tech* 2:497–507.

Dollhopf, F. L., G. Lechner, K. Neumann, and H. Sitte. 1972. The cryoultramicrotome Reichert. *J Microscopie* 13:152–53.

Fernandez-Moran, H. 1952. Application of the ultrathin freezing-sectioning technique to the study of cell structures with the electron microscope. *Ark Fys* 4:471–83.

Griffiths, G., A. McDowell, R. Back, and J. Dubochet. 1984. On the production of cryosections for immunocytochemistry. *J Ultrastructure Res* 89:65–84.

Parsons, D., D. J. Bellotto, W. W. Schulz, M. Buja, and H. K. Hagler. 1984. Towards routine cryoultramicrotomy. *Bull of Electron Microsc Soc of Am* 14(2):49–60.

Persson, A. 1972. Equipment for cryoultramicrotomy. The LKB Cryokit. *J Microsc* 13:162.

Roberts, I. M. 1975. Tungsten coating: a method of improving glass microtome knives for cutting ultrathin frozen sections. *J Microsc* 103:113–19.

Saubermann, A. J. 1986. Comparison of analytical methods for x-ray analysis of cryosectioned biological tissues. *Bull Electron Microsc Soc of Am* 16:65–69.

Somlyo, A. P., M. Bond, and A. V. Somlyo. 1985. Calcium content of mitochondria and endoplasmic reticulum in liver frozen rapidly in vivo. *Nature* 314:622–25.

Tokuyasu, K. T. 1973. A technique for ultracryotomy of cell suspensions and tissues. *J Cell Biol* 57:551–65.

———. 1989. Use of poly(vinylpyrrolidone) and poly(vinyl alcohol) for cryoultramicrotomy. *Histochemical Journal* 21:163–71.

Zierold, K. 1986. Preparation of cryosections for biological microanalysis. In *The science of biological specimen preparation,* M. Muller, R. P. Becker, A. Boyde, and J. J. Wolosewick, eds. Scanning Electron Microscopy, Inc., AMF O'Hare, IL 60666, pp. 119–27.

Reference Books

Hayat, M. A. 1970. *Principles and techniques of electron microscopy: Biological applications,* Vol. 1. New York: Van Nostrand Reinhold Co., pp. 183–237.

Reid, N. 1975. *Ultramicrotomy.* A. M. Glauert, ed. New York: Elsevier/North Holland Biomedical Press.

Reid, N., and J. E. Beesley. 1991. *Practical methods in electron microscopy, vol. 13: Sectioning and cryosectioning for electron microscopy.* A. M. Glauert, ed. Amsterdam, The Netherlands: Elsevier/North Holland Publishing Co.

Robards, A. W., and U. B. Sleytr. 1985. *Low temperature methods in biological electron microscopy.* A. M. Glauert, ed. New York: Elsevier/North Holland Biomedical Press.

Sjostrand, F. S. 1953. A new microtome for ultrathin sectioning for high resolution electron microscopy. *Experientia* 9:114–121.

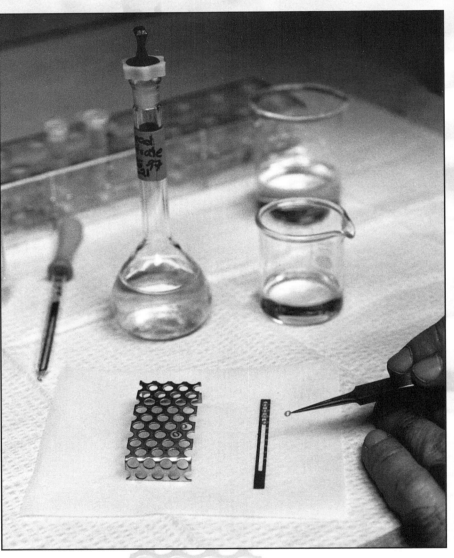

Chapter 5

Specimen Staining and Contrast Methods for Transmission Electron Microscopy

IN THE LIGHT MICROSCOPE, the "white" light that illuminates the specimen usually consists of a spectrum of colors. These colors become readily visible when the light strikes a specimen that has been stained with various dyes. Biological stains absorb certain colors in the spectrum, but transmit or reflect others to the eye. For example, in a specimen that has been stained with two commonly used biological stains, safranin and fast green, certain parts of the cell will appear red (safranin will stain acidic components of the cell such as DNA in the nucleus), while others will stain green (fast green stains basic components of the cytoplasm).

Colors do not exist in any electron microscope, since the illumination source is of a single wavelength determined by the accelerating voltage. In fact, a single wavelength helps to minimize the phenomenon termed *chromatic aberration,* which will degrade the resolution of the microscope (see Chapter 6). Since the illumination used in the electron microscope is of a single wavelength beyond the range of sensitivity of the human eye, one must examine images on viewing screens coated with phosphorescent materials. Such phosphors convert the kinetic energy of the invisible short wavelength electrons into longer wavelength (green to yellow-green) light that can be perceived by the eye. The images generated on such screens consist of various brightness levels ranging from very bright to quite dark so that the electron microscopist sees contrasting shades on a yellow-green screen. The darker shades are associated with areas of the specimen that have greater density, whereas brighter areas of the specimen have less density. Dense areas in the specimen deflect beam electrons to such an angle that they are eliminated from the image and generate various tonal ranges or contrast (see Figure 6.33). Since unstained biological tissues have little density differences—even compared to the surrounding environment—contrast is usually quite low.

It is important to increase the image contrast of most biological specimens by reacting selective cellular components with heavy metals. For instance, the fixative osmium tetroxide reacts with many organic molecules of the tissue and is reduced to metallic osmium. Such osmicated molecules are increased in density and will appear darker on the viewing screen. Although there is an improvement in image contrast with osmium, only occasionally will this impart sufficient contrast to render an adequate image. In the majority of cases, it is necessary to further increase overall contrast by using other heavy metals. Salts of heavy metals (such as lead citrate and uranyl acetate) further increase contrast of osmicated tissues when the metal ions of the salt react with various cellular components. Heavy metal salts that are used to increase contrast levels are therefore termed *stains.* This chapter covers the application of the more commonly used stains and metal shadowing techniques to enhance contrast of specimens for viewing by TEM. Techniques for enhancing the contrast of SEM specimens are covered in Chapter 3.

Positive Staining

An electron stain is said to exhibit *positive contrast* when it increases the density of a particular biological *structure* as opposed to the adjoining areas. In positive staining, the heavy metal salts attach to various organelles or macromolecules within the specimen to increase their density and thereby increase contrast differentially. This differs from the *negative contrast* situation where the background area surrounding the specimen is made dense by a heavy metal salt so that the specimen appears lighter in contrast to the darkly stained background. Figure 5.1 illustrates the positive staining of ultrathin, sectioned membrane vesicles, while Figure 5.2 demonstrates negatively stained whole intact vesicles.

The Physical Basis of Contrast

The heavy metal salts used as stains in the electron microscope consist of ions of a high atomic number with a large number of protons and electrons that scatter the beam electrons. When beam electrons encounter the atoms of a heavy metal ion, they will be deflected with minimal energy loss (elastic scattering) at such a wide angle that they will not enter the imaging lenses. This subtractive action (see Figure 6.33) gives rise to the various tonalities or shades evident on the screen. This contrast, which is dependent on the elimination of a number of electrons, is termed *amplitude contrast.*

In a second situation involving *phase contrast,* a low-angle deflection of the beam electrons may also occur so that the beam electrons lose some energy (inelastic scattering), but are still able to enter the imaging lenses. In thin specimens ($<$ 60 nm), these lower energy electrons have shorter focal lengths than do higher energy electrons and give rise to phase contrast, which may be seen as a light halo surrounding a dense object. In thick specimens, too many different energy levels of electrons enter the imaging lenses, so that resolution is degraded due to chromatic aberration (see Chapter 6).

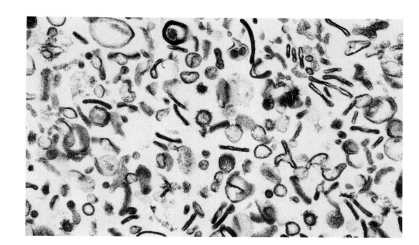

FIGURE 5.1 Ultrathin section through membrane vesicles from disrupted sperm cells. Sections were stained with uranyl acetate and lead citrate.

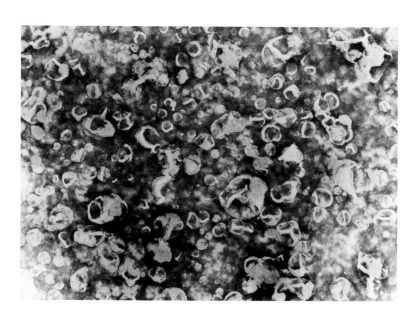

FIGURE 5.2 Negatively stained membrane vesicles from sperm cells. In this method, intact vesicles were mixed with negative stain (2% sodium phosphotungstate) and deposited onto Formvar and carbon-coated grids.

The two most commonly used positive stains are uranyl acetate (MW = 422) and lead citrate (MW = 1054). The exact mode of action of these stains is not completely understood. It is known that uranyl ions react strongly with phosphate and amino groups so that nucleic acids and certain proteins are highly stained. With lead stains, it is thought that lead ions bind to negatively charged components such as hydroxyl groups and osmium-reacted areas. Phosphate groups may also be involved in this phenomenon, since the use of phosphate buffers often enhances overall staining with lead ions.

Uranyl and lead stains are termed *general or nonspecific stains* when used in the routine manner since they will stain many different cellular compo- nents. This contrasts to their use in certain *cyto- chemical* and *immunocytochemical* techniques where prior treatment of the tissues may render macro- molecules reactive with these stains. For a discus- sion of cytochemistry and immunocytochemistry, consult Chapters 9 and 10.

Preembedding, Positive Staining with Uranyl Salts

Reaction of the tissues with uranyl stains may take place either before or after embedding in plastic. There are certain advantages to preembedding, or *en bloc,* uranyl staining. In fact, membrane preserva- tion is so enhanced that many cytologists consider uranyl salts to have fixative properties.

Adjustment of the uranyl stain to a pH of 5.2 also leads to the fine structural preservation of DNA as well as cell junctions, mitochondria, myofibrils, nucleoproteins, and phospholipids. Depending on the specimen, ultrastructural preservation is so much better when uranyl ions are used following aldehyde/osmium fixation, that it should be considered for routine use prior to embedding.

Preembedding Staining Procedure

1. After fixation of the tissues in an aldehyde and osmium fixative (the exact concentrations, buffers, times, etc., vary with the specimen type—see Chapter 2), the specimen is rinsed three times for 15 minutes each in double-distilled water.
   NOTE: This rinse must be extended even longer if a phosphate buffer was used for the earlier fixations, since phosphate ions will cause precipitation of the uranyl ions in the tissues.

2. Immerse specimens in the freshly prepared uranyl stain consisting of 0.5 to 2% uranyl acetate in double-distilled water for 1 to 2 hours at room temperature.
   NOTE: Uranyl salts are photolabile and must be protected from bright light. The stain is good for about one week and should be centrifuged or micropore-filtered prior to use. It is possible to refrigerate specimens overnight in the more dilute (0.05 to 0.2% w/v) aqueous solutions of uranyl acetate.
   CAUTION: Salts of uranium are toxic and may contain radioactive isotopes. Take precautions to avoid breathing the dust when weighing the powder. It is highly recommended to do all weighing in an enclosed glove box or a fume hood, or to wear a dust mask or filter. The scale and area where the weighing was done must be cleaned with a wet paper towel. Dispose of all uranyl salts and solutions as recommended by the local pollution control authorities. Radioactive uranyl salts are low-level radiation emitters; a radiological control office should be consulted for recommended disposal procedures. No special permits are required to use uranyl acetate salts.

3. Transfer specimen through dehydration series and embed in appropriate plastic prior to sectioning.
   NOTE: The dehydration should be as expeditious as possible since uranyl stains are slowly removed by organic solvents. Prolonged storage in alcohols, acetone, or propylene oxide may remove most of the stain. Small tissue blocks are highly recommended in order to speed the dehydration steps.

Even if tissues have been stained prior to embedding, it is still possible (and advisable) to stain the sections with uranyl salts followed by lead stain as described in the following sections.

Postembedding Staining with Uranyl Salts

Aqueous Solutions

One of the more common methods of staining thin sections is with 2% aqueous solutions of uranyl acetate. Usually, staining is accomplished in a simple device constructed from a petri dish, a sheet of dental wax, and a piece of filter paper (Figure 5.3). The filter paper is placed in the bottom of the petri dish, and a small amount of distilled water is added to soak the paper. A piece of dental wax (or Parafilm) is placed over the wet filter paper, and drops of the stain are placed on the wax surface. Individual grids are placed onto the droplets to float, section side down (Figure 5.4), and the petri dish lid is installed.

The moist atmosphere created by the damp filter paper prevents the stain from evaporating, so that staining times of 15 minutes to several hours at room temperature can be safely carried out. It is possible to accelerate the staining process by increasing the temperature to 40 to 60° C. If elevated temperatures are used, one must take care not to

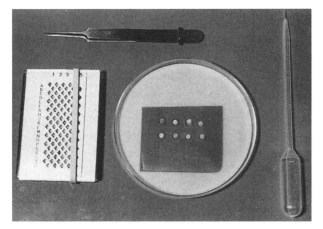

FIGURE 5.3 Setup required for staining of thin sections using aqueous uranyl acetate. Fine-tipped forceps (*top*) are used to manipulate the grids on the stain droplets on the rectangular piece of dental wax in the petri dish. A storage and filing system for the grids is shown on the left while a plastic pipette for depositing the stain droplets on the dental wax is shown to the right in the photograph.

melt the dental wax or paraffin sheeting. As with all uranyl stains, the solution should be protected from the light by placing the apparatus inside a drawer or simply by covering it with aluminum foil. In addition, the stain should be centrifuged (5,000 × g for 20 minutes) or forced through a syringe microfilter prior to use to prevent precipitates from attaching to grids. Cloudy solutions should be discarded.

After staining has been accomplished, the grids are individually removed using a fine-tipped forceps and rinsed with drops of distilled water from a plastic squeeze bottle held within 1 cm of the grid. Grids must be rinsed gently so that the supporting films are not broken or the sections lost. Alternatively, the grids can be washed by floating them, section side down, on large droplets of distilled water on dental wax or on distilled water contained in small beakers. One good method involves gently dunking the grids into several beakers of distilled water using 10 to 25 gentle dunks per beaker (Figure 5.5).

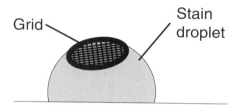

FIGURE 5.4 Sections may be stained by floating the grids on droplets of uranyl acetate on the wax surface.

FIGURE 5.5 Series of beakers of distilled water used for rinsing stained grids by dipping the grids into each of the beakers.

After rinsing, the grids may be stained immediately with lead stains or dried and stored until lead staining is desired.

If one wishes to store the grids, it is essential to remove the small amount of water remaining between the prongs of the forceps before placing the grid onto a clean surface. If this water remains in the forceps, the grid will be pulled between the prongs when the forceps is spread apart. To prevent this, a wedge-shaped piece of filter paper is placed between the prongs to soak up the water prior to releasing the grid (Figure 5.6). *NOTE: It is important to minimize the carryover of the various staining reagents from one staining solution to another by using the filter paper wedge between the prongs of the forceps.* Normally, one would store the grid on a clean filter paper in a petri dish or in a special grid storage box (Figure 5.3, left).

Helpful Suggestions when Staining

If the grid still adheres to the forceps when one attempts to transfer it onto a filter paper, it is helpful to place a drop of distilled water on the filter paper before placing the grid (section side up) on the moistened spot. The points of the forceps should be kept clean and free of burrs to prevent the adherence of the grid. Forceps points can be cleaned with filter paper and deburred by pulling a fingernail file between the closed forceps prongs. When working with nickel grids, it may be necessary to demagnetize both the forceps and grid by passing them through a demagnetizing loop (Figure 5.7).

Another method of staining using aqueous solutions of uranyl salts is by submerging the grids in the stain using one of the holders available for

FIGURE 5.6 A small wedge of filter paper may be used to remove liquids trapped between prongs of forceps that hold grid. This will prevent contamination of grid surface with the liquid.

FIGURE 5.7 Certain types of grids (nickel) as well as the stainless steel forceps often must be demagnetized by passage through a demagnetizing loop. Otherwise, grids will be attracted to the forceps and difficult to handle.

FIGURE 5.8 Molten dental wax was poured into the depressions in this glass spot plate to form a flat platform. The grids were submerged in the stain and arrayed along the wax surface.

multiple staining (Figure 5.12) or special glass containers (Figure 5.8).

Uranyl acetate is the most commonly used aqueous uranyl salt; however, 7.5% solutions of magnesium uranyl acetate in triple-distilled water give comparable, or possibly even better results depending upon the specimen (Frasca and Parks, 1965). They are used in the same manner as previously described for uranyl acetate.

Alcoholic Solutions

Some types of tightly cross-linked resins (Spurr's epoxy resin, for example) may not be adequately stained using aqueous solutions. To enhance pene-

tration, it is possible to formulate a stain with the heavy metal salt dissolved in an alcohol such as ethanol or methanol.

To prevent excessive evaporation of the alcohol, a chamber is fabricated as described for aqueous uranyl staining (Figure 5.3) except that the filter paper is moistened with 100% alcohol. It is difficult to stain the sections by floating the grids on drops of stain on a wax base because the grids have a tendency to sink in the solutions. Safer methods for staining using alcoholic solutions are illustrated in Figures 5.8 and 5.12.

Staining Methods for Alcoholic Solutions of Uranyl Salts

1. **Ethanolic Solutions of Uranium Salts (Epstein and Holt, 1963).**
   Probably the most commonly used alcohol-based staining solutions range in uranyl acetate concentration from 2 to 4% (weight/volume) to a saturated solution of uranyl acetate in 50% ethanol. *NOTE: Saturated solutions should be prepared at least one day prior to use since uranyl salts dissolve slowly. The stain should be centrifuged immediately before use (5,000 × g for 15 to 20 minutes) in a tightly sealed tube to prevent evaporation.* Staining times of 5 to 15 minutes are used at room temperature for sections of most epoxy-embedded specimens, but it may be necessary to increase the staining time to 1 to 2 hours or longer when using Spurr's epoxy resin. If extended times and temperatures are used, then the grids should be submerged in the stain rather than being floated on small droplets since they will sink anyway.

   After the grids have been stained, they are removed individually using a fine-tipped forceps and quickly rinsed in several changes of 50% ethanol. The safest method involves dunking the grids into several small beakers of 50% alcohol, taking care to minimize the carryover of alcohol from one container to the other by using a filter paper to blot the area between the prongs of the forceps. It is very important to transfer the grids quickly into the alcohols to minimize evaporation of the alcohol, which will cause a precipitation of the uranium salts on the sections. After the last rinse, the grids are placed on filter papers to dry and then a second, usually lead, stain may be applied.

   If one intends to apply a second stain (usually aqueous solutions of lead salts) to the grids immediately after the last rinse in 50% ethanol, then the grids should be dipped into a beaker of 25% ethanol followed by several changes of double-distilled water. The hydrated sections may then be stained using aqueous solutions.

2. **Absolute Methanolic Solutions of Uranyl Salts (Stempak and Ward, 1964).**
In this method, a 25% solution of uranyl acetate is prepared in absolute methanol. *NOTE: Absolute methanol is prepared by placing 100% methanol over a drying agent such as Molecular Sieves for several weeks to remove traces of water. Methanol may dissolve certain plastic substrates such as Collodion, so supported sections cannot be used in this procedure. In addition, some types of embedding plastics (methacrylates, for instance) may be weakened or dissolved by methanol. Caution is in order with this method.* The grids are submerged in the solution (they will not float on concentrated alcohols) and stained for 10 to 15 minutes.

After staining, the grids are removed using a fine-tipped forceps and gently but rapidly dipped 10 to 25 times each in 3 or 4 changes of absolute methanol. The prongs of the forceps should be drained with filter paper after each rinse to minimize carryover of uranyl stain. Great care must be taken to transfer the grid as rapidly as possible to prevent drying, which will cause a precipitation of the uranyl salts onto the surface of the sections (Figure 5.9). After the final rinse, the prongs are blotted and the grid gently placed on a filter paper and allowed to dry completely before staining with aqueous lead stains.

An alternate method involves the use of 2% uranyl acetate dissolved in absolute methanol containing 1% dimethylsulfoxide (DMSO). The DMSO is thought to enhance penetration of the stain into the plastic.

WARNING: This solution must be handled very carefully since DMSO will permit the rapid penetration of uranium ions through the skin. Gloves must be worn, and the solution should be handled in a fume hood.

Postembedding Lead Staining

Reynolds' Lead Citrate

After the specimen has been contrasted by uranyl staining, it is customary to apply a second stain, usually lead citrate dissolved in double-distilled water. The most commonly used lead citrate stain is

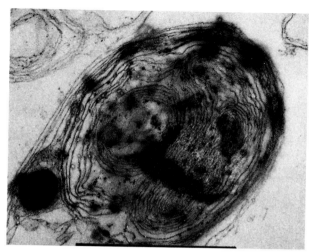

FIGURE 5.9A Grids contaminated with uranyl acetate salts that have precipitated due to slow transfer of the grids from one rinse to another.

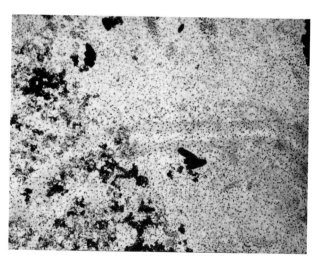

FIGURE 5.9B Lead precipitate usually appears more granular than does the uranyl acetate precipitate.

that described by Reynolds in 1963. Although time consuming to prepare, the stain is stable for many months if kept tightly sealed and protected from carbon dioxide.

Preparation and Use of Reynolds' Lead Citrate Stain

1. In a scrupulously clean, 50 ml volumetric flask, combine 1.33 g lead nitrate, 1.76 g sodium citrate, and 30 ml of CO_2-free double-distilled water (see next section). Shake the solution vigorously for several minutes and then 5 or 6

times over a 30-minute period. This will generate a milky white suspension of lead stain with no large particles. If one sees large chunks, continue shaking until the particles are dissociated, or start over.

2. Add 8.0 ml of a CO_2-free, commercially prepared, titrated solution of 1.0 N NaOH and swirl to mix the two solutions. The milky solution should turn clear. If the solution does not turn clear, something is wrong and the stain should be discarded.

    NOTES: Although it is possible to prepare 1.0 N NaOH using CO_2-free water and NaOH pellets, problems will result unless one uses NaOH pellets that are completely dry and CO_2-free (not having a powdery crust on the pellet surface). Since this is often difficult to ascertain, we recommend using commercially titrated solutions. For more information, see the articles by Heinrich (1985) and Reynolds (1963). A proper pH is extremely important with this stain. If the pH varies by more than 0.1 unit from pH 12.0, poor staining or precipitation will occur.

    CAUTION: Lead salts are extremely toxic. Exercise care in weighing the powders: wear gloves, use a fume hood or filter over face and mouth, wash hands after using, and clean work area thoroughly. Dispose of all lead salts and stains in a manner outlined by pollution control authorities.

3. After the pH has been verified, add CO_2-free water to the volumetric flask to bring the solution to a final volume of 50 ml. To prevent entry of CO_2 and possible precipitation of the staining solution, the volumetric should be tightly stoppered with a plastic (rather than ground glass) stopper.

4. Grids may be stained by flotation on the concentrated stain for 30 seconds to 15 minutes. Methacrylate sections must be stained for shorter times than epoxy plastics. In fact, it may be necessary to dilute the stain 1:10 to 1:100 using 0.01N NaOH to prevent overstaining of some acrylic plastics.

Unfortunately, lead stains are easily precipitated upon contact with CO_2. In order to prevent this, it is necessary to protect the staining solution during storage and especially during the staining process. The latter is easily accomplished by constructing a staining apparatus similar to that used for staining with uranyl acetate. In this case, the filter paper is wetted with 1N NaOH and several pellets of NaOH are placed on the filter paper and around the periphery of the dental wax (Figure 5.10). The NaOH will rapidly scavenge any CO_2 that may waft into the chamber.

Preparing CO_2-Free Water

All double-distilled water used in lead citrate staining should be free of CO_2. Such water may be conveniently prepared by autoclaving screw-capped bottles or flasks of water and then sealing them immediately upon removal from the autoclave (while still very hot). Alternatively, the flasks of water may be boiled 10 minutes and then sealed while hot. When opened, such flasks should emit a hissing sound, indicating that they were airtight. Opened flasks of water rapidly absorb CO_2 and should not be used after 10 to 15 minutes. They may, however, be reheated, resealed, and then cooled to room temperature prior to use.

Reynold's lead citrate stain has a rather long shelf life (up to 6 months) if sealed tightly. Even when slight turbidity is noted, the stain is often still useable after centrifugation (5,000 × g for 10 minutes) or passage through a microfilter. One particularly convenient method of clarification involves placing 1 ml of stain in a polypropylene, sealable microtube (Figure 5.11) and centrifuging the sample at high speeds for 4 or 5 minutes. The sealed tube is

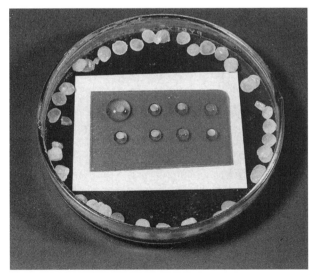

FIGURE 5.10 Setup for staining grids with lead citrate. Sodium hydroxide pellets are placed along the periphery of the dish to scavenge any carbon dioxide that may cause precipitation of the lead stain. In addition, the white filter paper just inside the NaOH pellets is soaked with a saturated aqueous NaOH solution.

impervious to CO_2 for many hours and may readily be recentrifuged as needed. If microfiltration is used, a few milliliters of stain or 0.01N NaOH must be run through the filter to remove any wetting agents used in the manufacture of the filter.

Staining in lead citrate is accomplished in the protected environment of the petri dish apparatus by first floating the grid, section side down, on a large drop of CO_2-free water for several seconds. The grid is then removed so that a drop of water remains over the sections, and the grid is transferred to a freshly deposited drop of lead citrate stain. The stain should not be used after standing exposed in the petri dish for longer than 15 to 30 minutes and it should not be reused. After use, the stain droplets should be discarded in an appropriately labeled container for removal by a pollution control authority.

After 30 seconds to 15 minutes, the grid is removed from the droplet of lead citrate so that a large drop remains covering the sections, and the grid is immediately dipped gently 25 times in a small beaker of either CO_2-free double-distilled water or a solution of freshly prepared 0.01 N NaOH. After this, the grid is dipped for the same number of times in CO_2-free double-distilled water and rinsed in several changes of plain double-distilled water. The prongs of the forceps must be blotted with filter paper at each transfer step to prevent carryover of

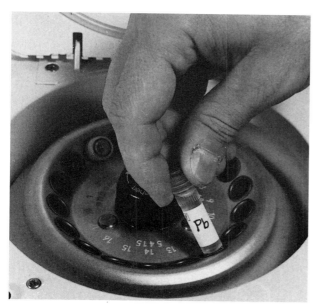

FIGURE 5.11   Lead stains may be centrifuged before use by placing a small volume in a microcentrifuge tube and spinning the sealed tube at high speeds (15,000 × g) for 5 minutes. Conventional centrifugation or filtration of the stain is also acceptable.

stain. After the final rinse, the grid is placed on a filter paper and allowed to dry prior to examination in the transmission electron microscope.

Other Methods for Lead Citrate Staining

Many other methods for lead staining are described in the literature and appear to work as well as Reynolds' lead citrate stain. Each method has its own advantages and problems, so that some trial and error should be expected if attempting some other method. The staining procedures are similar to those for the Reynolds' method and staining times of 2 to 15 minutes are to be expected, depending on the specimen.

It is possible to purchase lead citrate from a number of chemical companies, rather than making it by combining lead nitrate and sodium citrate as in the Reynolds' method. In a popular method described by Venable and Coggeshall (1965), 0.01 to 0.04 g of lead citrate is added to 10 ml of double-distilled water in a sealable centrifuge tube. After adding 0.1 ml of 10 N NaOH, the tube is capped and shaken vigorously until the lead citrate goes into solution. Prior to use, the sealed tube is centrifuged and used in the same manner as described for Reynolds' lead citrate.

In another method described by Fahmy (1967), one pellet of NaOH (0.1 to 0.2 g) is placed into 50 ml of autoclaved, CO_2-free water in a sealable centrifuge tube. After adding 0.25 g of lead citrate, the sealed container is shaken until the stain is dissolved. This solution has a long storage life, but should be centrifuged before use.

Microwave Staining

Just as microwaving can be used to accelerate fixation and embedding of specimens, it may be used to accelerate the staining process (Estrada et al., 1985; van de Kant et al., 1993). Grids may be microwaved for 15 seconds in uranyl acetate followed by 15 seconds in lead citrate to give results that might normally take over 1 hour to achieve using conventional procedures. It is necessary to calibrate the microwave oven to determine proper placement of specimens (Login and Dvorak, 1994).

Staining Many Grids

Until experienced in using the various staining methods described, researchers must be careful to prevent ruining valuable sections. At every step, it is possible to introduce contamination and precipitation on the sections (Figure 5.9) or possibly even to wash away the sections in the various solutions.

Consequently, the conservative approach is to stain only some of the grids at one time. As confidence in the methodology is gained, then more grids can be stained at one time. When stains are freshly prepared, it is very important to test the stains on some expendible sections in order to quality-control the batch of stain. In the same manner if the stain is thought to be too old, caution dictates that it should be discarded or tested prior to use. To evaluate a batch of stain for proper timing, stain three separate grids for 5, 10, and 15 minutes each and then examine them in the TEM for best contrast.

Normally, one can expect to stain 5 to 10 grids per session so that the manual methods described above are quite adequate. When larger numbers of grids (over 100) must be stained on a regular basis, it becomes necessary to resort to multiple or automated grid staining devices. Such systems must be thoroughly evaluated prior to use on valuable specimens or many hours of work may be lost. Some typical multiple-staining devices are shown in Figure 5.12. In addition, programmable machines are available that automate the entire staining process.

Removing Stain Precipitates

Most microscopists eventually encounter contamination of sectioned materials due to stain precipitation. It is sometimes possible to remove the precipitates by floating the grids on droplets of 0.5 N HCl for 0.5 to 1 minute. After several rinses in distilled water, the sections are restained with uranyl acetate and lead citrate. For more details, see Mollenhauer (1987).

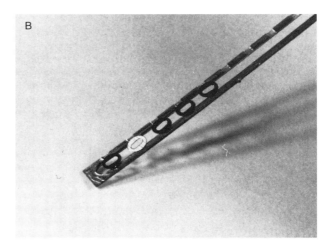

Negative Staining

In the positive staining methods described previously, the heavy metal ions react with macromolecules resulting in an increase in the density and contrast of the molecules. By comparison, in negative staining the macromolecule itself is usually unstained but is instead surrounded by the dense stain (Figure 5.2). As a result, the specimen appears in negative contrast (lighter in tone against a dark background). Negative stains are not used on sectioned materials, but are used to contrast whole, intact biological structures (viruses, bacteria, cellular organelles, etc.) that have been deposited on a supporting plastic or carbon film. The principle of this procedure is illustrated in Figure 5.13.

Because whole specimens are deposited on a grid, several conditions must be satisfied in order to

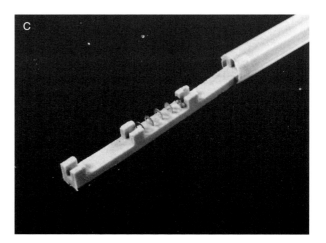

FIGURE 5.12   Devices for holding multiple grids in position during the staining process. After placing the grids into the various holders, the devices are placed into stain-filled receptacles for staining. Pictured from top left are the holders designed by (A) Hiraoka, (B) Synaptek GridStik™, and (C) Giammara.

A

SIDE
VIEW

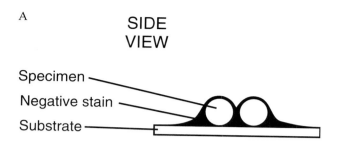

FIGURE 5.13 Illustration of negative stain and virus particles that have been negatively stained.
(A) Principle of negative staining showing side view of specimen surrounded by stain. The electron-dense stain fills indepressions in the specimen and is held tightly around the specimen by surface tension forces.
(B) Negatively stained virus particles from *Escherichia coli* bacteria. Darker areas represent local concentrations of stain.

B

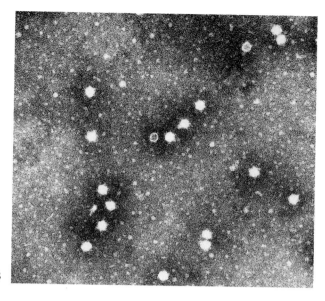

achieve optimal results. A firm, structureless substrate is essential to support the specimen. Collodion or Formvar substrates stabilized with carbon are satisfactory (see Chapter 4). Negatively stained specimens are usually examined at high magnifications, so that a double condenser lens and an anticontaminator in the specimen area are needed to minimize specimen damage. A minimum of beam current must strike the fragile specimen, so the bias must be adjusted to achieve minimal, yet adequate, illumination. Very often, the specimens have not been fixed, but are simply surrounded in the dried stain. In spite of this, the resolutions obtainable with the negative staining method are often better than with sectioned materials since thick plastic embedments are absent. Unlike fixatives, which may chemically combine with the specimen, negative stains enrobe fine structures in a firm, amorphous supporting matrix.

The negative staining technique is quite simple, rapid, and requires a minimum of experience and equipment. It may be used to evaluate cell fractionations (ribosomes, vesicles, tubules, etc.) and even to provide a rapid clinical diagnosis with certain viral infections. A good negative stain reveals the outline and form of structures that are spread out on a support film. The stain surrounds and fills in depressions in structures providing a hint of three-dimensionality. When the stain is fine and without graininess, the ultrastructure is then most adequately resolved.

Commonly Used Negative Stains

As was the case with positive stains, negative stains are salts of heavy metals such as uranium, tungsten, and molybdenum. Usually, 1 or 2% aqueous solutions are used, and the specimens can be viewed within minutes of staining. Some of the more popular negative stains are listed in Table 5.1. Uranyl acetate and phosphotungstic acid are the two most commonly used negative stains.

Preparation of Three Commonly Used Negative Stains

Uranyl acetate is dissolved with stirring in double-distilled water to give a 1 or 2% (w/v) final concentration. The solution should be prepared one day in advance since uranyl acetate dissolves slowly in water. Uranyl solutions should be yellow, clear, protected from light, and discarded if turbid. Contact with phosphate buffers must be avoided, since they will precipitate uranyl salts. Cacodylate and tris buffers are compatible with uranyl acetate.

Phosphotungstic acid is prepared as a 1 or 2% aqueous solution, and the pH is adjusted to 5.0 to 7.0 using 1 N KOH to yield potassium phosphotungstate (K-PTA). Different pH conditions will affect the staining, so some experimentation is in order with each specimen. When the pH is adjusted above 7.0, the stain becomes unstable and is more prone to precipitation. Below pH 6.0, the stain is

TABLE 5.1 Some Commonly Used Negative Stains

Salt	Preparation	Uses	Reference
Ammonium molybdate	1–3% aqueous	Membranes, enzyme subunits, cell fractions	Muscatello & Horne, 1968
Phosphotungstic acid (K-PTA)	0.5–2% aqueous, adjust pH to 5.5–8.0 with 1 M KOH	Viruses, bacteria, cell fractions, frozen sections, macromolecules (DNA, actin, enzymes, etc.)	Valentine & Horne, 1962; Horne, 1967
Uranyl acetate	0.5–2% aqueous	Same as above	Van Bruggen et al., 1960
Uranyl magnesium acetate	1% aqueous	Same as above	Valentine & Horne, 1962; Horne, 1967
Uranyl oxylate	12 mM uranyl oxylate + 12 mM oxalic acid. Mix equal parts and titrate to pH 6.5–6.8 with ammonium hydroxide	Small macromolecules	Mellama et al., 1967
Uranyl formate	0.5–2% aqueous solution adjust pH to 4.5–5.2 with ammonium hydroxide	Same as above	Leberman, 1965

stable for many weeks if refrigerated. To improve the spreadability of the stain (to overcome hydrophobic forces on the plastic/carbon substrate), one can add several components to the K-PTA: several drops of fetal bovine serum, 3 or 4 drops of a 10% solution of soluble starch, or bovine serum albumin to a final concentration of 0.01% (w/v). For osmotic balance, 0.4% sucrose may be added to K-PTA; however, this will decompose in the electron beam and contaminate the microscope unless anti-contaminators are in place over the specimen. Phosphate buffers are compatible with K-PTA.

Ammonium molybdate, 1 or 2% in double-distilled water, is a useful negative stain when examining certain enzyme subunits, membranes, or cell fractions. The contrast obtained with this stain is not as great as with uranyl acetate, but is comparable to K-PTA. Ammonium molybdate has a finer background than uranyl acetate and so may permit better resolution of extremely fine details. It does not appear to be as stable as K-PTA, but has a shelf life similar to uranyl acetate. Compatibility with various buffer systems and osmotic agents should be evaluated prior to use.

Negative Staining Procedures

Drop Method. In this procedure, a Formvar/carbon-coated grid is clamped into a locking forceps, and a drop of the sample is placed on the grid (Figure 5.14). After waiting 30 seconds to permit absorption of the specimen onto the substrate, a drop of negative stain is placed onto the grid and the excess is removed by touching a filter paper to the edge of the grid. After drying for 15 to 30 minutes, the sample can be examined in the electron microscope. It is also possible to mix the sample with an equal portion of stain (1 drop of each) on a Parafilm sheet, to apply the sample/stain mixture onto the grid, and then to blot and examine the grid as usual. Still another variation involves simply mixing equal portions of sample and stain and then using a pipette or clean platinum loop to transfer the sample/stain mixture onto coated grids arrayed on a piece of filter paper (Figure 5.15).

Flotation Method. A coated grid is floated on a droplet of sample on Parafilm or dental wax for 1 minute to permit adsorption of the specimen. The

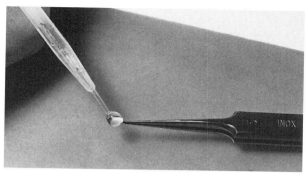

FIGURE 5.14 One method of negative staining involves deposition of the specimen onto a coated grid that has been locked into a forceps (as shown). After allowing the specimen to adsorb to the substrate, a drop of negative stain is applied to the grid.

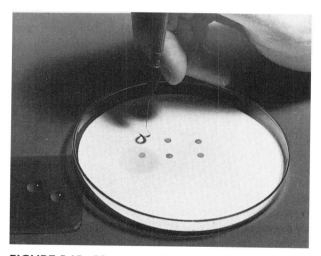

FIGURE 5.15 Negative staining may also be accomplished by mixing equal volumes of specimen and negative stain and depositing the mixture on a grid using either a loop (as shown) or a pipette.

grid is then transferred onto a nearby drop of negative stain for 30 seconds, blotted with a filter paper, and then air-dried for 30 minutes. Alternatively, the sample can be mixed with stain and the grid floated on the mixture.

Imaging Subcellular Components by Negative Staining

When cells have been fractionated into subcellular components by a variety of procedures, it may be necessary to examine the components to determine purity of the preparation or possibly to study the ultrastructure of a specific constituent.

The structure of the ribosome, for instance, was elucidated largely by examination of thousands of negatively stained ribosomal particles to generate models that represented three-dimensional views of the organelle. Figure 5.16 shows a panel of such views and a model that was generated from them by Dr. James Lake.

It is possible to view any subcellular component (membrane fractions, miscellaneous cytoplasmic filaments such as actin and myosin, mitochondria, etc.) using this procedure. The only limitation is that the structure not be so thick that it impedes the passage of electrons.

Revealing Viruses Using Negative Staining

It is possible to use negative staining in clinical situations as well as in research settings. Virologists have used the negative staining technique for over thirty years to study viral preparations in the laboratory, but the procedure of examining clinical isolates for the presence of virus is only occasionally used today—despite the potential for rapid diagnosis. In the laboratory, one method involves mixing a drop of viral preparation with a drop of negative stain and floating a coated grid on the surface of the mixture for 1 minute. After blotting and drying, the grid is examined in the electron microscope. Figures 20.175 to 20.184 show some virus images obtained using this procedure.

A slightly different approach is needed when clinical specimens are involved. Some infectious samples (biopsy tissues, feces, urine, pus, etc.) may require clarification and concentration by high-speed centrifugation in order to increase the number of viral particles, whereas others (vesicle fluid, for example) may be examined directly. It may prove beneficial to fashion a portable sampling kit as described in Figure 5.17 for use in a clinical setting. Such devices are small, easily used, and mailable if an electron microscope is not available on site. For more information regarding the use of electron microscopy in viral diagnosis, consult the references by Hsiung and Fong (1982) and Miller and Howell (1997).

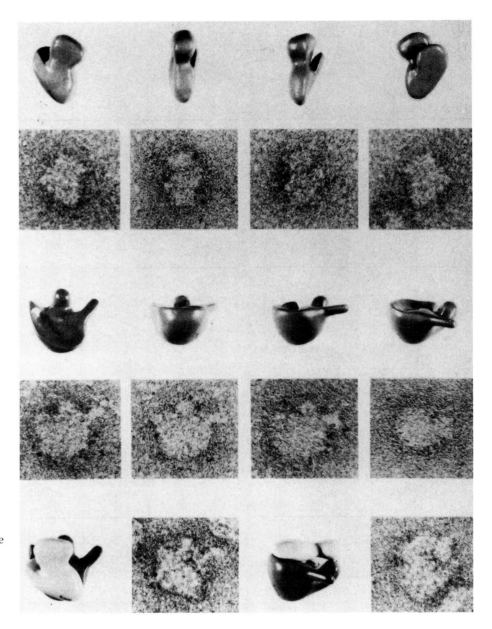

FIGURE 5.16 Three-dimensional model of ribosome that was constructed by examining many different views of negatively stained ribosomes.
(Courtesy of J. Lake.)

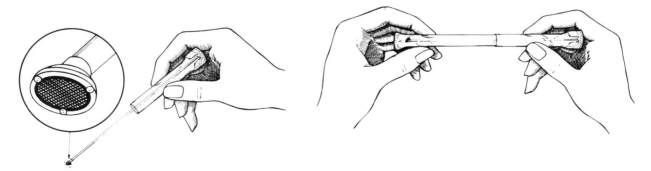

FIGURE 5.17 Diagram of portable device for collecting viral specimens from patients. A Formvar-coated grid is attached to the end of a modified cotton swab shaft and placed in contact with the specimen.
(Courtesy of Ted Pella, Inc.)

process, the screw prevents some areas of the card from receiving carbon. These uncoated, shielded areas serve as a relative gauge of thickness when compared to the areas of the card that received a darker coating of carbon. For more accurate measurements, digital resistance and quartz crystal monitors are available.

Due to its high melting point (3650° C), carbon electrodes can serve as a heat source to vaporize lower melting point metals. For instance, platinum and gold-palladium wires may be evaporated from heated carbon electrodes. Often carbon is evaporated along with the metals. This simultaneous evaporation is advantageous since carbon tends to "wet" specimen surfaces and thereby permits the metals to form thinner, more continuous films. To evaporate the metals, a proper length of acetone-cleaned metal wire is tightly wound around the tip of a pointed carbon rod that is then heated under high vacuum.

(b) Carbon Braid. Some EM supply houses sell braided carbon fibers in 3 or 4 foot lengths. This type of carbon is easier to evaporate and is suitable for making filmed carbon grids or for general coating purposes. To use it, a short length of braid is cut with a scissors and placed into special clamp-type electrodes (Figure 5.26). Although less amperage is needed to evaporate this type of carbon, enough heat may still be evolved to damage delicate specimens.

(c) Tungsten Filaments. One may make these filaments from annealed tungsten wire purchased from EM or scientific supply houses. A V-shaped, downward pointing filament is desirable since it maintains the molten metal near the tip to help form a point source. It is also possible to purchase multistranded tungsten filaments preformed into the V-shape (Figure 5.27A). These filaments are highly reliable, may be reused several times, and are less likely to break if overheated. One disadvantage is that they require more power to melt the metal (some vacuum evaporators may not be able to achieve the power levels needed) and more heat will be generated. It is also possible to purchase "tornado-shaped" tungsten filaments or baskets (Figure 5.27B) to hold chunks of metal or metal filings and powders (chromium chips, for example).

(d) Molybdenum Boats. These trough-shaped strips of metal (Figure 5.28) are also used to evaporate powders, metal filings, and chips of metal. They require a great deal of power, generating high heat that may affect delicate specimens—so caution is in order. They are useful for cleaning certain components of electron microscopes such as aperture strips. Since the metal tends to flow out into the trough as it melts, it is more difficult to achieve a point

source of evaporation so that shadows may be less sharp with this method of evaporation.

CAUTION: The electrodes will become extremely bright as evaporation commences. Do not look directly at the electrodes or damage to the eyes may result. Instead, use welder's goggles or a piece of overexposed and developed essentially black TEM film. Conventional sunglasses will not adequately protect the eyesight.

FIGURE 5.26 Carbon braid may be used instead of carbon rods as a source of carbon coating. A short length of carbon braid is clamped into the special holder as shown.

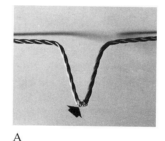 

A   B

FIGURE 5.27 (A) Multistranded tungsten filament with platinum wire wound around it (*arrow*). Heating the tungsten filament will melt and vaporize the platinum at the tip. (B) Spiral basket of tungsten wire used to melt chunks of metals such as chromium for evaporation.

FIGURE 5.28 Molybdenum trough clamped into electrode holder. Chunks of metal (*arrow*) may be placed into the trough for evaporation.

Equation 5.1: Estimating Amount of Metal Needed

It is useful to know approximately how much of the metal is needed to achieve a particular thickness of evaporated film. The following formula will help in this estimation:

$$M = \frac{4\pi r^2 t d}{\sin \alpha}$$

where M = weight of metal in grams
r = distance from source to specimen in cm
t = thickness of deposit in Angstroms
d = density of metal in gm/cm^2
α = angle of shadowing

Some Applications of Metal Shadowing and Negative Staining

Making Height Measurements Using Metal Shadowing

Metal shadowing can be used not only to reveal morphological features present in a specimen, but it may be used to measure the height of a specimen based on basic geometric principles. Figure 5.29 illustrates the relationship of the specimen to the evaporating electrodes. If one knows several parameters, it is possible to calculate the height of a particle or specimen based on measurement of the length of its shadow.

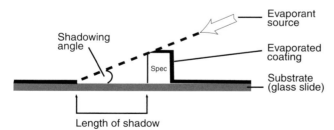

FIGURE 5.29 Principle of metal shadowing procedure showing deposition of metal alongside facing filament source. The specimen physically prevents the vaporized metal from reaching certain areas of the substrate. These "shaded" areas will show up as a white shadow.

Equation 5.2: How to Calculate Height of Shadowed Specimens

When small particles or organelles have been deposited onto a coated grid, it is possible to determine the width by directly measuring the image on

the negative. The height of the specimen can be calculated using the following equation.

$$H = \frac{b}{c} (l)$$

where H = height of specimen
b = height to filament from level of specimen
c = distance from point directly under filament to specimen
l = length of shadow

For example, suppose that the filament is situated 6 cm above the level of the specimen and that the distance measured from a point directly under the filament to the specimen is 12 cm. If one measured the length of the shadow and determined it to be l = 100 nm, then:

The height of the particle, H = 6/12 × 100 nm = 50 nm.

In this example, it may be noted that the shadow is twice as long as the specimen height.

One may recognize the term b/c as being equivalent to the geometric equation for calculating the tangent of an angle formed by a line connecting the filament to the edge of the shadow

$$\tan \Theta = \frac{height}{length}$$

$$height = (\tan \Theta)(length)$$

Consequently, on some vacuum evaporators equipped with a specimen stage that is graduated in degrees, one needs only to measure the length of the shadow and multiply this by the tangent of the angle to determine the height of the particle.

Replication of Biological Surfaces for Transmission Electron Microscopy

The SEM has made possible the investigation of the topographic features of a variety of biological structures. If a SEM is not available or if one needs better resolution than can be obtained in a particular SEM, it is possible to make a metal replica of a biological surface and to examine the replica in the TEM. (Replication methods for viewing biological specimens in the SEM are covered in Chapter 3, Replication Procedures). The resolution of this technique is approximately 2 to 3 nm; however, specimens with high relief can not be replicated easily. Several different methodologies are available, depending on the specimen and the nature of the information desired. Only the more commonly used methods will be covered in this book. For more possibilities, consult the reference book by Willison and Rowe (1980).

Platinum-Shadowed, Single-Stage Replicas. These types of replicas offer the greatest resolution and are to be considered before other methods, if the specimen is suitable. This is the same method used in preparing replicas in freeze fracture and freeze etching, except that the specimen is not frozen when the replica is made. The specimen must be well-fixed, dried, and capable of withstanding the conditions inside a vacuum evaporator.

Preparing Single-Stage Replicas

1. Fix and dry the specimen in a manner consistent with the preservation of the details that are to be studied. In some specimens (insects, plant materials such as leaves and seeds, or other hard specimens), this may mean simply allowing the samples to air-dry. Other samples (mammalian, bacterial cells, or other soft tissues) must undergo fixation, dehydration, and drying (critical point or freeze-drying) prior to replication. This technique works best with specimens that have little relief and can be dissolved using appropriate solvents (e.g., the sample should be expendable).

2. Place specimen inside vacuum evaporator and obtain vacuums in the 10^{-4} Pa (10^{-6} Torr) range.

3. Evaporate platinum onto the specimen surface at a 45° angle followed by carbon evaporated at a 90° angle. The thickness of the two evaporated materials must be determined by trial and error. Overly thick coatings will lack detail, while thinner ones will be fragile and easily broken during the processing steps.

4. Remove sample and, if possible, score the surface into 2 × 2 mm squares that will fit conveniently onto a TEM specimen grid. Razor blades or scalpels are normally used in this step.

5. Float off the replica squares from the surface of the specimen by slowly submerging the specimen into a container of distilled water. *NOTE: Difficult to strip coatings may sometimes be removed by pressing Scotch Brand tape onto the coated surface and pulling back the tape. The tape is then placed in chloroform to dissolve the adhesive and to free the replicas into the chloroform. After several changes in chloroform, the small replica squares are then placed into the proper organic solvent, as in step 6.*

6. Scoop up the replica squares using 30 mesh stainless steel screening (Small Parts, Inc., Florida) and transfer the replicas into an appropriate solvent to dissolve the organic materials: (a) full strength chlorine bleach (sodium hypochlorite) for several hours to overnight, or (b) 50% chromic acid (prepared by dissolving 5 gm sodium dichromate in 100 ml of 50% sulfuric acid). Either of these two solvents should remove any biological materials trapped in the replica; however, other solvents or enzymes may be needed initially to dissolve some types of organic materials.

7. After the organic material has dissolved (as determined by examining one of the processed replica squares in the TEM), scoop up the replica squares with the screening and transfer the replicas through at least four changes of double-distilled water.

8. Pick up individual replica squares by either: (a) lifting the replicas onto uncoated 200 mesh copper grids, or (b) picking up the replicas with a loop and transferring them onto a Formvar-coated grid. Experience is needed in both of these procedures since the replicas will tend to roll and fold over onto themselves.

9. Blot the grids, allow them to dry, and examine them in the TEM.

Figure 5.30 shows a platinum/carbon single-stage replica made of some bacterial cells. Figure 5.30A shows the presence of the bacterial cells in the replica (prior to dissolution in chromic acid) while Figure 5.30B shows only the replica remaining after dissolution of the organic material. Compare Figure 5.31, which shows bacterial cells viewed in the SEM, with Figure 5.30A and B, which shows bacterial cells viewed in the TEM.

Two-Stage, Negative Replicas. In bulk specimens that cannot be sacrificed, it may be possible to make a replica of the surface using the two-stage, negative replica procedure. The resolution is not as good with this technique, however, since one first makes a plastic replica of the biological surface and then coats the plastic with platinum/carbon. A schematic comparison of the single- and two-stage replication methods is given in Figure 5.32.

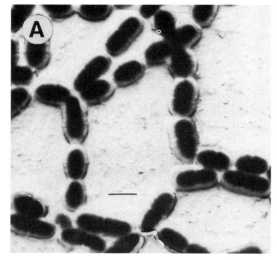

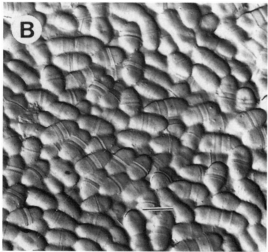

FIGURE 5.30 Platinum/carbon-coated bacterial cells (black). (A) *Streptococcus mutans* cells are still present in the preparation so that they obstruct fine details of the light gray metal coating. (B) The bacterial cells were dissolved using chromic acid. Only the platinum/carbon replica remains. Marker bars = 0.5 μm.

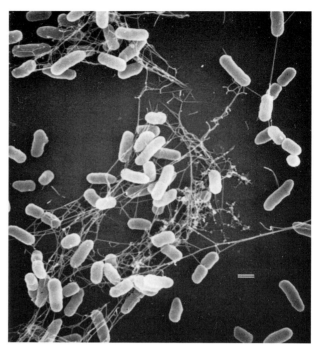

FIGURE 5.31 Some bacterial cells as in Figure 5.30 only viewed in the SEM. Marker bar = 0.5 μm.

4. Place the grids in the vacuum evaporator with the side that was in contact with the specimen surface facing up.
5. Prepare the platinum and carbon coatings as described in step 3 of the single-step replication procedure.
6. Gently lower the grids into chloroform to dissolve the Collodion (as evidenced by the platinum/carbon replica becoming free of the grid).
7. Scoop up the replica onto the grid and allow the chloroform to evaporate prior to placing the grid in the TEM.

Preparing Two-Stage, Negative Replicas

1. Onto a suitable specimen (fixed, dried, etc.) apply a thin layer of 1% Collodion in amyl acetate. When dry, apply several more layers of 2% Collodion.
2. With a blade, score the surface into squares and strip the plastic film from the surface as described in step 5 of the single-stage replication procedure. If adhesive tapes are used to assist stripping of the Collodion replica, avoid chloroform solvents since they will dissolve the plastic replica.
3. Pick up the plastic replica, place it on an uncoated grid, and blot it dry.

Extraction Replicas. This procedure is useful to physically extract insoluble components from the biological specimen for further study in the analytical electron microscope. It would be useful, for example, in the extraction and identification of carbon particles, asbestos, talc, etc., from lung or other tissues. Briefly, the tissues are covered with a layer of 5 to 10% polyvinyl alcohol (PVA) in water. After drying, the thin film is stripped from the specimen (pulling with it any particles in the tissues). After coating with carbon, the PVA is dissolved in water, leaving behind the carbon replica containing the particles which are examined in the electron microscope. For more detail, see the reference by Henderson (1975).

Single-Stage Replication

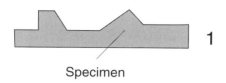

Specimen

1

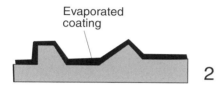

Evaporated coating

2

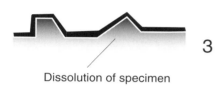

Dissolution of specimen

3

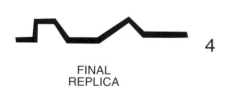

FINAL
REPLICA

4

Two-Stage Replication

Specimen

1

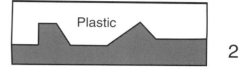

Plastic

2

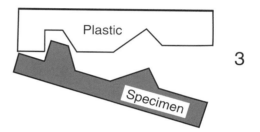

Plastic

Specimen

3

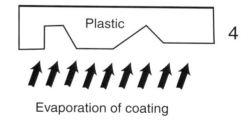

Plastic

4

Evaporation of coating

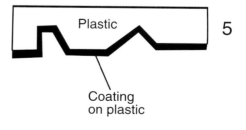

Plastic

Coating
on plastic

5

FINAL
REPLICA

6

FIGURE 5.32 Comparison of the single- and two-stage replication procedures. As may be expected by examination of the steps involved, the single-stage procedure yields replicas with better resolution.

Plasma Polymerization Replicas.   Tanaka and colleagues (1978) described a novel procedure for producing a high resolution replica by polymerizing naphthalene gas onto the surface of a specimen. The naphthalene gas is introduced into an evacuated bell jar and exposed to a high voltage discharge for 10 seconds. This polymerizes the gas into a thin film that clings to the surface of the specimen. A replica may then be produced by dissolving the biological material using a 1% solution of sodium hypochlorite for 5 to 10 minutes. The replicas may also be produced from the surfaces of frozen-hydrated or dried specimens, and replicas may be made from cells that have been labeled with colloidal gold probes (Figure 5.33). (See Chapter 9 and reference by Yamaguchi and Kondo, 1989.)

Visualizing Macromolecules

The imaging of the nucleic acids, DNA and RNA, is routinely done using either negative staining or metal shadowing techniques. Although better contrast may be obtained using metal shadowing, resolution will generally be better with negative staining techniques.

Rotary metal shadowing is preferred to static shadowing of nucleic acid molecules. In this procedure, negatively charged nucleic acids are usually suspended in a stabilizing *hyperphase* solution (consisting of ammonium acetate, EDTA [ethylene-diamine- tetraacetate], cytochrome c, and about 0.1 to 0.5 µg nucleic acid / 100 µl of total solution). The hyperphase is allowed to flow down a scrupulously clean microscope slide onto either a distilled water surface or the surface of a dilute solution of salts (the *hypophase*). The molecules spread over the surface of the hypophase, and the cytochrome c holds the nucleic acids firmly in a monomolecular film of denatured protein.

DNA Spreading and Shadowing Technique

The hyperphase solution containing the nucleic acid is allowed to spread over the surface of the hypophase (0.25M ammonium acetate, for instance) that has been dusted with talcum powder or powdered graphite. The nucleic acid/cytochrome c molecules will push the talc aside so that cleared areas containing specimen will be outlined by the talc/graphite. Collodion-coated grids are gently touched to the clear zones of nucleic acid to permit adsorption of the nucleic acid/cytochrome c and then blotted gen-

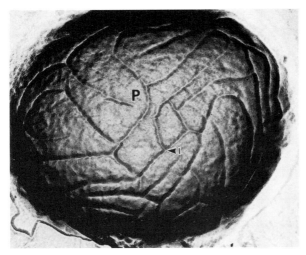

FIGURE 5.33A   Replica of the surface structure in a yeast cell showing invaginations (I) of the plasma membrane (P). The replica was produced using the plasma polymerization method of Tanaka, Sekiguchi, and Kuroda (1978).
(Courtesy of T. Hirano.)

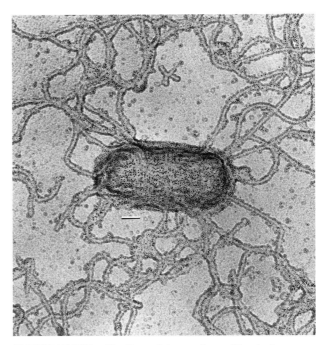

FIGURE 5.33B   Replica of the surface of bacterium, *Proteus vulgaris*, shown in stereoview. Flagellar and somatic antigens are recognized by the 5 nm and 10 nm colloidal gold particles, respectively. The replica was produced using the plasma polymerization method of Tanaka, Sekiguchi, and Kuroda (1978). Marker bar = 200 nm.
(Courtesy of M. Yamaguchi.)

tly by a filter paper touched to the edge of the grid. The grids can then be floated on a stain consisting of 50 mM uranyl acetate and 50 mM HCl in 95% ethanol. After rinsing in 95% ethanol, the grid is placed on a filter paper and allowed to air-dry. The dried grid is then placed on the stage of a vacuum evaporator, pumped to high vacuum, and rotated as a shadowing metal is evaporated. Figure 1-5 shows some DNA molecules that have been prepared using this technique. For more procedural detail, see the book edited by Sommerville and Scheer (1987).

Negative staining is conveniently used to reveal fine structural details of nucleic acids, enzyme subunits, and a number of other macromolecules (Figure 5.34). In the simplest approach, the macromolecule-containing sample is mixed with the negative stain and deposited onto a Collodion- or Formvar/carbon-coated grid as described previously in this chapter. After blotting with a filter paper, the grid is allowed to dry completely prior to examination in the electron microscope. A minimum of technical skills and equipment is needed with this procedure.

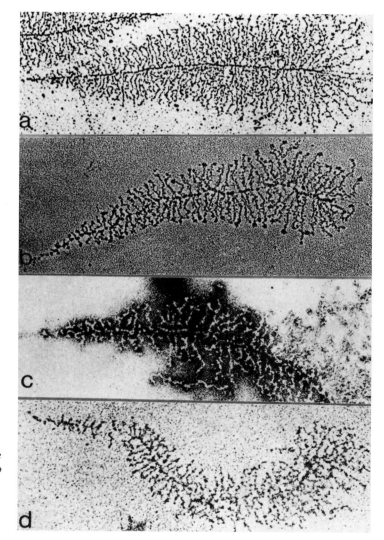

FIGURE 5.34 Different modes for visualization of spread amphibian genes in the act of transcription. The central dark line is DNA while the fibrils jutting out from the DNA are pre-RNA transcriptional units. (A) The most frequently used staining method for spread chromatin is phosphotungstic acid in 70% ethanol. (B) Phosphotungstic acid-stained preparations may be enhanced for contrast by additional shadow-casting using a heavy metal. (C) When phosphotungstic acid at a neutral pH is used, contrast reversal may be encountered. (D) In some instances, if the spread nucleic acids are picked up on extremely thin carbon films, the genes may be visualized without any additional heavy metal staining. Marker bars = 0.5 μm.
(Courtesy of M. F. Trendelenburg and IRL Press.)

REFERENCES

Bozzola, J. J. 1987. Clinical sampling device for rapid viral diagnosis by transmission electron microscopy. *J Electron Microscopy Tech* 5:243–48.

Epstein, M. A. and S. J. Holt. 1963. The localization by electron microscopy of HeLa cell surface enzymes splitting adenosine triphosphate. *J Cell Biol* 19:325–26.

Estrada, J. C., N. T. Brinn, and E. H. Bossen. 1985. A rapid method of staining ultrathin sections for surgical pathology TEM with the use of the microwave oven. *Am J Clin Pathol* 83:639–41.

Fahmy, A. 1967. An extemporaneous lead citrate stain for electron microscopy. *Proc 25th Annu EMSA Meeting,* pp. 148–49.

Frasca, J. M., and V. R. Parks. 1965. A routine technique for double-staining ultrathin sections using uranyl and lead salts. *J Cell Biol* 25 (No.1, Pt. 1):157–61.

Hayat, M. A. 1972. *Principles and techniques of electron microscopy: Biological applications,* Vol. 2. New York: Van Nostrand Reinhold Co.

Hayat, M. A., and S. E. Miller. 1990. *Negative Staining.* McGraw-Hill Publishing Co., New York.

Heinrich, H. 1985. Precipitate-free lead citrate staining of thin sections. *J Electron Microscop Tech* 2:275–81.

Henderson, W. J., et al. 1975. Analysis of particles in stomach tumors from Japanese males. *Environ Res* 9:240–49.

Horne, R. W. 1967. Electron microscopy of isolated virus particles and their components. In *Methods in virology,* Vol. 5. K. Maramorosch, and H. Koprowski, eds. New York: Academic Press, pp. 521–74.

Hsiung, G. D., and C. K. Y. Fong. 1982. *Diagnostic virology illustrated by light and electron microscopy.* New York: Yale University Press.

Karnovsky, M. J. 1967. The ultrastructural basis of capillary permeability studied with peroxidase as a tracer. *J Cell Biol* 35:213–36.

Leberman, R. 1965. Use of uranyl formate as a negative stain. *J Molecular Biol* 13:606.

Lewis, P. R. and D. P. Knight. 1992. *Cytochemical staining methods for electron microscopy.*

A. M. Glauert, ed. Amsterdam, The Netherlands: Elsevier/North-Holland Publishing Co.

Login, G. R., and A. M. Dvorak. 1994. *The microwave tool book.* Beth Israel Corp. ISBN 0–9642675–0–0.

Mellama, J. E., E. F. J. Van Bruggen, and M. Gruber. 1967. An assessment of negative staining in the electron microscopy of low molecular weight proteins. *Biochim Biophys Acta* 140:180–82.

Miller, S. E., and D. N. Howell. 1997. Concerted use of immunologic and ultrastructural analyses in diagnostic medicine—immunoelectron microscopy and correlative microscopy. *Immunol Investigations* 26:29–38.

Mollenhauer, H. H. 1987. Contamination of thin sections: Some observations on the cause and elimination of "embedding pepper." *J Electron Microsc Tech* 5:59–63.

Muscatello, U., and R. W. Horne. 1968. Effect of the tonicity of some negative-staining solutions on the elementary structure of membrane-bound systems. *J Ultrastructure Res* 25:73–79.

Reynolds, E. S. 1963. The use of lead citrate at high pH as an electron-opaque stain in electron microscopy. *J Cell Biol* 17:208–12.

Sommerville, J., and U. Scheer. eds., 1987. *Electron microscopy in molecular biology: A practical approach.* Oxford: IRL Press.

Springer, M. 1974. A simple holder for efficient mass staining of thin sections for electron microscopy. *Stain Tech* 49:43–46.

Stempak, J. G., and R. T. Ward. 1964. An improved staining method for electron microscopy. *J Cell Biol* 22: 697–701.

Tanaka, A., Y. Sekiguchi, and S. Kuroda. 1978. *J Electron Microsc* 27:378–81.

Trendelenburg, M. F., and F. Puvion-Dutilleul. 1987. Visualizing active genes. In *Electron microscopy in molecular biology: A practical approach.* J. Sommerville and U. Scheer, eds. IRL Press, Oxford, England. pp. 101–146.

Valentine, R. C., and R. W. Horne. 1962. An assessment of negative staining techniques for revealing ultrastructure. In *The interpretation of ultrastructure.* R. J. C. Harris, ed. New York: Academic Press, pp 263–78.

Van Bruggen, E. F. J., E. H. Wiebenger, and M. Gruber. 1960. Negative-staining electron microscopy of proteins at pH values below their isoelectric points. Its application to hemocyanin. *Biochim Biophy Acta* 42: 171–72.

van de Kant, H. J. G., Boon, M. E., de Rooij, D. G. 1993. Microwave applications before and during immunogold-silver staining. *J Histotechnol* 16:209–15.

Venable, J. H., and R. Coggeshall. 1965. A simplified lead citrate stain for use in electron microscopy. *J Cell Biol* 25(No. 2, Pt. 2):407–8.

Willison, J. H. M. and A. J. Rowe. 1980. *Practical methods in electron microscopy, Vol. 8: Replica, shadowing and freeze-etching techniques.* A. M. Glauert, ed. Amsterdam, The Netherlands: Elsevier/North-Holland Publishing Co.

Yamaguchi, M., and I. Kondo. 1989. Immunoelectron microscopy of *Proteus vulgaris* by the plasma polymerization metal-extraction replica method: Differential staining of flagellar (H) and somatic (O) antigens by colloidal golds. *J Electron Microsc* 5:382–88.

Chapter 6

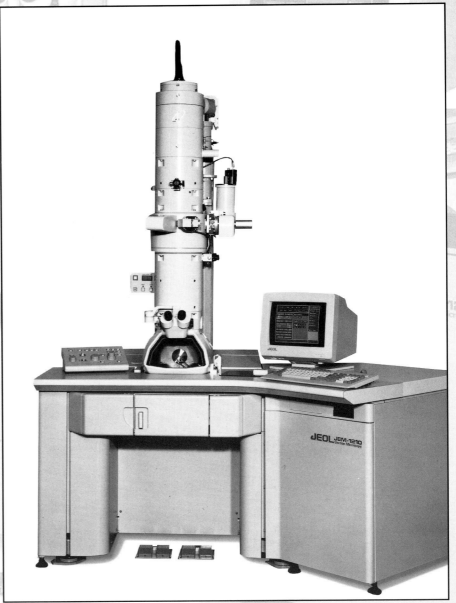

Courtesy of JEOL.

The Transmission Electron Microscope

A TRANSMISSION ELECTRON MICROSCOPE, or TEM, has magnification and resolution capabilities that are over a thousand times beyond that offered by the light microscope. It is an instrument that is used to reveal the *ultrastructure* of plant and animal cells as well as viruses and may provide an image of the very macromolecules that make up these biological entities. The TEM is a complex viewing system equipped with a set of electromagnetic lenses used to control the imaging electrons in order to generate the extremely fine structural details that are usually recorded on photographic film. Since the illuminating electrons pass *through* the specimens, the information is said to be a *transmitted* image. The modern TEM can achieve magnifications of one million times with resolutions of 0.1 nm.

The concepts of magnification and resolution may be understood as one views the letters on this printed page. If one moves away from the page, the distinct image of various letters slowly blurs into one indistinct object, as if the letters had merged. Conversely, if one moves closer to the page, two phenomena occur: the letters become larger and sufficient detail becomes apparent to distinguish the individual letters again. We have, in fact, "zoomed in" on the letters to both magnify and resolve finer detail. In a strict sense, *magnification* is a measure of the increase in the diameter of a structure (Figure 6.1A), and *resolution* is the ability to discriminate two closely placed structures that might otherwise appear as one and to see more details within the objects (Figure 6.1B).

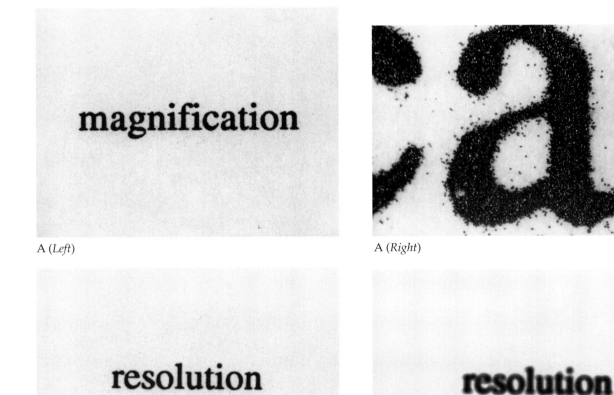

A (*Left*)

A (*Right*)

B (*Left*)

B (*Right*)

FIGURE 6.1   Illustration of the concepts of magnification and resolution. (A) (*left*) The word "magnification" shown magnified ten times (*right*). Both show good resolution. (B) (*left*) An image showing good resolution compared to an image (*right*) showing poor resolution. Both were recorded at the same magnification. If the image with poor resolution were to be magnified, the resulting image would be of poor quality due to empty magnification.

Higher magnification does not necessarily mean better resolution or vice versa. It is possible to enlarge an object without being able to see or resolve any more detail. For example, advertisements in comic or adventure books sell various "microscopes" with magnifications of over a thousand times. While such toys are able to magnify objects as claimed, the magnified images produced are very inferior because the overextended or *empty magnification* lacks resolution. Both magnification and resolution are necessary to produce a quality image.

Small objects are normally brought close to the eye in an attempt to magnify and resolve finer detail. However, the human eye cannot resolve two points that are any closer than 0.1 to 0.2 mm, no matter how close to the eye the object is placed. A simple hand lens may be used to project both a magnified and resolved image of an object on the retina of the eye so that the object can be seen in greater detail. The lenses of the electron microscope follow the same optical principles of magnification and resolution to permit one to observe the ultrastructure of biological specimens. Unlike glass lenses, which use visible light to form magnified and resolved images on the retina, the lenses of the electron microscope use shorter wavelengths of electromagnetic radiation that are not directly visible to the human eye. How the electron microscope uses such electromagnetic lenses to generate visible images on the human retina is the subject of this chapter.

Visible Light, Electrons, and Lenses

Electromagnetic Radiation and the Diffraction Phenomenon

Visible light represents a very small segment of a spectrum of waves making up the family known as *electromagnetic radiation* (Table 6.1). All of these

TABLE 6.1 Electromagnetic Radiations

Radiation Type	Wavelength Ranges (meters)	Usual Source	Usual Detector
Radio		Transmitter	Receiver
AM	545–188		
FM	3.40–2.79		
Television	5.55–0.34		
Microwaves	3×10^{-1}–3×10^{-3}		
Infrared (heat)	3×10^{-3}–8×10^{-7}	Hot Objects	Thermometers, Thermocouples, Nerve Cells
Visible Light	8×10^{-7}–4×10^{-7}	Electric Filament, Hot Objects	Photocell, Photographic Film, Eye
Ultraviolet	4×10^{-7}–1×10^{-9}	Electric Filament	Photocell, Photographic Film
X Rays	1×10^{-9}–1×10^{-11}	Impact of electrons on metal target	Photographic Film, Ionization Chamber, Geiger Counter
Gamma Rays	$< 1 \times 10^{-11}$	Radioactive Nuclei	Ionization Chamber, Geiger Counter
Accelerated, 60 kV electrons in microscope	5×10^{-12}	Electric Filament	Viewing Screen, Photographic Film

waves travel at the speed of light and differ from each other only in the distance from the top of one wave to the next (Figure 6.2). Starting at the long wavelength end of the spectrum and proceeding toward the shorter, there are the following waves: radio, infrared or heat, visible light, ultraviolet, X rays, gamma rays, and cosmic rays. Radio waves may have wavelengths ranging from several miles to several millimeters, while cosmic rays have wavelengths measured in femtometers (10^{-20} m).

Electromagnetic waves radiate from a source. For radio waves, the source is the transmitter, while a possible source of light waves may be a tungsten filament in a light bulb. The radiating waves emanate in ever-widening circles from the source until they come in contact with a solid object. When the waves strike the solid object, another series of waves is radiated from the edge of the object. The result is a new source of waves that merges with the original waves so that the light now appears to bend around the corner (Figure 6.3). This bending phenomenon is called *diffraction.*

Both the wavelike nature of electromagnetic radiation and the diffraction phenomenon may be readily demonstrated by observing the action of waves in a pool of water. A stone thrown into the water will generate a series of waves that radiate from the point where the stone entered the water. When the waves encounter a solid object such as a wall with an opening or aperture, another series of waves will be generated from the free edges of

the wall, creating the diffraction phenomenon (Figure 6.4).

When electromagnetic radiations undergo the diffraction phenomenon along the edge of a solid object (such as a cellular organelle), the diffracted waves *interfere* with the initial wavefront. This result is in an *unsharp image of the edge of the object being irradiated.* The edge appears to have a series of bands, or fringes, called *Fresnel fringes* (named after the French physicist Augustin Fresnel and pronounced *fre-nell*) running parallel to the edge. Unless magnified, these bands will not be seen individually but will appear to blend together to give

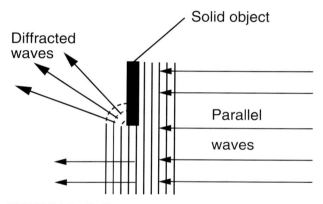

FIGURE 6.3 Diffraction phenomenon demonstrated by a series of parallel waves that strike the edge of a solid object. From the edge, a new series of waves (*dashed lines*) are generated that merge with the original front.

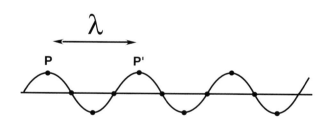

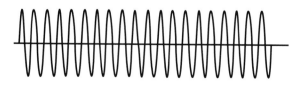

FIGURE 6.2 Wavelength (λ, *top*) is measured from the top of wave P to the top of wave P'. A short wavelength (*bottom*) has a smaller distance between waves.

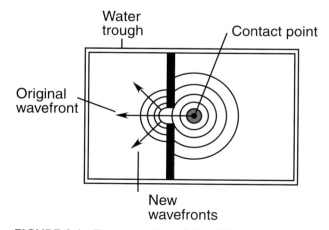

FIGURE 6.4 Demonstration of the diffraction phenomenon in a pool of water. A stone striking the water surface gives rise to waves that generate two more series of waves when they contact the solid walls extending into the water. These two new wavefronts merge with the original front that continues through the opening, or aperture, in the walls.

the impression of a fuzzy rather than a distinct edge. The net result is that the object appears slightly larger and indistinct (Figure 6.5).

Effect of Diffraction on Resolution

Resolution is degraded due to the diffraction phenomenon. Classically, loss of resolution has been illustrated by viewing a series of pinholes made in a thin metal foil. If the foil is held up to a strong light source and viewed with a lens, sharp points of light are not seen. Instead, one sees a bright central area surrounded by a series of Fresnel fringes originating from the edges of the foil (Figure 6.6A). In fact, even if one could generate a true point-source of light (as opposed to looking at light through holes), the Fresnel fringes would still be generated

since the waves would diffract from the edges of the lens or from any *apertures* in the lens system.

Apertures are simply holes of various sizes that may be used to control the amount of light passing through a lens. Most 35 mm cameras, for instance, have adjustable apertures that control the exposure by varying the amount of light striking the photographic film. The f/stop markings on the camera refer to the ratio of the diameter of the aperture to the focal length of the lens. Thus, a 5 mm aperture opening in a 55 mm lens has an f/stop of 55 mm/5 mm, or f/11. The amount of illumination passing through a lens is inversely related to f/stop: smaller numbered f/stops allow more light to pass. The f/stop of a transmission electron microscope lens with a focal length of 2 mm and a 50 μm aperture would be f/40 (i.e., 50 μm/ 2,000 μm).

Figure 6.6B is a photograph of some backlighted pinholes viewed in a light microscope. These ringed patterns are termed *Airy discs* after the 19th century astronomer Sir George Airy, who first described this pattern. The ringed patterns, or *circles of confusion,* tend to increase the apparent size of the holes so as to cause an overlap in some places. If the holes are too close, the overlap would obscure the distinct nature of the holes. Thus, *resolution is decreased by the diffraction phenomenon.*

To determine resolving power, it is important to know the radius of the Airy disc. The radius of the Airy disc as measured to the first dark ring is expressed by Equation 6.1:

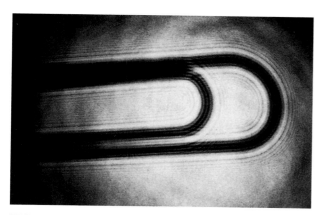

FIGURE 6.5 Diffraction of light using a solid object illuminated by a point source of light. The paper clip was illuminated by a laser and the shadow image recorded on a piece of photographic film. The diffracted waves interfere with the original waves, giving rise to the series of bands or Fresnel fringes around the object.

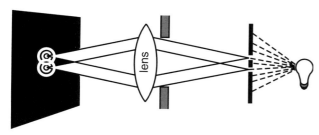

FIGURE 6.6A Degradation of resolution due to diffraction. Two pinholes held in front of a light source are viewed in a lens and the image is projected onto a flat surface. Instead of two sharp, bright spots, one sees two spots surrounded by diffuse rings (*left*). These enlarged, indistinct spots are caused by Airy discs. The thick barrier in front of the lens is an aperture.

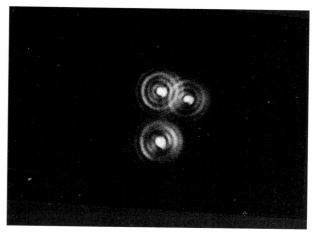

FIGURE 6.6B Airy discs generated by viewing three pinholes in a light microscope. A thin film of palladium/gold was deposited onto a glass slide and the slide was examined for naturally occurring pinholes in the film. Magnification of micrograph is 1,000×.

Equation 6.1: Radius of an Airy Disc

$$r = \frac{0.612\ \lambda}{n(\sin\ \alpha)}$$

where 0.612 = a constant
λ = wavelength of illumination
n = refractive index
α = aperture angle of the lens

The constant, 0.612, is based on an astronomical equation derived from the observation of self-luminous points (e.g., stars). The use of this constant in microscopy is problematic, since microscopy deals with non-self-luminous objects that generate diffraction effects when illumination strikes them.

The term *refractive index* (n in the equation) is a measure of the optical density of a medium. Light passing through a medium with a high refractive index is slowed down. When light passes through two media with dissimilar refractive indices, it will bend at the interface of the two rather than continue in a straight line. If one wishes light to continue in a straight line, the refractive indices of all of the media through which it passes must be similar. Some typical refractive indices are: air = 1.000, water = 1.333, glass = 1.5–1.6, standard immersion oil = 1.515. A vacuum is considered to have a refractive index of one.

Aperture angle (α in the equation) refers to the half angle of the illumination a lens can accept. The larger this acceptance angle, the more information will enter the lens. Consequently, as shown in Figure 6.7, a lens with a large aperture angle will accept more information from the object being viewed. In Figure 6.8, we see that the use of immersion oil in the light microscope increases the aperture angle and allows more information to effectively enter the lens, thereby increasing resolution.

The term *numerical aperture* or N.A. is equivalent to n(*sin* α) and is often engraved on the barrel of light microscope objective lenses. The higher this value, the greater is the information-gathering ability (and resolution) of the glass lens. The highest N.A. currently obtainable in glass lenses is around 1.5.

As early as 1896, Rayleigh established the criterion that if Airy discs are placed in such close proximity that their first dark rings contact each other, overlap will occur to the extent that they can no longer be resolved as two distinct units. Therefore, it is apparent that *the equation for the radius of the Airy disc is the equation for resolving power.* We can formally define *resolving power* as the minimum distance that two objects can be placed apart and still be seen as separate entities. Consequently, the shorter the

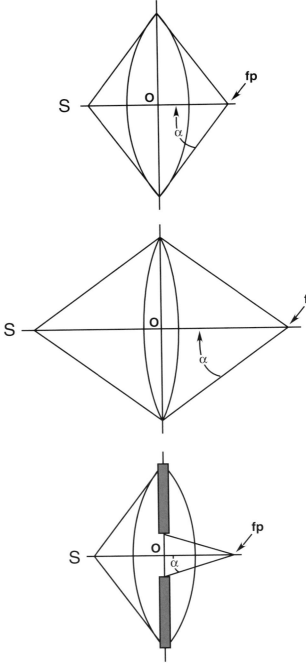

FIGURE 6.7 As aperture angle (α) increases, resolution improves. A high resolution lens (*top*) is a strong, short focal length lens with a large aperture angle formed when the lens is brought close to the specimen(s). Lower resolution lenses (*middle*) are generally weaker, with longer focal lengths and smaller aperture angles. An aperture placed into a high resolution lens (*bottom*) may diminish resolution, but contrast will be improved. Note how the aperture has decreased the aperture angle of the lens. α = aperture angle or half the angle of light leaving the lens, O = center of lens, fp = focal point where rays converge. The focal length of lens is expressed as the length of the line from point O to fp.

FIGURE 6.8 Refraction or bending of light to different degrees as it passes through media of different densities. On the left side of this combined drawing, one sees the situation when light travels through an object (O), glass slide/coverglass, and through air into an imaging lens (L). On the right side, oil has been added to fill the gap between the lens and the slide/coverglass. A lens brought close to a specimen on a glass slide will accept more information if oil (with a similar refractive index as the lens and slide) is used in the gap between the lens and slide (compare α′ angle without oil to α″ angle with oil). Since more information enters the oil-immersion lens, (at a wider α angle), greater resolution is possible. *(Courtesy of Carl Zeiss, Inc.)*

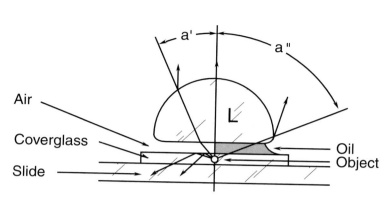

TABLE 6.2	Wavelengths of Visible Light
Color	**Wavelength in nm**
red	760–630
orange	630–590
yellow	590–560
green	560–490
blue	490–450
violet	450–380

distance, the better (or higher) is the resolving power of the system.

In order to obtain the best resolving power in an optical system, it is necessary to use the shortest possible wavelength and the largest aperture angle. For example, with a light microscope one would specify violet light (wavelength = 380 nm, from Table 6.2), immersion oil with a high refractive index, and a lens with a wide acceptance angle to obtain a resolving power of approximately 0.2 μm, as shown in Equation 6.2.

Equation 6.2: Calculation of Optimal Resolving Power of Light Microscope

$$r = \frac{0.612 \times 380 \text{ nm}}{1.5 \times 0.9} \qquad r = 172 \text{ nm}$$

Note: 0.9 is the value of the sine of a 64° angle, representing half the 128° acceptance angle of a lens (a typical value for a glass lens). Thus, the theoretical best resolving power of the light

microscope is 0.172 μm. The figure usually quoted is 0.2 μm.

In order to achieve the widest possible acceptance angle with any lens, the focal length of the lens must be as short as possible (Figure 6.7). This explains why the high resolution, oil-immersion objective lens on the light microscope is so close to the specimen (Figure 6.8).

Electrons, Waves, and Resolution

Physicists have demonstrated that, besides being discrete particles having a negative charge and a mass of 9.1×10^{-23} kg, *electrons also have wave properties*. In fact, the wavelength (λ) of an electron is expressed by the equation of the French physicist de Broglie as shown in Equation 6.3.

Equation 6.3: de Broglie Equation for Wavelength of an Electron

$$\lambda = h/mv \quad \text{where } h = \text{Planck's constant}$$
$$(6.626 \times 10^{-23} \text{ ergs/sec})$$
$$m = \text{mass of the electron}$$
$$v = \text{electron velocity}$$

After appropriate substitutions associating kinetic energy to mass, velocity, and accelerating voltage, the equation may be expressed:

$$\lambda = \frac{1.23}{\sqrt{V}} \text{ nm} \quad \text{where: } V = \text{accelerating voltage}$$

Therefore, if one were operating a transmission electron microscope at an accelerating voltage of 60 kV, the wavelength of the electron would be 0.005 nm, and the resolving power of the system—after substitution of these values into Equation 6.2—should be approximately 0.003 nm. In fact, the actual resolution of a modern high resolution transmission electron microscope is closer to 0.1 nm. The reason we are not able to achieve the nearly 100-fold better resolution of 0.003 nm is due to the extremely narrow aperture angles (about 1,000 times smaller than that of the light microscope) needed by the electron microscope lenses to overcome a major resolution-limiting phenomenon called spherical aberration. In addition, the *diffraction phenomenon* as well as chromatic aberration and astigmatism (to be discussed later) all degrade the resolution capabilities of the TEM. To appreciate these problems, it is necessary to understand how lenses function.

General Design of Lenses

An electromagnetic lens in the TEM may be thought of as a device that refracts or bends electromagnetic radiations to converge at a certain distance from the lens. Since it is impossible to represent waves adequately when drawing lens diagrams, one normally uses straight lines to represent rays of light (Figure 6.9A).

By convention, most drawings show only the outermost rays entering the lens. Be aware that there are numerous other rays between the peripheral ones that are not being represented in most drawings (Figure 6.9B shows only four rays).

It is important to note that both light and electron waves behave similarly as they pass through lenses and so both observe the same optical principles.

In studying Figure 6.10, it must be realized that the object being imaged is composed of many points, of which we are labeling only two, A and B or A' and B'. Furthermore, each point radiates information that passes through the entire curvature of the lens. The focal point (F) at which the rays that run parallel to the optical axis (line WXY in Figure 6.10) converge or

cross over defines the *focal length* of the lens (i.e., the distance between points X and F = focal length of lens). The type of lens shown is a double-convex or converging lens, the principle type of lens in all electron and most light microscopes. The point of entry of the electron into the lens field (central or peripheral) will determine the extent to which the ray will be refracted. Rays that pass far away from the exact center of the lens, or the optical axis, will be bent or refracted to a great degree, whereas those rays that pass along the optical axis will not be refracted at all.

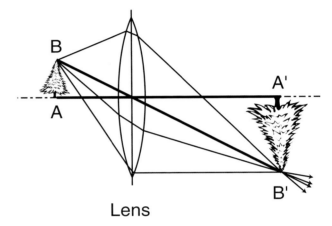

FIGURE 6.9B A single object point B is shown imaged at B' after passing through the lens. The reader should realize that each object point radiates many rays that are not shown in this drawing. Only four rays are shown. In practice only the outer rays are drawn in a lens diagram.

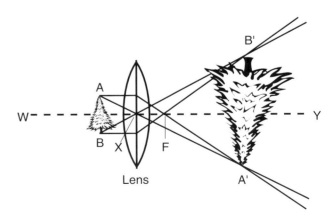

FIGURE 6.10 Image forming lens. An object, AB, is shown magnified to A'B'. Dashed line WXY marks the optical axis of the lens. The enlargement or magnification of the final image may be calculated by dividing the distance between points A' and B' by the distance between A and B. Focal length of this lens is the distance from point X to F (the focal point).

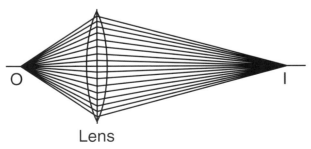

FIGURE 6.9A Diagram illustrating how multiple illuminating rays may be drawn passing through the lens. Object O radiates information that enters the lens and is imaged at I.

The manner that the lenses of an electron microscope refract electrons is discussed in the next section.

Design of Electromagnetic Lenses

Since electrons are particles with such small mass that they will be stopped even by gas molecules present in the air, glass lenses are of no value in an electron microscope. However, *since electrons have a charge, they can be affected by magnetic fields.* For example, an electron accelerated through a vacuum will follow a helical path when it passes through a magnetic field generated by a coil of wire with a direct current (DC) running through it (see Figure 6.11). Such simple electromagnetic coils are termed *solenoids.*

Suppose one illuminates a specimen (the arrow shown in Figure 6.12) with a beam of electrons such that some of the electrons that interact with a specimen point (A in Figure 6.12) are transmitted through the specimen and enter the electromagnetic lens. Depending on their precise trajectories as they enter the magnetic field, they will assume various helical paths as they speed through the lens. After leaving the lens, the electrons will focus at point A′ to generate an image point of the specimen. The distance from the center of the lens to where the electrons converge at A′ represents the focal length of the electromagnetic lens.

It is possible to change the focal length of an electromagnetic lens by changing the amount of DC current running through the coil of wire. This relationship is expressed in Equation 6.4:

Equation 6.4: Focal Length of Electromagnetic Lens

$$f = K\,(V/i^2)$$

where K = constant based on number of turns in lens coil wire and geometry of lens
V = accelerating voltage
i = milliamps of current put through coil

As the accelerating voltage of the electron is increased, the focal length is also increased since the electrons pass much more rapidly through the lens and assume looser helical routes. An increase in current put through the lens coil, however, results in a shorter focal length by forcing the electrons to assume tighter helical trajectories.

Being able to change the focal length of a lens is of practical importance, because this is how one can focus an image formed by a lens as well as change the magnification. In the light microscope, where the glass lenses are of a fixed focal length, focussing is done by physically moving the specimen into the proper plane of focus for each objective lens or vice versa. Similarly, magnifications are changed by removing an objective lens of one fixed focal length and replacing it with another. Obviously, the electromagnetic lenses of the electron microscope are

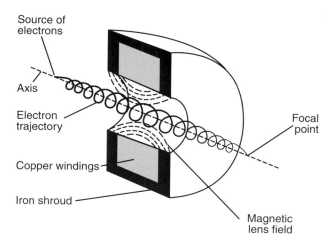

FIGURE 6.11 Single electron passing through electromagnetic lens. Instead of traveling in a straight line along the axis of the lens, the electron is forced by the magnetic field to follow a helical trajectory that will converge at a defined focal point after it emerges from the lens. Therefore, electromagnets, which are DC powered, behave similar to converging glass lenses.

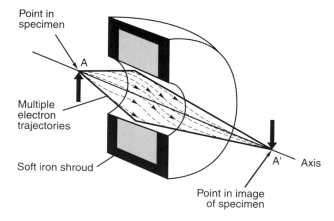

FIGURE 6.12 A group of electrons originating from point A in the specimen plane pass through an electromagnetic lens to all be focused at an appropriate point (A′) in the image plane. The specimen is represented as the heavy arrow in this drawing. The electromagnetic lens behaves as a thin biconvex glass lens as shown in Figure 6.9.

advantageous because they permit one to change focal lengths (e.g., change focus and magnification) by varying the current running through the lens coil without having to move the specimen or physically change lenses.

The efficiency of the electromagnetic lens can be greatly improved by concentrating the magnetic field strength close to the path of the electrons. This is accomplished by shrouding the coil on top, bottom, and side with a soft-iron casing so that the magnetism will run through the shroud (Figure 6.13A). The strength of the lens is thereby increased. (The term *soft iron* refers not only to the hardness of the metal, but also indicates that the iron is magnetized only when the electromagnetic field is conducted through it.)

The strength of the lens can be further increased by concentrating the magnetism to an even smaller area inside the lens bore by means of a liner termed a *polepiece* (so named because it sits in the north and south poles of the magnet). The cylindrical polepiece (Figure 6.13A and B) consists of upper and lower cores of soft iron held apart by a nonmagnetic brass spacer. The magnetic field is now concentrated between the top and bottom (north and south) iron components of the polepiece. These north and south cores of the polepiece are bored much smaller than the polepiece liner and must be as symmetrical as is mechanically possible in order to achieve high resolution. In practice, they are rarely perfect and may possess a number of defects that may degrade resolving power.

Defects in Lenses

A number of imperfections in lenses may reduce resolution. *Astigmatism* results when a lens field is not symmetrical in strength, but is stronger in one plane (north and south, for example) and weaker in another (east and west) so that only part of the image will be in focus at one time (Figure 6.14). A point would not be imaged as such, but would appear elliptical in shape; a cross would be imaged with either the vertical or horizontal arm, but not both, in focus at one time.

Some *causes of astigmatism* are an imperfectly ground polepiece bore, nonhomogeneous blending

FIGURE 6.13B (*top*) Photograph of a polepiece that fits into the Electromagnetic lens coil shown on the left. The soft-iron casing of the lens coil is removed to reveal the wire windings around the brass spool. (*bottom*) In the polepiece the north pole is arbitrarily on top, followed by a nonmagnetic brass spacer that holds north and south poles apart.

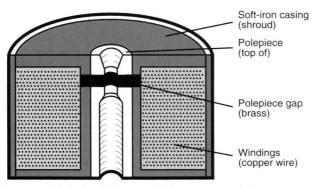

FIGURE 6.13A   Diagram of electromagnetic lens showing soft-iron casing (shroud) and soft-iron polepiece that slips down inside bore of lens.

of the polepiece metals, and dirt on parts of the column such as polepieces, apertures, and specimen holders. Because it is impossible to fabricate and maintain a lens with a perfectly symmetrical lens field, it is necessary to correct astigmatism by applying a correcting field of the appropriate strength in the proper direction to counteract the asymmetry. Such a device is called a *stigmator* and can be found in the condenser, objective, and intermediate lenses of the electron microscope (see Figure 6.35B).

Astigmatism in a glass lens could be corrected by regrinding the curvature of the lens so that the strength is symmetrical, or by imposing another lens field of the appropriate strength over one of the aberrant fields of the original lens—as is done with correcting eyeglasses.

Chromatic aberration results when electromagnetic radiations of different energies converge at different focal planes. With a glass lens, shorter wavelength radiations are slowed down and refracted more than are longer wavelengths of light. Effectively, the shorter, more energetic wavelengths of light come to a shorter focal point than do the longer wavelengths (Figure 6.15). In an electromagnetic lens, the reverse is true: *shorter wavelength, more energetic electrons have a longer focal point than do the longer wavelength electrons.* In both cases, however, chromatic aberration results in the enlargement of the focal point (similar to the Airy disc phenomenon caused by diffraction effects) with a consequential loss of resolution. Chromatic aberration can be corrected by using a monochromatic source of electromagnetic radiation. With glass lenses, one would use a monochromatic light (possibly by using a shorter wavelength blue filter). In an electromagnetic lens, one would insure that the electrons were of the same energy level by carefully stabilizing the accelerating voltage and having a good vacuum to minimize the energy loss of the electrons as they passed through the column. Thicker specimens give rise to a spectrum of electrons with varied energy levels and consequently worsen chromatic aberration (Figure 6.16A). Thin specimens are therefore essential for high resolution studies.

Chromatic change in magnification occurs when thick specimens are viewed at low magnifications using a low accelerating voltage. The *image appears to be sharp in the center, but becomes progressively out of focus as one moves toward the periphery* (Figure 6.16B). This is because the lower energy electrons are imaged at a different plane than the higher energy electrons. The effect is maximal at the periphery of the image, since these electrons are closer to the lens coils and, thus, are more affected by the magnetic field. This problem may be minimized by using thinner specimens, higher accelerating voltages, higher magnifications, and by correcting any other distortions that may be present in the lens.

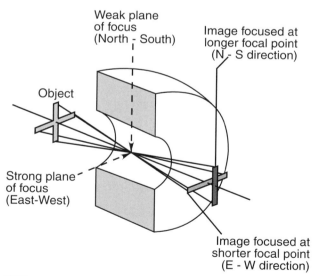

FIGURE 6.14 Astigmatism in a lens. Since the lens field is asymmetrically weaker in the north/south plane, objects oriented along the north/south axis will focus at a longer distance. By contrast, due to a stronger east/west lens field, objects oriented east/west will come to focus at a shorter distance from the lens. The effect is that only some portions of the image (either north/south or east/west) will be in focus at one time. Obviously, resolution will be degraded since the image will be focused in only one plane.

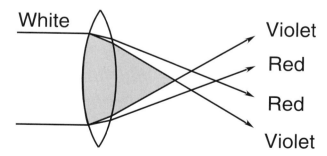

FIGURE 6.15 Chromatic aberration in a glass lens. Different wavelengths do not come to focus at the same point. Note how the violet part of the spectrum (*gray area*) focuses at a shorter distance than does the red part of the spectrum. This results in an enlarged, unsharp point rather than a smaller, focused one. Resolution of the point will be degraded.

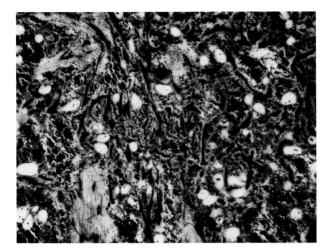

FIGURE 6.16A   Chromatic aberration in an overly thick section is evidenced by an image that is blurred overall due to degraded resolution.

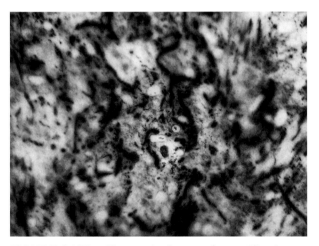

FIGURE 6.16B   Chromatic change of magnification occurs when an overly thick specimen is viewed at low magnifications with a low accelerating voltage. Only the central part of the image is sharp since the effect is maximal at the periphery.

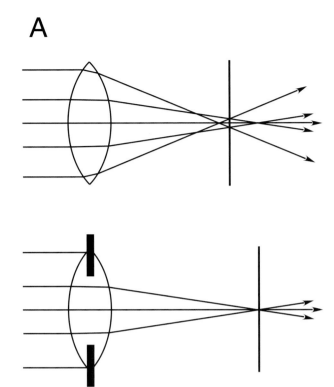

FIGURE 6.17   (A) Spherical aberration in a lens. Peripheral rays are refracted more than central rays, so that all rays do not converge to a common, small focal point. Instead, an enlarged, diffuse spot like the Airy disc will be generated. The vertical line indicates the one point where the point will be smallest (i.e., having the smallest circle of confusion).
(B) Correction of spherical aberration with an aperture (here shown inside the lens) to cut out peripheral rays and thereby permit remaining rays to focus at a common small imaging point. Resolution will be improved since individual image points in the specimen will be smaller.

Spherical aberration is due to the geometry of both glass and electromagnetic lenses such that rays passing through the periphery of the lens are refracted more than rays passing along the axis. Unfortunately, the various rays do not come to a common focal point, resulting in an enlarged, unsharp point (Figure 6.17A). At some distance, however, one should encounter the sharpest possible point that would constitute the *circle of minimum confusion* (i.e., the smallest Airy disc) and the practical focal point of the lens. *Spherical aberration may be reduced by using an aperture to eliminate some of the peripheral rays* (Figure 6.17B). Although apertures must be used in the electron microscope to reduce spherical aberration as much as possible, they decrease the aperture angle and thereby prevent the electron microscope from achieving the ultimate resolving power specified in the equation for resolution (Equation 6.1). In addition, the worsening of resolution as a result of using a longer focal length lens is shown in Equation 6.5.

Equation 6.5: Limit of Resolution, d_s, Imposed by Spherical Aberration

$$d_s = k_s \bullet f \bullet \alpha_o^3$$

where k = a constant related to lens characteristics
   f = focal length of lens
   α = aperture angle of lens, normally the objective lens

From Equation 6.5, we see why a short focal length lens combined with a smaller aperture (to generate a smaller aperture angle) will help to reduce the degradation of resolution caused by spherical aberration. Consequently, smaller apertures are generally more desirable than larger ones to improve resolution—in spite of the theoretical advantage offered by large apertures as indicated in Equation 6.1.

Certain types of image *distortions* may arise when spherical aberration occurs in the final imaging (or projector) lenses of the transmission electron microscope. Since peripheral electrons are refracted to a greater extent than central rays, the image formed by these peripheral electrons will be at a greater magnification and in a different focal plane than the image generated from more centrally positioned electrons. A grid of lines would not be imaged as square (Figure 6.18A), but would assume the shape of a sunken pillow, hence the name *pincushion distortion* (Figure 6.18B). This type of distortion may occur when attempting to operate the transmission electron microscope at excessively low magnifications. Another type of imperfection, *barrel distortion*, occurs when one attempts to use an elec-

tromagnetic lens in a demagnifying mode rather than the normal magnifying mode. In this case, the central part of the image is magnified more than the periphery so that the gridwork assumes a swollen or barrel shape (Figure 6.18C).

Fortunately, it is possible to neutralize one type of distortion with the other. For instance, if one projector lens is displaying excessive pincushion distortion, it is possible to operate another projector lens in the demagnifying mode to introduce an opposing barrel distortion. The lens systems of modern electron microscopes are designed to automatically counterbalance the various types of distortions throughout a wide magnification range. In older microscopes, however, one must take care not to introduce these distortions in the lower magnification range.

As one begins to use the electron microscope, a curious phenomenon called *image rotation* will be noticed as one changes magnification in older microscopes. This occurs because the electrons follow a spiral path through the lenses, and the spiral shifts as the strength of the lens is varied. Image rotation not only results in rotation of the image on the viewing screen as one increases magnification, but also exaggerates the effects of distortion. It is possible to minimize or eliminate image rotation entirely by ensuring that a series of lenses have opposing rotations rather than all having rotations in the same direction. This is accomplished by running the lens current through the coil in the opposite direction (i.e., reversing polarity) and is a principle utilized in some of the newer transmission electron microscopes.

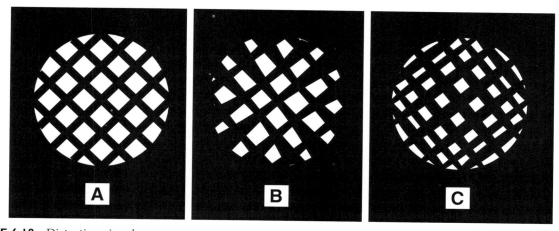

FIGURE 6.18 Distortions in a lens.
(A) Normal image of grid pattern. (B) Image with pincushion distortion. (C) Image with barrel distortion.

Magnification

Besides forming images with high resolution, the lenses of the electron microscope are able to further magnify these images. Magnification refers to the degree of enlargement of the diameter of a final image compared to the original (Figure 6.1). In practice, magnification equals a distance measured between two points on an image divided by the distance measured between these same two points on the original object, or

$$\text{Mag} = \frac{\text{image distance}}{\text{object distance}}$$

Consequently, if the image distance between two points measures 25.5 mm while the distance between these same two points on the object measures 5 mm, then the magnification is

$$\text{Mag} = \frac{25.5 \text{ mm}}{5.0 \text{ mm}} = 5.1$$

As will be discussed later, there are at least three magnifying lenses in an electron microscope: the *objective, intermediate,* and *projector lenses.* The final magnification is calculated as the product of the individual magnifying powers of all of the lenses in the system as shown in Equation 6.6.

Equation 6.6: Calculation of Total Magnification, M_T, of the TEM

$$M_T = M_O \times M_I \times M_P$$

where: M_T = total magnification or mag
M_O = mag of objective lens
M_I = mag of intermediate lens
MP = mag of projector lens(es)

For example, if the transmission electron microscope is operating in the high magnification mode, typical values for the respective lenses might be: 200 × 50 × 20 = 200,000×. If one were to operate the microscope in the low magnification mode, perhaps the values would be: 50 × 0.5 × 50 = 1,250×.

Intermediate magnifications may be produced by varying the current to the various lenses. Sometimes, it is desirable to view as much of the specimen as possible in order to evaluate quickly the quality of the preparation or to locate a particular portion of the specimen. In this case, an extremely low magnification is obtained by placing the microscope in the *scan magnification* mode, which can be accomplished by shutting off the objective

lens and using the next lens (the intermediate lens) as the imaging lens as follows: 1 × 0.5 × 100 = 50×. Of all the lenses used to change magnification, the objective lens is used the least. Normally it is maintained at around 50 or 100× while the other lenses are varied.

Although it is theoretically possible to increase the magnification indefinitely, the quality of the image magnified is dependent on the resolving power of the lenses in the system. Consequently, the term *useful magnification* is used to define the maximum magnification that should be used for a particular optical system. It is defined by the formula in Equation 6.7.

Equation 6.7: Useful Magnification

$$\text{Useful Magnification} = \frac{\text{resolution of the human eye}}{\text{resolution of the lens system}}$$

In the case of the light microscope, a typical value would be 1,000× because the resolving power of the human eye is about 0.2 mm, while the resolving power of the light microscope is approximately 0.2 μm. An electron microscope with a resolving power of 0.2 nm could be expected to have a top magnification of approximately 1,000,000×, or a thousand times greater than the light microscope. In practice, due to the diminished illumination at such high magnifications, microscopists would probably take the micrograph at a magnification of 250,000× and photographically enlarge the negative to the needed magnification. However, only rarely do biologists need such high magnifications.

Depending on the model of TEM, the magnification changes may occur as a series of discrete steps or as an infinitely variable or "zoom" magnification series.

Modern electron microscopes have digital displays that give the approximate total magnification when one varies the magnification control. Older microscopes usually have an analog gauge that may either read the current to one lens, or they may have a magnification knob with a series of click stops that may be correlated to a particular magnification. All microscopes (including light microscopes) must be calibrated in order to determine more accurately the total magnification, since a number of variables may cause the magnification to vary by as much as 20 to 30% over a short period of time. Even modern instruments with direct reading digital displays are guaranteed to be accurate to only ± 5 to 10% of the stated values. The procedure for magnification calibration is discussed later in this chapter.

Design of the Transmission Electron Microscope

Comparison of Light Microscope to Transmission Electron Microscope

The transmission electron microscope is similar in many ways to the compound light microscope. For instance, in both microscopes, electromagnetic radiations originating from a tungsten filament are converged onto a thin specimen by means of a condenser lens system. The illumination transmitted through the specimen is focused into an image and magnified first by an objective lens and then further magnified by a series of intermediate and projector lenses until the final image is viewed (Figure 6.19). Both kinds of microscope may record images using a silver-based photographic emulsion since it is sensitive to both types of radiations.

Of course, the lenses of light microscopes are composed of glass or quartz rather than the electromagnetic solenoids used in electron microscopes. Electron microscopes require an elaborate vacuum system to remove interfering air molecules that would impede the flow of electrons down the column. Such high vacuums are necessary from the point of origin of the electrons (the filament) up to, and usually including, the photographic film. Since the specimen is also subjected to these high vacuums, living specimens would be rapidly dehydrated if placed directly into the electron microscope. Nonetheless, it is possible to view rapidly frozen, hydrated, thin specimens by using cryostages that maintain the frozen state of cellular water even under bombardment by the electron beam.

Basic Systems Making Up a Transmission Electron Microscope

The transmission electron microscope (Figures 6.20A, B and C) is made up of a number of different systems that are integrated to form one functional unit capable of orienting and imaging extremely thin specimens. The *illuminating system* consists of the electron gun and condenser lenses that give rise to and control the amount of radiation striking the specimen. A *specimen manipulation system* composed of the specimen stage, specimen holders, and related hardware is necessary for orienting the thin specimen outside and inside the microscope. The *imaging system* includes the objective, intermediate, and projector lenses that are involved in forming,

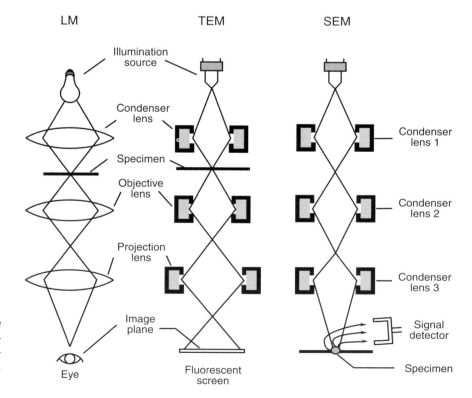

FIGURE 6.19 Comparison of light microscope (LM) to transmission (TEM) and scanning (SEM) electron microscopes. *(Redrawn with modifications from Postek, et al., 1980. Scanning Electron Microscopy: A Student's Handbook. Ladd Research Industries, Inc. Burlington, VT.)*

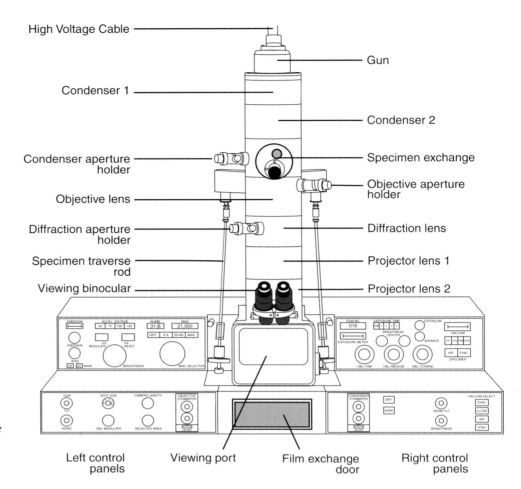

FIGURE 6.20A

Diagram of a modern transmission electron microscope with major components labeled. The function of each of the components is listed in Tables 6.3 and 6.4.

High Voltage Cable

Condenser 1

Condenser aperture holder

Objective lens

Diffraction aperture holder

Specimen traverse rod

Viewing binocular

Gun

Condenser 2

Specimen exchange

Objective aperture holder

Diffraction lens

Projector lens 1

Projector lens 2

Left control panels

Viewing port

Film exchange door

Right control panels

focusing, and magnifying the image on the viewing screen as well as the *camera* that is used to record the image. A *vacuum system* is necessary to remove interfering air molecules from the column of the electron microscope. In the descriptions that follow, the systems will be considered from the top of the microscope to the bottom. See Tables 6.3 and 6.4.

Illuminating System

This system is situated at the top of the microscope column and consists of the electron gun (composed of the filament, shield, and anode) and the condenser lenses.

Electron Gun. Within the electron gun (Figure 6.21), the *filament* serves as the source of electrons. The standard filament, or *cathode* (Figure 6.22), is composed of a V-shaped tungsten wire approximately 0.1 mm in diameter (about the thickness of a human hair). Being a metal, tungsten contains positive ions and free electrons that are strongly attracted to the positive ions. Fortunately, it is possible to

entice the outermost orbital, or valence, electrons out of the tungsten by first applying a high voltage to the filament and then heating the metal by running a small amount of DC electrical current through the filament while operating within a vacuum. In other words, a certain amount of energy must be put into the system to cause the electrons to leave the filament. The amount of energy necessary to bring about electron emission is termed the *work function* of the metal. Although tungsten has a relatively high work function, it has an excellent yield of electrons just below its rather high melting point of 3,653° K.

In practical terms, one first applies a fixed amount of negative high voltage (typically 50, 75 or 100 kV) and then slowly increases the amount of direct current running through the filament to heat it to achieve the emission of electrons (*thermionic emission*). As one applies more heat to the filament, the yield of electrons increases until the filament begins to melt and evaporate in the high vacuum of the microscope (Figure 6.23). At some optimal

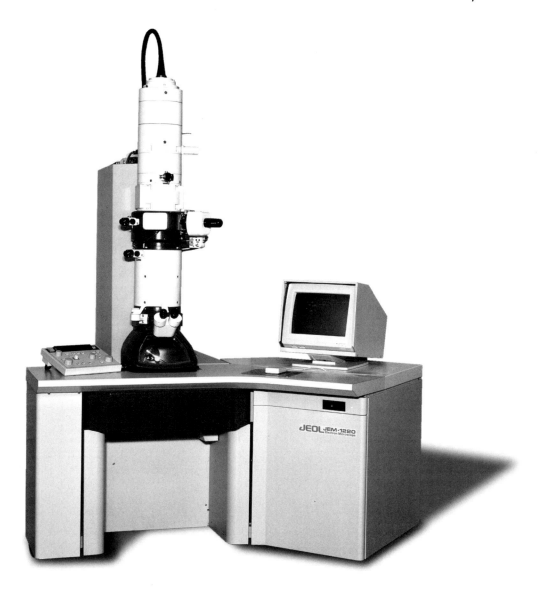

FIGURE 6.20B
Photograph of a
transmission electron
microscope.
(Courtesy of JEOL.)

temperature, the gun achieves good electron emission as well as an acceptable filament life: this is termed the *saturation point* (discussed later). By examining Figure 6.23, it becomes apparent that a good electron yield is achieved at 2,600° K where the filament life is around 100 hours. Oversaturating the gun by heating the filament even 200° K beyond 2,600° K results in a dramatic decrease in filament life. The average life of a filament ranges from 25 hours in older electron microscopes to over 200 hours in microscopes with good vacuum systems and scrupulously maintained gun areas. The *major causes of premature filament failure* are: oversaturation, high voltage discharge caused by dirt in the gun region, poor vacuum, and air leaks or outgassing from contaminants in the gun cham-

ber. The two controls on the panel of the microscope that are used to initiate the flow of electrons are usually labeled *accelerating voltage* (or possibly kV, HV or HT) and *emission* or *saturation* (or sometimes *filament*).

If one induces the emission of electrons from a filament as described above, this will result in the emanation of electrons in all directions. Without a mechanism for guiding them, most of the electrons would not enter the illuminating system. A second part of the electron gun, the *shield* (also called Wehnelt cylinder, bias shield, or grid cap), is involved in assuring that the majority of the electrons go in the proper direction. The shield is a caplike structure that covers the filament and is maintained at a slightly more negative voltage potential than the

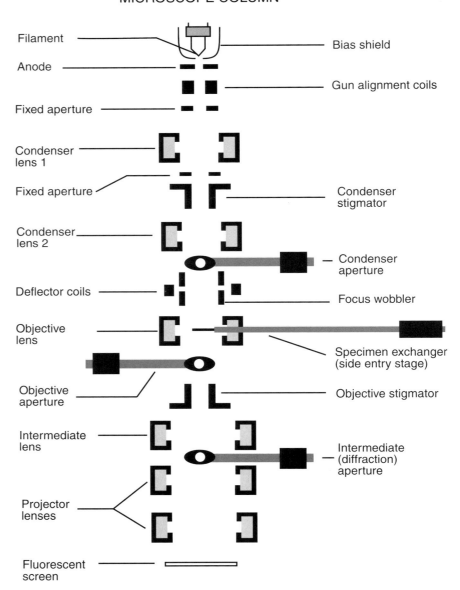

COMPONENTS OF TYPICAL TRANSMISSION ELECTRON MICROSCOPE COLUMN

Filament — Bias shield

Anode — Gun alignment coils

Fixed aperture —

Condenser lens 1 — Condenser stigmator

Fixed aperture —

Condenser lens 2 — Condenser aperture

Deflector coils — Focus wobbler

Objective lens — Specimen exchanger (side entry stage)

Objective aperture — Objective stigmator

Intermediate lens — Intermediate (diffraction) aperture

Projector lenses

Fluorescent screen

FIGURE 6.20C Diagram of major components making up the column of a modern transmission electron microscope.

filament. Because it is several hundred volts more negative than the 50 to 100 kV electrons, the shield surrounds the electrons with a repulsive field that is breachable only through a 2 to 3 mm aperture directly in front of the filament tip. Electrons exit the shield aperture and are drawn toward an apertured disc, or anode, the third part of the electron gun (Figure 6.21). The anode is connected to ground so that the highly negative electrons are strongly attracted to it. Thus it is positive with respect to the gun. In fact, the highly attractive pull of the anode in combination with the

negative surface of the shield act as an electrostatic "lens" to generate a crossover image of the electron source near the anode (Figure 6.24).

> Just as electromagnetic lenses have two forces or poles, *electrostatic* lenses have positively and negatively charged surfaces to attract or repel and, thereby, focus electrons. The term *crossover* refers to the point where the electrons focus or converge and cross over each other's paths.

TABLE 6.3	Major Column Components of the TEM*	
Component	**Synonyms**	**Function of Components**
Illumination System		
Electron Gun	Gun, Source	Generates electrons and provides first coherent crossover of electron beam
Condenser Lens 1	C1, Spot Size	Determines smallest illumination spot size on specimen (see Spot Size in Table 6.4)
Condenser Lens 2	C2, Brightness	Varies amount of illumination on specimen—in combination with C1 (see Brightness in Table 6.4)
Condenser Aperture	C2 Aperture	Reduces spherical aberration, helps control amount of illumination striking specimen
Specimen Manipulation System		
Specimen Exchanger	Specimen Air Lock	Chamber and mechanism for inserting specimen holder
Specimen Stage	Stage	Mechanism for moving specimen inside column of microscope
Imaging System		
Objective Lens	—	Forms, magnifies, and focuses first image (see Focus in Table 6.4)
Objective Aperture	—	Controls contrast and spherical aberration
Intermediate Lens	Diffraction Lens	Normally used to help magnify image from objective lens and to focus diffraction pattern
Intermediate Aperture	Diffraction Aperture, Field Limiting Aperture	Selects area to be diffracted
Projector Lens 1	P1	Helps magnify image, possibly used in some diffraction work
Projector Lens 2	P2	Same as P1
Observation and Camera Systems		
Viewing Chamber	—	Contains viewing screen for final image
Binocular Microscope	Focusing Scope	Magnifies image on viewing screen for accurate focusing
Camera	—	Contains film for recording image

*Not included are the mechanical adjustment screws for centering and tilting various lenses or apertures.

Variable Self-Biased Gun. It was stated earlier that the high voltage shield is slightly more negative than the filament. This difference in negative potential, or *bias,* is established by connecting the shield directly to the negative high voltage line while placing a variable resistor in the high voltage line to the filament (Figure 6.21A). By varying the value of this resistor, the filament may be made less negative than the shield (usually by 100 to 200 volts). The greater the value of the variable bias resistor, the less negative the filament will become. As the filament becomes less negative, fewer electrons will be able to pass through the shield aperture since they are now repulsed to a greater degree by the shield. The overall effect of the *variable bias,* therefore, is to regulate the escape of electrons through the shield aperture.

In addition to high voltage, one applies a certain amount of direct current to the filament in order to heat up the filament and enhance electron emission. As this current passes through the variable bias resistor, a certain amount of voltage is generated and applied to the shield in order to make it

TABLE 6.4 Major Components on Control Panels of the TEM*

Component	Synonyms	Functions of Component
Filament	Emission	Effects emission of electrons upon heating
Bias	—	Adjusts voltage differential between filament and shield to regulate yield of electrons
High Voltage Reset	HV, kV Reset	Activates high voltage to gun
High Voltage Select	HV, kV Select	Selects amount of high voltage applied to gun
Magnification Control	MAG	Controls final magnification of image by activating combinations of imaging lenses
Brightness	C2	Controls current to second condenser lens
Gun Tilt	—	Electronically tilts electron beam beneath gun
Gun Horizontal	—	Electronically translates electron beam beneath gun
Spot Size	C1	Controls final illumination spot size on specimen
Objective Stigmator	OBJ STIG	Corrects astigmatism in objective lens
Focus Wobbler	Focus Aid	Helps focus accurately at low magnifications
Exposure Meter	—	Monitors illumination for accurate exposures
Vacuum Meter	VAC	Monitors vacuum levels in various parts of scope
Focusing Control	Focus—fine, medium, coarse	Controls current to objective lens for accurate focusing of image
Brightness Center	Illumination Centration	Translates entire illumination system onto screen center
Condenser Stigmator	COND STIG	Corrects astigmatism in condenser lenses
Intermediate Stigmator	INT STIG	Corrects astigmatism in intermediate lens
Bright/Dark	—	Selects brightfield or darkfield operating mode
Main	—	Main power switch to console
Main Evac	EVAC	Main switch to vacuum system
HV Wobbler	HV Modulate	Wobbles high voltage to locate voltage center for alignment
Objective Wobbler	OBJ MODUL	Wobbles current to objective lens for alignment

*Not included are less frequently used controls as lens current switches or meters, current normalizers, selector switches for various magnification modes, diffraction controls, darkfield centration and stigmation controls, film counters and switches, vacuum status indicators, camera calibration switches, current and voltage stabilizer switches.

more negative. Therefore, as one continues to increase the heating current to the filament, the numbers of electrons coming off the filament will increase. But since the shield is becoming progressively negative, the total number of electrons actually passing through the shield aperture does not increase significantly. The so-called *saturation point* of the gun is the point where the number of electrons emitted from the gun no longer increases as the filament is heated. The gun is, therefore, said to be self-biasing, since it throttles back on electron emission as the heat is increased. *It is important that the operator realize that increasing the heat of the filament beyond the saturation point will not increase the*

brightness of the gun but will considerably shorten the filament life. On the other hand, *undersaturation* of the filament may lead to instabilities in the illumination of the specimen and cause problems if analytical procedures (such as X-ray analysis) are to be attempted. The arrangement for controlling electron emission in modern electron microscopes is termed the *variable self-biased gun.*

Controlling the Amount of Illumination Striking the Specimen. It is possible to make practical use of the variable bias to regulate the amount of illumination that strikes the specimen. For example, when operating at high magnifications with small

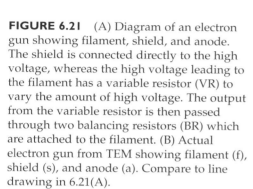
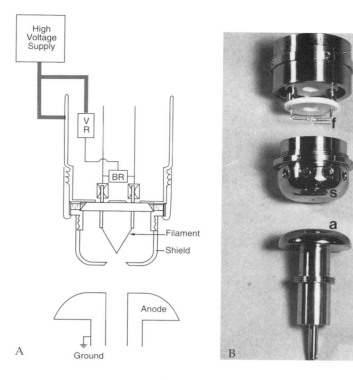

FIGURE 6.21 (A) Diagram of an electron gun showing filament, shield, and anode. The shield is connected directly to the high voltage, whereas the high voltage leading to the filament has a variable resistor (VR) to vary the amount of high voltage. The output from the variable resistor is then passed through two balancing resistors (BR) which are attached to the filament. (B) Actual electron gun from TEM showing filament (f), shield (s), and anode (a). Compare to line drawing in 6.21(A). (*A, modified from a drawing provided by Hitachi Scientific Instruments.*)

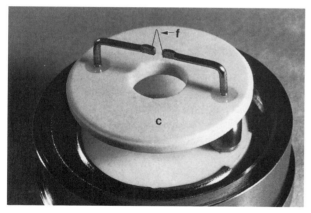

FIGURE 6.22 Standard V-shaped tungsten filament (f) used in most electron microscopes. The filament is spot-welded to the larger supporting arms, which pass through the ceramic (c) insulator and plug into the electrical leads of the gun.

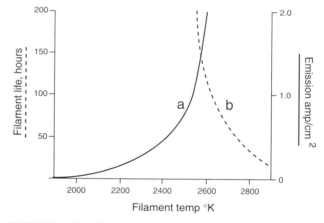

FIGURE 6.23 The importance of proper saturation of the filament is shown in these two curves. (a) Solid line shows the relationship between electron emission from a filament as a function of the temperature of the filament. The optimal temperature is 2,600° K. (b) When one exceeds 2,600° K filament temperature, the filament life drops dramatically (*dashed line*).

condenser spot sizes, it may be necessary to alter the bias to effect greater gun emissions. Of course, the filament life will be shortened, but this may be necessary in order to critically view and focus the specimen. It is also important to remember that the greater the beam current, the greater the specimen damage.

Moving the filament closer to the shield aperture will permit more electrons to pass through to the con-

denser lenses. However, if the filament is placed too close to the aperture, the bias control by the shield will be lost, and the emission will become excessive. Filaments placed too far away from the shield aperture, on the other hand, may never yield sufficient numbers of electrons from the gun. Therefore, careful

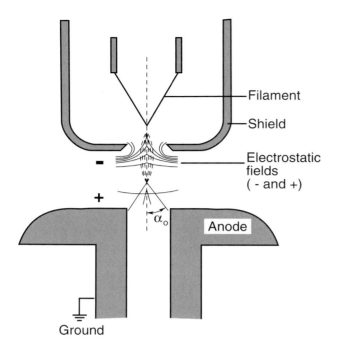

FIGURE 6.24 The self-biased electron gun. The shield (Wehnelt cylinder) is slightly more negative than the filament to control the release of electrons from the gun. A variable bias resistor (see Figure 6.21A) regulates the degree of negativity of the filament. The anode serves as a positive attracting force and serves as an electrostatic lens (in combination with the shield) to help focus the electrons into a crossover spot approximately 50 μm across.

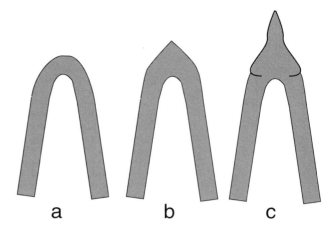

FIGURE 6.25 Drawing of three different filament tips.
(a) Standard V-shaped filament tip.
(b) Standard filament tip that was flattened and then sharpened to a fine point.
(c) Filament tip where a crystal of tungsten was spot-welded onto the curved end.

placement of the filament relative to the shield aperture is very important and should be in accordance with the manufacturer's specifications.

The distance of the anode from the filament and shield is also important. As one moves the anode closer to the filament, more electrons will be extracted from the gun. This becomes a consideration when using lower accelerating voltages where it may be necessary to move the anode closer to assist in the extraction of the lower energy electrons (50 kV, for example). Some electron microscopes have an external adjustment screw that will mechanically adjust the height of the anode, while other models have a pneumatically actuated "anode lifter" that changes in response to the kilovolt selected by the operator. Older microscopes may have no such adjustment.

The choice of kV should be considered carefully. Lower kVs such as 50 kV will generate an image with higher contrast but lower resolution, while higher kVs (100–125 kV) improve resolution but lower overall contrast. Less specimen damage will result at the higher kVs since the speedier electrons interact for a shorter period of time with the specimen.

Other Gun Designs. The majority of transmission electron microscopes use the gun design described in the previous section. However, some notable variations are also available, depending on the needs of the researcher. The filament shape may be altered as illustrated in Figure 6.25b, where the tip was first flattened and then sharpened to a point. It is also possible to purchase a pointed filament made by welding a single crystal of tungsten onto the curved tip of a standard filament (Figure 6.25c). Both types of *pointed filaments* have a considerably shorter lifetime than do standard filaments. However, since the initial gun crossover image is much smaller and the beam is highly coherent, they are necessary for high resolution studies where beam damage may be a consideration (e.g., viewing crystalline lattice planes).

Besides being made of tungsten, filaments may also be constructed of *lanthanum hexaboride*, which has a lower work function. Typically, these filaments operate at temperatures 1,000° K lower than tungsten and have a brightness several times greater than a standard tungsten source. The lifetime of such filaments ranges from 700 to 2,000 hours. This type of filament may be made from a single LaB_6 crystal with one end having a point measuring only several micrometers across (Figure 6.26). LaB_6 filaments are coming into use slowly, since they are considerably more expensive than tungsten filaments and are extremely chemically reactive when hot. For the lat-

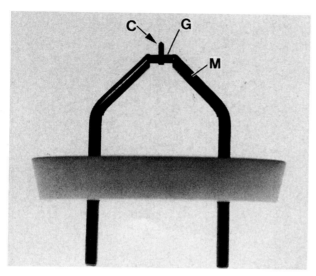

FIGURE 6.26 Lanthanum hexaboride cathode. The crystal (C) is held in place by means of pyrolytic graphite (G) blocks with compressive force generated by molybdenum (M) alloy posts designed to withstand extremely high temperatures.
(Cathode provided by FEI Company.)

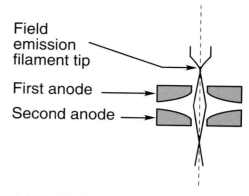

FIGURE 6.27 The field emission gun. Electrons are extracted from a single crystal of tungsten by a series of anodes that are made several thousand volts positive. It is not necessary to heat this type of filament.

ter reason, vacuums greater than 10^{-5} Pa are essential (see section "Vacuum System"), and special filament mounts must be constructed from such nonreactive elements as rhenium or vitreous carbon. LaB_6 filaments are useful when small beam crossover sizes containing large numbers of electrons are necessary—as in high magnification/resolution studies, for elemental analysis, or in high resolution scanning electron microscopy.

A totally different gun, nearly a thousand times brighter than the standard gun, may also be used under certain conditions. In the *cold field emission gun,* the filament is a single crystal of tungsten with its atomic crystalline lattice precisely oriented to maximize electron emission. Electrons are not generated by thermionic emission (heating), but are actually drawn out of the tungsten crystal by a series of positive high voltage anodes that act as electrostatic lenses to focus the gun crossover to a spot size of 10 nm (Figure 6.27). A major disadvantage of the cold field emission gun is the ultrahigh vacuum required (greater than 10^{-8} Pa) and the extreme susceptibility of the filament to contaminants. Cold field emission guns are very useful in high resolution scanning and scanning transmission electron microscopes and are now being incorporated into high resolution transmission electron microscopes. See Table 6.5 for a comparison of the three major filaments.

Condenser Lenses. This second major part of the illuminating system gathers the electrons of the first crossover image from the gun and *focuses electrons onto the specimen.* Modern transmission electron microscopes have two condenser lenses, unlike the first microscopes that had only one. The first condenser lens (designated C1) is a demagnifying lens that decreases the size of the 50 μm gun crossover to generate a range of spot sizes from 20 μm down to 1 μm. The second condenser lens (C2), on the other hand, enlarges the C1 spot. The overall effect of both lenses is to control precisely the amount of electron irradiation or illumination striking the specimen. *The operating principle for using C1 and C2 is to generate a spot on the specimen of the proper size to illuminate only the area being examined.* Therefore, at higher magnifications smaller spot sizes should be focused on the specimen (Figure 6.28B), while larger spots may be used at lower magnifications (Figure 6.28A). Because spot sizes are controlled, beam damage can be minimized to parts of the specimen not being viewed. This offered a great advantage when TEMs were made with two condenser lenses rather than one.

Suppose one is working at a magnification of 50,000×. At this high magnification, the C1 lens should be highly energized to demagnify the 50 μm illumination spot from the gun down to 1 to 2 μm. Next, the C2 lens should be used to adjust the size of the C1 illumination spot to cover only the specimen area being viewed. Since the average viewing screen is about 100 mm across, a 2 μm spot of illumination enlarged 50,000× would just cover the screen (2 μm × 50,000 = 100 mm). Therefore, the C2 lens should also be highly energized to generate a 2 μm spot on the specimen. At a magnification of

TABLE 6.5   Comparison of the Three Major Filaments in Terms of Brightness, Size of the Source Crossover, Energy Spread, Service Life, and Vacuum Required

	Cold Field Emission	Lanthanum Hexaboride	Tungsten Filament
Brightness (A/cm^2 • sr)	10^9	10^7	10^6
Source Diameter (nm)	<10	10^4	>10^4
Energy Spread (eV)	0.2–0.3	1.0–2.0	1.0–2.0
Service Life (hours)	>2,000	1,000–2,000	40–100
Vacuum Required (Pa)	10^{-8}	10^{-5}	10^{-3}

(Courtesy D. Rathkey, FEI Company.)

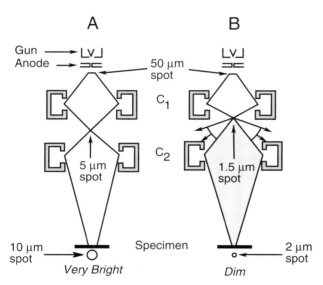

FIGURE 6.28   The condenser lens system. (A) In this mode, the 50 μm gun crossover is reduced to 5 μm by condenser lens 1, C1, and then slightly enlarged by condenser lens 2, C2, to yield a 10 μm spot on the specimen that is five times brighter than the initial gun crossover. (B) At higher magnifications, the 50 μm gun crossover is reduced to 1.5 μm by a highly energized C1. This refracts the peripheral electrons to such a great angle that they cannot enter C2 and are therefore lost. After C2 slightly enlarges the C1 spot, the resulting 2 μm spot is rather dim.

mize specimen damage but with enough illumination to focus.

If one studies Figure 6.28, it is apparent why smaller spot sizes are necessarily dimmer. If C1 is highly energized in order to generate a small spot (Figure 6.28B), the focal length is made so short and the aperture angle so great that many electrons are refracted to such an extent that they do not enter C2. On the other hand, if C1 is weakened to generate a larger spot, the focal length is longer and the aperture angle is smaller so that effectively all electrons may now enter C2 (Figure 6.28A). Therefore, as the C1 spot is made progressively smaller, overall illumination tends to diminish. This poses a problem at higher magnifications where very small spot sizes are needed. Illumination may become so dim that microscopists must allow 30 to 60 minutes for their eyes to adapt to working under such dark conditions. However, it may be possible to increase the illumination on the specimen using the techniques described previously in the chapter section entitled "Controlling the Amount of Illumination Striking the Specimen," or by using high sensitivity electronic cameras to view dimly illuminated specimens.

Apertures in Condenser Lenses. Depending on the design of the transmission electron microscope, one or both condenser lenses may have apertures of variable sizes. Generally, the C1 aperture is an internal aperture of a fixed size, while the C2 aperture is variable by inserting into the electron beam pathway apertures of different sizes attached to the end of a shaft. A popular method is to use a molybdenum foil strip containing 3 or 4 holes of 500, 300, 200, and 100 μm in diameter

10,000×, it is possible to keep C1 highly energized but to use C2 to magnify or spread out the 2 μm spot an additional 5× to just cover the 100 mm screen. However, the illumination will be about 5 times dimmer. It is important, however, to keep the C2 lens spread out away from crossover to mini-

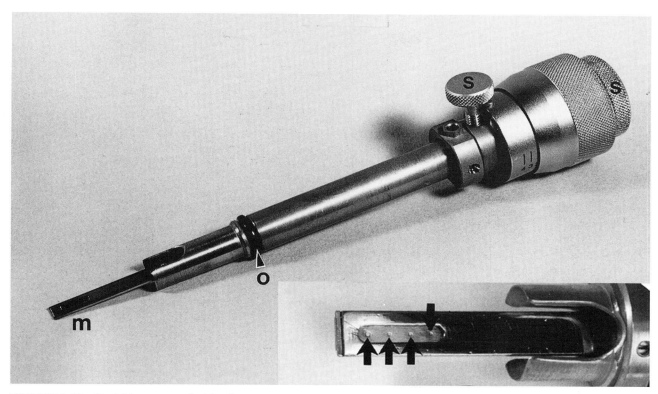

FIGURE 6.29 Variable aperture holder from a TEM. The rod contains a molybdenum strip (m) with apertures of various sizes. Positioning screws (s) permit the precise alignment of the apertures in the electron beam. An O-ring seal (o) permits the aperture to be sealed off inside the vacuum of the microscope column. Insert shows enlargement of the molybdenum aperture strip held in place by a brass retainer clip. Arrows point to apertures in the strip.

(Figure 6.29). The various sizes may be inserted and centered using external adjustment screws during the alignment procedure (described later in this chapter). Larger condenser apertures permit most of the electrons to pass through the lens and, therefore, yield a brighter spot on the specimen. Smaller apertures cut out more peripheral electrons and, hence, reduce the illumination on the specimen. However, since spherical aberration is concomitantly reduced, greater resolution is possible using smaller condenser apertures. The operational principle to remember is *larger condenser apertures give more illumination but with more spherical aberration.*

Specimen Manipulation System

Most biological specimens are mounted on a copper meshwork or *grid.* Grids are placed into a specimen holder and, after insertion into an air lock, the chamber is evacuated and the specimen holder is inserted into the stage of the microscope. (In very old microscopes, no air locks were provided, so it

was necessary to admit air to the entire column in order to insert a specimen. Such changes would take 5 to 10 minutes versus 30 or so seconds in modern air-locked microscopes.)

The *specimen stage* is a micromanipulator for moving the specimen in x and y directions in increments as small as 10 nm, the width of a cell membrane. Depending on the design of the specimen holder and stage, it may also be possible to tilt and rotate the specimen inside the column of the electron microscope. Some of the newer microprocessor-controlled TEMs have automated stage controls that permit motorized and precise movement of the specimen. It is possible to program the stage controller to scan specimens and even to systematically record images as, for example, when one wishes to make a montage of the numerous individual electron micrographs (see Chapter 8 for more information on montages). An important feature of such computer-controlled stages is the ability to memorize specified coordinates and to be able to return to these locations on command.

Top-Entry Stage. One type of *specimen holder* is a brass cartridge with a long cylinder that enters the objective lens as the holder is placed in the stage on top of the objective lens (Figure 6.30). The grid sits on the end of the cylinder and is held firmly in place by a tight-fitting sleeve that slips over the cylinder.

> Firm contact between grid and specimen holder is essential in order to dissipate the buildup of heat and static charges resulting from bombardment by the electron beam. To optimize this dissipation, insert the copper mesh facing the electron beam (i.e., specimen down). However, be careful that the specimen and supporting membrane are well adhered to the copper grid, or they may detach and fall onto the microscope stage or objective lens, necessitating time-consuming disassembly of the microscope.

After the grid has been inserted in the specimen holder, the cartridge is placed in an air lock where the air is removed by the vacuum system. The air lock is opened to the high vacuum of the microscope and the specimen holder transferred onto the movable stage of the microscope. It is again important that the contact between the specimen holder and brass stage be firm in order to dissipate any static charges. For instance, a dirty specimen stage or holder will prevent such contact and may lead to thermal or electrostatic drift in the specimen. *Dirty specimen holders are a major cause of specimen drift in the TEM. Holders must be examined and cleaned on a regular basis.*

> Some top-entry holders are designed to tilt mechanically several degrees, and some stages have gearing that will enable the specimen to be rotated 360 degrees. It is possible to vary the length of the cartridge nosepiece and construct holders suited to special purposes. For instance, short nosepiece holders are useful for high contrast and low magnification applications (longer focal length objective lens), while longer nosepieces are suitable for high resolution studies (short focal length objective lens). Although it is possible to insert only one cartridge holder into the stage, some microscopes have air locks that accommodate several cartridges so that it is not necessary to break vacuum in order to insert another specimen (Figure 6.31).

Side-Entry Stage. In this type of stage, the specimen grid is introduced into the microscope stage by entering through the side of the objective lens polepiece. The specimen holder resembles an aperture holder consisting of a rod with a flat plate on one end that has one or more recessed areas for holding grids (Figure 6.32).

FIGURE 6.30 (*left*) Short, top-entry grid holder for high contrast, low-magnification work. Resolution is not as good with this type of grid holder since the specimen is placed higher in the objective lens, necessitating a longer focal length of the lens. (*right*) Standard top-entry specimen grid holder for high resolution work. The specimen grid is placed on the end of the grid holder shaft and held in place with a sleeve that is slipped over the shaft (*arrow*). These holders are placed in the specimen stage with the grids in the downward position in the polepiece.

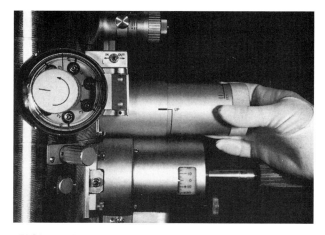

FIGURE 6.31 Multiple grid holder specimen air lock. In this design, it is possible to load six grid holders into the air lock, evacuate the sealed chamber, select the holder, and insert it into the column using the exchanger arm shown.

The grid is placed into the flanged recess and held in position with a clip or plate (Figure 6.32B). The specimen rod may then be evacuated in the air lock and the rod inserted into the microscope stage. With some types of rods, the end bearing the specimen grids may be inserted into the stage and detached while the carrier rod remains in the air lock. Most types of rods are inserted into the specimen stage and remain in the stage during the viewing process. Since the rod enters through the side of the pole-piece, it is necessary to design the polepiece with a large enough gap to permit entry and allow for tilting of the specimen by as much as 65 degrees from the horizontal.

Side-entry stages provide much more versatile manipulation of the specimen. Besides the standard x and y horizontal movements, the specimen holder may permit tilting, rotation, a second axis of tilt (double-tilt stage), and special modifications as described in the next paragraph. Since it is also necessary to accurately set the specimen in the correct focal plane

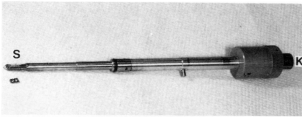

A

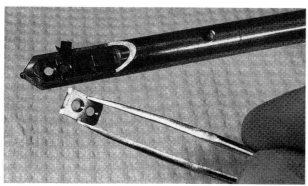

B

FIGURE 6.32 (A) Side-entry, multiple specimen grid holder. k = specimen selection knob for positioning proper grid, s = specimen holder area where grids are inserted and held. (B) Close-up view of side-entry, multiple grid holder showing a specimen grid in the countersunk depression on the right (*arrow*). The grids are held in place by means of the plate shown in the prongs of the forceps. The plate is positioned over the grids and held in place by a spring-loaded latch.

of the objective lens, a z-axis or vertical movement is always provided to allow accurate eucentric positioning. Modern side-entry stages offer high resolution capabilities nearly comparable to top-entry stages and permit more versatility for specimen manipulation and orientation for analytical purposes. For these reasons, the side-entry stage is currently favored over the top-entry stage in the latest generation of TEMs.

Special Stages.
It is possible to manipulate the specimen in the electron microscope in a number of ways using special specimen stages or holders. For instance, the specimen may be subjected to stretching and compression in a *tensile stage*, and heating or cooling in specially modified *thermal stages*. Of particular interest to biologists is the *cold stage*, since it permits the examination of rapidly frozen specimens (such as live virus preparations) that are still hydrated and have not been exposed to chemical fixation or staining. Besides examination of fluid specimens, it is also possible to study ultrathin frozen, hydrated sections of unprocessed biological materials for elemental analysis. Although specimen preparatory techniques are still being refined, cold stages offer tremendous potential when combined with the analytical capabilities of the TEM.

Imaging System

This part of the microscope includes the objective, intermediate, and projector lenses. It is involved in the generation of the image and the magnification and projection of the final image onto a viewing screen or camera system of the microscope.

Objective Lens. By far, this is the single most important lens in the transmission electron microscope, since it forms the initial image that is further magnified by the other imaging lenses. In order to achieve such high resolutions, the lens must be highly energized to obtain the short, 1 to 2 mm focal lengths necessary. The lens must be free of astigmatism and have minimal aberrations. This means that the polepieces must be constructed from homogeneously blended metals, be as symmetrical as possible, and contain devices, or stigmators, for correcting astigmatism. The objective lens is used primarily to focus the image. The objective lens also initially magnifies the image whereas other lenses are used to magnify the image further. Of all of the lenses used in the magnification of an image, the objective lens is the least variable so that it can maintain the very short focal lengths necessary for high resolution and still be convenient to focus (i.e., if the strength of the objective lens were varied over a wide range, refocusing would require major adjustments of the lens current to the lens). Currently, as magnifications are changed, the adjustments to the objective lens needed to bring the image into focus are not excessive.

Because any fluctuations in either lens current or high voltage would affect the focus of the objective lens, both must be made extremely stable. Since contamination may introduce astigmatism into any lens system, some way of minimizing contamination in the objective lens is needed. Such devices, called *anticontaminators,* are now essential for high quality work.

> Images are formed in the objective lens by a "subtractive" action. Depending on specimen thickness and density of various parts of the specimen, some electrons (inelastically scattered ones) will pass through the specimen and into subsequent lenses with a loss of some energy. Other electrons may be deflected upon contact with parts of the specimen and rendered unable to enter the objective and other imaging lenses. Still other electrons may lose all of their energy upon impact with the specimen and are likewise lost. Most of the electrons that enter the objective lens are ultimately projected onto the phosphorescent viewing

screen to cause a certain level of brightness. The more electrons passing through any one point on the specimen, the brighter the image generated.

Regardless of how an electron is lost, this loss is evidenced on the screen as a darker region. Hence, areas of high density/thickness will appear darker than areas of less density/thickness resulting in various "gray levels" on the screen. Since biological specimens have inherently small density differences between the various parts of the cells, it is necessary to enhance these differences by reacting high density metals (osmium, lead, uranium, etc.) with specific subcellular structures. These heavy metals are introduced during the specimen preparation processes of fixation and staining (see Chapters 2 and 5).

Apertures in Objective Lens.   As will be illustrated in subsequent chapters, obtaining a thin specimen with good contrast is not always easily done. *The function of the objective aperture is primarily to enhance contrast* by trapping more of the peripherally deflected electrons (Figure 6.33). Apertures of various sizes may be positioned in the polepiece gap in the back focal plane just under the specimen. Arranged on a similar positioning rod as the condenser apertures, these apertures are much smaller in size (70, 50, 30, and 20 μm, for example) and

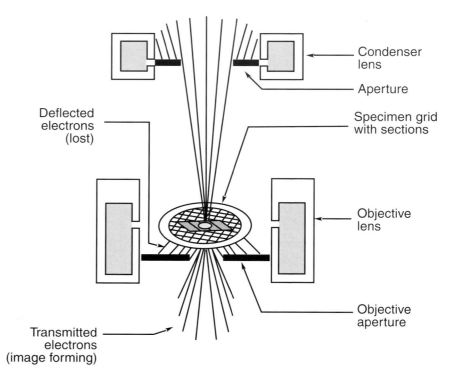

FIGURE 6.33 Objective aperture located between upper and lower parts of polepiece, just under the specimen. The major function of the aperture is to help remove peripherally deflected electrons to enhance image contrast. In addition to the specimen and objective aperture, a chilled anticontaminator blade (see Figure 6.34) may also be inserted just above the specimen (or sometimes above and below the specimen) to prevent contaminants from condensing on specimen.

Condenser lens

Aperture

Specimen grid with sections

Objective lens

Objective aperture

Deflected electrons (lost)

Transmitted electrons (image forming)

more prone to contamination. Small objective apertures give increased contrast, although at the expense of overall illumination and resolution.

Photographers, as well as electron microscopists, make use of apertures not only to control the amount of illumination entering a lens, but further to control the depth of field. It is well known that "stopping down" or decreasing the size of the aperture results in bringing more of the foreground and background into focus, whereas wide open apertures result in only a narrow zone being in focus. *Depth of field, therefore, refers to the depth in the specimen plane that is in focus.* As is demonstrated in Equation 6.8, *smaller apertures increase the depth in the specimen that is in focus.*

Equation 6.8: Depth of Field

$$D_{fi} = \frac{\lambda}{\sin \alpha^2}$$

where λ = wavelength of radiation
α = aperture angle

If we are using an accelerating voltage of 60 kV, the wavelength of the electron is 0.005 nm. A large 200 μm aperture would generate an aperture angle of illumination in the objective lens of approximately 10^{-2} radians. Upon substitution in the equation, we obtain a depth of field of approximately 50 nm. Now, if a 100 μm aperture is used, the aperture angle is approximately 10^{-3} radians, which yields a 100 times greater depth of field of 5 μm. Consequently, when using smaller apertures in both the objective and condenser lenses to generate narrow aperture angles, the entire depth of the specimen is in focus. This is in contrast to the light microscope, where larger aperture angles result in rather narrow depths of field, making it necessary to focus through the various levels to view the entire depth in the specimen. Depth of field and depth of focus are illustrated in Figure 6.36, later in this chapter.

Apertures are usually constructed from high melting point metals such as molybdenum or platinum. Although they may be configured as single discs with a central hole, they are more commonly fabricated from thin foils cut into a strip. Individual holes are precisely drilled through the metal and scrutinized for burrs that would affect the symmetry of the field. Using the external controls on the TEM column, a single hole may be selected and positioned symmetrically around the axis of the electron beam.

It is possible to purchase standard strips containing holes commonly used in a particular instrument, or one can order customized aperture strips with specific types and thickness of metal and specified hole diameters. For optimum performance, apertures must be cleaned periodically, the frequency depending on the types of specimens and general cleanliness of the microscope itself. Molybdenum apertures are cleaned by placing the strip into a flat holder of tungsten or molybdenum and heating the strip in a vacuum evaporator (to avoid oxidation). After reaching a cherry red color, most contamination will be evaporated and removed by the vacuum system. However, it may be necessary to repeat this process several times. One should not exceed the cherry red color, since the molybdenum may weld onto the holder. Platinum strips may be cleaned by passage through a propane gas flame followed by immersion in hydrofluoric acid and ammonium hydroxide. Aperture strips are easier to clean if contamination has not built up over time.

A *thin foil aperture* may also be used in the objective or condenser lenses. Such apertures are made of an extremely thin layer of gold and are "self-cleaning." In practice, one must regularly run the focused electron beam over the rim of the aperture in order to bake off the contaminants. When such foils can no longer be cleaned using such a procedure, they must be replaced.

Anticontaminators in Specimen Area. *Anticontaminators* are found in close proximity to the specimen, aperture, and polepiece of the objective lens. They are essentially metal surfaces that are chilled with liquid nitrogen from a reservoir outside the column of the microscope. Most contaminants originating from the specimen or the microscope will condense onto the extremely cold anticontaminator and be removed from the system.

Anticontaminators are sometimes called *cold fingers*. Some anticontaminators may resemble an aperture holder, except that the brass plate is much thicker and the aperture much larger (Figure 6.34). Other anticontaminators are ring-shaped and encircle the specimen. Since anticontaminators must fit into the same cramped space that also accommodates the specimen and objective aperture, they must be designed very carefully. Anticontaminators must be polished clean periodically to remove condensed materials and then carefully positioned to avoid contact with the specimen holder or objective aperture.

FIGURE 6.34   Specimen anticontaminator or cold finger. The large container (c) is filled with liquid nitrogen to chill the cold finger blade (b) that is located just above and below the specimen. An O-ring seals the apparatus from the atmosphere. *(Courtesy of Gatan, Inc.)*

Astigmatism Correction. *Stigmators* are located beneath not only the objective but also the condenser and intermediate lenses. They function to correct the radial lens asymmetries that prevent one from focusing the image in all directions and generating circular illumination spots. Since an astigmatic lens is stronger in one direction (north-south, for instance) than another, one creates a compensating field of equivalent strength in the opposite direction (east-west). Two parameters must be considered: direction of the astigmatism (*azimuth*) and strength of the astigmatism (*amplitude*). One must be able to adjust both variables to suit the particular situation.

Older stigmators were composed of pairs of magnetic slugs that could be mechanically rotated into position to compensate for astigmatism. Newer microscopes use primarily electromagnetic stigmators since they are less expensive to build, easier to use, and somewhat more precise in their correction. Electromagnetic stigmators may consist of eight tiny electromagnets encircling the lens field. By varying the strength and polarity of various sets of magnets, one can control both amplitude and azimuth in order to generate a symmetrical magnetic field (Figure 6.35). When stigmators become dirty, they will no longer effectively compensate for astigmatism and must be withdrawn from the microscope and cleaned.

Intermediate (Diffraction) Lens. As one proceeds down the column, this lens immediately fol-

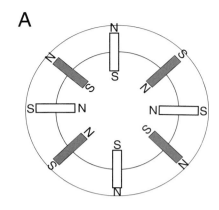

FIGURE 6.35   (A) Conceptual drawing of electromagnetic stigmator showing orientation of eight electromagnets around the lens axis. Strength and direction are controlled by adjusting appropriate combinations of magnets to generate a symmetrical field. The stigmator is located under the condenser and the objective lens polepieces. (B) Actual stigmator apparatus taken from an electron microscope. The large arrow indicates one of the eight electromagnetic iron slugs oriented around the central axis. The entire apparatus fits up into the bore of the objective lens so that the area indicated in the large arrow is positioned just under the specimen. The smaller arrow points out individual electrical contacts through which current flows to energize the electromagnets. The close-up photograph (*bottom*) shows some of the electromagnets that are positioned near the specimen (*arrow*).

lows and is constructed similarly to the objective lens. In older, simpler microscopes, magnification is altered by varying the current to this lens, while in newer microscopes the preferred method is to use combinations of several lenses to allow a wider, distortion-free magnification range. The major function of this lens is to assist in the magnification of the image from the objective lens. At very low magnifications, the objective lens is shut off and the intermediate lens used in its place to generate the primary image. Although the image produced by the very long focal length intermediate lens is poor compared to that generated by using all three lenses, it is adequate for low magnification work. The intermediate lens may be equipped with an aperture that is used when operating the microscope in the diffraction mode (Chapter 15).

Projector Lens. Most modern transmission electron microscopes have two projector lenses (P1 and P2) that follow the intermediate lens. Both P1 and P2 are used to further magnify images from the intermediate or diffraction lens. Except for very high magnifications, only three of the four imaging lenses are normally energized at any one time, and various triplet combinations are used to achieve the magnification range desired. In a microscope with four imaging lenses, the first projector lens can also be used as a diffraction lens, and it may be possible to insert a specimen into a specially modified holder located either between P1 and P2 or below P2 for specialized, low angle diffraction studies. As with intermediate lenses, projector lenses suffer from distortions that have less effect on resolution than do aberrations occurring in the objective lens.

Projector lenses are said to have great *depth of focus,* meaning that the final image remains in focus for a long distance along the optical axis. This is determined by Equation 6.9.

Equation 6.9: Depth of Focus

$$D_{fo} = \frac{M^2 \cdot RP}{\alpha}$$

where: M = total magnification
RP = resolving power of instrument being used
α = aperture angle established by objective lens

At a magnification of 100,000×, in an instrument with resolving power of 0.2 nm and having an aperture angle of 10^{-2} radians, the depth of

focus of the projector lens may be calculated to be 200 meters. This becomes important when one realizes that the photographic film is not in the same plane as the viewing screen. For the same reason, it is possible to locate multiple image recording devices at various points beyond the projector lens, since they will all be in focus. However, the magnification will increase as one moves farther away from the projector lens. The relationship between depth of field and depth of focus relative to aperture angle is shown in Figure 6.36.

Viewing System and Camera. The final image is projected onto a viewing screen coated with a phosphorescent zinc-activated cadmium sulfide powder attached to the screen with a binder such as cellulose nitrate. Most electron microscopes provide for an inclination of the viewing screen so that the image may be conveniently examined either with the unaided eye or with a *stereomicroscope* called the binoculars. With the stereomicroscope, although the image may appear to be rough due to the 100 μm-sized grains of phosphorescent particles making up the screen, it is necessary to view a magnified image in order to focus accurately. Some microscopes may provide a second, smaller screen that is brought into position for focusing. In this case, the main screen remains horizontal, except during exposure of the film. All viewing screens will have areas marked to indicate where to position the image so that it will be properly situated on the film.

Pre-evacuated films are placed into an air lock (*camera chamber*) under the viewing screen and the

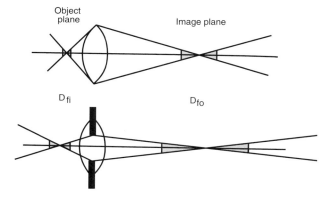

FIGURE 6.36 Depth of field (D_{fi}) occurs in the object plane, while depth of focus (D_{fo}) refers to the depth in the image plane that is in focus. In the bottom figure, note that an aperture increases both the depth of field and depth of focus.

chamber evacuated to high vacuum. The chamber is then opened to the column to permit exposure of the film. It is necessary to have a supply of pre-evacuated films for insertion into the microscope, since it may take several hours to remove residual moisture or other contaminating gases from the film. Cameras may hold from one to fifty individual 3 ¼" × 4" films, or one may use a variety of roll cameras loaded with 35 mm or larger films. Most films used in electron microscopy are orthochromatic and thus may be handled using an appropriate safelight (see Chapter 8).

All modern electron microscopes provide for metering the exposure and have a shutter to time the exposure accurately. Since the screen is coated with the same type of material as are camera light meters, it is possible to monitor the current generated by the electron beam striking the screen.

Some electron microscopes provide for an integrated reading when the image is viewed on the large main screen and for a localized or "spot" reading when a smaller viewing screen is used for imaging. Normally, integrated readings are adequate; however, if the image has extremes of contrast, the spot meter is very valuable to insure proper exposure. It is usually necessary to calibrate the exposure meter when different film is being used or a new microscope is installed. This is done by taking a series of micrographs with the meter set for different sensitivity levels (similar to the ASA or film speed). After the developed negatives are evaluated, the proper setting may be recorded and the sensitivity properly adjusted.

A shutter is provided to time the exposure so that the proper negative density (as determined by the previous calibration) may be obtained. Most electron microscopes have timers that vary from a fraction of a second to "hold" positions in which a timer may be used for very long manual exposures. Electron micrographs are exposed for 0.5 to 2 seconds in order to record all density levels and to minimize image shift or *drift* (i.e., slow movement of the image after exposure to the beam). Once the time has been selected, the illumination level is adjusted with the C1 and C2 lens controls until the exposure meter reaches the calibration point. The film is then advanced under the viewing screen, and the screen is moved to permit electrons to pass onto the film. As one begins to raise the viewing screen, the beam is blocked by the shutter until the screen is totally raised. The shutter is then opened for the proper interval, after which the beam is again blocked until the screen is repositioned.

Vacuum System

The vacuum system is a major assemblage of pumps, switches, and valves that are involved in evacuating air primarily from the pathway of the electron beam, but also from other selected areas of the microscope. It may be thought of as a plumbing system designed to connect the vacuum pumps to chambers (specimen, camera, gun, column) that need evacuation. Although earlier electron microscopes had vacuum systems that were more manual in operation, the standard automated vacuum plant in present-day electron microscopes is little changed from its predecessors. Besides the obvious advantage of convenience, the major advantage of current systems is the inclusion of extensive safeguards against contamination, misvalving, and power and water failure (although these safeguards were sometimes incorporated even in the earliest of microscopes).

A vacuum is needed in the electron microscope not only to increase the *mean free path* of an electron (i.e., the distance an electron must travel without encountering an interfering gas molecule), but also to prevent high voltage discharges between the filament/shield and the anode. Such discharges are one of the major causes of filament failure. In addition, all filaments (especially field emission) are extremely sensitive to oxidation and must be protected. A fourth reason why a vacuum is needed involves the removal of contaminating gases (especially water vapor and organics present in laboratories) that are broken down under high energy electron bombardment and generate corrosive radicals that combine with the specimen to destroy fine structure.

Vacuum Terminology

Since the terminology used to describe vacuum situations may be confusing even to the practicing electron microscopist, it is important to be familiar with certain basic concepts and phrases. Confusion arises since vacuum is expressed in units of pressure: a high vacuum is low pressure. In addition, a general acceptance of a standardized nomenclature (*bar* versus *torr* versus *Pascal*) has met with resistance. It is not uncommon to see all three terms used even today in the literature. For that reason, one should be aware of the relationship between the various terms to fully understand the discussion.

If one takes a long tube sealed on one end, fills it with mercury, and inverts the tube in a bowl, some of the mercury will flow out into the bowl. At sea level, the height of the column of mercury remaining in the tube, measured from top to fluid level in the bowl, would be 760 mm. The 760 mm of mercury is supported in the glass tube by the pressure of the atmosphere pressing down into the bowl. This simple barometer was developed in the mid-1600s by Evangelista Torricelli and is the basis for many pressure measurements. If the atmospheric pressure is lowered, the mercury begins to flow out into the bowl, and the height of the column becomes lower. At some point, only 1 mm of mercury will be supported. This 1 mm of mercury, supported by 1/760 atmosphere, is equivalent to one *torr* (named in Torricelli's honor) and would be the atmospheric pressure one might encounter 28 miles above sea level. One torr = 1.32×10^{-3} atmosphere.

The international standard for pressure measurement is the metric unit termed the *Pascal* (named in honor of the French philosopher and scientist, Blaise Pascal, who lived from 1623 to 1662), and is abbreviated *Pa*. One Pascal equals 9.92×10^{-6} atmosphere. Since there are 133 Pa per torr:

$$1 \text{ Pa} = 1/133 \text{ torr}$$
$$= 7.52 \times 10^{-3} \text{ torr}$$

To convert a pressure stated in torr into Pascal, multiply by 133; to convert a figure stated in Pascal into torr, divide by 133. Since these figures may not be easily remembered, a crude approximation may be obtained if we add 2 to the torr exponent to obtain Pa. So, a vacuum of 10^{-7} torr is very roughly equivalent to 10^{-5} Pa (actually 1.33×10^{-5} Pa).

The relationship between the three major pressure measurement terms is given in Table 6.6.

Vacuum ranges for the TEM vary from the so-called "high vacuums" of 10^{-1} to 10^{-4} Pa, to "very high vacuums" in the 10^{-4} to 10^{-7} Pa range, and "ultrahigh vacuums" in the 10^{-7} to 10^{-9} Pa range. It may be of interest to note that even in electron microscopes

operating at a high vacuum of 10^{-4} Pa, there are still over 10^{10} air molecules per cubic centimeter. However, the mean free path of an electron under such conditions is over 50 meters.

Rotary and Diffusion Pumps. The majority of electron microscopes achieve operating vacuum conditions using two types of pumps: *rotary pump* and *diffusion pump*. The rotary pump is used first to lower the pressure into the 10^{0} to 10^{-1} Pa range (*rough pumping*), and then the diffusion pump is used to achieve higher vacuums in the 10^{-3} to 10^{-4} Pa range. These pumps use different principles to achieve their ultimate vacuums, and certain limitations in each should be realized.

Rotary Pump. This **mechanical pump** is also sometimes called a **rough pump** or **forepump** since it is often used to establish a rough starting vacuum before the diffusion pump is used. A typical **rotary pump** is shown in Figure 6.37. A rotating cylinder with sliding spring-mounted vanes is mounted off center inside a larger cylindrical space. As the rotor turns clockwise from position A to B, an enlarged space is created with a lowered pressure that draws air into B. (A similar phenomenon occurs when one releases a squeezed rubber bulb of an eyedropper pipette to draw fluid or air into the bulb.) As the rotor continues turning, space B is sealed off and becomes smaller as the rotor moves the contents into space C. The gas in space C is compressed to the extent that a spring-loaded valve is forced open and the air is exhausted from the system (similar to squeezing a rubber bulb filled with fluid). The pump has thereby moved some of the air from inside the closed microscope system to the outside. It is possible to link two such systems in series (exhaust of one

TABLE 6.6   Relationships Between Commonly Used Units of Pressure

	Pascal	Torr	Millibar
1 Pascal	1	7.5×10^{-3}	1.0×10^{-2}
1 Torr	133	1	1.33
1 Millitorr	1.3×10^{-1}	1×10^{-3}	1.3×10^{-3}
1 Millibar	100	0.75	1
1 Atmosphere	1.01×10^{5}	760	1.01×10^{3}

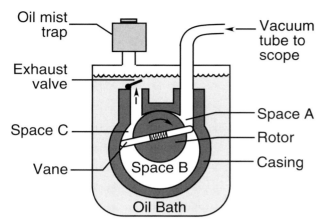

FIGURE 6.37A Rotary pump. Illustration of principle of operation of pump module. As the rotor turns, space A becomes enlarged, creating a vacuum that sucks air into the space. When the rotor rotates further and seals off the space by means of the spring-loaded vane, a large volume of air has been removed from the TEM and into the pump. Upon further rotation of the rotor, the sealed space designated B becomes smaller and the air is compressed. Eventually, the compressed air is moved over to space C by the rotor and a spring-loaded valve opens to exhaust the air to the outside of the system. The oil is used to lubricate the moving rotor and vanes and to carry frictional heat to the outside of the case.

FIGURE 6.37B Photo of a belt-driven, two-stage rotary pump. The electric motor driving the pump is to the left. The cylindrical cannister on top of the pump module is a high-efficiency filter to remove oil fumes from the exhaust of the pump. The port through which the vacuum is applied is indicated by an arrow.

FIGURE 6.37C Photograph of a direct-drive, two-stage rotary pump found on most modern electron microscopes.

into the inlet of a second pump) to achieve a two-stage pump, the most commonly used type of rotary pump. Two-stage pumps do not necessarily pump faster, but do achieve better vacuums.

Since friction is a problem as the vanes rub along the inside of the larger cylinder, it is necessary to lubricate the system. This is accomplished by immersing the pump module in an oil bath so that some oil is constantly coating the two rubbing surfaces. The oil serves not only to lubricate the system and provide a vacuum seal between the two surfaces, but also to transfer any frictional heat to the outside of the pump casing. Being mechanical pumps, the seal between the rotor and casing will, after extensive use, wear so that good vacuums can no longer be obtained. In addition, the oil absorbs moisture and other laboratory solvents, and the lubrication and pumping efficiency falls off. Depending on the pumping loads and atmospheric moisture, the rotary pump oil must be replaced at regular intervals (1 to 4 times a year). *One must be aware that used oils are carcinogenic, so gloves must be worn and proper disposal procedures must be followed* (some service stations will take the oil). Since all rotary pumps release a fine mist of oil out of the exhaust port, it is very important that the exhaust be vented to the outside or, at the very least, fitted with efficient oil mist traps. The major causes of rotary pump demise are failure to maintain the proper oil level and infrequent oil changes.

Diffusion Pump. In contrast to the rotary pump, diffusion pumps contain no moving parts but are able to complete the pumping operation started by the rotary pump. Diffusion pumps were developed nearly 70 years ago and still form an important component in most vacuum systems. The basic design, shown in Figure 6.38, consists of a series of stacked towers with umbrella-like caps with highly polished surfaces. The towers are placed inside of a water cooled cylindrical enclosure that is open by way of a main valve on the end connected to the microscope and has an electric heater and side-mounted outlet at the base. Special oils in the bottom of the pump housing are boiled, and the pressurized vapor is forced up the towers to be deflected downwards by the umbrellas or jet assemblies. Since the oil molecules are traveling at supersonic speeds, any air molecules that have diffused into the pump will be knocked to the bottom of the diffusion pump chamber where they will be removed by a rotary pump connected to the side port. *Diffusion pumps cannot function unless they are "backed" by or connected to operating rotary pumps to remove the accumulated air molecules.* The hot oil vapors strike the sides of the pump housing, condense on the water-cooled surface, and flow to the bottom of the pump to be boiled and recycled. In an effort to slow down the creep of diffusion pump oil into the column, water- or liquid nitrogen-chilled baffle plates are often installed at the mouth of the diffusion pump to trap contaminants. Such baffles not only provide a cleaner vacuum environment, but also act as cryopumps to improve the pumping efficiency of the diffusion pumps.

The throat size as well as the number of umbrella towers determines the number of stages in a diffusion pump, with each stage capable of lowering the pressure by nearly one Pascal. Therefore, a three-stage diffusion pump should improve the vacuum by a factor of 10^{-3} Pa. If a vacuum system had been previously rough pumped to a vacuum of 10^{-1}, the final vacuum would be in the 10^{-4} Pa range.

Different types of oils, with very *low vapor pressures* (i.e., not easily vaporized), are used in diffusion pumps. Silicone-based oils (Dow Corning 704 or 705)

FIGURE 6.38A Diffusion pump. Photograph of exterior of a diffusion pump showing cooling coils surrounding body. Stacked, umbrella-like caps (*lower, right*) fit down into the body of the pump.

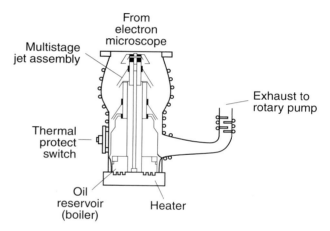

FIGURE 6.38B Cutaway diagram of diffusion pump interior.

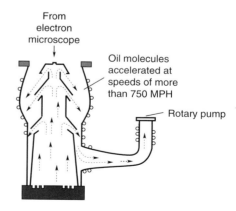

FIGURE 6.38C Operating principle of pump: supersonic oil particles drive air molecules to base of pump to be withdrawn by a rotary pump (not shown) attached to the side arm.
(B and C, courtesy of Varian Associates—Vacuum Products Division.)

may be used in some vacuum systems. However, since they are very difficult contaminants to remove, polyphenyl ethers (Santovac 5, Convalex 10) or per-fluoropolyethers (Fomblin Y VAC) are preferred in clean systems. One simply cannot replace one oil with another, however, since the boiler temperatures and vapor pressures of the various oils are different and must be matched to the specific system.

Diffusion pumps have no moving parts and thus require little maintenance and may function for many years without problems. However, should they be exposed to pressures greater than 1 Pa, some oil may be blown into the rotary pump. When the diffusion pump oil level is sufficiently lowered, pumping efficiency will be diminished and the boiler may overheat and burn out the heater. Therefore, the oil level should be checked if pumping efficiency drops significantly and no other causes may be found. The pump should be unbolted, drained of oil, cleaned with the proper solvent, dried, and refilled with the precise amount of proper oil. The Viton, heat resistant, sealing gaskets should be replaced, if deformed.

Reading Vacuum Levels. The level of vacuum may be measured using several different monitors. Vacuums generated by the rotary pump are evaluated using either a thermocouple or Pirani gauge.

The *thermocouple gauge* (Figure 6.39) consists of a filament that is heated with a constant amount of current. The temperature of the wire is measured directly by an attached thermocouple (thermometer). Large numbers of air molecules conduct away heat and lower the temperature of the wire. Instead of reading temperature, the meter scale is replaced with one expressed in vacuum units. As air molecules are

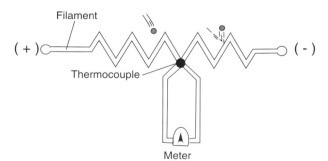

FIGURE 6.39 Principle of operation of thermocouple gauge. A thermocouple senses the temperature of a heated wire. As air molecules are removed, the temperature of the wire rises so that temperature may be correlated to vacuum level.
(Courtesy of Varian Associates—Vacuum Products Division.)

removed due to an improved vacuum, the wire gets hotter and the needle reads better vacuum conditions rather than higher temperatures. Therefore, in this gauge, the amount of voltage fed to the wire is the constant, while the temperature is the variable that is measured on the surrogate scale.

The *Pirani gauge* consists of a filament that is maintained at a constant level of electrical resistance by running the proper voltage through it. The resistance is established by heating the filament to a certain temperature. Air molecules conduct heat away from the filament, causing a decrease in temperature and resistance. A Wheatstone bridge circuit senses the resistance drop and applies more voltage to the filament to maintain the resistance level or temperature. A voltmeter is used to sense the amount of voltage being put into the circuit. As air molecules are removed from a system, the filament loses less of its heat and less voltage is needed to maintain the resistance. The voltmeter scale may be replaced with a scale calibrated in vacuum units rather than volts, so that one reads the vacuum directly. Lower voltage is read as better vacuum. Therefore, in the Pirani gauge, the temperature and resistance are maintained at a constant level, while the voltage is the variable that is measured (and converted into vacuum units).

As the pressure decreases to 10^{-1} Pa or less, the Pirani and thermocouple gauges are no longer sensitive. Instead, a cold cathode gauge or an ionization gauge may be used.

The *cold cathode gauge* (Figure 6.40A) consists of cathodes and an anode or collector wire that are maintained at several thousand volts relative to each other. The high voltage field ionizes gas molecules present in the gauge and causes the electrons to travel to the anode. The electrons collide with gas molecules and ionize them producing positive ions. A powerful external magnet increases the sensitivity of the meter by causing the electrons to assume long spiral trajectories, rather than straight lines. The positive ions travel to the negatively charged cathode and the resulting ion current is measured and expressed in vacuum units rather than amperage. This gauge is used in the 10^{0} to 10^{-6} Pa ranges.

For wider measurements of vacuums in the 10^{-2} to 10^{-12} Pa ranges, an *ionization gauge* (Figure 6.40B) is desirable. This gauge consists of a hot filament that generates electrons that travel towards a positively charged coil of wire called a grid. In their travels to the grid, the electrons ionize air molecules and cause a flow of positively charged gas molecules that are attracted to a collector. The ion current collected is proportional to the pressure in the chamber.

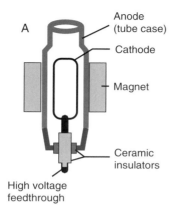

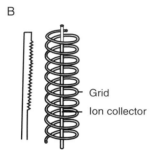

FIGURE 6.40 (A) Diagram of cold cathode vacuum gauge. Air molecules are ionized between the high voltage cathode and anode. The flow of ions is measured as current by a meter. The magnet increases the path lengths of the electrons to increase ionization and boost sensitivity of the gauge. (B) Diagram of an ionization gauge. A heated filament (*left*) generates electrons that travel toward the positively charged grid. As they travel to the grid, the electrons ionize any gas molecules present and generate gas ions that are collected by the central wire.
(*Courtesy of Varian Associates—Vacuum Products Division.*)

Total Vacuum Systems. The principles behind a complete vacuum system are more readily understood by viewing a simplified diagram as shown in Figure 6.41. The system consists of the rotary and diffusion pumps, switching valves, connecting pipes, air locks, and vacuum gauges. In all vacuum systems, certain operational principles or rules must be followed: (a) sealed chambers must be evacuated first with a rotary pump followed by a diffusion pump; (b) diffusion pumps must be backed by a rotary pump; (c) air may be admitted to chambers only if closed off from all pumps; (d) rough pumps and diffusion pumps must not pump on the same sealed chamber at the same time. Descriptions of several operational situations follow.

Turning On Cold System. When one first turns on a cold electron microscope, there is a brief period of rough pumping of the column area. Water is run through the diffusion pump cooling coils while the oil is brought to a boil. The diffusion pumps are sealed from the column but are being backed by a rotary pump. This usually takes 15 to 20 minutes to heat the oil.

Column Pump Down Sequence. After pumping for a period of time (perhaps 20 to 30 minutes), the microscope has reached the proper operating vacuum so that the filament may be activated for viewing.

Specimen Insertion. A specimen grid is loaded into the holder, inserted into the sealed specimen air lock, and rough pumped into the proper range. The rough pump is disconnected and the specimen inserted into the high vacuum environment inside the microscope for examination.

Specimen Removal. The specimen retrieval system is used to move the specimen back into the sealed air lock. Air is admitted to the air lock and the specimen removed.

Removal of Exposed Films. After electron micrographs have been taken, the camera chamber is sealed off from all vacuum pumps and air is admitted to the camera chamber. After removal of exposed films and replenishment with unexposed, prepumped films, the sealed-off camera chamber is first pumped by rotary pump and then properly connected to the diffusion pump and the column.

Shut Down Sequence. After shutting off the lenses and high voltage, all air-locked chambers are evacuated and sealed. Power to the diffusion pump heater is shut off, but cooling water and rotary pump backing are maintained for 15 minutes to permit the oil to condense into the bottom of the diffusion pump boiler. The backing pump is sealed from the diffusion pump and power to the backing pump is shut off. Air may be bled into the vacuum line between pumps and sealed. Cooling water is shut off.

For further readings on the theory and practical aspects of operating a vacuum system, see the reference volume by Bigelow (1994).

Vacuum Problems and Safety Features

A number of accidental situations may disable a vacuum system and possibly contaminate the column of

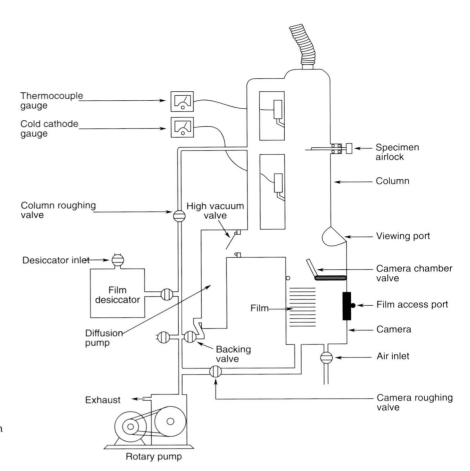

Thermocouple gauge

Cold cathode gauge

Column roughing valve

High vacuum valve

Specimen airlock

Column

Viewing port

Camera chamber valve

Desiccator inlet

Film desiccator

Film

Film access port

Diffusion pump

Camera

Backing valve

Air inlet

Exhaust

Camera roughing valve

Rotary pump

FIGURE 6.41 Simplified diagram of a vacuum system in a transmission electron microscope.

the electron microscope. Descriptions of the more common types of accidents follow.

If the diffusion pump is connected to the column and the flow of cooling water is interrupted, the diffusion pump oil will not condense and the vapors may drift in the reverse direction to contaminate the column. This is usually prevented either by installing an overtemperature cutoff switch on the diffusion pump or a flow detector in the water line that will shut off power to the heater if water flow stops. The use of liquid nitrogen-chilled baffles or traps (anticontaminators) placed above the diffusion pumps will also lessen or prevent the contamination. Diffusion pump oil that has been severely overheated may undergo cracking and must be replaced.

Another problem may result when the backing rotary pump is disabled, perhaps due to a broken drive belt or power outage. If the diffusion pump is connected to the column when this occurs, then the lower pressure of the column will draw the oil from the rotary pump through the diffusion pump and into the column creating a major disaster requiring extensive cleaning. Most automated vacuum systems will

detect the deterioration of vacuum in the column and close all valves. Since the valves are pneumatically activated from a pressurized tank, no electrical power is needed. In a manually valved system, it is poor practice to leave an unattended microscope in high vacuum. In manual systems, it is possible to build a battery-powered warning bell that will be activated by a loss of vacuum in either the column or the backing line of the rotary pump.

Misvalving of the microscope may also lead to messy situations. If one opens the diffusion pump to a column containing too much air, the rush of air into the diffusion pump may blow the diffusion pump oil out into the rotary pump or possibly back up into the lower part of the column. Worst of all, if the column is connected to high vacuum and one improperly opens the bleed valve to vent the lines between the rotary pump and diffusion pump (the last step of system shut down), air will rush through the bottom of the diffusion pump and blow vaporized diffusion pump oil into the column with great force. Probably the diffusion pump towers will be displaced and damaged by impact with the cooling baffles at the top of the diffusion pump.

Obviously, a major cleaning of the column, as well as disassembly and cleaning of the vacuum system, is needed. The only safety measure to prevent misvalving of a manually operated vacuum system is a thorough understanding of the proper valving sequences of the vacuum system.

Other Types of Vacuum Pumps

Besides the rotary and diffusion pumps, several other specialized vacuum pumps may be more suited to particular electron microscope applications as, for example, when ultraclean vacuums are needed. The more commonly used types of specialized pumps are:

Turbomolecular pumps (Figure 6.42) are oil-less systems that consist of a series of rapidly spinning rotors with blades or vanes inclined at an angle. Between the rotors are a series of static blades (stators) slanted in the opposite direction. Individual rotors and stators are arranged alternately in 8 to 12 layers or stages. The rotors spin at 20,000 to 50,000 rpm and strike gas molecules with such impact that they are driven down to the next stage when the next rotor knocks them deeper into the pump with increasing momentum. The stators encourage the downward movement due to their inclination. Upon reaching the bottom of the pump, the gas molecules are removed usually by a mechanical rotary pump. Unlike diffusion pumps, turbo pumps do not require water cooling and may pump a sealed chamber at atmospheric pressures. The 10^{-7} to 10^{-8} Pa attainable vacuums are better than diffusion pumps,

although the rate will be slower. Turbo pumps are more expensive and require more frequent maintenance than diffusion pumps, but are much cleaner and less susceptible to misvalving problems. Never move an instrument even slightly when a TMP is operational because the pump may be damaged.

Entrainment pumps (ion, cryogenic) work on a principle totally different than other pumps that move gas molecules by impact (diffusion, turbomolecular) or positive displacement (rotary) methods. These pumps entrain or hold gas molecules onto chemically reactive or extremely cold surfaces rather than remove the molecules totally from the system. They are used in applications requiring extremely clean, ultrahigh vacuums in the 10^{-9} Pa or better range as, for instance, in lanthanum hexaboride and field emission electron guns. Having no moving parts, these pumps are less prone to breakdown but must be regenerated or purged of the entrained molecules periodically. A major disadvantage of these types of pumps is the differential pumping speeds for various molecular species. For instance, cryogenic pumps remove water molecules quite efficiently but other gas species as nitrogen or methane may be less efficiently entrained by the pump. It may be necessary to combine two different types of entrainment pumps to cover the range of molecules likely to be encountered in any one research situation. Cryosorption pumps entrain molecules on a cooled surface by means of van der Waals forces. The pumping mechanism is based on cryocondensation and cryotrapping onto such cooled surfaces as molecular sieves, porous silver, and activated charcoal.

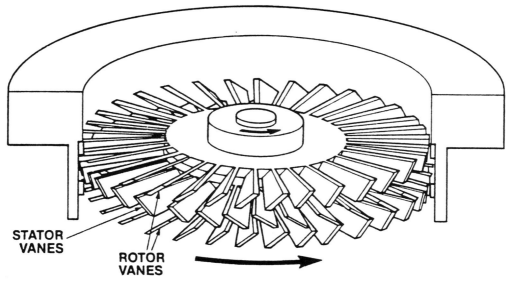

STATOR VANES

ROTOR VANES

FIGURE 6.42 Illustration of moving rotor vanes and nonmoving stator vanes of a turbomolecular pump. *(Courtesy of Varian Associates—Vacuum Products Division.)*

Ion pumps, such as the sputter ion pump, operate on the same principle as the cold cathode ionization gauge used in measuring vacuum status. High voltage cathodes generate a flow of electrons toward high voltage anodes. The electrons collide with and ionize gas molecules causing the positive ions to flow to the cathode while the electrons continue their collision path toward the anode. The gas molecules collide with the surface of the cathode and are chemically held to the surface by a reactive coating of titanium or zirconium.

Preparing the Transmission Electron Microscope for Use

Alignment Theory

After the microscope has reached high vacuum status, one must prepare the electron optics for optimal results. This usually takes place at the beginning of the microscopy session or after maintenance has been performed on any of the systems associated with image formation. Although the details of this process, called *alignment,* may vary significantly from one transmission electron microscope to another, certain basic conditions must be satisfied in order to obtain high quality micrographs. This section will explain and outline a general procedure for fulfilling these conditions. More detailed protocols are furnished by the various microscope manufacturers.

The major reasons for aligning any optical system are convenience of operation and improved resolution. For instance, when increasing the magnification, it is very annoying to have the image and even the illumination move off the viewing screen. These phenomena are caused by misaligned imaging and illumination systems, respectively. *In alignment, the optical axes of the illuminating and imaging lenses are lined up along a common axis; the various apertures are centered relative to this axis, and any astigmatism present in the condenser and objective lenses is corrected.* In practice, the common axis is aligned to the center of the viewing screen. The corrections may involve certain mechanical and electronic adjustments that may take anywhere from 15 minutes for a routine "touch up" alignment to several hours if major systems have been moved, for example after cleaning the components (lens polepieces, stigmators, etc.) inside the TEM column.

Since it is impossible to fabricate a perfectly symmetrical electromagnetic lens, the optical axes of the various lenses seldom correspond to the physical center of the lens but rather correspond to either a voltage center or current center. The voltage and current centers are axes around which the image is seen to expand or rotate as either the high voltage or current to a lens is changed, respectively. One way to find the optical axis would be to vary or "wobble the filament high voltage" (i.e., vary it by several hundred volts) and observe the rotation of the image around some point. The appropriate lens would then be shifted so that the rotation point is conveniently centered on the viewing screen. This is called *voltage centering* and is commonly used in the alignment procedure. A second way to locate the optical axis of a lens, *current centering,* is to vary the current to that lens and observe the rotation of the image around the current center. The lens is then moved so that this rotation is centered on the viewing screen. The current and voltage centers will not coincide exactly, but will be very close to each other in a well-constructed lens.

The optical axis is placed on the screen center because any electrical disturbances are minimal at the optical axis. For instance, any high voltage fluctuations will cause the image to expand and rotate around the screen center. If the important part of the image is centered on the screen, then it will show the least effect, whereas the movements are more exaggerated toward the edge of the field (Figure 6.43). Similarly, any current instabilities or image drift will be minimal at the optical axis, which was placed at the screen center.

The movements necessary to situate the optical axis of an electromagnetic lens along the common alignment axis may involve a horizontal x and y movement, called *translation, shift, or traverse,* and an inclination along a vertical axis, called *tilt.* Some lenses have provisions for both types of movements, others may have only a translation capability, while some may be factory aligned and not conveniently moved at all. There has been a tendency in newer microscopes to factory align major lenses mechanically and to encourage users to trim this alignment electronically using electromagnetic deflection coils. In these systems, alignment is much faster and easier to accomplish.

Alignment Practice

1. **Optional steps (when replacing burnt-out filament).** Routine alignment normally would begin as described in the next paragraph.

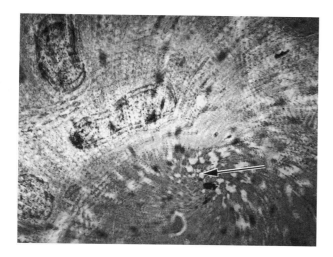

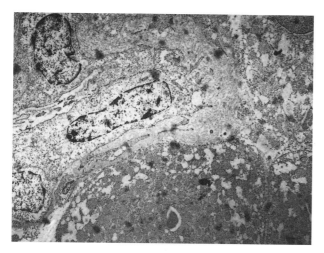

FIGURE 6.43 (*top*) Instabilities in a lens are minimal at the optical axis (*arrow*). Notice how the image is less smeared at the optical axis compared to the region that is off axis. This image was obtained by wobbling the current to the objective lens. (*bottom*) Normal micrograph without instabilities in the objective lens current.

However, when the filament burns out, it will be necessary to remove the cathode cartridge, clean the gun, and install a new filament. After setting the new filament in the cathode cartridge and replacing the shield, the filament height and centration are carefully adjusted relative to the shield aperture. The cathode cartridge is then attached to the high voltage assembly, the system sealed, and the microscope evacuated. During the pumpdown, one should withdraw the smaller objective and diffraction apertures from the column, while the condenser apertures may remain in position if they were previously

centered. One may now proceed to the next step and begin alignment of the column.

2. **Alignment begins at the top of the microscope column with the illumination system.**
   If the microscope was very much misaligned (perhaps in an attempt to locate the illumination after the filament failed), then it may be necessary to switch off some of the imaging lenses in order to find the beam. Otherwise, the lenses are left on and set to the recommended positions for alignment.

3. **Filament emission and rough saturation.**
   The high voltage is applied to the filament and the filament heated to the approximate *saturation point* by watching the viewing screen with the beam at crossover to see that the dark striations on the filament tip just disappear. If one has an accurate emission meter, heat up the filament until the emission meter no longer continues to rise. If no reading is obtained on the meter, or if the reading is excessive, then the bias to the shield should be changed until the recommended emission is obtained (usually 10 to 25 µA). If this fails, then the filament distance from the shield aperture should be rechecked and the procedure repeated.

4. **Filament-cloud centering.**
   Vary the C2 lens current or brightness controls until some illumination is seen on the screen center. If possible, decrease the spot size using C2 control and maintain the central location on the screen by translating the condenser lens. Maximize the illumination by translating the gun, but take care not to burn the screen with excessive beam irradiation by spreading the beam, as needed, using the C2 control. Slowly desaturate the gun using filament emission control and focus the filament image using C2. Tilt the gun relative to the entire illumination system until a symmetrical image of the undersaturated filament halo is obtained; then translate the illumination spot to the center of the screen with gun translation controls (Figure 6.44).

5. **C2 aperture centration.**
   Start with crossover at center of screen. Check the centration of the C2 aperture by taking the C2 control through the crossover point and back. If the illumination spot expands and contracts symmetrically at the screen center, then the aperture is centered. If the illumination sweeps off the screen, the two aperture centration controls should be adjusted until the illumination spot stays centered.

6. **Condenser lens 1 alignment.** Since the illumination spot size may be varied to suit the magnification, the C1 lens must also be aligned. This is accomplished by bringing C2 to crossover at the screen center and varying the C1 lens control through the various spot sizes (1 through 10 μm, for example) and observing the positioning of the spots on the screen. Focus the largest C1 spot to crossover and center the spot on the viewing screen using gun translation controls. Switch the C1 to the smallest spot, focus to crossover and center the spot using illumination translation control. When the illumination spot stays centered when switching the C1 through the various spot sizes, the C1 is considered aligned. Desaturate the filament slightly and recheck the cloud symmetry (using gun tilt) and cloud centering (using gun translation).

7. **Condenser lens stigmation.**
   Astigmatism in the condenser lenses results in an illuminating spot that is not circular and generates an asymmetrical aperture angle that degrades resolution. One method of astigmatism correction is to focus the image of the undersaturated filament on the screen using C2 and then sharpen up the image of the filament using the condenser stigmator (Figure 6.45).

8. **Specimen height adjustment.**
   On specimen stages with adjustable specimen height settings (z-axis controls), it is important to place the specimen in the proper focal plane of the objective lens. In this location, optimal resolution will be obtained and the image will remain centered if the specimen is tilted. To accomplish this, center a distinctive part of the test specimen and focus it. Tilt the stage ± 10° from the horizontal and observe if the focused image shifts off center. If so, return the image to

center using the z-axis control (and possibly other stage centration controls, depending on design of the TEM). This setting is critical not only for optimal resolutions, but also for convenience of operation if tilting is to be done. In addition, a misplaced z-axis will also affect accuracy of the magnification settings.

9. **Imaging lens alignment.**
   a. *Objective lens.* Depending on the microscope, the imaging lenses may have various combinations of translate and tilt adjustments. In some instruments, the objective lens may have one or both movements, while in other microscopes this lens is fixed and other lenses are aligned relative to it. If the objective lens is adjustable, this is usually accomplished by wobbling the current through its lens coil (or by reversing the polarity of the lens) and observing the motion of a focused image like a holey film (see Chapter 4). The objective lens is tilted and translated until the rotation of the image occurs about the screen center.

   b. *Diffraction lens.* The intermediate (diffraction) lens may be centered by switching the TEM to the diffraction mode and centering the focused spot using the diffraction spot centering controls. This very bright diffraction or crossover spot is sometimes called a *caustic figure,* since it may actually burn a mark on the viewing screen. If the microscope is to be used for diffraction purposes, then the diffraction spot must also be centered at the various camera lengths to be used.

   c. *Projector lens.* The projector lens is usually centered by varying the current to the

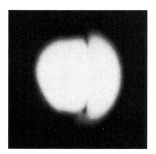
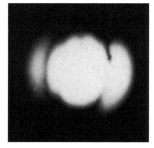

FIGURE 6.44 Nonsymmetrical filament image (*left*) prior to alignment. A nearly symmetrical filament image (*right*).

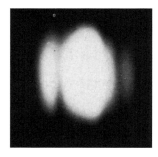
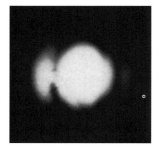

FIGURE 6.45 Astigmatism in the condenser lens is evidenced by a smearing of the image of the undersaturated filament in the north-south direction (*left panel*). Stigmated condenser shows a more symmetrical undersaturated filament image.

projector lens and translating the system until the image rotates about the center of the viewing screen. In some microscopes with two projector lenses, the projector alignment is accomplished in two stages. After turning off the first projector lens (P1), the second projector (P2) lens is first aligned by observing the two caustic figures generated by reversing polarity of the intermediate lens and moving the most peripheral caustic figure to screen center using the P2 controls. The P1 lens is then turned on and centered in the same way.

10. **Voltage centering.**

Most electron microscopes provide a tilt of the illuminating system that must also be accomplished after the other lens alignments. This is an important correction since the illumination may be inclined at a wrong angle to properly enter the optical axis of the objective lens. Voltage centering is achieved by first moving the focused image of a distinctive object such as a holey film to the center of the screen using the specimen traverse controls and activating the high voltage wobbler device to vary the high voltage. The image will rotate and expand around some point that should be returned to the center of the viewing screen using the illumination system tilt controls. When properly tilted, the image will rotate and expand around the screen center as the high voltage is wobbled (Figure 6.46).

11. **Objective aperture centration.**

With the specimen in place, vary the lens current to the intermediate (diffraction) to form the caustic figure. A diffraction aperture is inserted and centered using the aperture centering controls. The diffraction image is sharpened; the objective aperture is inserted and moved until its outline is centered around the bright central spot (Figure 6.47). The diffraction aperture may then be removed and the TEM returned to the normal viewing mode.

12. **Objective lens stigmation.**

A holey film is inserted into the microscope and focused at high magnification (at least 50,000× or twice the magnification that one will be using). As the image of the hole is focused, a bright line, or Fresnel fringe, will be seen around the hole. Focusing is carefully accomplished and the symmetry of the white fringe around the hole is observed. In this just

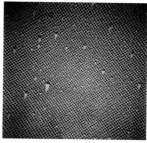

FIGURE 6.46 Expansion of the image around the voltage center of an electromagnetic lens (*left*). This image was obtained by modulating the high voltage and recording the image shifting during the modulation. The more distinct micrograph (*right*) shows the image after centering of the voltage and termination of the modulation. (*Courtesy of S. Schmitt.*)

FIGURE 6.47 Off-centered objective aperture (*left*) compared to the image of a properly centered objective aperture (*right*). (*Courtesy of Scott Pelok.*)

overfocused position, the white fringe will be readily seen against the background structure of the film. The stigmator is adjusted until the white overfocused fringe is symmetrically arranged around the edge of the hole (Figure 6.48). Note that the background details become sharper and more contrasted when astigmatism is corrected. In fact, this is another way to correct for astigmatism. Namely, a carbon or other finely structured film is focused as sharply as possible at a very high magnification. The stigmator controls are then used to fine focus the image to maximize contrast (Figure 6.49).

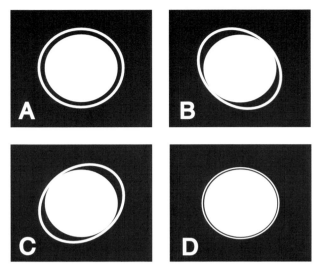

FIGURE 6.48 Correction of astigmatism. (A) Establish overfocused fringe. (B) Change focus until direction of astigmatism becomes apparent. (C) Place the stigmator direction or azimuth so that it is 90 degrees to the astigmatism. (D) Adjust strength or amplitude to generate a symmetrical field.

Major Operational Modes of the Transmission Electron Microscope

During the alignment procedure, one should be aware that the conventional transmission electron microscope may be set up for operation in several different operational modes. Depending on the design of the microscope, this may involve relatively few or many mutually exclusive adjustments. In addition, certain specimen preparation techniques may be utilized to further enhance these operational modes.

High Contrast

A constant problem with biological specimens is their low contrast. In the high contrast mode, the instrument is adjusted to give contrast at the expense of high resolution. As a result, this mode is generally used at magnifications under 50,000×. The conditions that may be changed to enhance contrast are summarized below.

How to Obtain High Contrast

1. **The focal length of the objective lens is increased.** This necessitates using shorter specimen holder cartridges (Figure 6.30, left) in a top entry stage to position the specimen higher in the objective lens. In a side entry stage,

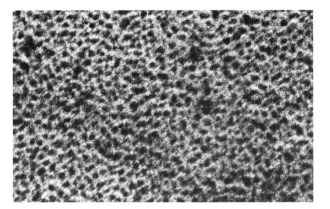

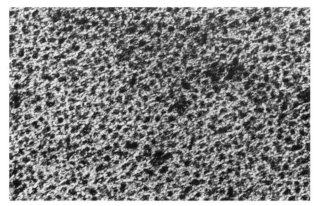

FIGURE 6.49 Alternate method for correcting astigmatism using Pt/Ir or carbon evaporated film. (*Top*) Focus grain image as much as possible using objective lens. (*Bottom*) Sharpen focus and contrast using stigmator controls.

adjustment of the z-axis or specimen positioning may also be needed if a special holder is not provided. It may be recalled that longer focal lengths result in narrower aperture angles, a worsening of chromatic aberration, and a loss of resolution.

2. **Lower accelerating voltages are used.** The resulting lower energy electrons are more readily affected by differences in specimen density and thickness, and contrast will be thereby increased. Unfortunately, this interaction with the specimen generates a population of imaging electrons with a wide range of energies, resulting in an increase in chromatic aberration. Lower accelerating voltages are also more damaging to the specimen, since the electrons are slowed down more and transfer more energy to the specimen, resulting in excessive heating. Lower energy electrons are more susceptible to poor vacuum conditions, with the

exacerbation of chromatic aberration. Clean, high vacuums are needed to minimize electron energy losses, and the microscope itself should be clean, since these electrons are more easily affected by astigmatism. Lastly, it will be recalled that lower energy electrons have longer wavelengths, so that the resolving power will be degraded.

3. **Smaller objective apertures should be utilized.** These apertures will remove more of the peripherally deflected electrons from the specimen, so that the subtractive image from the objective lens will be accentuated in contrast (i.e., the signal-to-noise ratio is increased). Small apertures are more prone to astigmatism problems, making clean vacuums and specimen anticontaminators essential.

4. **Photographic procedures may be employed.** Most images generated in the transmission electron microscope are enhanced for contrast using photographic techniques. During exposure of the electron micrograph, the sensitivity of the exposure meter may be adjusted to slightly overexpose the film. Underdevelopment will then enhance the contrast range in the final negative. Details will necessarily be lost in the intermediate density ranges. Of course, during the printing of the negative, one may use higher contrast photographic papers (see Chapter 8).

5. **The specimen may be prepared to enhance contrast.** Standard fixation and staining techniques will increase density by depositing the heavy metals along various organelles. Certain embedding media (polyethylene glycol) that may be dissolved or etched away will help boost contrast, or one may utilize stained, frozen sections without any embedding media. The easiest approach is simply to cut thicker sections; however, the resulting chromatic aberration and superimposition of structure will degrade resolution.

High Resolution

Most of the conditions used to achieve high resolution in the electron microscope are the opposite conditions discussed above for the high contrast mode. Since contrast will be lacking in these specimens, efforts should be made to boost contrast using appropriate specimen preparation and darkroom techniques, as described in the previous section.

How to Obtain High Resolution

1. **The objective lens should be adjusted to give the shortest possible focal length** and the proper specimen holders used. In some systems, this is simply a matter of pressing a single button, whereas, in certain microscopes several lens currents must be changed concomitantly. Perhaps it may even be necessary to insert a different polepiece in the objective lens.

2. **Adjustments to the gun, such as the use of higher accelerating voltages,** will result in higher resolution for the reasons already mentioned in the discussion on high contrast. Chromatic aberration may be further lessened by using field emission guns since the energy spread of electrons generated from such guns is considerably narrower. (The energy spread for tungsten = 2 eV while field emission = 0.2–0.5 eV.) In an electron microscope equipped with a conventional gun, a pointed tungsten filament will generate a more coherent, point source of electrons with better resolution capabilities.

3. **Use apertures of appropriate size.** For most specimens, larger objective lens apertures should be used to minimize diffraction effects. If contrast is too low due to the larger objective aperture, smaller apertures may be used but resolution will be diminished. In addition, they must be kept clean since dirt will have a more pronounced effect on astigmatism. Small condenser lens apertures will diminish spherical aberration, but this will be at the expense of overall illumination. The illumination levels may be improved by altering the bias to effect greater gun emissions; however, this may thermally damage the specimen.

4. **Specimen preparation techniques** may also enhance the resolution capability. Extremely thin sections, for instance, will diminish chromatic aberration. Whenever possible, no supporting substrates should be used on the grid. To achieve adequate support, this may require the use of holey films with a larger than normal number of holes (holey nets, see Chapter 4). The areas viewed are limited to those over the holes.

5. **Miscellaneous conditions** such as shorter viewing and exposure times will minimize contamination, drift, and specimen damage, and help to preserve fine structural details. Some of the newest microscopes have special accessories for minimal electron dose observation of the

specimen and may even utilize electronic image intensifiers to enhance the brightness and contrast of the image. Anticontaminators over the diffusion pumps and specimen area will diminish contamination and resolution loss. High magnifications will be necessary, so careful adjustment of the illuminating system is important. It may take nearly an hour for the eyes to totally adapt to the low light levels, and this adaption will be lost if one must leave the microscope room. Alignment must be well done and *stigmation must be checked periodically* during the viewing session. The circuitry of the microscope should be stabilized by allowing the lens currents and high voltage to warm up for 1 to 2 hours before use. Bent specimen grids should be avoided since they may place the specimen in an improper focal plane for optimum resolution. In addition, they prevent accurate magnification determination and are more prone to drift since the support films are often detached.

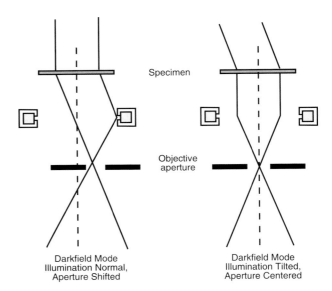

FIGURE 6.50 Schematic diagram showing two ways of setting up microscope for darkfield imaging: (*left*) displacement of objective aperture off-axis; (*right*) tilt of illumination system into on-axis objective aperture.

Darkfield

In the normal operating mode of the transmission electron microscope, the unscattered rays of the beam are combined with some of the deflected electrons to form a brightfield image. As more of the deflected or scattered electrons are eliminated using smaller objective lens apertures, contrast will increase. If one moves the objective aperture off axis, as shown in Figure 6.50, left, the unscattered electrons are now eliminated while more of the scattered electrons enter the aperture. This is a crude form of darkfield illumination. Unfortunately, the off-axis electrons have more aberrations and the image is of poor quality.

Higher resolution darkfield may be obtained by tilting the illumination system so that the beam strikes the specimen at an angle. If the objective aperture is left normally centered, it will now accept only the scattered, on-axis electrons and the image will be of high quality (Figure 6.50, right). Most microscopes now have a dual set of beam tilt controls that will permit one to adjust the tilt for either brightfield or darkfield operation. After alignment of the tilt for brightfield followed by a darkfield alignment, one may rapidly shift from one mode to the other with the flip of a switch. Both sets of controls also provide for separate stigmation controls to correct for any astigmatism introduced by the tilting of the beam to large angles.

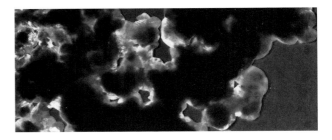

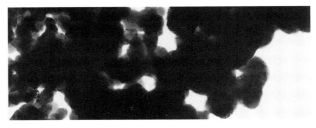

FIGURE 6.51 (*top*) Darkfield image obtained by tilting illumination system. (*bottom*) Same specimen viewed in standard brightfield mode. Specimen consists of inorganic salt crystals.

The darkfield mode can be used to enhance contrast in certain types of unstained specimens (thin frozen sections) or in negatively stained specimens. An example of a darkfield image is shown in Figure 6.51.

Diffraction

In specimens that contain crystals of unknown composition, the diffraction technique may be used to measure the spacing of the atomic crystalline lattice and determine the composition of the crystal, since different crystals have unique spacings of their lattices. *The diffraction phenomenon is based on the reflection or diffraction of the electron beam to certain angles by a crystalline lattice.* Instead of focusing a conventional image of the crystal on the viewing screen using the objective lens, one uses the intermediate or diffraction lens to focus on the back focal plane to see the selected area diffraction (SAD) on the screen. Since the crystalline lattice diffracts electrons to form bright spots on the viewing screen (similar to the mirrored rotating sphere sometimes used in ballrooms to reflect a light source onto the walls), the image will consist of a central, bright spot surrounded by a series of spots, which are the reflections. The central bright spot represents nondiffracted rays while the peripheral spots represent rays diffracted at various angles. The distance of these spots from the bright central spot is inversely proportional to the spacing of the crystalline lattice. A crystal with small lattice spacings will diffract the electrons to greater angles to give spots that are spaced far from the central spot. This is unfortunate for biologists, since organic crystals, such as protein, with large lattice spacings will diffract the beam so little that the spots will be crowded around the central bright spot and engulfed by its brilliance. With organic crystals, the specialized technique of high dispersion electron diffraction must be used.

Diffraction Practices

After the crystal is located (using the standard bright-field imaging mode), the crystal is centered on the viewing screen and the objective aperture removed. After placing the TEM in the selected area diffraction (SAD) mode, an SAD aperture of the appropriate size is inserted to select the area of the crystal one wishes to diffract. Focus sharply on the edge of the SAD aperture using the SAD (intermediate lens) control. Refocus the image using the objective lens focus controls to bring the image into the same plane as the intermediate aperture. (If contrast is inadequate at this point, temporarily reinsert the objective aperture to check focus and then remove it before proceeding.) Place the TEM into the diffraction mode (usually a button labeled "D" or "DIFF") and ensure that the second condenser (C2) lens is spread to prevent

burning of the viewing screen. For photography, adjust the size of the diffraction pattern using the camera length control, readjust the C2 lens so that the pattern is very dim, and focus the central bright spot as small as possible using the intermediate lens.

In order to cut down on the glare from the bright central spot, a physical beam stopper is inserted to cover it. Exposures are usually made for 30 to 60 seconds in the manual mode since the illumination levels will be very low. Single crystals will generate separate spots while polycrystalline specimens will produce so many spots around the central point that they will blend to form a series of concentric rings (Figure 6.52). Some biological applications of diffraction may be to confirm that a crystal present in human lung tissue is a form of asbestos, or to identify an unknown crystal in a plant or bacterial cell. See also Chapter 15 and the reference sources at the end of this chapter.

Checking Performance

Alignment

It is possible to check and make minor corrections to the alignment of the electron microscope in less than 10 minutes, as follows:

Without a specimen in the TEM, verify that the filament cloud is symmetrical, the condenser lens is stigmated, and the C1 aperture is centered. Check that the C2 aperture is centered by varying the current to the C2 lens and observing that the illumination stays centered. Insert a holey film and set the microscope at a low magnification. Bring the illumination to the small-

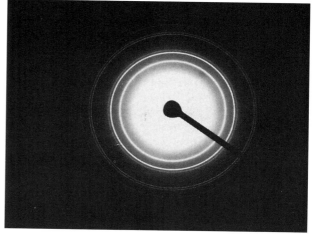

FIGURE 6.52 Diffraction pattern obtained from polycrystalline specimen showing characteristic ring pattern.

est possible crossover spot size and then increase to the top magnification. Both the illumination spot and the image should stay close to the center of the screen. If either move off screen, then adjustments will be needed. Now, repeat this process using the largest spot selectable by C1 and make corrections if the illumination does not stay centered. Insert a specimen and tilt the stage $\pm10^0$ to see that the specimen stays centered. Wobble the high voltage at a high magnification and observe that the focused image rotates around the center of the screen. Go to the diffraction mode with the intermediate lens and observe that the caustic pattern is centered. Check centration of the objective aperture while in the diffraction mode. Finally, stigmate the microscope if necessary.

Electrical Stability

The microscopist should be aware that a recently turned on microscope will be somewhat unstable until the high voltage and lens circuits have warmed up for perhaps 30 to 60 minutes. If imaging problems are still encountered and instabilities in the microscope are suspected, several areas may be checked as follows:

Problems with the filament may be checked by undersaturating the filament and carefully observing its focused image on the screen. Changes in the size or brightness of the halo may indicate problems with the vacuum, bias controls, contamination in the gun area, or it may indicate that a weak filament is about to burn out. Instabilities in the high voltage, diffraction, or projector lenses may be detected by observing the caustic image. If the caustic image changes in sharpness or moves, then instabilities in these circuits should be pursued further. Objective lens instabilities may be detected by observing an overfocused Fresnel fringe in a holey film. Should the fringe change in any way, objective lens circuits should be checked. All electron microscopes have built-in gauges or test points in circuit boards that may be monitored using appropriate testing equipment. However, unless one has a thorough understanding of the process and an appreciation of the dangers involved, this is best left to trained personnel. At least one will have diagnosed the problem as a microscope—not a specimen—problem.

Image Drift

Gradual shifting of the image, usually in one direction, is a common and annoying problem, encountered especially when support films are not used.

The most common cause of image drift is a dirty specimen holder. All holders should be checked and cleaned on a routine basis. Drift might also be caused by heating of the specimen due to excessive beam irradiation, in which case a smaller spot size, condenser aperture, or less filament emission may be tried. If the section or plastic substrate is not firmly attached to the grid, it will move as the grid heats up—the amount of movement is related to the beam intensity and the area illuminated. This may be confirmed by examining a test specimen known to be thermally stable. Contamination in the area above the specimen may lead to charging and shifting of the image as the static is discharged. This should be suspected if the image shifts as the C2 illumination level is changed. Moderate drift of the specimen on the screen may be seen through the viewing binoculars. An easy way to test a specimen for drift is to place a recognizable specimen structure next to a fixed point on the viewing screen and observe the movement of the specimen over an interval 2 to 3 times greater than the exposure time for the negative. Drift may be documented by taking an exposure of an overfocused hole, waiting for 1 to 2 minutes with the specimen still being irradiated, and then taking a second exposure on the film. This double-exposed film is then developed and the distance traveled by the hole may be converted into nanometers traveled per second (Figure 6.53). For example, if one is hoping to resolve 1 nm and the exposure time is 4 seconds, then the specimen should not drift any more than 1 nm per 4 seconds. At a magnification of 500,000×, the drift should not exceed 5 mm over 4 seconds. (See the section "Magnification Calibration" later in this chapter.)

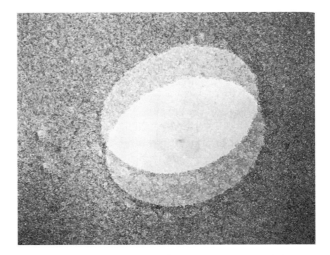

FIGURE 6.53 Drift in an electron microscope. Two exposures are made on the same negative with an interval of 1 to 2 minutes between exposures. The rate of drift may be calculated based on the distance moved over the intervening time period.

Contamination

Deposition of contaminants onto the specimen that is being subjected to electron bombardment will degrade resolution. One should be prepared to quantitate the rate of contamination in order to determine when it has become unacceptable for the resolution level needed.

The contamination rate may be quantitated by double-exposing a hole in a test specimen at a magnification of 250,000× as described in the previous paragraph. The decrease in the diameter of the hole, by deposition of contamination along the rim, is an excellent indicator of the rate of contamination. A 2 to 3 minute exposure should indicate a contamination rate of less than 0.01 nm per second with all anticontaminators in operation. Without a specimen anticontaminator, the rates may be ten times higher even in a clean microscope. It should be realized that a recently cleaned microscope column is actually quite contaminated with organic vapors used in the cleaning procedure, so that the microscope should be allowed to remain in high vacuum for one full day in order to remove these molecules. The anticontaminators on the diffusion pumps should be filled with liquid nitrogen, but the specimen anticontaminator should not be chilled during that time, since it will quickly load up with contaminants that may later be released onto the specimen.

Magnification

It is necessary to calibrate the magnification settings of all electron microscopes, since different mechanical and electronic alterations will result in significant variations. Even in the most modern of microscopes, some manufacturers warrant the figure displayed to be within only ± 10% of the actual value.

Although it is possible to calibrate the entire magnification range in the microscope, this may not be necessary if one uses only certain settings. The frequency of calibration varies with the individual need for accuracy, as well as with the servicing intervals for major systems in the microscope. Once or twice a year may be sufficient for routine work where high accuracy is not required. In critical operations, calibration should take place during each use of the microscope or, alternately, one may include a standard of known size on the same grid as the specimen. A number of situations may cause a change in previously determined magnification figures.

After *cleaning* operations involving the polepieces, specimen stage, or specimen holders, magnifications should be recalibrated. *Servicing* of electronics may affect magnifications if lens current or reference circuits were repaired.

A major cause of erroneous figures arises from lens *hysteresis* (also called *remanence*), or the residual magnetism left in the soft-iron polepieces even after the electromagnetic field strength has been changed. This may be minimized by starting the magnification calibration series at a low magnification and taking micrographs at increasingly higher magnifications. Similarly, the low- to high-magnification scheme should be followed when viewing and recording images. One should also verify that the calibration is accurate at different accelerating voltages.

The placement of the *specimen height* in the objective lens is very critical. Since the position of the grid must be within 40 μm for accurate replication of a particular magnification setting, the grid must be perfectly flat and the z-axis (specimen height) must be accurately set. This is a significant consideration with a side-entry stage since the z-axis is adjustable over a very wide range. If the stage is tilted, the x and y traversement will result in a change of specimen height. Bent grids should be avoided, and the specimen should always be on the same side of the grid relative to the beam.

Magnification Calibration

Calibration may be accomplished by first acquiring accurate standards with established sizes and properties. One convenient method that may be used to calibrate the entire magnification range in the TEM involves the use of a commercially prepared, cross-section of a single-crystal, semiconductor wafer. The standard, called MAG*I*CAL™, covers magnifications from 1,000× to 1,000,000×. When viewed in the TEM, the standard appears as a series of alternating light (silicon) and dark (SiGe alloy) bands of accurately determined thicknesses. Since the series of bands range from 10 nm to over 5 μm, calibrations are accurate over the wide range mentioned. To determine the magnification (M) based on the standard, one uses the simple equation below:

$$M = \frac{\text{distance measured on negative}}{\text{known distance on standard}}$$

For example, on a negative, if one measured the spacing between the bands to be 10 mm, while the distance is known to be 10 nm on the standard, then the magnification is one million times, or:

$$M = \frac{1 \times 10^{-3} \text{ meter}}{1 \times 10^{-9} \text{ meter}}$$

Using conventional methods of calibration, however, more than one standard will be needed to cover the entire magnification range. For instance, from the lowest magnifications up to perhaps 40,000 to 50,000×, one may use commercially available cross-ruled *diffraction grating replicas* with 2,160 lines/mm (Figure 6.54). The procedure followed is to focus carefully on the granularity of the grating and take an electron micrograph at the desired settings. After development, the negatives are placed on an illuminated view box and distances measured in millimeters. Measurements should be made from the same relative positions in each line (e.g., middle of the dark band of one line to a similar position on the second line). One then counts the number of spaces included in the specific millimeter distance measured. These values are substituted into Equation 6.10.

Equation 6.10: Magnification Calculation from Diffraction Grating Replica

$$M = \frac{2.16\,(A)}{B} \times 10^3$$

where: M = magnification
  A = distance in mm between lines on electron micrograph
  B = number of spaces between lines

Several *sources of error* exist with the diffraction grating method. The grating is a platinum/carbon replica of a standard optical grating and may have undergone some distortion during the mounting on the grid, thereby affecting the accuracy of measurements. Generally "waffle" gratings are preferred to the parallel line gratings, since the crossed lines permit measurements in both directions to increase the accuracy. In order to have an accuracy of ± 2%, it is necessary to include at least 10 spaces in the distance measured. In practical terms, this means that gratings give a ± 2% error only up to a magnification of 25,000×. However, if a ± 5% variability is acceptable, then they may be used up to 50,000 to 60,000×. At the higher magnifications, in order to minimize the error involved in counting only a fragment of a space, one measures a small piece of debris at a known accurate magnification and then remeasures this same object at the unknown magnification. The amount of enlargement between the two is the factor by which the known setting is multiplied. An object measuring 5 mm at 20,000× and now measuring 12.5 mm is 2.5× larger, so that the new magnification is 2.5× 20,000 = 50,000×.

Above 50,000×, one must resort to *organometallic crystals* with established lattice spacings. A good standard is beef liver catalase prepared by placing a drop of the catalase suspension on a coated grid for 5 to 10 seconds and then negatively staining the specimen by floating the grid on a drop of 2% aqueous solution of ammonium molybdate or potassium phosphotungstate (see Chapter 5). The largest lattice, beef catalase, has spacings of 8.8 nm. Alternative standards might include bacteriophage or tobacco mosaic virus particles that may be obtained from colleagues or purchased from the American Type Culture Collection.

Resolution

Most electron microscopes have a guaranteed optimum resolution figure that was verified by the manufacturer usually upon installation of the microscope. As long as one obtains satisfactory micrographs, it may be mistakenly assumed that the resolving power has not changed. One should be aware of several methods to verify this, since the degradation of resolving power may be so insidious

FIGURE 6.54 Magnification calibration standard. This series of micrographs are of a standard diffraction grating containing 2,160 lines/mm. The magnifications were calculated to be 10,400×; 21,000×; and 30,200×, respectively.

as to go unnoticed until the quality of work is brought into question, perhaps to the embarrassment of the microscopist.

The *point-to-point method* is the most readily accepted method since it graphically demonstrates the ability of the microscope to visualize two fine points separated by a specific distance. As a test specimen, one may use either a thin carbon film alone or with a thin film of platinum-irridium alloy evaporated onto the carbon. A series of micrographs are made at the top magnification needed to demonstrate the resolution (300,000 to 500,000×). Even if all conditions are perfect, it is very difficult to obtain satisfactory results from a single micrograph, since precise focusing is essential. Consequently, a *through-focus series* is made by first focusing the image as best as possible and then backing off the finest focus knob counterclockwise by two click stops. Micrographs are then rapidly taken, advancing the focus clockwise for each exposure until a total of five shots have been made. The five micrographs are later examined on a light box and several separated points are located on the two films that are closest to focus. The smallest distances between two points are located on both negatives and converted into actual nanometer distances based on an accurate knowledge of the magnification (Figure 6.55). It is important to locate the points on two films, since electron noise ("snowy" background) may give an impression of two points that do not physically exist. *A through-focus series may be used with some specimens in order to obtain the most accurate focus with the highest possible resolution.*

The *lattice test* is based on demonstrating a crystalline lattice of known spacings. Simply stated, if one sees the lattice structure in a graphitized carbon particle or a single crystal gold foil (Figure 6.56), then distances of 0.34 nm and 0.20 nm, respectively, are being resolved. There are several problems with taking such figures as being strictly accurate, since astigmatism and electron noise may artificially enhance the resolving capabilities supposedly being demonstrated. Most microscope manufacturers will provide two figures for resolving power, for example, lattice = 0.14 nm and point-to-point = 0.30 nm.

The *Fresnel fringe method* (Reisner, 1956) (Figures 6.57 and 6.58) for calibrating resolution is undoubtedly the most convenient method since the test specimen, a holey film, is readily available to all microscopists. After correcting for astigmatism, the fringe is focused as accurately as possible and a series of micrographs is taken by varying the focus slightly between the various exposures. The films are examined for that negative that shows the finest fringe by being slightly overfocused. The fringe width is then evaluated by measuring the distance from the center of the overfocus fringe

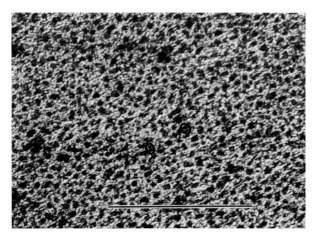

FIGURE 6.55 Resolution standard, evaporated Pt/Ir film showing a resolution of 0.4 nm at points circled. Magnification bar = 50 nm.

FIGURE 6.56 Resolution standard, gold foil. The lattice spacings show a resolution better than 0.204 nm. Final magnification of print is 2.4 million times.

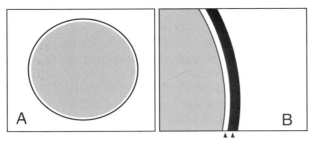

FIGURE 6.57 (A) Convenient resolution standard, Fresnel fringe method. The fringe width is measured, based on known magnification, and the finest fringe width measured is equal to resolution. (B) Enlargement showing distance to be measured indicated by arrowheads.

FIGURE 6.58 Positive print showing edge of holey film with Fresnel fringe (*white line, arrow*) in overfocused condition.

(e.g., the dark line on the negative) to the center of the white space just inside of the dark line (Figure 6.57B). This distance in nanometers is the resolving power of the instrument. Of course, one must have a good magnification calibration in order to determine the fringe width accurately. Since this method is based on accurate stigmation and focusing of the microscope, an error of 20% is not unusual. However, with experience this simple method may prove adequate for all but the most precise situations.

Levels of Usage of the Transmission Electron Microscope

The preceding chapter was directed toward giving the investigator a sound background in the fundamentals of the transmission electron microscope. For serious, committed electron microscopists maintaining their own equipment, this level of detail—and even much more—is a necessity. However, there are several other levels of experience that might be appropriate for individuals using an electron microscope.

The experienced user and maintainer. This is an individual who has responsibility for the operation and maintenance of the microscope. Considerable experience in the theoretical basis of microscopy, electronics, and a mechanical inclination are necessary. If the microscope fails, and even new ones do fail, this individual must have the capability to repair it. Rarely do electron microscope users have full maintenance capability. If the microscope breaks down or if specialized replacement parts are needed, then service representatives may be called upon.

The experienced user with a maintenance contract. Such researchers are involved in the routine operation of the microscope and the recording of images. The level of familiarity with the instrument may include routine alignment and perhaps routine servicing (cleaning of apertures and specimen anticontaminators, replacement of burnt-out filaments, calibration of magnification and resolving power, etc.). Major repairs and routine cleaning of sensitive microscope parts (polepieces, specimen stages, column liners, etc.) may be left to service engineers as part of an annual service contract. Such yearly contracts may vary from 5 to 10% of the original cost of the instrument. In situations where productivity cannot be interrupted by a researcher becoming involved in microscope maintenance, such contracts are probably quite cost effective. In addition, the contracts may include the replacement of extremely expensive parts (high voltage tanks, vacuum system components, electronic circuit boards, stages and specimen holders, etc).

The assisted viewer. Individuals who have only a temporary need for electron microscopy may rely on more experienced personnel to assist them in specimen preparation and the viewing of specimens in the electron microscope. Often, researchers in need of electron microscopy will contact a microscopist at their own or perhaps at nearby institutions to arrange for the needed services. In this case, the researcher may simply view the specimens while a trained individual oversees the operation of the microscope. With time, and a continued need for electron microscopy, the researcher may opt to become more personally involved in the operational procedures. However, a more efficient use of the researcher's time may still involve assistance by an experienced microscopist.

A user contracting for service. Occasionally a researcher may need electron microscopy as a small part of a study, perhaps to confirm some data obtained by other techniques. The researcher may have little experience in image interpretation and no experience or interest in learning even the basics of microscope operation. Such individuals usually contact *consultants* who are experienced in the area of research in which the investigator is involved and who have access to an electron microscope. The researcher may send specimens to the consultant, who then studies the specimens in the electron microscope, records images, and provides the researcher with photographic prints and perhaps even an analysis of the micrographs. Such consultants generally command premium fees since they are experienced in both electron microscopy as well as the interpretation of the images.

Shared Facilities

Due to the high cost of acquiring and maintaining an electron microscope and associated support equipment, shared facilities are becoming more common than individual researchers having such units. A shared or centralized facility may involve only one electron microscope serving several researchers with similar research interests or it may be quite extensive with many electron microscopes equipped for various specialized needs. Users of such facilities may be operating at different levels of experience, so that trained microscopists and technologists may be involved in various service and training activities. Central facilities are usually financially supported to various extents by a central administration and/or by a system of external grants and contracts. Often there is an established fee structure for the various levels of usage of the facility.

REFERENCES

Agar, A. W., R. H. Alderson, D. Chescoe. 1974. Principles and practice of electron microscope operation. In *Practical methods in electron microscopy,* Vol. 2., A. M. Glauert, ed. Amsterdam, The Netherlands: Elsevier/North Holland Biomedical Press.

Beeston, B. E. P., R. W. Horne, and R. Markham. 1972. Electron diffraction and optical diffraction techniques. In *Practical methods in electron microscopy,* Vol. 1 of *Practical methods in electron microscopy* Series, A. M. Glauert, ed. Amsterdam, The Netherlands: Elsevier/North Holland Biomedical Press.

Bigelow, W. C. 1994. Vacuum methods in electron microscopy. In *Practical methods in electron microscopy,* Vol. 15 of *Practical methods in electron microscopy* Series, A. M. Glauert, ed. London: Portland Press.

Chapman, S. K. 1986. Maintaining and monitoring the transmission electron microscope. In *Practical methods in electron microscopy,* Vol. 8 of *Practical methods in electron microscopy* Series, A. M. Glauert, ed. Herndon, VA: BIOS Scientific Publishers.

Chescoe, D., and P. J. Goodhew. 1990. The Operation of transmission and scanning electron microscopes. In *Practical methods in electron microscopy,* Vol. 20 of *Practical methods in electron microscopy* Series, A. M. Glauert, ed. Herndon, VA: BIOS Scientific Publishers.

Goldstein, J. I., D. E. Newberry, P. Echlin, D. C. Joy, C. Fiori, and E. Lifshin. 1981. *Scanning electron microscopy and x-ray microanalysis. A textbook for biologists, materials scientists and geologists.* New York: Plenum Press.

Meek, G. A. 1976. *Practical electron microscopy for biologists.* New York: John Wiley and Sons.

Misell, D. L., and E. B. Brown. 1987. *Electron microscopy for biologists.* New York: John Wiley and Sons.

Misell, D. L., and E. B. Brown. 1987. Electron diffraction: An introduction for biologists, Vol. 12 of *Practical methods in electron microscopy* Series, A. M. Glauert, ed. Amsterdam, The Netherlands: Elsevier/North Holland Biomedical Press.

O'Hanlon, J. F. 1980. *A user's guide to vacuum technology.* New York: John Wiley and Sons.

Reisner, J. H. 1956. Practical aspects of lens correction for astigmatism. Parts I and II. *Sci Instr News* 1:5–12.

Varian Vacuum Products Division. 1986. *Basic vacuum practice.* Palo Alto: Varian Associates Vacuum Products Division.

Wischnitzer, S. 1981. *Introduction to electron microscopy.* New York: Pergamon Press.

Chapter 7

Courtesy of AMRAY.

The Scanning Electron Microscope

ABOUT THE SAME TIME the first transmission electron microscope (TEM) was nearing completion in the 1930s, a prototype scanning electron microscope (SEM) was constructed by Knoll and von Ardenne, in Germany. Unfortunately, the resolution of this first instrument was no better than the light microscope. Following several refinements made by Zworykin at the RCA laboratories in the United States, as well as improvements made by McMullan and Oatley at Cambridge University in England, a commercial SEM became available in 1963. A later version, the Cambridge instrument, shown in Figure 7.1A, had resolving powers of about 20 to 50 nm, useful magnifications of 20 to 75,000×, and a *depth of field* (e.g., depth in the specimen that was in focus) that was 300 times greater than the light microscope. A recent model SEM is shown in Figure 7.1B. Modern instruments resolve typically 2.0 nm with magnifications up to 200,000×.

FIGURE 7.1A One of the first commercially produced SEMs, the Cambridge Mark II Stereoscan.

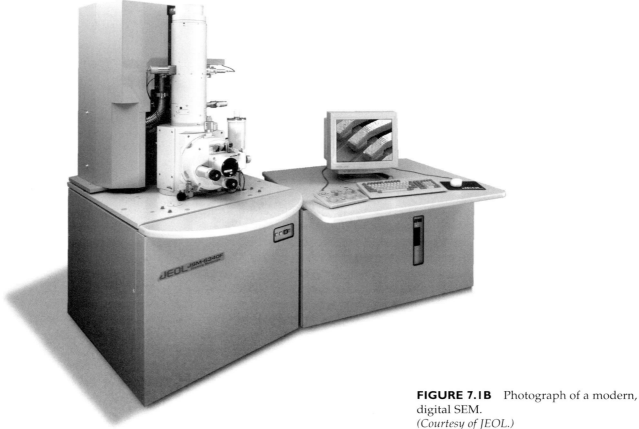

FIGURE 7.1B Photograph of a modern, digital SEM.
(Courtesy of JEOL.)

FIGURE 7.1 C, D, and E Some SEMs have a small broadfield optical microscope to help locate the area of interest in the specimen. Once located using light optics, the electron optics may be utilized to achieve high resolution with great depth of field. (D) Low power light optical image of stamens in flower. (E) SEM image obtained from same area.
(Courtesy of Topcon Technologies, Inc.)

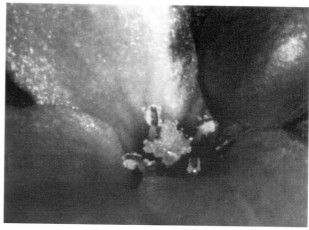

D

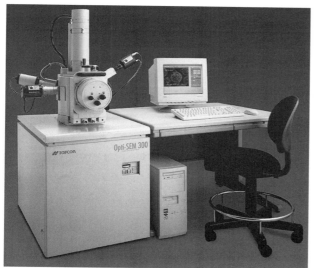

C

E

The SEM received much notice in the popular press due to the readily recognizable, greatly enlarged, three-dimensional images of insects, flowers, pollen, etc. Even now it is not unusual to find popular books dedicated to the beautiful, often surrealistic images observed with these instruments (Figure 7.2). In contrast to the TEM, that is used to view *thin* slices of biological specimens, the SEM can be used by biologists to study the *three-dimensional* features of individual cells and even whole organisms. Some SEMs permit one to insert specimens 3 to 12 inches in size into the specimen chamber and view them with a depth of field of several millimeters. In addition, the SEM is a standard instrument in the electronics industry where it is often situated in the fabrication area to control the quality of microcircuitry.

Even though the lenses of the SEM are constructed similarly to those of the TEM (Chapter 6),

these electromagnetic lenses do not form an image of the specimen according to the optical principles used by the more conventional light and transmission electron microscopes. Instead, the lenses of the SEM are used to generate a demagnified, focused spot of electrons that is scanned over the surface of an electrically conductive specimen (Figure 7.3). As these impinging electrons strike the specimen, they give rise to a variety of signals (see Chapter 15), including low energy *secondary electrons* from the uppermost layers of the specimen. Some of the secondary electrons are collected, processed, and eventually translated as a series of *pixels* (picture elements) on a cathode ray tube or monitor. For each point where the electron beam strikes the specimen and generates secondary electrons, a corresponding pixel is displayed on the viewing monitor. The brightness of the pixel is directly proportional to the number of secondary electrons generated from the

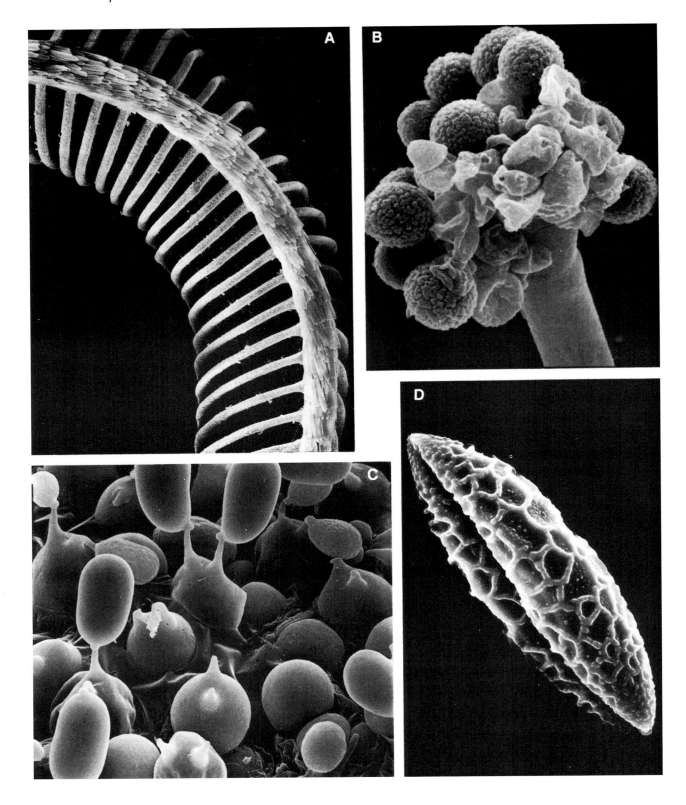

FIGURE 7.2   (A) Antenna of moth, 50×. (B) Fungal fruiting body on soybean leaf, 1,700×. (C) Spore structures on underside of mushroom, 2,500×. (D) Day lily pollen grain, 750×.
(A and D, courtesy of S. Schmitt. C, courtesy of A Krajec.)

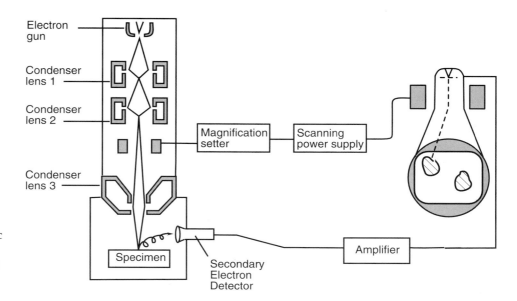

FIGURE 7.3A Schematic diagram of SEM showing basic components making up the complete system.

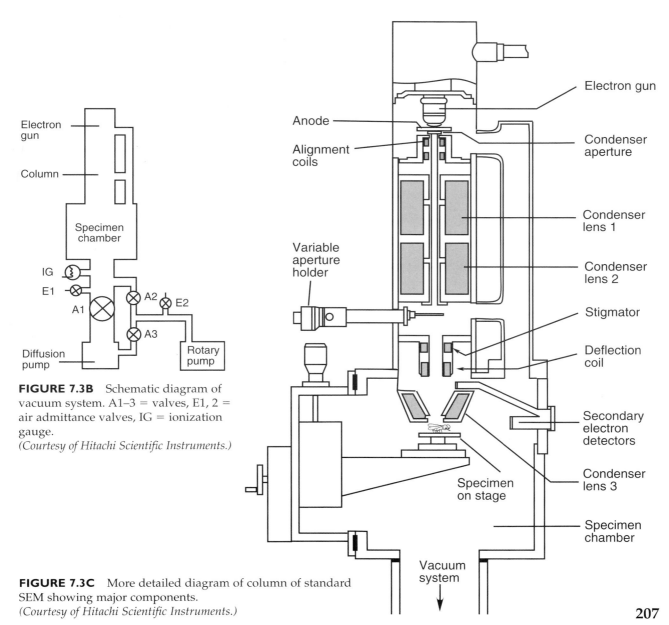

FIGURE 7.3B Schematic diagram of vacuum system. A1–3 = valves, E1, 2 = air admittance valves, IG = ionization gauge.
(Courtesy of Hitachi Scientific Instruments.)

FIGURE 7.3C More detailed diagram of column of standard SEM showing major components.
(Courtesy of Hitachi Scientific Instruments.)

specimen surface. Since the electron beam is scanned rapidly over the specimen, the numerous, minute points appear to blend into a continuous-tone image composed of many density levels or shades of gray. The shading is similar to an ordinary black and white photograph in which light and dark areas give the impression of depth.

Basic Systems of the SEM

The SEM may be subdivided into a number of component systems that carry out various functions. Among these systems, a *lens system* is involved in producing a small, focused spot of electrons that are then rastered over a specimen surface by means of a *scan deflection system*. A *specimen stage* is needed so that the specimen may be inserted and situated relative to the beam. A *secondary electron detector* is used to collect the electrons and to generate a signal that is processed by electronics and ultimately displayed on *viewing* and *recording monitors*. Modern SEMs also have the capability for storing and processing digital images. A *vacuum system* is necessary to remove air molecules that might impede the passage of the high energy electrons down the column—as well as to permit the low energy secondary electrons to travel to the detector.

Certain components of the SEM are identical to those found in the TEM. Because the quality of the SEM image is dependent on the signal (i.e., the numbers of electrons generated from the specimen surface), an intense electron source is essential to generate a good yield of secondary electrons from the specimen. Consequently, high resolution SEMs use lanthanum hexaboride or field emission guns. The construction of the electron gun, electromagnetic lenses, and vacuum systems are otherwise similar in both instruments. It is necessary to read the sections in Chapter 6 on the TEM prior to reading this chapter because Chapter 7 will deal only with those components and functions that differ in the SEM. An overall schematic representation of the SEM is shown in Figure 7.3.

Electron Optical and Beam Control Systems

This system of electromagnetic lenses, deflection coils, and stigmators is involved in the control and refinement of the electron beam after it leaves the electron gun and before it strikes the specimen. As in the TEM (Chapter 6), most SEMs use a V-shaped, tungsten filament that is heated to effect the thermionic emission of electrons, which are accelerated in the direction of the anode due to the application of negative high voltage. After leaving the bias shield and forming an initial focused spot of electrons of approximately 50 μm in diameter, a series of two to three condenser lenses are used to successively demagnify this spot sometimes down to 2 nm or less. As will be seen later, *small spot sizes are essential for the better resolutions required at high magnifications.*

Condenser Lenses 1 and 2

The number of condenser lenses may vary from two to three, depending on the resolving powers needed. The *first condenser lens, C1,* begins the demagnification (i.e., decrease in spot size) of the 50 μm focused spot of electrons formed in the area of the electron gun.

As the amount of current running through the first condenser lens (C1) is increased, the focal length of the lens becomes progressively shorter and the focused spot of electrons becomes smaller. It is apparent from studying Figure 6.28 that a short focal length C1 lens subsequently causes such a wide divergence of the electrons leaving this lens that many electrons are not able to enter the next condenser lens (C2). *The overall effect of increasing the strength of C1 is to decrease the spot size, but with a loss of beam electrons: Resolution improves, but the overall signal (number of secondary electrons) coming from the specimen will be weaker since fewer beam electrons strike the specimen.*

Apertures are placed in the lenses to help decrease the spot size and to reduce spherical aberration (Chapter 6) by excluding the more peripheral electrons. Each of the condenser lenses behave in a similar manner and possess apertures, some of which may be either fixed in size and placement in the column or which may be variable and adjustable using controls on the column of the SEM.

Final Condenser Lens

The *final condenser lens,* often inappropriately called the objective lens, is the strongest lens in the SEM and does the final demagnification of the focused spot of electrons. This final or third lens is used primarily to fine-tune the spot size without a loss of beam electrons. As will be discussed later, *the final lens is used to focus the image seen on the monitor or cathode ray tube (CRT).*

The final condenser lens usually contains two sets of *deflection coils* and a stigmator. The deflection coils are connected to a *scan generator* to raster the electron spot across the specimen. Rastering not only moves the spot in a straight line across the specimen, but also moves the spot down the specimen as well (i.e., possesses both x and y movements) (Figure 7.4A). A *change of magnification* is achieved by varying the length that the beam is scanned over the specimen versus the length displayed on the viewing screen. For instance, if the electron probe is scanned over a 10 mm distance on the specimen, and displayed on the monitor at a final length of 10 cm, this represents a magnification of 10×. Going to a smaller scan length of 1 μm would give a final magnification of 100,000× on the 10 cm viewing screen. Therefore:

$$\text{Magnification} = \frac{\text{length displayed on screen}}{\text{length scanned on specimen}}$$

Dual Magnification Mode

A useful feature present on some SEMs is the *dual magnification* mode that permits the simultaneous viewing of two different magnifications either on two separate viewing screens (some SEMs have more than one) or on a single screen that is split into two areas. This is accomplished by sending alternate scans taken at different magnifications to different display screens. For instance, the first scan across the specimen may be at 100×, while the next scan would be at a magnification of 1,000×. If all of the odd-numbered scans are then displayed on the 100× area, while the alternate even-numbered scans are displayed on the 1,000× area of the screen, one sees the two different magnifications at the same time (Figures 7.4A and 7.4B.) This feature is useful for rapidly locating an area of interest at a low magnification while still being able to scrutinize detail at a higher magnification.

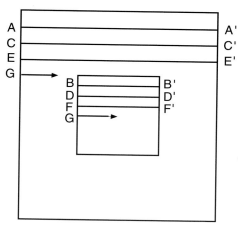

FIGURE 7.4B In the dual magnification mode, one set of lines (A to A′, C to C′, etc.) is scanned a particular length over the specimen; while alternating lines (B to B′, D to D′, etc.) are scanned a shorter distance. *(Redrawn from Postek, et al, 1980.)*

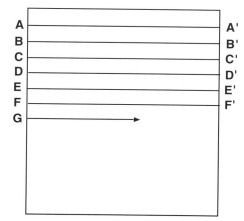

FIGURE 7.4A The focused beam of electrons is scanned in a raster pattern over the specimen surface. The first scan is from A to A′, with the beam moving down and then scanning line B to B′, etc. *(Redrawn from Postek, et al, 1980.)*

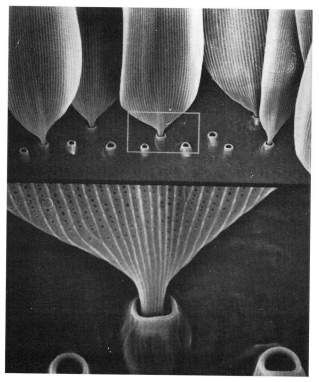

FIGURE 7.4C Scanning electron micrograph of scales on butterfly wing taken using the dual magnification function. The small box in the top half of the micrograph indicates the area enlarged in the bottom half of the micrograph. Both images are viewed simultaneously in the SEM. *(Courtesy of S. Schmitt.)*

Stigmator Apparatus. The *stigmator (stigma* means "mark" or "spot" in Latin) is a device that is used to control any distortions in the roundness of the spot formed by the electron probe that is scanned over the specimen. Since the spot on the viewing monitor is round, it is important that secondary electrons from the corresponding point on the specimen emanate from a round spot. For instance, if the electron beam spot is not round but elliptical, then extraneous information (i.e., too many secondary electrons) is being put into the round spot on the viewing monitor (Figure 7.5A). Beam spots that are not round will generate an image on the viewing monitor that is smeared in one direction (Figure 7.5B). This phenomenon, called *astigmatism,* is one of the major causes of loss of resolution in the SEM. The major reason for astigmatism is contamination on one of the apertures (usually the final aperture), since this causes a distortion of the symmetrical electromagnetic field.

Stigmators usually consist of 6 to 8 small electromagnetic coils inside the lens bore of the final condenser lens. As described in Chapter 6, stigmators are able to correct the asymmetrical distortions to the electromagnetic field by introducing an opposing field of appropriate strength (amplitude) in the proper direction (azimuth) so as to counterbalance the offending field (Figure 7.5C). This is usually accomplished manually; however, the newest SEMs have *semiautomatic stigmation* mod-

ules that permit the rapid and precise correction of astigmatism, which saves time and assures accurate stigmation.

Apertures in the Final Condenser Lens. The final lens usually has externally adjustable apertures that may be readily varied in size. Normally, final apertures on the order of 50 to 70 μm are used to generate smaller, less electron dense spots for secondary electron generation and imaging. Larger apertures (200 μm or so) are used to generate larger spots with greater numbers of electrons. These large spots con-

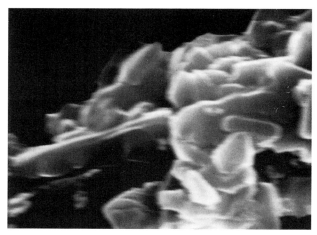

FIGURE 7.5B   Scanning electron micrograph showing astigmatism in the EW direction due to elongated spot shown as "b" in Figure 7.5A.

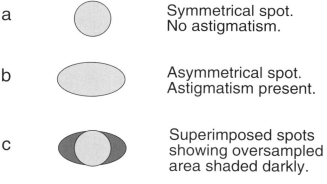

a Symmetrical spot. No astigmatism.

b Asymmetrical spot. Astigmatism present.

c Superimposed spots showing oversampled area shaded darkly.

FIGURE 7.5A   Comparison of electron beam spots focused on specimen. A properly sized symmetrical spot (*a, above*) samples only the specimen areas appropriate for the magnification and resolution desired (See Equation 7-2). In b, the asymmetrical spot is appropriately sized in the NS direction but excessively large in the EW direction. The image will therefore appear in focus in the NS direction but smeared in the EW direction. In c, one sees the oversampled area darkly shaded.

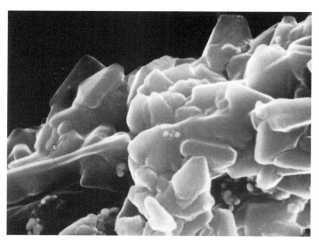

FIGURE 7.5C   Scanning electron micrograph showing correction of astigmatism (shown in Figure 7.5B, *above*). Stigmators were used to force the previously elliptical spot in Figure 7.5B into a symmetrical shape.

tain a great deal of energy and may damage fragile specimens. They are used primarily to generate X-rays for elemental analysis rather than for imaging purposes (Chapter 15).

Apertures are important parts of electron microscopes because they not only affect spot size and beam current, but also affect depth of field and help diminish spherical aberration.

Apertures and Depth of Field

Apertures can be used to control the *depth of field* in the specimen. Depth of field refers to the depth in the specimen that appears to be in focus. As discussed in Chapter 6, depth of field is expressed as Equation 7.1.

Equation 7.1: Depth of Field

$$D_{fi} = \frac{\lambda}{NA^2}$$

where λ = wavelength of illumination
NA = numerical aperture

Variation in the depth of field as a result of changes in aperture size is shown in Table 7.1. As illustrated in Figures 6.36 and 7.6, smaller apertures generate narrower beams with smaller aperture angles. The diameter of the beam (i.e., spot size) varies less along the length of such narrow beams compared to wider angled beams. Consequently, as the beam is scanned along the contours of the specimen, the spot size varies less at the various levels, so that the specimen will appear sharply in focus at the various levels.

NOTE: *As the working distance increases, the aperture angle becomes narrower (Figure 7.7). Consequently, the depth of field figures given in Table 7.1 are valid only at a working distance of 10 mm.*

Depth of field is also affected by the distance the specimen is situated from the final condenser lens, the so-called *working distance*. From Figure 7.7 it can be seen that the aperture angle decreases as one increases the working distance. Consequently, the depth of field will increase as one increases the working distance. Unfortunately, this increase in depth of field is at the expense of resolution, since the numerical aperture decreases and since long focal length lenses are more susceptible to chromatic aberration (see Chapter 6). The relationship between working distance and image quality is summarized in Table 7.2.

TABLE 7.1 Effect of Aperture Size on Depth of Field at Various Magnifications and at a 10 mm Working Distance

| Mag | Depth of Field | | |
	100 μm Aperture	200 μm Aperture	600 μm Aperture
10×	4 mm	2 mm	670 μm
50×	800 μm	400 μm	133 μm
100×	400 μm	200 μm	67 μm
100,000×	0.4 μm	0.2 μm	0.067 μm

Source: Goldstein, et al., 1981. Scanning Electron Microscopy and X-ray Analysis: A Text for Biologists, Materials Scientists, and Geologists. New York: Plenum Publishing Corp., 134.

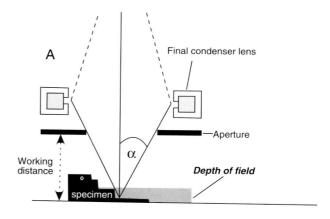

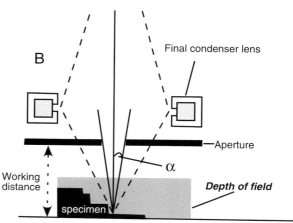

FIGURE 7.6 Depth of field (the depth that is in focus in the specimen) is increased by using smaller apertures as shown in B.
(Redrawn from Postek et al, 1980.)

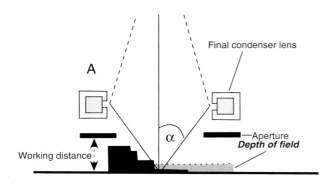

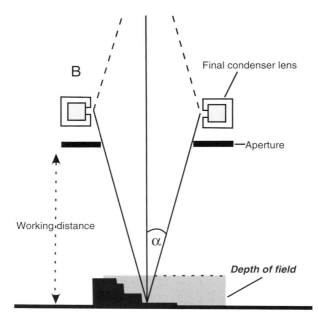

FIGURE 7.7 When the working distance is increased as shown in B, this decreases the aperture angle alpha so that the depth of field is also increased. *(Redrawn from Postek et al., 1980.)*

Resolution in the SEM. As with any magnifying instrument, *resolution* in the SEM refers to the ability of the instrument to image two closely placed objects as two entities rather than a single object. The size of the final spot is related to the resolution of the SEM: *Smaller beam spot sizes permit better resolution.* The final lens is used to focus the size of the illuminating beam spot to match the magnification used. Since the secondary electrons arising from the beam spot striking the specimen are additively displayed as a spot of a fixed size (usually around 100 μm) on the viewing monitor, the diameter of the beam spot on the specimen must not exceed a certain size as defined by Equation 7.2:

Equation 7.2: Maximum Allowable Spot Size of Beam

$$\text{Maximum Spot Size} = \frac{100 \ \mu\text{m}}{\text{Magnification}}$$

For example, if one is working at a magnification of 10×, the beam spot size on the specimen must not exceed 10 μm. A magnification of 100,000× would require a beam spot size of 1 nm or less on the specimen. If the beam spot size goes beyond this size, secondary electrons are generated from areas outside of what is being summarized on the spot on the monitor. This results in an unsharp image, since extraneous information is present in the display spot. The final lens may be used to decrease the beam spot and thereby "focus" the image. Obviously, the final lens can reduce the beam spot only so much, so that one must rely on the other condenser lenses to further assist in the demagnification. Therefore, if one can not achieve a satisfactory focus using the final lens, it may be necessary to increase the strength of the other condenser lenses to achieve smaller spot sizes. The relationship between beam spot size and image quality is summarized in Table 7.3.

From the preceding discussions, it is apparent that smaller apertures not only give better resolu-

TABLE 7.2 Working Distance and Quality of Image					
Working Distance (mm)	**5**	**10**	**20**	**35**	
Resolution	Best		⟶	Worst	
Depth of Field	Shallow		⟶	Deep	
Signal Strength	Strong		⟶	Weak	

TABLE 7.3 Beam Spot Size and Quality of Image			
Beam Spot Diameter (nm)	**1**	**100**	**500**
Resolution	Best	⟶	Worst
Signal Strength	Weak	⟶	Strong

tions due to decreased spherical aberration and spot size, but they also increase the depth of field. Unfortunately, these gains are at the expense of a decreased signal coming from the specimen since fewer electrons are present in the impinging beam. The relationship of aperture size to image quality is given in Table 7.4.

Interaction of Electron Beam with Specimen

When the accelerated 15 to 25 KV electrons of the SEM strike the specimen, they give rise to a number of emanations (see Figure 15.1). Depending on the speed of the electrons, as well as the density of the specimen, the beam may penetrate to a variable depth in the specimen. The initial point of entry may be at a 2 nm diameter spot; however, the electrons scatter randomly throughout the specimen until their energy is dissipated by interaction with atoms of the specimen. A longitudinal section across the point of entry of the electron beam will reveal a teardrop-shaped area (see Figure 15.17) where the electrons have spread. Two types of electron scattering result from the interaction of the primary beam electrons with the atoms of the specimen: elastic and inelastic scattering.

Elastic Scattering. An *elastically scattered electron* is one that has changed direction without losing velocity or energy. This type of scattering results when a beam electron collides with or passes close to the nucleus of an atom of the specimen. When such an elastically scattered electron exits the specimen back in the direction from which it came, it is termed a *backscattered electron.* Such high energy electrons (on the same order as the beam electrons) may interact with atoms of the specimen prior to their exit to generate secondary electrons some distance from their initial point of entry in the specimen (Figure 7.8). If backscattered electrons strike

parts of the microscope, this may also generate extraneous secondary electrons that are summed with the secondaries from the specimen to give rise to *noise* in the final image (see the section "Signal Versus Noise," in this chapter). It is possible to detect backscattered electrons using special detectors as described later.

Inelastic Scattering. During *inelastic scattering,* some beam electrons interact with the atoms of the specimen to produce low energy or *secondary electrons.* Secondary electrons have energy ranges of 0 to 50 eV and are the electrons most commonly used to generate the three-dimensional image. Occasionally, tightly bound, inner shell electrons may interact with a beam electron to effect the ejection of the inner shell electron. This event ionizes the atom until outer orbital electrons fill the inner void and energy is dissipated in the form of characteristic X-

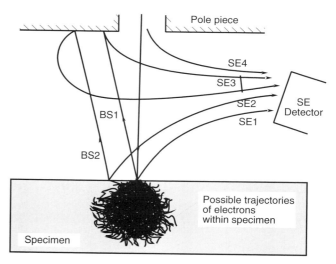

FIGURE 7.8 Sources of signal and noise. True signal is represented by secondary electrons that are generated from the spot of electrons focused on the specimen. True signal is represented by secondary electrons SE1 since they originate from the point where the beam strikes. SE2 are secondaries that originate some distance from the focused spot and represent spreading of the beam upon interaction with the surface of the specimen. Noise is generated when backscattered electrons BS1 and BS2 strike parts of the lens polepiece to generate SE3 electrons. SE4 secondaries originate if beam strikes parts of the lens such as aperture or polepiece liner. *(Courtesy of David C. Joy and The Royal Microscopical Society.)*

TABLE 7.4 Aperture Size and Quality of Image				
Aperture Diameter (μm)	**30**	**200**	**400**	**600**
Resolution	Best ─────────→ Worst			
Depth of Field	Deep ─────────→ Shallow			
Signal Strength	Weak ─────────→ Strong			

rays and an Auger electron. These emanations can be detected by specialized X-ray or electron spectrometers. Auger spectroscopy is rarely used in the biological sciences. It is an analytical method for determining the elemental composition of the uppermost atomic layers only. Other types of emanations resulting from the primary electron beam striking the specimen include heat, continuum X-rays, and light. All of these emanations may be detected and displayed or recorded as described in Chapter 15.

Specimen Manipulation

The specimen is normally secured to a metal stub (usually aluminum) and is grounded to prevent the buildup of static high voltage charges when the beam electrons strike the specimen. In order to orient the specimen precisely, relative to the electron beam and electron detectors, all SEMs have controls for rotating and traversing the specimen in x, y, and z (height) directions. Besides *rotational, lateral, and height adjustments,* it is also possible to *tilt* the specimen in order to enhance the collection of electrons by a particular detector. Judicious combinations of these movements not only permit accurate location of desired areas of the specimen, but they may have a large effect upon magnification, contrast, resolution, and depth of field. Consequently, when poor images are encountered, some improvement may be gained by a simple reorientation of the specimen.

Electron Detector, Signal Processing, and Recording Systems

Signal Versus Noise

The emanations, or signals, generated as a result of the electron beam striking a specimen are used to convey different types of information about the specimen. In the usual imaging mode in the SEM, *signal* consists of the secondary electrons generated from the spot struck by the beam electrons. *Noise,* on the other hand, consists of secondary electrons originating at locations away from where the beam struck the specimen.

The Origin of Noise in the SEM

In the case of secondary electrons (Figure 7.8), SE1 and SE2 represent signal coming from the beam spot. Noise is represented by SE3 and SE4 electrons that were generated as a result of backscattered electrons BS1 and BS2 striking metal parts of the SEM. Noise is

usually evidenced as a "snowy" image devoid of clear details. Noise may also arise when the beam strikes parts of the microscope column (e.g., final apertures) to generate spurious secondary electrons. Faulty electronics may also generate noise during the processing of the signal. For more details, see the article by Joy (1984).

A way of expressing the relationship between true versus extraneous signal is the *signal to noise ratio* (S/N). Whenever noise rises to an unacceptable level (based on a poor quality image), the signal to noise ratio is said to be low. One must therefore either reduce the noise or raise the signal to achieve a satisfactory image. Since it is more difficult to reduce the noise level, most microscopists attempt to raise the amount of signal from the specimen. Ways of increasing signal from the specimen are described later in this chapter (Major Operational Modes of the SEM).

Secondary Electron Detector

After the beam electrons strike the specimen, low energy secondary electrons leave the specimen from many different angles. Because they are weakly negative, the secondary electrons will be attracted to any positive source. The secondary electron detector uses this phenomenon to gather electrons. The most common type of secondary electron detector is based on the original 1960 *scintillator-photomultiplier* design of Everhart and Thornley (Figures 7.9 and 7.10).

In this system, the secondary electrons are attracted to a Faraday cage that is maintained at up to +300 V and surrounds the secondary electron detector. Upon reaching the cage, the electrons are more strongly attracted to the end of the detector (Figure 7.9) since its thin aluminum coating is placed at +12,000 V. The electrons strike the several nanometer thick aluminum coating with such impact that they pass through the aluminum layer, strike a phosphorescent *scintillator* material, and generate a brief burst of light, a scintilla, that travels down the lucite or quartz light guide. This burst of light then strikes a photocathode surface on the end of a photomultiplier. The *photocathode* is coated with a material that generates electrons on contact with light.

After being generated in the photocathode, the photoelectrons then enter the *photomultiplier* and travel down a series of dynodes or electrodes that proportionally increase the number of electrons at each stage (Figure 7.11). The yield of secondary

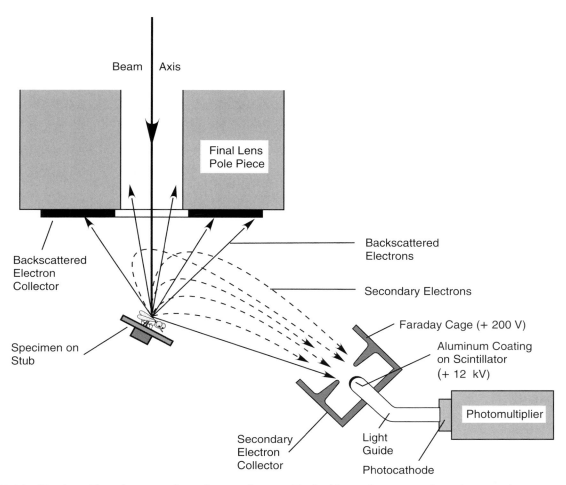

FIGURE 7.9 Everhart-Thornley secondary electron detector. Dashed lines show secondary electrons that are attracted to an aluminum-coated scintillator due to the positive pull exerted by the Faraday cage and the +12,000 V potential of the aluminum coating over the scintillator on the end of the light pipe. After striking a photocathode, the signal enters the photomultiplier. Backscattered electrons are not attracted by the detector.

FIGURE 7.10 Photograph of Everhart-Thornley secondary electron detector from a Kent-Cambridge SEM. The arrowheads show the paths secondary electrons might travel from the specimen (S) to the detector (not shown) housed inside the Faraday cage (F). After striking the scintillator of the detector, photons of light travel down the plastic light guide (L) to the photocathode of the photomultiplier (not shown).

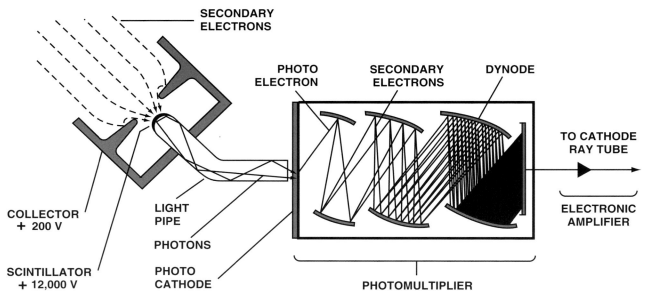

FIGURE 7.11 Schematic of photomultiplier where photoelectrons (generated by photons from the light pipe striking the photocathode) are multiplied by striking a series of high voltage dynodes to generate more secondary electrons.

electrons may be increased by increasing the voltage to the dynodes to effect a higher *gain* (on the order of 10^5 to 10^6). The SEM control that changes the gain of the photomultiplier is sometimes called the gain or, more commonly, the *contrast* control, since it affects the overall contrast of the image on the display monitor. *Increasing the photomultiplier gain will increase the contrast of the image* (the highlights are increased more than the shadow areas resulting in a greater brightening of the light areas compared to the dark areas).

Signal Processing

The small amount of voltage generated by the photomultiplier now enters a preamplifier-amplifier component of the SEM where the weak signal is amplified electronically. When one increases the output from the *preamplifier,* the *brightness* of the image increases overall (both the highlights and shadow areas are boosted).

Most SEMs have a control termed *gamma* that may be used to nonlinearly and selectively extend the contrast range in either the highlights or the shadows without loss of information in the nonamplified component. For instance, at a gamma setting of 0.5, contrast in the highlights is expanded while the shadows are compressed in contrast. At gamma 1.0, contrast levels in both the highlights and shadows are expanded. At gamma 2.0, contrast in the shadows is expanded while the highlights are com-

pressed. Gamma is most useful in situations where the specimen has excessive contrast (Figure 7.12).

Some Specialized Features of the SEM

A special control termed *dynamic focusing* may be used on specimens that are tilted relative to the beam. Without dynamic focusing, the beam spot would enlarge as it was scanned from the top to the bottom of the specimen (Figure 7.13). To maintain the spot size within the appropriate range, electronic modules are utilized to change the focal length of the final condenser lens as it is scanned over the specimen. In this manner, the spot size is the same at the top of the specimen as it is when it reaches the low points at the bottom of the tilted specimen.

Tilt correction is necessary on tilted specimens since the top of the specimen is at a higher magnification (shorter scan length) than the bottom of the specimen (longer scan length). This physical phenomenon may be corrected by reducing the scan lengths of the beam as it traverses from top to bottom. The factor by which the scan length is reduced depends on the degree of tilt and is adjustable by a control on the panel of the SEM. This correction must be used carefully since specimens with a great deal of topography may be distorted by activating this control (Figure 7.14). Flat specimens work best with this mode.

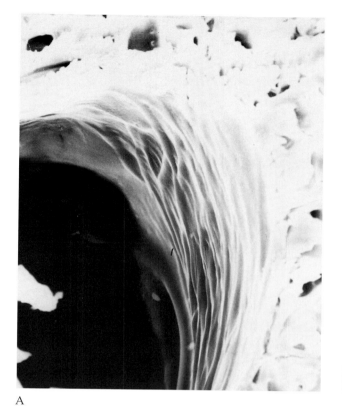 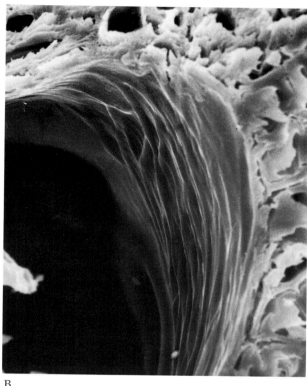

A   B

FIGURE 7.12 Demonstration of the use of gamma to diminish excessive contrast. The left-hand image is without gamma applied while the right-hand image shows the use of gamma. *(Courtesy of S. Schmitt.)*

Normally, the secondary electrons are displayed on the viewing monitor in the *brightness modulation* mode (i.e., the brightness of the spot on the viewing monitor is proportional to the number of secondary electrons from the corresponding point on the specimen). It is also possible to display the secondary electrons in the *Y-modulation* mode. In this mode, the scan on the viewing monitor is proportionally displaced in the y direction on the viewing monitor depending on the number of secondaries detected at each location on the scan (Figure 7.15, left). This image is similar to one displayed on an oscilloscope. If one accumulates a number of these scans on the viewing monitor, three-dimensionality of the specimen is enhanced (Figure 7.15, right). This mode is particularly useful for specimens that are flat and devoid of much topographic contrast. Interpretation of images generated using Y-modulation is difficult since bright areas of the specimen (which give rise to deflections in the y direction) may be due to other than topographic features (see "Contrast and Three-Dimensionality of the SEM Image," later in this chapter).

Image Recording

Recording of the image displayed on the cathode ray tube or monitor differs from the methods followed in transmission electron microscopy. Unlike TEM, where the electrons interact directly with the photographic medium, SEM images are most often photographed directly from a monitor through the lens of either a 35 mm roll film camera or a larger format 4″ × 5″ sheet film camera. The shutter of the camera is not used. Instead, the camera shutter remains open as the electron beam is slowly scanned across the specimen for 90 seconds or so.

Films for Use in the SEM

Several types of conventional black and white films may be used to record the image (Kodak Commercial Film 4127, Kodak PLUS-X Panchromatic Professional sheet films 4147, 2147, Kodak Technical Pan Films 2415 in 35 mm rolls, 6415 in 120-size rolls, and 4415 in sheet films). Most often, researchers use a 4 × 5

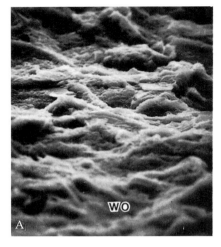
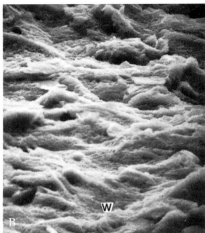

FIGURE.7.13 Tilted sample without (WO) and with (W) dynamic focusing applied.
(Courtesy of S. Schmitt.)

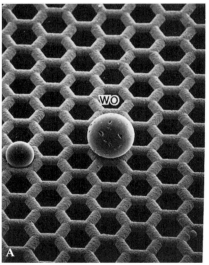
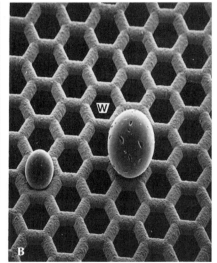

FIGURE 7.14 Tilted sample without (WO) and with (W) tilt correction applied. In these two micrographs, note how the hexagonal background pattern has been corrected with tilt correction applied and how the spherical objects have been distorted. Specimen consists of two glass beads deposited onto a TEM grid with hexagonal mesh pattern.
(Courtesy of S. Schmitt.)

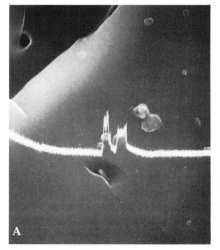
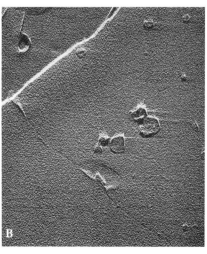

FIGURE 7.15 Standard brightness modulated secondary electron image (*left*). Compare to Y-modulated image on right.
(Courtesy of R. Tindall.)

instant type film, such as Polaroid Type 55, that gives both a 4 × 5 positive and a high-quality negative. Such films are 4 to 5 times more expensive than standard films, but are favored since darkroom time is saved and the quality of the image may be evaluated in less than one minute. When numerous images are to be recorded, roll films may provide a significant savings but at the loss of immediate gratification. Additional information on films and negative handling is found in Chapter 8.

Some Useful Features in Modern SEMs

A valuable addition to most SEMs is the *automatic data display* that permits the generation of informational data on the viewing and recording monitors. With this accessory, experiment numbers, dates, accelerating voltages, magnifications, and magnification scales may be displayed (see Figure 7.19). Keyboards permit the insertion of text and numbers anywhere on the image to be permanently recorded on the negatives.

Automatic brightness and contrast modules permit calibration of the exposure parameters for each type of film used in the SEM. Once calibrated, perfectly exposed negatives are generally the rule and very little recording medium is wasted. In addition, *autofocus* controls permit the precise focusing of the SEM even at very low magnifications.

Contrast and Three-Dimensionality of the SEM Image

The three-dimensional appearance of SEM images is due to differences in contrast between various structural features of the specimen when they are displayed on the viewing monitor. Contrast arises when different parts of the specimen generate differing amounts of secondary electrons when the electron beam strikes them. Areas which generate large numbers of secondary electrons will appear brighter than areas that generate fewer secondary electrons. The yield of secondary electrons by these various areas may be influenced by several conditions.

The *orientation of the specimen topography* relative to the electron beam and secondary electron detector greatly affects the yield of secondary electrons. As illustrated in Figure 7.16, certain areas of the specimen (designated "D" in the figure) will

not be struck by the beam and will not yield any secondary electrons. These areas will appear dark on the display monitor. Areas such as "I" in Figure 7.16 will be struck by the beam, but since they face away from the detector, fewer secondary electrons will be collected and intermediate levels of brightness will be displayed. Optimal yields of secondary electrons would come from areas that are struck by the beam and face the detector ("B" in the figure). These areas would appear as highlights in the image.

A second condition that affects the yield of secondary electrons is the *angle that the beam enters the specimen surface*. If the beam enters a specimen at a 90 degree angle, the beam penetrates directly into the specimen and any secondaries generated below a certain depth will not be able to escape. On the other hand, if the beam strikes the specimen in a grazing manner, then the beam does not penetrate to a great depth and more secondaries will be able to escape since they are closer to the surface. Since rounded objects are more likely to be grazed by the electron beam than would flat objects, round areas usually appear to have a sharp bright line around

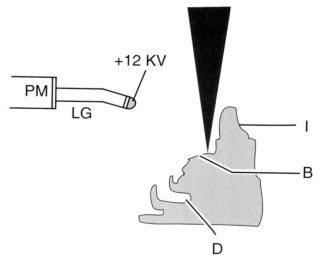

FIGURE 7.16 Three-dimensionality and contrast are due to the yield of secondary electrons from various parts of the specimen. Areas marked B face the beam and are in line of sight with the detector so that they will appear bright, I (intermediate brightness) faces the beam but fewer secondaries reach the detector since it is not in line of sight, D is dark in appearance since the beam does not strike this area and no secondaries are generated.

them due to the enhanced yield of secondaries (see Figure 7.17).

In a third situation, thin, raised areas of the specimen usually appear much brighter than broad, flat areas. This phenomenon is termed the *edge effect* since it takes place along sharp edges or peaks in the specimen. These areas appear brighter because the secondary electrons are able to escape from all sides of the thin areas in the projection (Figure 7.18). An example of a specimen demonstrating pronounced edge effect is also shown in Figure 7.19.

Other conditions that may affect contrast in a specimen include:

- The distribution of elements with *different atomic numbers*. Higher atomic numbered elements have a greater yield of secondary and backscattered electrons than do elements with lower atomic numbers. Higher atomic numbered elements therefore appear brighter in the SEM.
- Higher *accelerating voltages* result in lower contrast due to greater beam penetration and enhanced secondary yield from all parts of the topography. If more contrast is needed than can be obtained using the SEM contrast controls, then lower accelerating voltages should be used.
- *Charge accumulation* (*charging*) on incompletely coated or nongrounded areas of the specimen will result in an increase in contrast. For instance, large areas that are suspended by a thin stalk tend to build up a static charge from

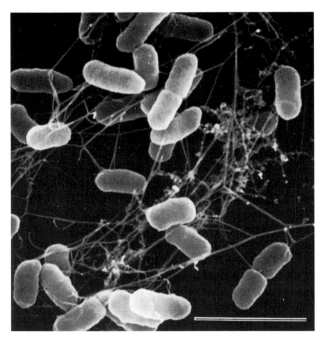

FIGURE 7.17 Rounded specimens demonstrate an enhanced emission of secondary electrons on their periphery since the electron beam grazes, rather than penetrates the surface coating on the specimen. Note bright periphery on all of the rounded bacterial cells. Marker bar = 2.3 μm.

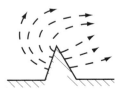 

FIGURE 7.18 The edge effect, or enhanced electron emission, occurs along the edges of thin raised areas since secondary electrons may exit from both sides of the structure.

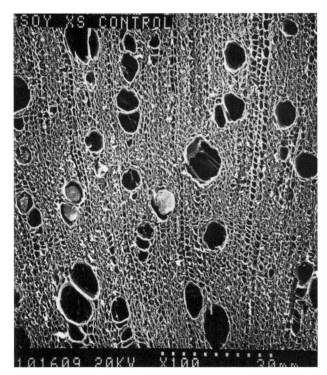

FIGURE 7.19 In this specimen of soybean stem, the cell walls are very bright due to enhanced emission of secondary electrons. The edge effect is diagrammed in Figure 7.18. The information bar on the bottom of the micrograph indicates that the image was recorded at 20 kV accelerating voltage and at a magnification of 100×. In addition, a magnification scale is recorded onto the negative to show that the total distance between the 11 white dots above the 100× mark is 0.30 mm.

the electron beam and cannot dissipate the charge rapidly enough through the thinned portion. This may cause the deflection of the beam so that it strikes other areas to generate an excessive amount of secondary electrons. Likewise, naturally *magnetic areas* in a specimen may either deflect or attract the beam to affect the yield of secondary electrons.

• When *crystals* are oriented along certain lattice planes relative to the beam, an enhanced yield of secondaries may result in an increase in brightness along these lattice planes so that certain crystals will appear much brighter than others.

Stereo Imaging with the SEM

The conditions described previously contribute to the generation of an image that appears to have depth even though the image is recorded on a piece of photographic paper that has only two dimensions. It is possible to generate images with even more three-dimensionality by a variety of other methods. The reader has undoubtedly experienced slide shows or even motion pictures in which special glasses were used by the audience to enhance the three dimensionality of the images on screen. In the SEM, the enhancement of three-dimensionality not only serves an esthetic function but it assists the viewer in discriminating between projections and depressions, and it elucidates spatial relationships to help determine distances between two objects (i.e., are two cells actually close enough to touch or is one simply in front of another one). In addition, some investigators feel that since one usually combines two separate micrographs in order to generate the single image (see next section) then the resolution and signal-to-noise ratio of the stereo image is enhanced.

The perception of depth is due to the *parallax phenomenon* wherein separate views from the right and left eyes are merged by the brain into one image. Since the left eye sees more of the left side of an object and less of the right side than does the right eye, the brain interprets this variation as depth. It is possible to achieve this same effect in the SEM by taking two separate micrographs of an object so that one of the micrographs shows more of the left side of the object while the other micrograph shows more of the right side. When the two micrographs are placed side by side so that the left eye views the micrograph showing more of the left side

while the right eye views the other micrograph, and when the images are merged by the brain (usually with aid of an optical device), the perception of depth occurs in the micrograph. The key to the procedure is to (1) generate two different micrographs from separate viewpoints and (2) merge these separate views using the appropriate tools.

Generating Two Micrographs with Separate Views

Two micrographs (termed *stereo pairs*) with the appropriate separate views may be generated in several different ways. One would take one micrograph followed by a second that had a different view due to (a) shifting the specimen slightly to the left or right, (b) rotating the specimen, (c) deflecting the electron beam electronically, or (d) tilting the specimen before taking the second micrograph.

The first two methods are less desirable since they are only effective at low magnifications and require a eucentric goniometer stage, respectively. (Eucentric stages maintain the central location of the specimen when the stage is tilted.)

Not all SEMs have the capability to tilt the beam between micrographs so that the third method may not always be possible. On the other hand, since it is possible to tilt the specimen in all SEMs, the last method is probably the most prevalent one. A description of the general steps involved in generating the two micrographs by this method follows.

Using Tilt to Generate Stereo Views in the SEM

1. Take the first micrograph in the usual manner. The specimen should show reasonable contrast and be set up to show good depth of field (i.e., proper working distance and aperture sizes).
2. Without moving the specimen, take a wax marking pencil and trace over the outline of the specimen on the viewing CRT. Rather than marking directly on the CRT, carefully tape a piece of clear acetate plastic over the CRT and mark on this sheet.
3. Tilt the specimen slightly. The amount of tilt depends upon the magnification (more tilt at lower magnifications) and the topography of the specimen (more tilt for specimens with flat topographies). In general, most researchers use tilts ranging from 4 to 15 degrees.
4. Refocus the specimen using the Z control (specimen height) adjustment of the stage. One must not focus using the final condenser lens, as

is the normal procedure, since this would affect the magnification of the subsequent micrograph.

5. Realign the image so that is coincides as closely as possible with the image traced on the acetate sheet using the wax pencil.

6. Adjust the brightness and contrast of the image in the SEM to match the first micrograph and take the second micrograph.

Merging Two Micrographs with Separate Views to Generate a Stereo Image

The most common method of presenting stereo images is to print the two separate images at the same contrast and density levels in the darkroom. The micrographs usually are small contact prints no larger than 4″ × 5″. The micrographs are then placed alongside each other and viewed with a special set of glasses (stereo viewers) placed directly over the two prints. By moving the two prints carefully, while looking through the viewers, it is possible to get the two images to converge to obtain a striking stereo effect. Relatively few individuals are able to place an index card between the two prints and by crossing their eyes force the two images to converge into the stereoscopic view. Figure 7.20 shows the setup and simple tools needed for viewing micrographs in stereo. This type of stereo viewer may be obtained from most EM supply houses at a relatively modest cost. An example of two images that were prepared by tilting a specimen are shown in Figure 7.21. The stereo glasses shown in the preceding figure are needed to produce the desired image.

A very simple procedure requiring only a small mirror was described by Harrington and Welford (1990). The two contact prints are produced in the usual manner except that one of the prints (produced from the negative that was tilted closer to the detector) is printed with the emulsion side up so that a mirror image is generated. After aligning the two prints side by side with a gap of approximately ½″, a mirror approximately 6″ high is placed between the two images with the reflective side toward the print that was reversed (i.e., mirror facing the mirror image). After placing one's nose at the edge of the mirror and carefully orienting the mirror until the two images blend into one, it is possible to obtain a stereo image quite easily. The key to this procedure is the careful production of the mirror-imaged print. An example of these types of prints is given in Figure 7.22. A 6″ high mirror is recommended for viewing the two micrographs as indicated in Figure 7.23.

FIGURE 7.20 Setup for viewing stereo pairs. The two micrographs are placed alongside each other under the special stereo viewing glasses, which merge the two slightly different views into one image that has three-dimensionality.

Preparing Stereo Images for Projection

If one wishes to project stereo images for an audience, several different (but equally demanding procedures) may be used. One method involves the use of glasses made up of lenses of complementary colors such that the left eye is covered by a red filter while the right eye is covered by a green filter. One then prepares two separate positive slides from the two stereo pairs and projects them using two separate projectors (the projector on the left with a red filter over its lens and the projector on the right with a green filter). The slide projected by the left-handed (red) projector can only be seen by the left (red lensed) eye, while the right eye (green lens) can see only the image from the right (green) projector. This technique can also be accomplished using polarizing, instead of colored filters, except that alignment of the two projectors is more critical and special screens are needed.

Another method of producing a slide for stereo imaging involves the production of a single colored

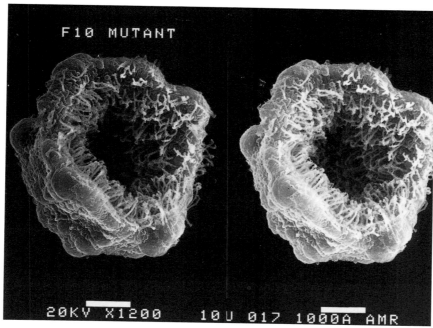

FIGURE 7.21 Stereo pairs ready for viewing using the special glasses shown in Figure 7.20. This illustration shows an inverting *Volvox* embryo magnified 1200×. The left-hand view was tilted +7 degrees relative to the right-hand view. *(Courtesy of G. M. Veith.)*

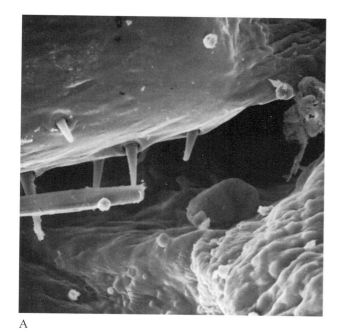
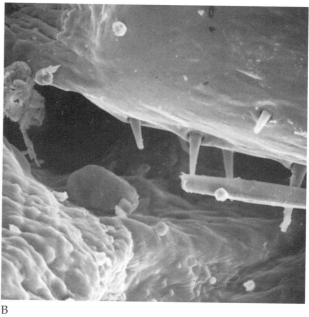

A   B

FIGURE 7.22 Stereo pairs ready for viewing using only a six inch high common mirror. The mirror is positioned between the micrographs and one views down the mirror as shown in Figure 7.23. *(Courtesy of D. Harrington and A. Welford and Wiley-Liss Publishers.)*

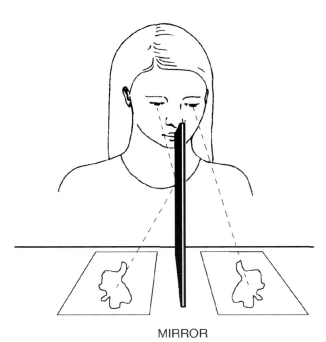

MIRROR

FIGURE 7.23 Method of viewing mirror image stereo pairs shown in Figure 7.22.
(Redrawn from Harrington and Welford, 1990.)

(rather than black and white) slide that has merged the red (left) and green (right) images onto one slide. In this instance, one needs only the colored glasses and a single projector for viewing purposes. The disadvantage to this technique is that the production of the merged image on the colored slide is difficult, except in the latest generation of SEMs. Newer SEMs have color viewing screens and electronic beam tilting capabilities so that it is possible to automatically tilt and record appropriately colored images of the stereo pairs on a single slide. Such colored slides that contain the stereo image encoded in color are termed *anaglyphs*. For more information on stereo imaging in the SEM, see the references by Barber and Emerson (1980), Peachey (1978), and Wergin (1984).

Major Operational Modes of the SEM

The SEM may be operated in two major modes: high resolution or great depth of field. As will be seen, these modes require largely conflicting parameters so that it is impossible to optimize both conditions simultaneously.

High Resolution

High resolutions demand a small, coherent spot with minimal aberrations and good signal-to-noise ratios. A summary of some of the conditions necessary to achieve high resolutions follows.

Conditions for High Resolution

1. *Small spot sizes* are achieved by optimally energizing all condenser lenses. The first condenser lens is adjusted to give a small spot that will then be further demagnified by the remaining lenses. As described previously (see Figure 6.28), the small spot will have a diminished number of electrons (i.e., low current) due to the wide aperture angles needed to achieve diminished spot sizes, so take care not to over energize the first condenser lens (follow manufacturer's recommendations).
2. *Small apertures* should be used to help diminish the size of the spot as well as to minimize spherical aberration. Again, small apertures will exacerbate the loss of current in the beam spot and may lead to a fall off of signal.
3. *Good signal-to-noise ratios* are essential in order to reveal details present in the smaller spot. The operational principal here is: *Maximize the generation of secondaries from the specimen.* This may be accomplished by:
   a. Putting *more current in the beam* by getting higher emissions of electrons from the gun (alter bias settings, move filament closer to aperture in shield, move anode closer to filament, use lanthanum hexaboride filament or cold field emission gun).
   b. Using *slower scan rates* on the specimen. Longer dwell times of the beam on the specimen will generate more secondary electrons from the spot. Generally, when images are photographed, slow scan speeds are automatically utilized to maximize secondary electron emission (thus increasing the S/N ratio), but it may be possible to slow the scan even more. One must remember, however, that this increase in current may damage sensitive specimens.
4. *Proper accelerating voltages* should be selected based on the nature of the specimen. The principle to follow here is: *Use the lowest possible kV that gives an acceptable signal-to-noise ratio.* Since beam penetration and enlargement of the spot size result from using increasingly higher accelerating voltages, one seeks a voltage high enough to minimize chromatic aberration (see

Chapter 6), while not degrading resolution. If the specimen is dense, then lower kVs should be employed since the beam electrons tend to spread near the surface of the specimen, increasing the spot size. Most biological specimens are not overly dense so that accelerating voltages in the 10 to 15 kV range are normally employed. Some particularly fragile specimens may benefit from even lower kVs, but this puts severe demands on the cleanliness and vacuum requirements of the SEM in order to minimize chromatic aberration.

5. *Short working distance* is needed in order to achieve a small enough spot, as well as to minimize chromatic aberration. It may also be recalled from Chapter 6 that shorter focal length lenses have the wider aperture angles necessary to achieve higher resolutions.

6. *Proper stigmation* of the spot is necessary for the reasons described in the chapter section entitled "Condenser Lenses." Any contamination, especially on the final aperture, will cause astigmatism and a loss of resolution. Lower accelerating voltages are more susceptible to astigmatism, stray magnetic fields, and chromatic aberration, so a thoroughly clean and well-evacuated SEM is necessary to maintain proper stigmation at lower kVs. Since astigmatism builds up with time, one must continually check and correct for astigmatism during the microscopy session.

Great Depth of Field

Great depth of field may be achieved by using small apertures, long working distances, and lower magnifications. Since resolution is less evident due to the latter two conditions, one may now increase the spot size to achieve a better signal-to-noise ratio. This operational mode is used to view specimens with extremes of topography. It should be realized that, from an esthetic point of view, it may be desirable to have only the area of interest in sharp focus while throwing a cluttered background slightly out of focus.

Imaging Other Types of Specimen Signals

So far, we have discussed only the more commonly used detector for imaging secondary electrons. Other types of emanations may also be detected and displayed on the monitor to reveal information in addi-

tion to topography. These detectors include those for backscattered electrons and visible light. The analysis of X-ray emanations will be covered in Chapter 15 since the technique also applies to TEM and scanning transmission electron microscopy.

Backscattered Electrons

Backscattered electrons, as mentioned earlier, are elastically scattered electrons that exit from the specimen with energy levels similar to the primary or beam electrons. Being of high energy, they may escape from great depths in the specimen and exit some distance from the point of entry of the beam electrons. As they interact with the atoms of the specimen, they may generate secondary electrons that are quite some distance away from the secondaries generated by the beam spot (effectively increasing the size of the beam spot and degrading resolution). In addition, backscattered electrons may strike metallic parts of the SEM chamber to generate even more secondary electrons (see Figure 7.8). All of these extraneous electrons may be detected by the secondary electron detector to raise the background level in an image and ultimately degrade resolution.

Since backscattered electrons travel in straight lines at high velocities, most cannot be attracted over to the standard secondary electron detector. The few that do enter the secondary detector generate a high signal upon entering the photomultiplier. Consequently, some of the highlights seen in the standard secondary image may be due to backscattered electrons.

If one quantitates the yield of secondary and backscattered electrons in various atomic elements, it will be observed that as the atomic number increases, the yield of both secondary and backscattered electrons increases. This is because larger atomic-numbered elements have more orbital electrons available to interact with beam electrons. Interestingly, a comparison of the ratio of secondary to backscattered electrons finds that there is a significantly higher proportion of backscattered electrons generated compared to secondaries as the atomic number increases. Both of these phenomena have practical implications to the operator: *Higher atomic numbered elements appear slightly brighter than lower atomic numbered elements in a secondary electron detector and significantly brighter in a backscattered electron detector.*

In practice one should be aware of a few basic principles when using backscatter imaging modes on biological specimens. Since the contrast in this mode is the result of differences in atomic number (so called *Z-contrast* where Z refers to the atomic number), there should be an atomic number difference of 3 between the various elements in the sample, otherwise contrast differences may not be discernible. The specimen should be as flat as possible in order to reduce interfering topographic contrast caused by specimen terrain.

Backscattered Electron Detection

Secondary electrons are used to generate an image based on topographic contrast, whereas backscattered electrons are not normally used to study topographies. Instead, contrast is based on detecting areas of different atomic numbered elements (Z-contrast). Unfortunately, in the biological sciences most of the elements present in specimens are of a relatively low atomic number and yield few backscattered electrons. However, if one applies selective stains of heavy metals such as silver or lead, the stained areas will appear very bright due to the high numbers of backscattered electrons (Figure 7.24).

Although relatively few backscattered electrons enter the secondary electron detector, it is possible to filter out secondary electrons by shutting off the positive voltages to the detector or by making the detector slightly negative to repel secondaries. Since only a relatively low number of backscattered electrons are collected using this method, the image will be rather noisy. The yield may be increased if the first condenser lens is adjusted to give a larger spot containing more primary beam electrons. Unfortunately, these highly energetic spots may damage some specimens and still not effect a good yield of backscattered electrons.

Specially designed detectors may be installed into most SEMs to collect backscattered electrons. All of these detectors use one principle to enhance efficiency of collection: *Increase the surface area of the detector and place it high above the specimen where backscattered electrons most likely will be encountered.* Several different types of detectors are available, but the two main types used in biological sciences are the wide angled scintillator-photomultiplier and the solid state detector.

Types of Backscattered Electron Detectors

The *wide angled scintillator-photomultiplier* uses a technology similar to the conventional secondary detector, but with some important differences. The detector is located high and above the specimen rather than off to one side, and the scintillator surface is considerably larger than the conventional secondary detector (Figure 7.25). Since a photomultiplier is utilized, this type of detector has high overall performance with a high signal-to-noise ratio, and good resolution and discrimination capabilities for different atomic numbered elements. A variation of this basic design involves placing multiple scintillator-photomultiplier detectors high above the specimen to increase the detection area. Such multiple units are quite expensive due to the large amount of electronics involved.

FIGURE 7.24 Comparison of secondary electron image to backscattered image. (A) Secondary electron image of *Xenopus* (frog) optic nerve tract. (B) Same specimen as in (A), except viewed in the backscattered imaging mode. Since the nerves have been stained with silver, they appear much brighter than the background so that it is much easier to trace them throughout the tissue.
(Courtesy of J. S. J. Taylor and The Williams and Wilkins Co.)

Solid state detectors are probably the most common type of backscatter detector used since they are less expensive, reasonably sensitive to differences in atomic number, easily maintained, and take up less valuable space inside the specimen chamber. In one type of solid state detector, four quadrants of a circular silicon diode are mounted directly under the final condenser lens (Figure 7.26B). Backscattered electrons that strike the diodes cause the ejection of electrons in the silicon and generate a flow of current proportional to the number of backscattered electrons striking it. This small amount of current is then amplified by electronics, and the signal is ultimately sent to the display mon-

itor where large amounts of current are displayed as bright spots on the viewing screen. Some disadvantages include lower sensitivities (since photomultipliers are not involved) and a lower resolution than the scintillator types.

Important Considerations when Using Backscattered Imaging with Biological Specimens

The beginning microscopist should be aware that backscattered electrons are most useful for discriminating areas of different atomic numbered elements in the

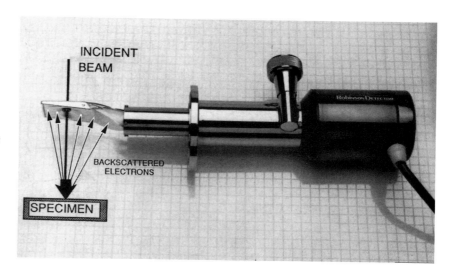

FIGURE 7.25 Photograph of a wide angled scintillator-photomultiplier backscattered electron detector. The electron beam passes through the hole in the center of the flattened portion (*left*) to strike the specimen and generate backscattered electrons. These electrons are detected when they strike the scintillator coating on the end of the wedge-shaped lucite light guide.
(*Courtesy of Electron Detectors, Inc.*)

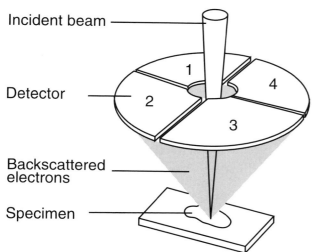

FIGURE 7.26A Diagram of solid state detector placed under final lens. Backscattered electrons strike the silicon diode and the current is detected and related to numbers of electrons striking it.
(*Courtesy of JEOL.*)

FIGURE 7.26B Photograph of backscattered electron detector showing the circular silicon diode that is divided into quadrants.
(*Courtesy of G.W. Electronics.*)

specimen. For this reason, one should not coat specimens with heavy metals such as palladium or gold in order to render them conductive. Instead, if a conductive coating is needed, a layer of carbon is usually deposited over the specimen using a vacuum evaporator (see Chapter 5).

Backscattered electrons are not very plentiful, so one may wish to increase the yield using larger spot sizes with more current (i.e., having more electrons) or possibly higher accelerating voltages. Both of these conditions will improve the signal, but at the expense of resolving power. More backscattered electrons will be detected if the detector is placed high and above the specimen and the specimen is in line of sight with the overhead detector.

Excessive topography in a specimen may interfere with the discrimination of areas of different atomic number. Consequently, specimens with low topographies are better suited for such Z-contrast investigations. This is not to exclude using the backscatter electron detector to improve topographic contrast. However, it is generally true that *in the conventional SEM secondary detectors are used primarily for conventional imaging using topographic contrast, while backscatter detectors are best used for differentiating different elements in a specimen based on Z-contrast.*

Cathodoluminescence

Upon bombardment by beam electrons, certain materials such as phosphors, semiconductors, and insulators may emit photons in the ultraviolet and visible energy ranges. It is possible to detect this very low level of light by using appropriate equipment. A simple detector may be fashioned from some types of secondary electron detectors by removing the scintillator disc. In this manner, the light guide will pick up light emanating from the sample and photomultiply and process this signal in the same manner as it would the photons generated by secondaries striking the scintillator.

One Type of Cathodoluminescence Detector

Since a modified secondary electron detector only detects 4% of the light emitted from the specimen with a spatial resolution of approximately 0.1 μm (the light microscope is 0.2 μm!), this method is adequate for cursory investigation. Several specially designed cathodoluminescence detectors have been designed to enhance sensitivity. One such device is diagrammed in Figure 7.27. In this system, the collector consists of a parabolic mirror placed over the specimen. The par-

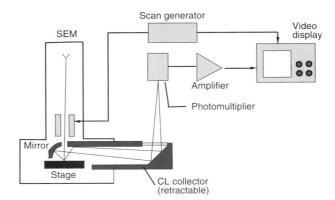

FIGURE 7.27 Diagram of one type of cathodoluminescence detector. The specimen is surrounded by a front-surfaced parabolic mirror. Light is reflected by the mirrors into the detector, a light pipe/photomultiplier system.
(Redrawn from original provided by Oxford Instruments North America, Inc.)

abolic mirror reflects light emitted from the specimen into a reflecting light guide and into a photomultiplier for signal amplification. In addition, an analysis system may be added to analyze the spectrum of light emitted by the specimen to aid in the identification of the chemical nature of the luminescing areas.

Only a few biological compounds cathodoluminesce, so there may be limited need for this type of detector. However, certain stains such as Calcofluor (a water-soluble dye that reacts with newly deposited cell walls), Brilliant Yellow 6G, and fluorescein isothiocyanate exhibit cathodoluminescence and may be used to stain cells to tag certain structures. In addition, certain biological compounds naturally exhibit the phenomenon. These include: tryptophan, tyrosine, adenine, guanine, thymine, polymerized DNA, polyurethane, some herbicides, pesticides, cotton, certain areas of adrenal and lung tissues, and leukocytes treated with formaldehyde. Most SEM manufacturers are able to supply (or recommend a manufacturer of) such detectors. A bibliography of articles dealing with cathodoluminescence was published in 1976 by Brocker and Pfefferkorn.

Using Secondary and Backscattered Electrons in Biological Studies

As was discussed in the previous sections, secondary electrons are used to reveal the three-dimensional features of specimens, whereas backscattered electrons are generally used to differentiate areas that differ in atomic number. In some studies, it may be useful to combine both types of signal in order to study the distribution of a particular material beneath or on the surface of a specimen.

The value of using backscattered electrons to image histochemically labeled *subsurface* components was first described by Abraham and DeNee in 1973 and further developed by Soligo and DeHarven (1981) who studied the distribution of heavy metal localized enzymes in human leukocytes from normal and leukemic patients. In 1984, DeHarven et al. described a technique for localizing *surface* antigens in the SEM using anitbody-labeled, 20 and 45 nm colloidal gold particles. Antigenic sites appeared as brightly labeled points using a solid state backscattered electron detector. When the secondary electron signal was combined with the backscattered signal, the three-dimensional distribution of the particles over the cell surface was revealed as shown in Figure 7.28.

In the following three studies, the investigators used both types of detectors in such capacities. For more detail, refer to the original articles.

Study 1: Localization of Cell Surface Receptors Using Gold Labels.

In a 1990 study that used the technique described by DeHarven et al. (1984), Helinski and coworker localized cell surface receptors for interleukin-2 in human T leukocytes. Cells were initially fixed using an aldehyde fixative and rinsed extensively to remove reactive aldehyde groups. Mouse monoclonal antibody against the cell receptor was reacted with the fixed cells. After rinsing to remove the unattached antibody, the cells were reacted with gold particles labeled with goat anti-mouse antibody. After an overnight fixation in glutaraldehyde to stabilize the label on the cell surfaces, the cells were dehydrated and the specimen was coated with carbon for conductivity. The sample was examined in the SEM using a standard secondary electron detector as well as a solid state backscatter detector. In the example shown in Figure 7.28, the secondary electron detector was used to give an overall image of the cells, whereas the backscattered electron detector was used to reveal the sites where the antibody gold is attached.

Study 2: Detection of Subsurface Structures.

Since backscattered electrons may emerge from deep inside biological tissues, this study by Horiguchi, et al. (1984) took advantage of the penetrating power of the primary electron beam to examine nuclei inside of intact cells. Small pieces of tadpole tail muscle were fixed in an aldehyde fixative and rinsed extensively

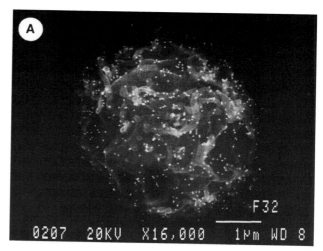

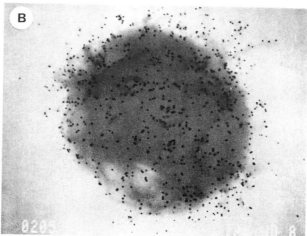

FIGURE 7.28 Normal human blood T cells that have been labeled with colloidal gold particles. The cells were first reacted with a purified mouse monoclonal antibody against a specific cell surface marker for interleukin-2 followed by gold particles coated with goat anti-mouse IgG. (A) This image is generated by combining the secondary signal to give the overall image of the cell with the backscattered image of the bright gold particles. (B) This image of the same cell is taken only in the backscattered electron imaging mode with reversed polarity so that the 40 nm gold particles appear dark rather than bright. Marker bar = 1 μm.
(Courtesy of J. L. Pauly and Wiley-Liss, Inc.)

prior to submersion of the specimens in an ammoniacal silver solution. After rinsing in distilled water, the tissue was again fixed in glutaraldehyde, rinsed in acetic acid, dehydrated up to 70% alcohol, and frozen in liquid nitrogen. Tissues were fractured to expose blood vessels, further dehydrated, critical point dried, and coated with carbon for conductivity. A standard secondary electron detector was used for three-dimensional imaging purposes, whereas a solid state detector was used to image the silver-stained organelles. Cell nuclei were specifically stained with the silver and exhibited high contrast against an unstained background (Figure 7.29).

Study 3: Selective Staining of Neuronal Tissues. Some of the selective silver stains used in light microscopy may be employed for SEM. In a study by Taylor, Fawcett, and Hirst (1984), central nervous tissues of *Xenopus laevis* (newt) that had been fixed in a variety of light microscope fixatives were dehydrated, embedded in paraffin, sectioned at 7 μm, and mounted on coverslips. The sections were stained with a variety of silver stains for neuronal tissues,

dehydrated, critical point dried, and carbon coated prior to examination in the SEM. The specimens were examined using both the standard secondary detector as well as a solid state high resolution backscattered electron detector using 20 to 25 kV accelerating voltage. In the secondary imaging mode, one was able to observe the surface features of such structures as the retina (Figure 7.30A). With the backscattered detector, one was able to observe the silver-stained nerve cells lying under the limiting membrane of the retina (Figure 7.30B). The procedure described in this paper makes it possible to more effectively distinguish and trace nerve fibers in a variety of tissues.

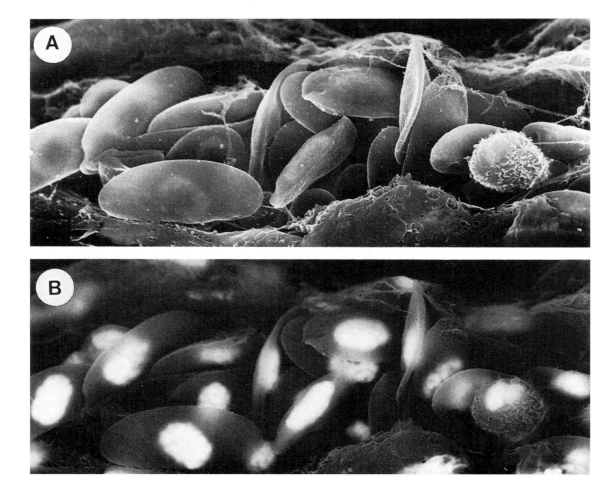

FIGURE 7.29   Red blood cells from tadpole as viewed in the (A) secondary electron mode and (B) backscattered electron mode. The nuclei were stained with a silver stain and appear very bright. Note that the backscattered image reveals structure beneath the surface of the cell membrane due to the penetrating power of the backscattered electrons. *(Courtesy of Kyozo Watanabe and The Williams and Wilkins Co.)*

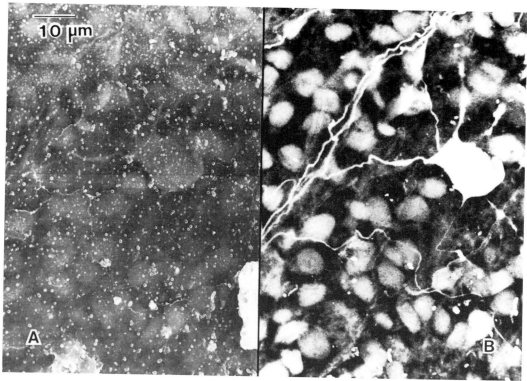

FIGURE 7.30   Secondary (A) and backscattered (B) views of identical specimen, optic nerve of frog. In the secondary mode, you see only surface features whereas in the backscattered mode it is possible to view through the limiting membrane of the retina and see the underlying nerve cells that have been stained with a silver compound. Since the axons were stained with silver, they have an enhanced emission of backscattered electrons and are therefore brighter than surrounding cells.

(Courtesy of J. S. J. Taylor and The Williams and Wilkins Co.)

Specialized Instrumentation for Observing Unfixed Tissues

Observation of Frozen Specimens

The majority of biological specimens that are examined in the SEM undergo the standard procedures of fixation, dehydration, drying, mounting, and coating with a metal prior to observation. Cold stages have been developed to fit on most standard SEMs and permit one to observe a fresh-frozen specimen using low accelerating voltages for a short period of time. Such stages would be useful with certain types of specimens that are not adequately preserved by the standard methods or whenever speed of viewing is necessary (i.e., to rapidly freeze a dynamic cellular process and to view it with a minimum of disturbance). For exam-

ple, the secretion of water-soluble polymers by an organism would be best observed using a rapidly frozen specimen. It may be obvious that certain types of specimens (snowflakes, ice cream, frozen biological specimens taken from Arctic areas) that occur in the frozen condition may be destroyed upon thawing.

In a typical situation, a specimen is first cryoprotected using glycerol or DMSO and rapidly frozen to liquid nitrogen temperatures below $-170°$ C (see Chapter 14). After transferring the frozen specimen onto a prechilled specimen stage, the specimen is evacuated and inserted into the SEM viewing chamber. A reservoir of liquid nitrogen is used to maintain the temperature of the cold stage throughout the viewing process. In many instances, after warming up the specimen slightly, any ice crystals that formed on the surface will be sublimed away by the vacuum system of the SEM. The specimen surfaces may then be observed at low to intermediate magnifications.

In order to minimize charging, accelerating voltages of 1 to 5 kV are used. Salts remaining in the frozen aqueous system of the specimen also impart some conductivity to the tissues. Metal coatings are still desirable; consequently, some cold stages contain a small set of electrodes or a sputter coater. Once metal coated, frozen specimens may be observed for much longer periods of time and with improved resolutions compared to uncoated specimens. Since such extremely cold specimens act as miniature cryopumps, they attract and condense contaminants onto the specimen surfaces. Clean vacuum systems or good cryotraps are needed to prevent specimen contamination. Some cryostages provide the operators with manipulators for fracturing the frozen specimen to expose underlying structures. A soil nematode observed on a cryo stage is shown in Figure 7.31.

Observation of Fresh Specimens

In 1989 a unique type of SEM, termed an environmental SEM (ESEM), became available commercially. It allowed the observation of unfixed, non-frozen, uncoated biological specimens at pressures of about one million times higher than the conventional SEM (e.g., 2.7×10^3 Pa, as described in Chapter 6). The instrument, manufactured by Philips Electron Optics (Figure 7.32), operates in the standard accelerating voltage ranges of 0 to 30 kV with claimed resolution capabilities of 10 nm at a pressure of 6×10^2 Pa. The ESEM is differentially pumped so that the gun area has vacuums in the 10^{-4} Pa range common in most conventional SEMs. The pressure then rises (due to a differential pumping system) as one approaches the specimen chamber. As a result, the ESEM column is at conventional vacuum levels while only the specimen chamber is at the increased pressures needed to observe fresh, hydrated tissues. The detector used in this system is unlike the conventional one developed by Everhart and Thornley, but is based on the gaseous discharge detector described by Danilatos in 1982. Some examples of biological specimens viewed in the ESEM are shown in Figure 7.33.

Several other manufacturers of SEMs now provide so-called low or variable pressure SEMs with differential pumping of the specimen area and column. In these instruments, a backscattered electron detector is used instead of a gaseous discharge detector as used in the environmental SEM. Images similar to those shown in Figure 7.34 may be easily obtained in these instruments. It is anticipated that continued development and exploitation of the capabilities of the environmental and variable pressure SEMs by researchers will permit the study of dynamic processes in living cells.

FIGURE 7.32   The environmental SEM (ESEM) for observing fresh, unfixed specimens. Since the instrument is able to operate at relatively low vacuum levels (in contrast to the conventional SEM), it is possible to observe live, hydrated specimens at claimed resolutions of 5 to 10 nm. A different type of detector than found in the conventional SEM is used in the ESEM.
(Courtesy of Philips Electron Optics.)

FIGURE 7.31   A soil nematode viewed in the frozen hydrated state using an SEM equipped with a cryo stage that was chilled with liquid nitrogen. No metal coating was used.
(Courtesy of Oxford Instruments.)

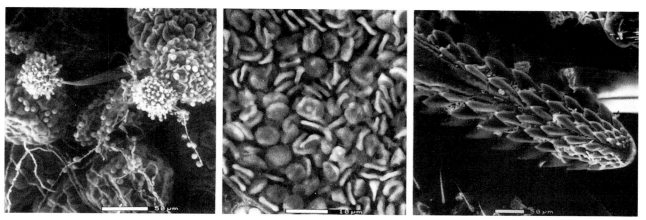

FIGURE 7.33 Examples of different types of unfixed, hydrated biological specimens observed in the Environmental SEM. (A) Bread mold viewed at 12 kV accelerating voltage, 490 Pa vacuum, −15° C. Marker bar = 50 μm. (B) Fresh-frozen red blood cells in lung tissue. 20 kV, 370 Pa, −15° C. Marker bar = 10 μm. (C) Mouthparts of tick. 20 kV, 490 Pa, ambient temperature. Marker bar = 50 μm.
(Courtesy of Philips Electron Optics.)

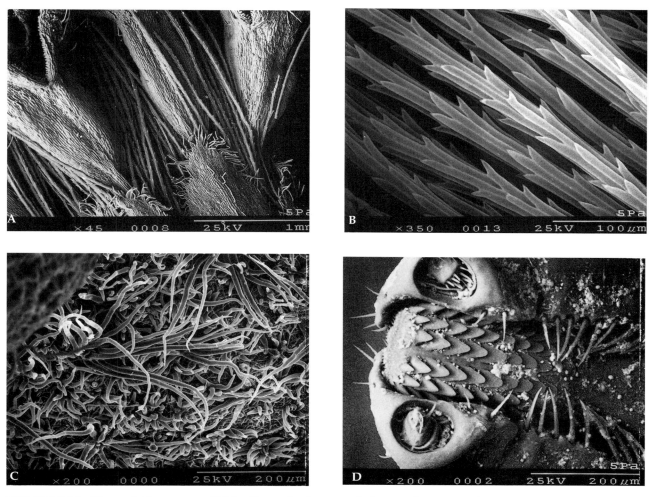

FIGURE 7.34 The variable pressure SEM utilizes a backscattered electron detector to obtain images of unfixed, hydrated biological specimens under higher vacuum conditions than used in the Environmental SEM. It may be necessary to cool (not freeze) the specimens in order to slow down dehydration under the partial vacuum. (A) Freshly picked composite flower. Marker bar = 1.0 μm. (B) Enlarged area of B showing detail on stamen hairs. Marker bar = 100 μm. (C) Detail of surface of flower petal. Marker bar = 200 μm. (D) Mouthparts of tick. Marker bar = 200 μm. All specimens were viewed at 25 kV, 5 Pa vacuum, and at ambient temperature.

Alignment and Operation of the Scanning Electron Microscope

Although the SEM has a considerably simpler alignment and operational procedure than the TEM, proper alignment and operation are essential for high-quality imaging. Even modern, digital SEMs require various degrees of manual alignment especially after filament changes, column cleaning, or electronic maintenance procedures that may affect lens or high voltage settings. Details may vary from one brand of SEM to another, but the basic steps are similar: filament saturation, electron gun alignment, aperture alignment, astigmatism correction, specimen placement, and image recording. If you make drastic changes in the accelerating voltages (from 10 to 20 kV, for example), recheck the saturation through the following stigmation steps.

Prior to beginning alignment of the SEM, a conductive specimen (for example, a magnification or resolution standard) should be inserted and set to a low magnification (around 100×) to readily obtain an image.

Filament Saturation

Even though most SEMs permit you to view the doughnut-shaped image during heating of the filament (Figure 6.46), most operators prefer to monitor the waveform to locate the saturation point. In practice, select the *waveform* instead of the regular image display mode. Doing so switches the scanning mode to single line scan and displays the information as a Y-modulated rather than brightness modulated image (see Figure 7.15, left). In this mode, the peaks represent bright areas of the specimen with high electron emission while the valleys represent areas of low emission. The difference between the peaks and valleys would translate as *contrast* in the normal viewing mode. If the peaks are excessive (due to an excessively contrasted specimen), adjust peaks back onto the screen. As you slowly heat the filament, the entire waveform will begin to move up the viewing screen. This represents an increase in the overall brightness of the image due to more beam electrons striking the specimen and generating increasing numbers of secondary electrons. At some point you may encounter a *false peak* (labeled "2" on Figure 7.35) and the waveform may actually begin to decrease even as you continue to increase the heating of the filament (labeled

"a" on Figure 7.35). Goldstein, et al. (1981) feel that this may be due to a localized area of the filament reaching the emission temperature before the whole filament tip does. It is easy to identify a false peak since a further increase in the filament heating will reverse the downward trend since an increased electron emission from the gun will lead to an increase in signal from specimen.

NOTE: As you continue heating the filament, the waveform may move off the top of the viewing screen due to increased number of secondary electrons from the specimen. Maintain the waveform on the lower third of the screen using the brightness *controls on the SEM console.*

At some point, the waveform will cease to move up the viewing screen (labeled "4" on Figure 7.35). This is the *saturation point.* Further increase in the heating of the filament will overheat it and lead to premature filament burnout. Normally, you can turn back the filament heater slightly (from 4 to 3 in Figure 7.35). At this point the filament is considered properly heated. Now, examine the *emission current* on the SEM display to ascertain that it is neither excessive nor inadequate. Depending on the SEM and type of filament (tungsten, LaB_6, cold field emission), the emission current (measured in μA) is adjusted to the proper setting using the *bias* control. For example, at an accelerating voltage of 30 kV, the emission of a tungsten or LaB_6 filament may be adjusted up to 100 μA whereas a cold field emission filament may be set for 5 to 10 μA emission.

CAUTION: Check with the manufacturer of the SEM to determine the proper filament emission settings to avoid damaging the filament. This is especially true of LaB_6 and cold field emission filaments.

Electron Gun Alignment

Observe the waveform on the viewing screen and adjust the horizontal x and y adjustment of the gun alignment control until the brightness is maximized—as evidenced by the maximum rise of the waveform. It may be necessary to use the brightness controls to keep the waveform on the viewing screen if the secondary electron emission on the specimen is excessive.

Aperture Alignment

Most SEMs have a variable aperture near the final condenser lens that can be centered externally using mechanical adjustment screws. After focusing the image at a magnification of approximately 500×, activate the *aperture align* switch that modulates or wobbles the current to the final lens. Observe the shifting movement of the image on the screen and center the

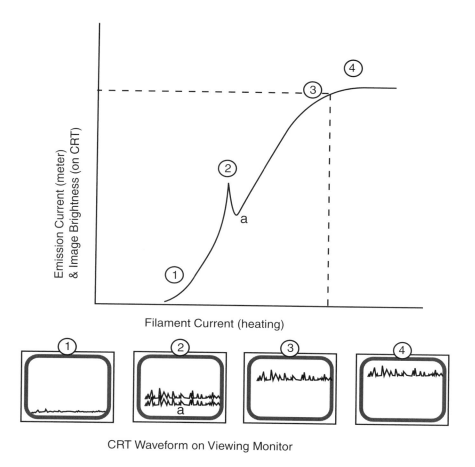

A CRT Waveform on Viewing Monitor

FIGURE 7.35 (A) Top diagram shows rise of brightness as the filament current is raised and the filament heated. A "false peak" is encountered at "2" followed by a dip in the brightness at "a" and then a continuation of the increase in brightness to the saturation point "4." Beyond the saturation point, the filament may be damaged. The bottom diagram shows a series of four images of the waveform one would see at the four points on the saturation curve shown in the top diagram. (B) Blown filament taken from an SEM. The presence of the rounded feature on the filament (*arrow*) is due to overheating or over saturation of the filament while the projections from the filament wire (*arrowheads*) indicate oxidation of the filament due to a vacuum leak.

B

aperture until the movement is stopped. Increase the magnification and perform further centering of the aperture at least twice the magnification that will be used to record the image. Most operators perform final aperture alignment at the top magnification of the SEM.

Gun Tilt

If it is necessary to adjust the tilt of the gun, this may be achieved by first focusing the specimen, then varying the first condenser lens focus control back and forth through crossover (normally the center position on the control). Observe the movement of the specimen (as in aperture alignment). Adjust the tilt controls so that when the condenser I focus is varied, the specimen movement is minimized.

Astigmatism Correction

Focus on a detailed area of the specimen magnified at least 5,000× or at twice the magnification that you will be recording the image. Adjust the astigmatism correction controls to minimize smearing in any one direction and to achieve a sharp image. Recheck the sharpness using the final condenser focus control and touch up the astigmatism correction if needed. Since astigmatism changes as you operate the SEM, it is important to recheck the correction periodically. The frequency of correction depends on the cleanliness of the specimen and the SEM column. Newer SEMs may have an automatic astigmatism correction; however, you should still periodically check that the correction is accurate.

Image Recording

Practically all SEMs now have the capability of recording numbers and alphanumeric characters directly onto the image. This may be by means of a conventional keyboard or, in some cases, by means of a small key pad. The information becomes part of the permanent photographic or digital image and may include magnification, scale markers, accelerating voltage, operator number, negative number, and operator-selected text. Many exposure parameters (brightness, contrast, time) as well as focus are automatic on most modern SEMs and the resulting images are accurately recorded most of the time.

However, you must be able to recognize properly focused and exposed images since the automatic controls (like automatic handheld cameras) must sometimes be overridden. All SEMs provide for such manual overrides.

Magnification Calibration

Just as it is questionable practice to always use the automatic modes of the SEM (autofocus, autobrightness and contrast, semiautomatic stigmation, etc.), it is also not safe to assume that the magnification stated on the display is absolutely accurate. Most manufacturers will guarantee the accuracy to ± 10%. If greater accuracy is required, then magnification standards must be used to calibrate the SEM just as in the

TEM. Since magnifications will change with working distances and accelerating voltages, and even though modern digital SEMs should correct for these changes, you should perform the calibration at the same settings that the images will be recorded for optimal results. One type of magnification standard that may be used in the SEM is waffle grating (Figure 7.36), which is similar to the replica grating used in TEM calibration (Figure 6.54). The procedure used for calibration of the magnification is exactly as described for the TEM (see page 198).

Resolution Checking

Specimens used for checking the resolution of the SEM are different from those used in the TEM. However, the criteria are the same: ability to resolve the shortest distance and visualize many levels of gray. One type of specimen consists of gold that has been thermally evaporated onto a polished carbon substrate (Figure 7.37A). Such specimens may be purchased commercially for $40 to nearly $200 but they will last for a long time if cared for properly (protect from dirt, mechanical abrasions, and airborne organics). If you have the instrumentation (vacuum evaporator with two independently adjustable DC heating supplies), the standards may also be prepared in your laboratory by following the procedure described by Humenansky (1987). A second type of specimen consists of gold that has been evaporated onto audiotape (Figure 7.37B). Use the old style of reel-to-reel recording tape that does not contain plastic binders. If you evaporate a light coating of gold onto a short piece of tape that has been glued onto an SEM stub and examine the coated tape in the SEM at high magnification, a series of fine fissures or cracks will be observed. Simply look for the smallest cracks resolvable and record the image using only moderate contrast so that many gray levels are evident. The reason for avoiding high contrast is that is distorts the visibility of the edges and gives an overly optimistic reading. Likewise, improper adjustment of the stigmation controls may artificially enhance the resolution by smearing the image in the direction of the fissures. Therefore a good test specimen should have many levels of gray, be in good focus, and be stigmated properly. When these parameters are met then you can measure the distances and calculate the resolution in exactly the same manner as described for the TEM using the *point-to-point* method (see page 199).

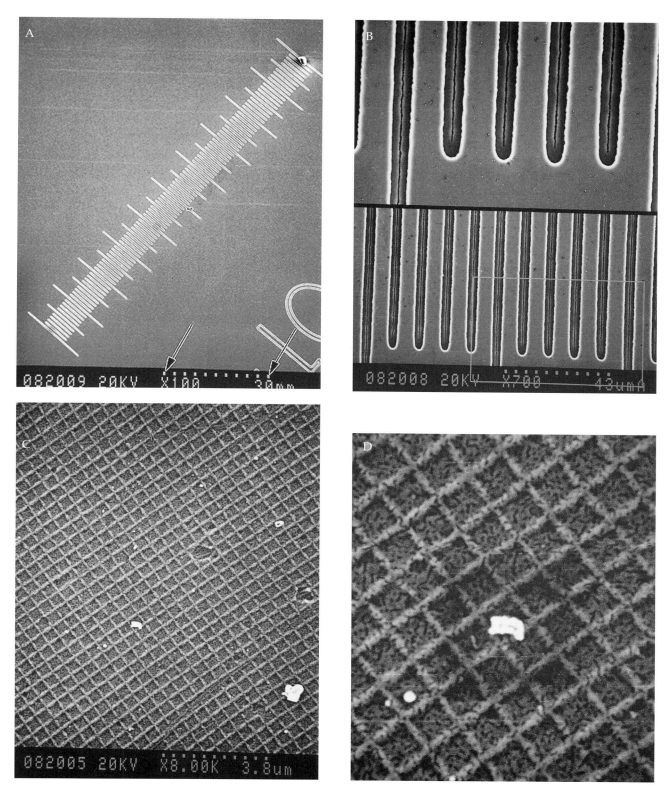

FIGURE 7.36 Magnification standards used to calibrate the SEM. (A) At a low magnification, a 1 mm ruled standard may be used. At a magnification of 100×, the SEM placed a marker bar indicating 0.30 mm from one point (*arrow, left*) to the end of the series of points (*arrow, right*). Calculations confirmed this magnification as 0.307 mm. (B) Higher magnification of ruled standard at a microscope setting of 700× and marker bar indicating 43 μm. Calculations indicated the marker bar is 43.86 μm. (C) Waffle grating imaged at 8,000× and marker bar indicating 3.8 μm. Calculations indicated the marker bar is 3.77 μm. (D) SEM magnification of 25,000× and marker bar indicating 1.20 μm. Calculations indicated the marker bar is 1.195 μm. All of the calibrations indicated that the SEM was well within specification for magnification settings.

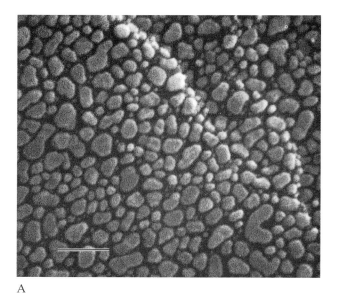

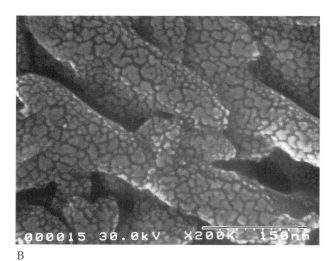

A

B

FIGURE 7.37 Resolution standards used to check resolving capabilities of the SEM. (A) Evaporated gold particles on carbon substrate. (B) Evaporated gold film on magnetic recording tape. Magnification marker = 150 nm in both images. *(Courtesy of Hitachi Scientific Instruments.)*

Digital Scanning Electron Microscopes

Most current generation SEMs are controlled by microprocessors that may be built into the SEM or operate from a separate personal computer using specialized software written for the SEM. All functions or selections that were formerly analog are digitally based (for example, kV, focus, magnification, image capture, stigmation, brightness and contrast, image processing). Images are digital and the information is encoded in a series of picture elements, or *pixels*, capable of being saved on a variety of recording media. One major advantage of the digital SEMs is the ability to carry out *frame averaging* in which a series of images or frames is compared and averaged so that any randomly occurring event such as noise or charging will be averaged out or removed. This gives an immediate improvement in the quality of the image on the viewing screen. Although the overall resolution of the image will suffer in comparison to an analog image, digital images are quickly and inexpensively captured, easily manipulated, transferred electronically, and may be used directly in most publishing programs. Unless the highest resolution photo-based image is required, digital images will fulfill the majority of microscopic imaging needs. It is possible to upgrade older analog SEMs with the image capturing capabilities of digital microscopes. For more information on digital microscopy and image analysis see Chapter 18.

REFERENCES

Abraham, J. L., and P. B. DeNee. 1973. Scanning electron microscopy histochemistry using backscattered electrons and metal stains. *Lancet* 1:1125.

Barber, V. C., and C. J. Emerson. 1980. Preparation of SEM anaglyph stereo material for use in teaching and research. *Scanning* 3:202–6.

Becker, R. P., and O. Johari, eds. 1979. *Cell surface labeling*. Scanning Electron Microscopy, Inc., AMF O'Hare, IL, 344 pp.

Brocker, W., and G. Pfefferkorn. 1976. Bibliography on cathodoluminescence. *Scanning Electron Microsc* IV: 725–37.

Danilatos, G. D. 1982. Foundations of environmental scanning electron microscopy. *Advances in electronics and electron physics* 71:109–250.

Danilatos, G. D., and R. Postle. 1982. The environmental scanning electron microscope and its applications. *Scanning Electron Microsc* I:1–16.

DeHarven, E., R. Leung, and H. Christensen. 1984. A novel approach for scanning electron microscopy of colloidal gold-labeled cell surfaces. *J Cell Biol* 99:53–57.

Everhart, T. E., and T. L. Hayes. 1972. The scanning electron microscope. *Scientific American* 226:54–68.

Everhart, T. E., and R. F. M. Thornley. 1960. Wide-band detector for microampere low-energy electron currents. *J Sci Inst* 37:246–48.

Goldstein, J. I., D. E. Newbury, P. Echlin, D. C. Joy, C. Fiori, and E. Lifshin. 1981. *Scanning electron microscopy and x-ray microanalysis: A text for biologists, materials scientists, and geologists.* New York: Plenum Publishing Corp., 673 pp.

Haggis, G. H., E. F. Bond, and R. G. Fulcher. 1976. Improved resolution in cathodoluminescent microscopy of biological material. *J. Microsc* 108:177–84.

Harrington, D. A., and A. Welford. 1990. An alternate method of viewing stereo pairs from the SEM. *J. Electron Microsc Tech* 15:101–2.

Helinski, E. H., G. H. Bootsma, R. J. McGroarty, G. M. Ovak, E. de Harven, and J. L. Pauly. 1990. Scanning electron microscopic study of immuno-gold-labeled human leukocytes. *J Electron Microsc Tech* 14:298–306.

Horiguchi, T., F. Sasaki, and H. Takahama. 1984. Identification of cells by backscattered electron imaging of silver stained bulk tissues in scanning electron microscopy. *Stain Technol* 59:143–48.

Humenansky, J. 1987. Manufacture and use of test samples to adjust and evaluate the SEM, TEM and STEM. *EMSA Bulletin* 17(1):68–72.

Joy, D. C. 1984. Beam interactions, contrast and resolution in the SEM. *J. Microsc* 136:241–58.

Kan, F. W. K. 1995. Backscattered electron imaging of colloidal gold-labeled fracture label preparations, chapter 10. In *Rapid freezing, freeze fracture, and deep etching,* N. J. Severs and D. M. Shotton, eds., Wiley-Liss, NY.

Lyman, C. E., D. E. Newbury, J. I. Goldstein, D. B. Williams, A. D. Romig, Jr., J. T. Armstrong, P. Echlin, C. E. Fiori, D. C. Joy, E. Lifshin, and K. R. Peters. 1990. *Scanning electron microscopy, X-ray microanalysis, and analytical electron microscopy: A laboratory workbook.* Plenum Press, NY.

Murphy, J. A., and G. M. Roomans, eds. 1984. *Preparation of biological specimens for scanning electron microscopy.* Scanning Electron Microscopy, Inc., AMF O'Hare, IL, 344 pp.

Peachey, L. D. 1978. Stereoscopic electron microscopy: principles and methods. *Bull Electron Microsc Soc Am* 8(1):15–21.

Postek, M. T., K. S. Howard, A. Johnson, and K. L. McMichael. 1980. *Scanning electron microscopy: A student's handbook.* Ladd Research Industries, Inc., Burlington, VT, 305 pp.

Soligo, D., and E. DeHarven. 1981. Ultrastructural cytochemical localizations by backscattered electron imaging of white blood cells. *J Histochem Cytochem* 29:1071–79.

Taylor, J. S. H., J. W. Fawcett, and L. Hirst. 1984. The use of backscattered electrons to examine selectively stained nerve fibers in the scanning electron microscope. *Stain Technol* 59:335–41.

Wells, O. C. 1974. *Scanning electron microscopy.* New York: McGraw-Hill Book Company, 421 pp.

Wergin, W. P. 1984. Importance of incorporating stereopsis and stereometry into a scanning electron microscopy course. *Scanning Electron Microsc* III:1225–35.

Chapter 8

Production of the Electron Micrograph

Photographic Principles

Although digital imaging (see Chapter 18) is rapidly replacing conventional photographic recording media, the final record of most electron microscope investigations is usually the positive photographic print that is produced from a black-and-white negative. This electron micrograph is the medium of presentation of one's work to the scientific community and serves as a basis for judging the quality of the research in electron microscopy. The production of a high-quality negative and print is as important as any of the preceding preparatory steps. Perhaps because this process may be regarded as a technologically simple endeavor, the photographic aspect of electron microscopy is often slighted. This is unfortunate since excellent morphological data may be lost unless it is recorded properly. An unsuitably exposed or processed negative usually yields a poor quality print, whereas a properly managed negative nearly always has the capability to produce a good print.

General Steps in Producing a Conventional Electron Micrograph

- Expose the sensitive photographic emulsion by means of electrons (TEM) or photons (SEM).
- Develop the exposed emulsion in the darkroom to generate a negative image.
- Treat the negative with fixer for stabilization.
- Wash and dry the negative.
- Project the negative onto a photographic paper using an enlarger.
- Develop the photographic paper.
- Stop development process and stabilize the image in photographic fixer.
- Wash and dry the photographic print.

A number of factors may result in the production of suboptimal negatives, necessitating rescue efforts in the darkroom or even the return to the electron microscope to re-record the images. It is very important, therefore, to be aware of what is necessary to produce both a high-quality negative as well as a good photographic print.

Negative Recording Medium

Currently, the most prevalent method of recording images in electron microscopes involves the transfer of energy from either an electron or a photon into a sensitive crystal (or grain) of silver halide (usually silver bromide). The AgBr grains (Figure 8.1) used in negative materials for the electron microscope are typically around 0.5 μm in diameter and are suspended in a matrix of gelatin.

During the manufacture of the recording material, the AgBr/gelatin suspension, or emulsion, is spread over the substrate (plastic film, glass, or paper) and allowed to dry as a 12 μm thick opalescent layer that is sensitive to both photons and electrons. In negative materials, the emulsion layer is usually overlayered with a 1 μm coating of plain gelatin to protect the underlying silver grains against abrasions that would physically render the grains developable (i.e., a black scratch on the negative). In some negative films, a second layer of plain gelatin is applied to the back side of the film to prevent curling of the developed film during the drying process. An antihalation dye may be added to the backing or substrate of certain negative materials such as Kodak 4489 film to prevent reflection of light back through the sensitive emulsion layer. The dye makes the back side darker in appearance under a safelight and facilitates the recognition of the shinier emulsion-coated side. The dye is removed during the development process. Other films such as Kodak SO-163 have no antihalation layer and appear the same color (off-white) on both sides.

The silver halide crystals in an emulsion may be manufactured in various sizes. The size of the

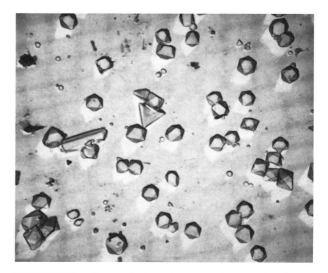

FIGURE 8.1   Transmission electron micrograph of a cubic silver bromide emulsion that may be used on photographic recording media. The silver grains have been shadowed with platinum to generate the white "shadows." Magnification = 20,000 ×.
(Courtesy of Eastman Kodak.)

grains will affect the resolution of the negative-recording medium and may affect the speed of the film (see the section "Speed and Resolution" later in the chapter). For instance, negative medium with large silver bromide grains will not have the resolving capabilities of a negative medium with very small silver halide grains. Electron microscope emulsions have small silver grains (e.g., fine grain) in order to enhance the resolution capabilities of the negative.

For the production of electron microscope negative-recording media, the emulsion is coated over either a film base of plastic (usually polyester or cellulose acetate) or glass. Plastic, rather than glass, films are used most often since they are less expensive, easier to store and transport, safer to handle, and have better adherence of the emulsion to the plastic. On the other hand, polyester films take longer to evacuate when placed inside of the microscope (cellulose acetate films take even longer) and the plastic substrate is more easily scratched. Glass substrates are dimensionally more stable than plastic and are occasionally used when the most critical of measurements must be made from the negative, as in electron diffraction studies. Glass substrates are only rarely used these days as polyester and cellulose acetate films have become dimensionally more stable. Figure 8.2 is a diagram of a typical recording medium used in electron microscopy.

Exposure Process in the SEM

In the SEM, images are usually recorded by photographing the image off a high resolution cathode ray tube (CRT). A camera, carrying 35 mm or sheet film of various sizes, is used in this process. For convenience, most researchers use an instant film such as Polaroid Type 55, which produces both a high quality negative and a positive print at the same time. Normally a 4" × 5" carrier is used for the individual packets that are slipped into the back of the camera. After development (by pulling the exposed film packet through two pressure rollers of the special film holder), the negative part of the packet is placed in a concentrated solution of sodium sulfite to remove the developer gel. After 30 seconds, the negative is washed in running tap water and dipped in a dilute wetting agent such as Photo Flo and dried in a dust-free environment. The positive print should be coated with an acidic polymer that neutralizes the developer and seals the print in a layer of plastic. Some types of instant print films do not have to be coated since they are covered with a thin layer of plastic during the manufacturing process.

Instant films are quite convenient and provide excellent negatives as well as acceptable prints. They are, however, expensive when one needs to take many micrographs. For economic reasons, it is possible to use other roll or sheet films to record the images from the CRT. Among the films that can be used are the slower speed panchromatic films (Kodak Ektapan 4162, Kodak Plus-X Pan 4147 and 2147, Technical Pan Films 2415, 4415, and 6415) as well as Kodak Commercial Film 4127, specially designed for recording SEM images from a CRT. Other types of **panchromatic** films by other manufacturers are also acceptable and may be tested for suitability. The development process is normally conducted in total darkness for all except the 4127 film, which is **orthochromatic.** (Panchromatic films are sensitive to nearly all wavelengths of light and must be handled in total darkness, whereas orthochromatic films are not sensitive to certain wavelengths and can be handled under so-called "safelight" conditions.) The processing of these films is similar to the one described for TEM (next section), except that the type of developer, as well as the final working dilution, may differ. Since developers and processing procedures are continually being improved by the manufacturers, one should follow the current recommendations included with the particular recording medium.

Exposure Process in the TEM

Although photographic emulsions are sensitive to both photons and electrons, the emulsion will react differently when exposed to a photon compared to

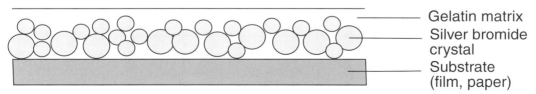

FIGURE 8.2 Diagram of a cross section of recording medium (film or paper) used to capture electron or photon images.

an electron. A photon from the light microscope typically has an energy of 2 to 3 electron volts. It takes approximately 10 of these low energy photons to expose a single silver halide grain. On the other hand, a single electron in a transmission electron microscope has energies several tens of thousands times greater (50 to 100 kV) than a photon, so that one electron will expose many different silver grains as it passes through the emulsion. For instance, a 10 kV electron will pass through about 1 μm of emulsion before its energy is dissipated, whereas a one million volt electron is capable of passing through approximately 2,000 μm of emulsion. The latter high energy electrons may pass through the emulsion, strike the negative substrate, and be backscattered through the emulsion to expose more grains. This results in a diffusion of the overall image so that it becomes less sharp. Fortunately, at the energy levels used in the standard TEM, the 50 to 75 kV electron is slowed down sufficiently by the emulsion so that only several silver grains are normally exposed by a single electron. Thus, photographic emulsions are exposed much more efficiently by electrons than by photons. In practice, the exposure of an emulsion by electrons may be considered to be 100% efficient.

When an electron strikes a silver halide grain, some of its energy is transferred to the crystal to generate aggregates of 3 to 4 metallic silver atoms, termed a *latent-image* speck. If the exposed silver halide crystal is placed in the proper developer, the latent-image specks serve as nucleation sites to rapidly convert the entire crystal from a silver halide (i.e., Ag+) to a reduced grain of metallic silver (Figure 8.3). In contrast, silver halide crystals that were not struck by an electron will develop at a considerably slower rate. One must stop the development process after a certain period of time in order to prevent these unexposed crystals from developing and causing background "fog," or an increase in the overall density of the negative. It is in the photographic "fixer" that these undeveloped silver halide crystals are selectively made soluble and removed, leaving behind only the metallic silver grains still embedded in the gelatin.

Grain Versus Noise

A negative film may be thought of as a carrier of several layers of silver halide crystals that become developable if struck by an appropriate particle. In the ideal situation, the electrons that form the image upon interaction with the specimen will expose all of the proper silver halide crystals to generate a clean,

noise-free image. However, unless enough electrons strike the emulsion (i.e., proper exposure), gaps or voids consisting of unexposed areas may be interspersed among exposed silver halide crystals. This will yield a "snowy" or noisy image when the negative is printed. The inexperienced microscopist may interpret this noise as a grainy picture (meaning that one is actually seeing overly large individual silver grains rather than gaps between groups of silver grains). One may mistakenly conclude that the resolution of the emulsion on the negative is inadequate, since the silver grains appear to be excessively large. In fact, one will not begin to see the actual silver grains until the negative is enlarged over 12 times.

Speed and Resolution

Photographic emulsions with large silver grains are said to be of *high speed*, since it takes fewer photons to expose a given area of emulsion. On the other hand, *high resolution* photographic emulsions have smaller silver bromide grains, requiring more pho-

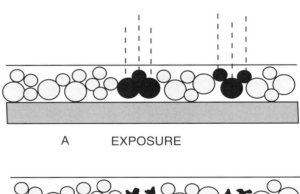

A   EXPOSURE

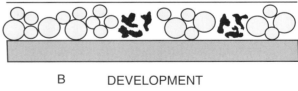

B   DEVELOPMENT

C   FIXING IN HYPO

FIGURE 8.3   During the development process, AgBr emulsion grains that were exposed to electrons or photons are developed into metallic Ag grains. Unexposed and undeveloped AgBr grains are removed by the photographic fixer, leaving behind only the developed Ag grains suspended in the gelatin.

tons to expose a given area, and are consequently said to be slower speed emulsions. Since the efficiency of exposure by electrons is essentially 100%, speed of emulsions per se is usually a minor consideration in transmission electron microscopy. *Resolution capability of the emulsion is the major concern.* For this reason, most electron microscope emulsions have small silver grains in order to be able to record fine details of the image.

Resolution is also affected by the accelerating voltage used. Higher energy electrons tend to backscatter more throughout the emulsion so that the image is spread out and resolution is slightly degraded. Some negative materials for use at higher accelerating voltages have an increased thickness of plain gelatin overlying the emulsion in order to slow down electrons and diminish the image spread. A few emulsions may utilize grains that require more than one electron hit to render them developable.

Since the silver grains used in negative emulsions for the TEM are relatively small, it is essential that one *collect enough electrons on the TEM negative (proper exposure)* to fill in the image details in order to avoid image gaps and improve the signal. *The benefits of collecting more electrons are an increase in the resolution capabilities and an increase in contrast of the TEM negative.* This is an important difference between emulsions exposed to photons versus electrons: Continued exposure to photons (as in the case of SEM images recorded from a CRT) will increase only the overall density and not the contrast. Because it is so important from a practical standpoint to collect more electrons to effect better resolution and contrast in TEM negatives, this is discussed in the next section.

Improving Resolution and Contrast in TEM Negatives

It is possible to collect more electrons on the negative by making several different adjustments to the TEM. The easiest way is to *increase the time of exposure* of the TEM negative to the electron beam. In other words, allow the beam to strike the specimen (and negative) for a longer period of time. In older microscopes, this might simply involve increasing the exposure time from 2 to 4 seconds, for example. In older instruments, this can be achieved by adjusting the electron beam intensity to the proper point for a 2 second exposure and then overriding the automatic exposure system by manually setting the exposure to 4 seconds.

Alternatively, in newer instruments, one may reset the sensitivity setting of the exposure metering system so that the film speed is effectively lower.

This will program the electronics to automatically allow more electrons to strike the negative. It will be necessary to calibrate the exposure meter by taking a number of trial exposures at different meter settings until the proper density/contrast levels are observed in the developed negative.

NOTE: To most photographers, electron microscope negatives that are properly exposed for contrast/resolution appear to be overexposed (overly dense and low in contrast due to the large number of gray tones). If there is any doubt about exposure of TEM negatives, it is better to err on the side of overexposure since an underexposed negative lacks informational detail.

A recalibration of the sensitivity setting of the exposure meter may force the microscopist to increase the number of electrons striking the specimen. This may be achieved by: (a) adjusting the condenser lenses to focus more electrons per unit area on the specimen (i.e., adjust illumination or brightness controls on the TEM), (b) using a larger condenser aperture, or (c) changing the gun bias or filament height relative to the shield to increase the yield of electrons out of the gun (see Chapter 6).

In addition, specimen contrast may be increased by using lower accelerating voltages, smaller objective apertures, and pointed filaments (see Chapter 6).

Besides using various instrumental manipulations described above, contrast may be affected in other ways as well. It is possible to increase negative density and contrast by increasing development activity through: (a) increasing developer concentration, (b) increasing developer temperature, and (c) increasing development time. Unfortunately, since signal and noise both increase proportionally using these methods, the resolution of the negative will not be as good compared to the procedure of gathering more electrons.

The best method to obtain a high-quality negative is to properly expose the negative in the electron microscope and to develop it according to the manufacturer's recommendations rather than resorting to rescue efforts in the darkroom.

Commercial Films, Handling, Developing, and Troubleshooting

Negative Recording Media for TEM

Several products exist for recording images in the TEM. Probably the most commonly used negative medium is Kodak 4489 Electron Microscope Film.

TEM Negative Media

Kodak 4489 film has a substrate of polyester of 0.18 mm in thickness and comes in four different sizes—with the 3¼″ × 4″ being the most prevalent. It is composed of an extremely fine grained emulsion that is highly efficient in the detection of electrons in the 40 to 100 kV range. The film is not sensitive to several wavelengths (i.e., it is orthochromatic) and can be conveniently handled under OA (green/yellow), IA (light red) or OC (light amber) safelights. It contains a dyed gel layer on the backside to facilitate recognition of the lighter emulsion side. This film is also available as a 35 or 70 mm roll film in 100 ft lengths, where it is designated as SO-281.

A second film, Kodak SO-163, has the same general characteristics as 4489, except that it is twice as fast (slightly larger grains) and requires longer to pump down. The recommended safelights are: I (red), GBX-2 (red), or 6B (brown). SO-163 is not available in the roll format.

Like the Electron Image glass plates, SO-163 is more versatile than 4489 in terms of the development and exposure conditions. Whereas 4489 offers only intermediate speed and signal to noise when developed, SO-163 and the Electron Image Plates can additionally be processed for maximum speed and low signal to noise (with unstable specimens) or for minimum speed and maximum signal to noise (with stable specimens).

In general, TEM negatives are placed within the evacuated microscope just below the viewing screen. However, not all negative materials are placed inside the vacuum of the TEM.

In some electron microscopes, the electrons strike a fiber optic faceplate to generate photons that travel outside the vacuum chamber to form a photonic image on a sensitive emulsion. Since the film is not inside the vacuum chamber, no evacuation of the film is necessary, and the film may be loaded directly from atmospheric conditions.

Handling of Negative Materials and Conventional Processing

Proper handling of film begins *prior* to opening the package. If the film was stored in a refrigerator or freezer (to extend its shelf life), it is necessary to allow several hours for the material to warm to room temperature prior to opening. Otherwise, moisture may condense on the chilled emulsion and ruin the recording medium. The film should be opened under appropriate safelight conditions as specified by the manufacturer. It is important that the proper size bulb be used in the safelight and that the safelight be checked periodically for leaks or cracks in the filters. This is best done by placing a test sheet of unexposed film (emulsion side facing the safelight) in the work area. Several coins (penny, nickel, quarter and half-dollar) are placed in several locations over the film and removed after various times (perhaps 1, 5, 10, and 20 minutes). The film is developed normally and the negative examined over a backlighted viewbox. The outlines of the coins should not be visible if the appropriate safelight conditions were followed.

The work area over which the negatives are to be loaded/unloaded into the cassette carriers of the electron microscope must be dust- and static-free. This is easily accomplished by thoroughly wiping the working surface with a damp, lint-free towel. The damp towel not only removes lint, but helps to dissipate any static generated during the cleaning process.

Removal of the films or plates from the package must be done carefully so as not to abrade the surface of the emulsion. For instance, in an attempt to separate closely packed films, one may be tempted to slide one film over another. This may not only generate an abrasion, but enough static electricity may be produced to cause local fogging of the film in the shape of a "lightning bolt" (Figure 8.4). It must

FIGURE 8.4 Static discharge artifact on a TEM negative. Due to low humidity in the TEM room, static electricity generated by sliding one negative over another caused the buildup of static electricity which, when discharged, gives off a miniature "lightning bolt" that exposes the sensitive emulsion.

be realized that the abrasions need not be deep enough to gouge the emulsion, since light pressure may be sufficient to mechanically expose the silver halide grains. In fact, most "scratches" seen on negatives are probably caused by slight contacts with the emulsion surfaces. Try not to contact the emulsion, but handle the film by the edges only.

Lintless nylon gloves should be worn when handling film to avoid transferring body oils and moisture to the medium. In fact, it is good practice to avoid skin contact with any object that is to go into the electron microscope. Lint and dust particles must be avoided, since they will be opaque to the electrons and recorded as transparent areas on the negative and black profiles on the final prints. One may remove the particles using either a camel's hair brush or special compressed air that is oil and moisture free. It is poor practice to blow on the films, since droplets of moisture (saliva) may be deposited onto the emulsions.

After removing the exposed negatives from the microscope and loading fresh films into cassettes, the unexposed film cassettes are placed into a chamber for preevacuation for several hours prior to insertion into the high vacuum of the TEM. Usually the films are evacuated with a rotary pump in a chamber near the TEM. Under the 10^{-1} to 10^{-2} Pa vacuums achieved, moisture and other contaminating gases are removed from the film (e.g., outgassing). Prepumping will greatly speed up subsequent film exchanges in the microscope column itself, since it may take several hours of evacuation to achieve operational vacuums in the microscope. In this manner, one has a supply of preevacuated film ready to insert into the TEM. Most TEMs have a built-in prepumping chamber in the microscope chasis for storing loaded film cassettes until needed.

The exposed negatives are usually transferred into plastic or metal racks or carriers prior to processing. Such carriers can accommodate an entire load of exposed films for passage through the development process (Figure 8.5).

NOTE: Although it is possible to develop negatives in photographic trays, this is not a consistent, safe, or efficient way to manage more than 3 or 4 films or plates.

Although not necessary, nitrogen burst agitation systems are preferred whereby pulsed bursts of nitrogen gas are forced from the bottoms of the various solutions to agitate the processing chemicals and maintain a fresh supply of active chemicals over the emulsion surface.

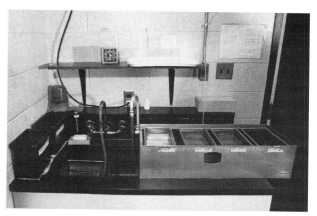

FIGURE 8.5 Typical setup used to develop sheet films (TEM or SEM). The containers of various solutions (developer, two water rinses, fixer) are placed into a large temperature controlled water bath and kept covered when not in use to prevent oxidation. Negatives are placed into plastic racks or holders to be developed in groups of 15 or more.

Most microscopists manually move the racks of negatives through the various containers of solutions (Figure 8.6). It is important that a *consistent manner of agitation* be established in order to generate a uniformly and properly processed negative. The protocol illustrated in Figure 8.6 is highly recommended for microscopists having problems developing negatives. Figure 8.7 shows a flow diagram for negative processing as used in a typical electron microscopy laboratory.

Standardization of Procedures

Once a satisfactory processing procedure has been established, it is important that this protocol be followed strictly. This becomes very important whenever several investigators are working on the same or related projects. Otherwise inconsistent negatives will be produced that will be difficult to match up during final darkroom printing for publication purposes. Several points to observe are:

1. Use the same type of negative film whenever possible.
2. Standardize the type and dilution of developer used, as well as the temperature, time, and agitation scheme followed.
3. Evaluate the status of the chemicals used for processing. Developers have a stated capacity for a certain number of films, after which quality will significantly deteriorate. In addition, certain developers have a stated shelf

Step 1 Step 2

Step 3 Step 4

FIGURE 8.6 Method of agitating sheet films for development of negatives. For uniform processing, follow this four-step agitation cycle. Each cycle should be completed in about 5 to 7 seconds and repeated every 30 seconds. Use this same agitation cycle for processing plates in a rack or holder.
(Courtesy of Eastman Kodak.)

life and should be discarded after that time—even though the full number of films may not have been run through the solution. Always record the date of preparation and numbers of negatives put through the process. Fixers are easily checked using a drop of hypo checker that turns turbid upon exhaustion of the fixer.

4. Clean vessels and tanks are essential to prevent contaminating the chemicals used in processing. Always rinse out measuring vessels before and after use to prevent cross-contamination. Better yet, use separate vessels to measure out the various solutions.

Occasionally, poor quality negatives may be produced due to either operator error or faulty chemicals. Table 8.1 should help to narrow down the causes of the problem.

Evaluation of Negatives

Routinely evaluate the dried negatives prior to making photographic prints rather than simply printing all of the negatives. It is probable that not all negatives will be needed because some may be redundant while others may be of a poor quality. Examine the negatives first for suitability for the project (appropriate area of specimen, proper magnification, etc.) and then for technical quality (good preservation and embedding, high-quality sections on a clean substrate, good staining and contrast,

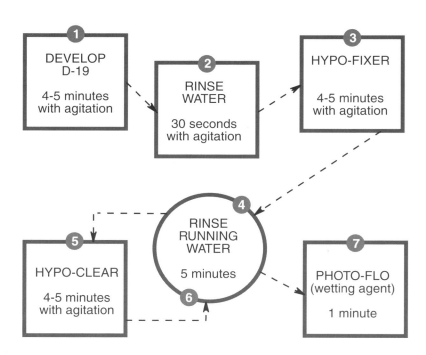

FIGURE 8.7 Flow diagram of typical process for developing sheet film negatives for TEM. SEM negatives are developed in a similar manner, except that a different developer would be used.

TABLE 8.1 Common Problems with TEM Negatives

Appearance of Negative	Possible Causes	Remedies
Low contrast	Incorrect sample preparation	Use stains or shadowing to enhance contrast
	Proper sized aperture not used	Insert smaller objective aperture
	Improper exposure	Increase exposure in TEM
	Improper development	Check recommended procedures
	Exhausted developer	Check developer
Hazy, unsharp image, lack of detail	Sample, thermal drift	Prepare samples correctly
	Instrumental instability	Check instruction manual
	Contaminated column	Clean EM column
	Improper processing of negatives	Use recommended procedure
	Incorrect focus	Refocus. Try auto-focus of TEM
Bright fringe around specimen	Incorrect focus	Refocus. Try auto-focus of TEM
Viewing screen not evenly illuminated	Beam deflectors not centered	Center condenser traverse/tilts
	Condenser lens too near crossover position	Adjust condenser lens away from crossover
Scratches and white spots on negatives	Dirt or dust on negatives prior to exposure in EM	Remove dust from unexposed negatives with brush or compressed air
	Handling abrasion in loading	Handle film more carefully
Overall film density with low contrast	Fogging of film or plate	Use proper safelights, check for light leak
	Overdevelopment of negative	
Streaks	Uneven development	See recommended processing protocols in this chapter
	Improper agitation	
Higher density along edges of negatives	Excessive agitation in developer	Agitate properly (see Figure 8.6)
	Insufficient agitation in stop bath	
Mottled or uneven appearance	Use of acid stop bath before fixing	Eliminate acid stop bath
	Insufficient water rinse before fixing	Rinse as recommended
Negatives adhere to processing racks	Too little agitation at beginning of processing	Agitate properly in developer (see Figure 8.6)
Water spots	Excess water not removed from negative surfaces	Use Photo-Flo or other wetting agent before drying

Source: Kodak Publication P-236.

proper exposure and development, good focus, absence of drift and astigmatism, absence of scratches and lint). The evaluation is normally done on a light box using an 8 to 10× magnifier (Figure 8.8). It is possible to rapidly eliminate undesirable negatives and thereby save both time and expense in the darkroom. Some negatives may have to be retaken in the electron microscope.

Automatic Processing of Negative Materials

In electron microscope laboratories where speed or large numbers of conventional negatives are required, automated processors (Figure 8.9) may be considered. Modern processors produce developed and dried negatives in 3 to 5 minutes and dry pho-

FIGURE 8.8 Evaluation of electron microscope negatives is best accomplished on a light box using an 8 to 10× magnifier.

FIGURE 8.9 Automatic processor for processing film or photographic paper. *(Courtesy of Mohr Industries.)*

tographic prints in about the same amount of time. In one hour, it is possible to generate 150 developed and dried negatives or 90 dry 8″ × 10″ prints. In some units, one developer may be used to process either negatives or prints so that it is not necessary to keep two separate sets of developers on hand.

Darkroom Printing

The last step in the production of an electron micrograph is the generation of a positive print on photographic paper. This is an exciting step because it is the first chance one has for a detailed analysis of the features revealed by the tissue section. Since electron microscope usage is expensive, generally only a preliminary survey is made of tissue features while one is operating the microscope. More detailed analysis can be accomplished at one's convenience using the printed electron micrograph. Electron micrographs represent the culmination of many hours of preparatory work and should be carefully prepared.

Micrograph enlargement or printing, as it is sometimes called, is carried out in a darkroom. Certain darkroom techniques can be used to enhance selected features of the negative and make the micrograph more pleasing to the eye. On the other hand, poor darkroom technique will be detrimental to the interpretation of the micrograph and will detract from the aesthetic appeal of the electron micrograph. In addition, poor darkroom techniques

will effect the acceptability of the micrograph for publication.

Work Prints and Final Prints

Micrograph production is usually handled in two steps. *Work prints* of all negatives are made initially. These are used to analyze specimen features and to determine which micrographs should be given more attention in the printing process and which should be enlarged. From work prints, a decision is made about which of the prints should be printed again for publication. Usually, great care is not exercised in making work prints. Work prints may be made in a small size to save photographic paper. Some investigators will make *contact prints* of their negatives by placing the negative directly on photographic paper and shining light through the negative onto the photographic paper. Contact prints are small (the same size as the negative), but may be used for analysis in a rough way. Selected micrographs may be printed at a larger size at a later time or may be printed in rapid succession to save time. The quality of work prints is usually slightly suboptimal, but of sufficient quality that their detail may be analyzed. Usually, the standard 3¼″ × 4″ negative is magnified about 2.6 to 2.7 times to obtain an 8″ × 10″ print. Prints of this size are generally considered adequate for survey purposes.

Final prints or *publication prints* are high-quality photographic prints made from selected negatives for the purposes of making a scientific presentation such as a printed publication, poster, or a slide talk. Usually several prints that vary only slightly in contrast and/or exposure are made from one negative in order to have the capability of selecting one that is considered optimal. Great care is exercised in making these prints. Well-focused negatives may be magnified up to ten times to obtain the maximum resolution from the negative, although such an enlargement is rare and will require special equipment and large size photographic paper. When the negative is magnified greatly, usually only a portion of the negative is enlarged.

The Enlarger and Accessories

The photographic enlarger is used to project a magnified image of the negative onto photographic paper coated with a photosensitive emulsion. Upon exposure, the emulsion, containing a *latent image*, will produce a positive image during development. What was a shadow (darkened) on the fluorescent screen of the transmission electron micrograph became a reversed image (light area) on the negative is now returned to its original shadowlike appearance on the print. The darkened images one sees on a transmission electron micrograph represent the areas where electrons were deflected to various degrees by tissue components that had been

reacted with osmium or heavy metal stains. In the long run, the density on the photographic paper is related to the number of electrons striking a particular area of the negative.

There are four types of enlargers used (Figure 8.10). The *diffuse light source* enlarger employs either a frosted light bulb or a fluorescent bulb as the illumination source. The light is diffused by a ground glass placed between the light source and the negative. The diffuse nature of the light source produces light rays that are scattered. Such enlargers are not used for printing electron micrographs. A *condenser* enlarger is similar to a diffuse light source enlarger in utilizing a light bulb, but in place of the ground glass are condenser lenses that focus the light onto the negative. Condenser enlargers are commonly used in the printing of scanning electron micrographs since they soften the image. The *point light source* enlarger employs an intense tungsten or halogen light source housed in a clear bulb. The source acts as if it were a single, concentrated, light source, allowing the condenser lenses of the enlarger to generate light that is more coherent than a condenser enlarger and not diffused. The point source, which is also more complex to use, is considered optimal for transmission electron microscopy, producing micrographs more representative of the negative image. The overall resolution and contrast of

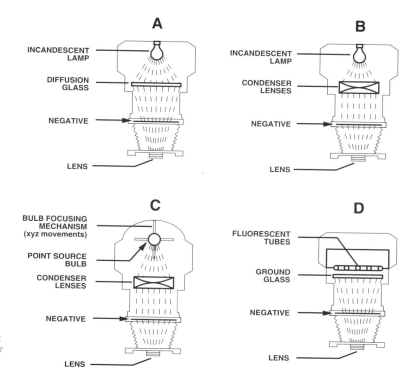

FIGURE 8.10 Drawings of (A) a diffuse light source enlarger, (B) a condenser enlarger, (C) a point light source enlarger, and (D) a cold light source enlarger. (*Diagram adapted from Postek et al., 1980,* SEM: A Students Handbook, *reprinted courtesy of M. T. Postek.*)

the print are increased using a point light source enlarger. Point source enlargers are the most commonly used type for printing TEM micrographs. A *cold light source* enlarger is similar to a diffuse light source enlarger, only the light source is usually a fluorescent tube. It is suitable for printing scanning electron micrographs, the result being an overall softened image.

Negatives used in most TEM investigations are commonly sized $3\frac{1}{4}'' \times 4''$. The negative is situated in an appropriate negative holder, usually without glass supporting windows in order to avoid dust and scratches, even though the negative sandwiched between glass will ensure a flatter image (all areas will be evenly focused).

The enlarging lens must be compatible with the size of the negative to be enlarged to achieve the appropriate field enlargement. For a $3\frac{1}{4}'' \times 4''$ electron microscope negative, the enlarging lens should have a focal length of around 135 mm. Lenses that will not enlarge the negative more than a few times or that cut out the corners of the negative are inappropriate. The enlarging lens on a condenser enlarger has a feature, the aperture, which will determine the amount of light that is projected. F/stop numbers indicated on the side of the enlarger lens regulate the amount of light passing through the lens. Each higher number indicated allows half the amount of light (as the previous smaller number) to pass through the lens. For example, an f8 aperture allows twice the amount of light as an f11 aperture. One focuses the projected negative image usually at the lowest numbered f/stop to obtain the maximum light possible and to minimize the depth of field (see Chapter 6). On a point light source enlarger, a variable transformer, instead of the aperture, is used to vary the illumination. Closing down the aperture will cause the point source filament to become visible on the enlarger stand.

The height of the enlarger on the support stand is the major practical factor that determines the enlargement. For each enlargement, the focus knob must be manipulated to bring the image into sharp focus. Critical focus is achieved using a small magnifying device to image the silver grain in the negative on the projected image until the grains of the negative are sharply focused by the enlarger controls. An automatic timer attached to the enlarger determines the length of time the light is exposed to the photographic paper.

An easel holds the photographic paper under the enlarger. Some easels may frame a print of any size or dimension up to the size of the photograph-ic paper used. Other easels frame only one size of photographic paper. Special orange or red safelights, which do not expose the photographic emulsion, are used in the darkroom to allow one to see the surroundings while working with the enlarger and while viewing the projected image.

Printing Papers

When choosing a photographic paper for electron microscopy, one has the following options:

Fiber base or resin coated. Fiber-based papers have a paper or polyester base, whereas resin coated (also known as RC) are fiber-based papers coated with a surface layer of clear plastic. Although fiber-based papers have been the traditional paper of choice, the newer resin-coated papers are more convenient to process and require shorter water rinsing times. RC papers may be air-dried in racks whereas fiber-based papers require special heated dryers. Both are acceptable and commonly used for electron microscopy.

Paper weight. The thickness of the base determines the weight. Less wrinkling of the paper is seen with heavier weight papers, although single weight paper is more convenient to process and less costly.

Tone. The warmness or coldness of papers is referred to as the tone of the paper. Warm papers are more brown in tone; cold papers are more blue-black in tone. A cold paper is often used for transmission electron microscopy, whereas a warm paper is often used for scanning electron microscopy. Common brands of fiber-based paper for electron microscopy are Kodabromide, Kodak Polycontrast, Ilfobrom, and Agfa Brovira. Comparable resin-coated papers are produced by Kodak, Ilford, and Agfa.

Surface. Paper surfaces extend from glossy to matte (flat). Glossy is customarily used for publication purposes; however, poster presentations are easier to view if nonreflecting matte surfaces are used.

Contrast. This refers to the range of blacks and whites in the micrograph. Sharp blacks along with sharp whites and fewer intermediate gray tones constitute *high contrast*. The presence of grays in the micrograph produces *low contrast* and gives the micrograph a "muddy" appearance. There are many possible contrast ranges between high and low contrast.

In general, stained tissue sections are of a low contrast. Thus, for transmission electron microscopy, high-contrast-producing photographic papers are usually required to obtain a pleasing micrograph. Within the overall constraint for the need to increase contrast, some electron microscopists prefer more gray tones and others opt for fewer gray tones, preferring instead sharp blacks and whites. This is simply a matter of taste. Within certain limits, both high contrast and low contrast micrographs are seen in publications. It is not aesthetically pleasing to publish a high contrast micrograph next to one of much lower contrast. Micrographs should match each other in contrast.

Contrast may be regulated by either the grade of paper employed or with filters and variable grade paper. *Graded papers* have numbers from #1 to #5 (or sometimes #6) with #1 producing the lowest contrast and #6 the highest contrast. Most transmission electron microscopists find that grades #3 to #6 give the desired results with the usual range of negatives produced by the electron microscope. A contrast series shows the differences in image contrast one may expect when graded papers are utilized (Figure 8.11). When using *variable contrast papers,* only one paper type is used. The contrast is regulated with filters inserted into the enlarger during the printing process. Both methods for varying contrast are in

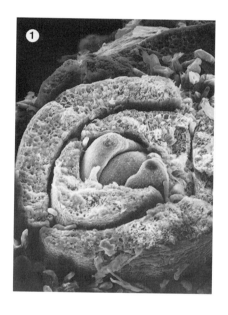
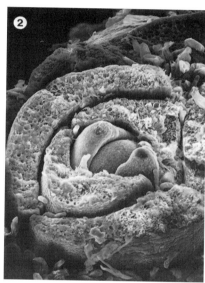
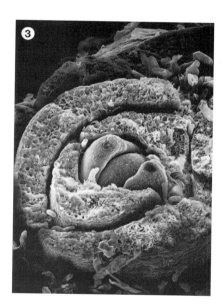
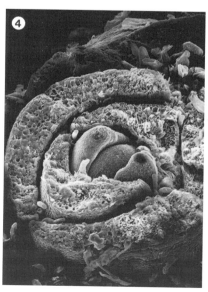

FIGURE 8.11 A contrast series showing prints made on resin-coated paper ranging from contrast grades #1 through #5.

common usage for electron microscopy. Expect that a higher contrast filter or graded paper will require more exposure than one of a lower contrast.

A typical problem that surfaces, especially with beginners, is determining how to manipulate contrast and exposure (see following) simultaneously to obtain a suitable micrograph. To some degree, exposure and contrast are a matter of individual taste. This problem is solved if one first learns to distinguish, through examination of micrographs, the difference between contrast and exposure, and one corrects micrograph problem(s) in a systematic manner. With experience the problem will be minimized.

Enlarging

Basic photography texts should be consulted for the detailed steps in enlarging the negative image to make a print (see references). Fortunately, many individuals are familiar with the enlarging process and help is usually available.

The intensity of light on the photographic paper is one factor that determines the *exposure* or the overall darkening of the photographic paper. The more total light, the darker the image. All other variables being constant, the amount of light coming from the enlarger at any instant may be regulated by the enlarging lens in a condenser enlarger and a variable rheostat in the point light source enlarger. The amount of light striking the photographic emulsion is also dependent on the time the photographic paper is exposed to the light. Figure 8.12 is a series of prints made on the same grade paper where only the exposure time is varied while the contrast is generally similar for all exposures. To determine if a problem is a contrast or exposure problem or both, one should analyze a micrograph systematically for differences in blacks and whites (contrast) and for the degree of general darkening of the paper (exposure). Problems may exist with both contrast and exposure.

Both contrast and exposure can be manipulated to produce an acceptable background density of a print. As a general rule, where there are no tissue elements (for example, intercellular space or clear vacuoles), these areas of the print should match the unexposed border of the micrograph or be dark-

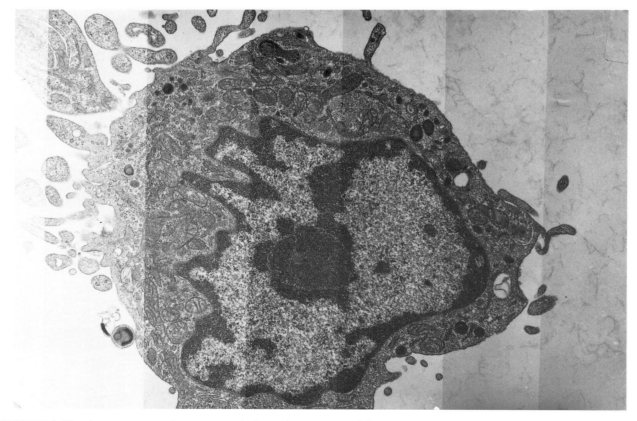

FIGURE 8.12   An exposure series or test strip is made on paper of the same grade, varying only the time the light was projected onto the photographic paper.

ened (grayed) slightly as compared with the border (Figure 8.11). When this is evident, the exposure is correct. If the contrast is too great or too little, use a photographic paper that will compensate.

A good way to determine how a print will appear is to expose a small piece of printing paper or *test strip* before the final print is made. The test strip can be covered partially by opaque cardboard and uncovered systematically in timed steps from one end to the other (e.g., at 3 second intervals) during the exposure to determine optimal exposure time. When the print is developed, it will show a series of exposures along the test strip. Examination of the strip will allow one to determine the appropriate exposure. Examining the area of the print with the proper exposure will then allow one to determine the suitability of the contrast.

Since considerable time and effort may go into producing a print, one should keep a record of the exposure time, light setting, and intensity of the light as determined by a rheostat on the enlarger. A soft pencil may be used to record this and other information on the back of the photographic paper used to make the print.

Print Processing

The standard print processing setup is shown in Figure 8.13. Recommended times and brief instructions follow.

Developer. Use the manufacturer's recommended developer. Dektol made by Kodak is a common developer. Develop 1 to 2.5 minutes for fiber-based papers and 1 minute for resin-coated papers. Agitate gently but constantly.

Stop bath. The stop bath will neutralize the developer and allow the fixer to last longer. A commercial acetic acid (4%) bath is recommended as a stop bath

for fiber-based papers. A water bath is used for resin-coated papers. About 15 seconds is needed in the stop bath.

Fixer. One or two baths of a product such as Kodak Rapid Fix are used. Fiber-based papers require about 5 minutes in fixer, while resin-coated papers require about 2 minutes. If two fixer baths are used, then as the first fixer bath becomes old, it is replaced with the second bath and a new bath is made to replace the second bath. An exhausted fixer solution is easily determined with a drop of test solution (Edwal or Kodak) that will turn cloudy if the fixer is in need of replenishment.

Water rinse. The micrographs are washed for 2 to 5 minutes in water.

Fixer remover. For fiber-based papers only, use one of several commercial solutions (for example, Perma Wash by Kodak) for 2 to 4 minutes.

Wash. Fiber-based papers should be washed from 8 to 15 minutes, whereas resin-coated papers can be washed for 4 to 5 minutes. It is preferable to use a system in which running water is used to agitate the prints. Special rotating washers or swirl washing sinks have been designed for this purpose. However, care must be exercised with the resin-coated papers or the emulsion will be scratched or abraded (i.e., do not wash resin-coated papers in a tumbling drum system).

Burning-in and Dodging

Negatives are not always evenly exposed. There are many causes of uneven exposure, some of which relate to variations in the thickness of the tissue section, others to unevenness of the electron illumination, and yet others to errors made in the developing

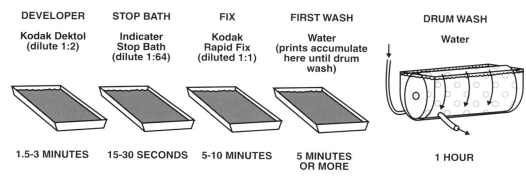

FIGURE 8.13 Paper print processing setup.

and enlarging process. Whatever the cause, the final print can be made to show more of an even exposure by the rescue processes of dodging and burning-in. When a print shows one underexposed, lighter area, an opaque material (for example, cardboard) with a hole cut in it (1″ to 3″) is held between the light source and the printing paper during a fraction of the entire exposure (for example, 5 seconds of 20 seconds). (The hole should be held close enough to the lens so that its edge is out of focus and not sharp.) The area that was underexposed is allowed to expose beyond the time that is optimal for the remainder of the print. The hole is moved slowly to cover all underexposed areas of the print and "feather" the burned-in edges. Thus, the underexposed area is "burned-in" to compensate for negative imperfections.

Often a portion of an electron microscope print will be overexposed (darker) as compared with the remainder of the print. This artifact may have the same causes as those mentioned previously for underexposure. The circular piece of cardboard, cut out from the opaque cardboard described above, can be mounted onto a narrow rod (for example, a long applicator stick) of some type and used to "dodge" the overexposed area. The cardboard is placed between the enlarging lens and the printing paper and moved slowly to cover the overexposed area. Usually dodging is effective if under 30 to 40% of the exposure time is blocked to a selected area. Several attempts may be necessary before the desired result is produced. It is possible that the same print may require both burning-in and dodging.

Techniques to Enhance Contrast

There are several procedures that allow the small increases in contrast one often seeks in a transmission electron micrograph. Use of a point source enlarger will increase contrast. Exposure times that are long (30 seconds to 1 minute) at low light levels will produce a slightly more contrasted micrograph. Development in undiluted developer will increase contrast. All of the above used in concert will provide the equivalent of one paper grade of higher contrast (e.g., grade #5 to #6).

Matte and Glossy Electron Micrographs

Even though a glossy paper is used to print micrographs, fiber-based prints can be made either glossy or matte (nonshiny) during the drying process. If heated drum dryers are used and the emulsion surface is placed in contact with the drum, the print will be glossy. Most work prints and publication prints are dried in this manner. A matte finish is obtained by placing the emulsion side toward the drying cloth. Prints displayed in a poster session of a scientific meeting are dried for a matte finish. They reduce glare and thus are easier to view. Resin-coated papers are purchased with specified matte or gloss finishes and never dried in a conventional drum dryer because the plastic coating will melt and adhere to the surface of the heated drum.

Preparing Micrographs for Publication

Electron micrographs are made for viewing. If one's goal is publishing, it is hoped the micrographs will contain new information that will add to the existing body of scientific literature. The goal of every biological electron microscopist is to publish data in the best form possible. The data for the microscopist *is* the electron micrograph. Therefore, it is imperative to produce micrographs of the highest quality for presentation at scientific meetings and for publication.

The Final Print

Choosing a Representative Print

A criticism commonly heard is that electron microscopists publish micrographs that are not representative of tissue features. This criticism should be taken into account when selecting prints to portray the findings of a study. The final print should not be misleading as regards the typical findings of the study. If one can easily avoid misrepresentation, then why is the criticism often leveled at microscopists? The answer lies in the microscopist's desire to publish the most aesthetically pleasing print possible. In doing so, there may be a tendency to select a beautiful print that may, at the same time, show an atypical or infrequently encountered structural feature. Clearly, this tendency should be avoided.

Matching Several Prints Placed Together

Final or publication prints are usually based on scrutiny of a work print and after a determination

REFERENCES

Crang, R. E. 1987. Montaging electron micrographs. *J Electron Microsc Tech* 7:53–60.

Electron microscopy and photography. 1973. Kodak Data Book No. P-236, Rochester, NY.

Farnell, G. C., and R. B. Flint. 1975. Photographic aspects of electron microscopy. In *Principles and techniques of electron microscopy. Biological applications,* vol. 5. M. A. Hayat, ed., New York: Van Nostrand Reinhold Co., pp. 19–61.

Horenstein, H. 1983. *Black & white photography. A basic manual.* 2d ed. Boston: Little, Brown and Co.

Postek, M. T., K. S. Howard, A. Johnson, and K. L. McMichael. 1980. *Scanning electron microscopy: A student's handbook.* Ladd Research Industries, Inc., Burlington, VT. 305 pp.

Shipman, C. 1974. *Understanding photography.* Tucson: H. P. Books.

Smith, R. F. 1990. *Microscopy and photomicroscopy: A working manual.* CRC Press, Inc., Boca Raton, FL.

Chapter 9

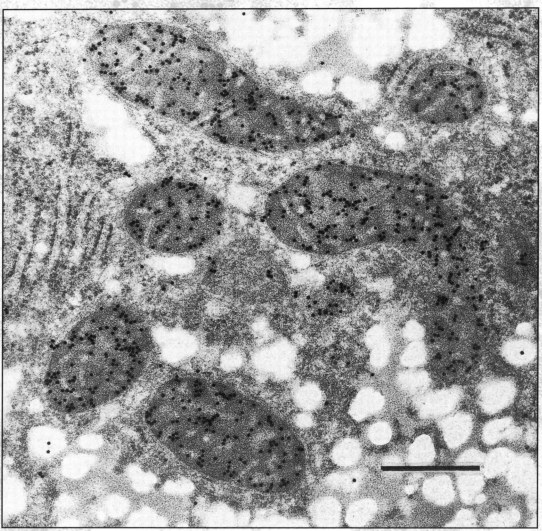

Courtesy of M. Bendayan.

Immunocytochemistry

THE PROCESS FOR DETECTING ANTIGENS using anti-bodies directed specifically against them is called *immunochemistry*. When the electron microscope is used to detect the localization, the technique is usually referred to as *immunocytochemistry*. Although sometimes immunocytochemistry is said to also be performed at the light microscope level, a more commonly used term is immunohistochemistry. The molecular recognition properties of the anti-body for the antigen in the *antigen-antibody reaction* are what provide the specificity to the localization process. An antigen is localized by antibodies that are applied to the cells. A *tag* recognizable under the light or electron microscope is attached to the last of the added antibodies. In theory, a single molecule of a particular antigen can be localized by a tagged antibody or series of antibodies.

Immunocytochemistry has made valuable contributions to our knowledge of the location of a variety of substances. It is the most powerful and potentially specific of all of the localization techniques described in this text. Immunocytochemistry has permitted the localization of substances where the only major requirement for localization is that a specific antibody be developed against the antigen. Of the various localization techniques currently available for electron microscopic investigations (see Chapters 10 through 12), immunocytochemistry is not only the most widely used, but holds the most promise for the future.

Immunocytochemistry was a natural out-growth of immunohistochemistry. Most of the dyes and fluorescent compounds used to visualize the immune complexes at the light microscope level are not appropriate tags for use with the electron microscope because they are neither electron dense nor do they produce an electron-dense product.

In 1959, Singer used an antibody tagged with the electron-dense protein ferritin to obtain ultra-structural localization. The tag, which was electron dense due to the presence of iron in the protein, was applied prior to embedding and came to be known as a *preembedding stain*. The first *postembedding stain* applied to a tissue section was the electron-dense heavy metal uranium (Sternberger et al., 1965). The development, by Graham and Karnovsky (1966), of cytochemical methods to localize enzyme (peroxidase) activity allowed Nakane and Pierce (1967) and others to label an antibody with an enzyme and to use the *enzyme-labeled tag* for electron microscopy. Subsequently, the peroxidase-antiperoxidase (PAP) technique was developed by Sternberger et al. (1970) to improve the sensitivity and resolution of the enzyme method. Ultrastructural tags of more

recent vintage include the *avidin-biotin system* (Heitzman and Richards, 1974) and the currently very popular *colloidal gold technique* (Faulk and Taylor, 1971).

The Antigen-Antibody Reaction

When higher organisms (most vertebrates) encounter a foreign molecule or *antigen* repeatedly, they will develop *antibodies*. The term immunogenesis refers to the production of antibodies in response to the antigen.

For instance, antibodies may be formed against antigens, such as might be on red blood cells from a different organism. Complex molecules under a molecular weight of about 5,000 are usually not immunogenic, but may be made so if conjugated, or chemically coupled, to a larger molecule. Proteins and complex branched chains of sugars are very immunogenic, whereas lipids, nucleic acids, and less complex molecules are poorly immunogenic. A *determinant* is the unique portion of the antigen, either molecular or conformational or both, which is recognized by the antibody.

Antibodies may be harvested from animals which have become immunogenic in the natural course of events to some environmental stimulus or disease. Humans with certain autoimmune diseases have a high level of antibodies to some of their own tissue components. For example, actin (a filamentous structural protein) antibodies are often present in the sera of patients with certain forms of hepatitis.

An *adjuvant* is a substance or treatment that may be used to boost the immune response. Antibodies may also be produced artificially by repeated injection of antigens with, or without, accompanying adjuvant. Repeated injection is, by far, the most common method to induce antibodies for use in immunocytochemistry. Animals that are commonly used to produce antibodies for immunocytochemistry are the rabbit, guinea pig, mouse, and goat. The response to antigens, with or without adjuvant, in these species is the production of *polyclonal antisera*, which is always a mixture of antibodies against different sites (determinants) on the same antigen. Polyclonal antibodies bind to an antigen at multiple determinants. There are many sites available on the antigen to allow binding of numerous antibodies. *Monoclonal antibodies* are highly specific in that they bind to a single determinant. Monoclonal antibodies are produced first by animal immunization and secondly by *in vitro* fusion of an individual antibody secreting cell from the immunized animal with specific tumor cells. The cell that is formed is called a *hybridoma cell*. It not only secretes clones of antibodies binding to a single determinant of the antigen, but divides in culture and may be frozen and thawed so that a continuous source of monoclonal antibody is available.

The advantage of using monoclonal antibodies is that they are highly specific. The disadvantages of using monoclonal antibodies for immunocytochemistry are that they are often of low affinity (binding strength) and they bind at one site only, a site which may be altered during tissue preparation or may be hidden and not accessible to the antibody.

Antibodies, while generally thought of as being specific for the molecule that elicited their formation, may *cross-react* with other molecules having the same or similar determinant. Cross-reactions may be with related antigens in another species or molecules that have little or no functional or structural similarity to the antigen except at the antibody binding site.

The antibodies present in a subfraction of the serum (noncellular portion of the blood) of mammals are collectively called *immunoglobulins*. Structural subclassifications of immunoglobulins, abbreviated IgG, IgA, IgM, IgD, and IgE, have a variety of immune and nonimmune functions. We will focus on the structure of the most common of these, the *IgG* molecule, since other classes of immunoglobulins have many structural similarities to IgG. IgG is the major immunoglobulin in serum of most mammals and the most common immunological probe for immunocytochemistry.

Figure 9.1 depicts the general structure of an IgG molecule, having a molecular weight of about 150,000. Of its four polypeptide chains, two are *heavy chains* (MW 50,000) and two are *light chains* (MW 25,000). The identical heavy chains are linked to each other by disulfide bridges, and the two light chains are also attached to each heavy chain by disulfide bridges. The molecule has the general appearance of a *Y*, which is usually depicted diagrammatically as an inverted *Y*. The two short limbs of the Y, composed of both light and heavy chains, may bend somewhat at the hinge region of the heavy chain (shown in Figure 9.1 as a coil). Two identical *antigen binding sites* are found at the ends of two short limbs of the Y on the light and heavy chains, whereas the stem of the Y serves, among

other functions, to notify the immune system that after binding to an antibody, it has recognized a foreign substance that needs to be eliminated. Enzymatic cleavage with papain severs the heavy chains, yielding three fragments (Figure 9.2): two *Fab* (antibody binding) fragments and a single *Fc* (crystallizable) fragment. Fab fragments may be used in place of IgG in immunocytochemical procedures. Subsequent figures will depict the IgG molecule in the simple Y configuration.

It is important for our purposes to realize that *antibodies themselves, being complex protein molecules of high molecular weight, are immunogenic* (capable of

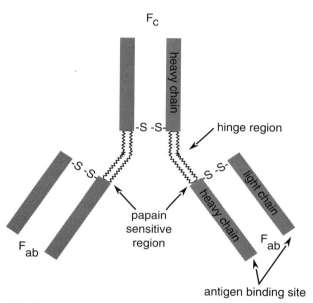

FIGURE 9.1 Structure of the IgG molecule. (Fab, antigen binding fragment; Fc, crystallizable fragment; S-S, disulfide bonds).

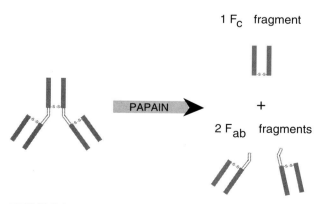

FIGURE 9.2 Cleavage products of IgG after treatments with papain, a proteolytic enzyme found in the papaya plant.

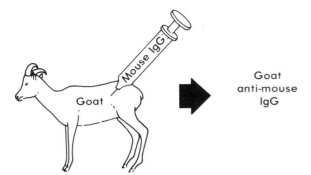

FIGURE 9.3 Goat anti-mouse IgG is produced after injection of a goat with mouse IgG. Antibodies produced in this manner are specifically directed against other antibodies (proteins), and thus are commonly used as secondary antibodies in immunocytochemical reactions.

eliciting an immune response). For example, a goat repeatedly injected with mouse IgG will form antibodies to this immunoglobulin fraction. The antibody produced is called goat anti-mouse IgG (Figure 9.3). This terminology for naming antibodies, by first specifying the species in which the antibody was produced and, second, specifying the species against which it was directed, will be employed throughout this chapter. Although technically incomplete, in some instances the name of the species in which the antibody was produced is omitted.

Approaches to Labeling

The general approach to immunocytochemistry is to attach *tags* to antibody molecules. The tagged antibodies are then exposed to the tissue antigen of interest to which they specifically attach. There are several ways in which this may be accomplished.

The *direct method* for antibody labeling is straightforward, in principle. An antibody is first prepared in another species against the antigen. This antibody is conjugated to a tag, visible (or later made visible) in the electron microscope. The *primary antibody,* as it is called, is then reacted with the tissue antigen (Figure 9.4). The tag on the 1°Ab (antibody) is then observed by electron microscopy.

By far, the most common strategy for antigen localization is the *indirect method* (Figure 9.5). The tissue antigen is first exposed to a primary antibody that will attach to it. After washing away unbound antibody, a *secondary antibody,* prepared in a different species and directed against the immunoglobulins of the class of the primary antibody (e.g., goat anti-mouse IgG), is allowed to bind. Usually the second antibody contains the tag that may be visualized by electron microscopy, although tertiary, and so forth, antibodies that contain the tag may be added.

The indirect method is the most commonly used procedure for several reasons. The primary antibody is usually difficult to obtain, since frequently only small amounts of purified antigen may be available to elicit its formation. Conjugation of any primary antibody to a tag is often difficult and inefficient; consequently it is often a "hit or miss" procedure that may exhaust the precious lit-

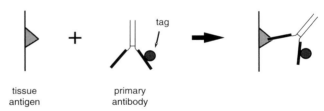

FIGURE 9.4 The *direct method* of antibody labeling. A tissue antigen is exposed to a primary antibody that has been previously conjugated to a tag.

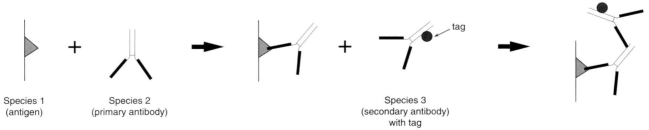

FIGURE 9.5 The *indirect method* for antibody labeling. A tissue antigen (from species 1) is exposed to a primary antibody (generally made in a second species) that has been made to bind the antigen. After binding of the primary antibody, a tagged secondary antibody (made in a third species) is exposed to the bound antigen-antibody complex. The secondary antibody was produced to react against all IgGs of the second species (Figure 9.4) and contains a tag. The result is a two-layered antibody sequence with an attached tag.

FIGURE 9.6
Amplification of localization using polyclonal antisera. Since the tagged secondary antibodies can react with multiple determinants on the primary antibody, the response is greatly amplified by the increased binding of secondary antibody to primary antibody and the presence of multiple tags.

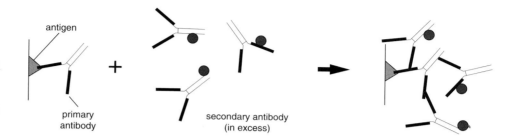

FIGURE 9.7 Localization using *protein A*. Tagged protein A is added to the antigen-antibody complex. Protein A binds to the Fc portion of the IgG molecule.

tle primary antibody available. Addition of tags may also severely interfere with the binding of the primary antibody to the antigen. Secondary antibodies, on the other hand, are usually plentiful since they are easily obtained in bulk from commercial sources. The ultrastructural tagging is, therefore, not as risky if plenty of secondary antibody is available. Furthermore, localization is amplified by the number of secondary antibodies containing tags, which can bind to a single antigen (the primary antibody) due to the polyclonal nature of most antisera (Figure 9.6). Several secondary antibodies may bind to a single primary antibody thus amplifying the reaction.

Protein A, produced by certain strains of *Staphylococcus aureus* bacteria, or related proteins such as protein G from a streptococcal organism, have the interesting ability to bind to the Fc portion of some immunoglobulins (notably IgG) from several species. A variety of ultrastructural tags may be conjugated to protein A. Sequential addition of primary antibody and protein A-tag represents a major ultrastructural form of localization, albeit based only partially on immunologic principles (Figure 9.7). The affinities of protein A and G for various classes of immunoglobulins are given in Table 9.1.

TABLE 9.1   The Relative Affinities of Protein A and G for Various Antibody Subclasses in Various Species

Antibody	Affinity for protein A	Affinity for protein G
Human IgG$_1$	+ + + +	+ + + +
Human IgG$_2$	+ + + +	+ + + +
Human IgG$_3$	—	+ + + +
Human IgG$_4$	+ + + +	+ + + +
Mouse IgG$_1$	+	+ + + +
Mouse IgG$_{2a}$	+ + + +	+ + + +
Mouse IgG$_{2b}$	+ + +	+ + +
Mouse IgG$_3$	+ +	+ + +
Rat IgG$_1$	—	+
Rat IgG$_{2a}$	—	+ + + +
Rat IgG$_{2b}$	—	+ +
Rat IgG$_{2c}$	+	+ +

Summarized data taken from Harlow and Lane (1988).

Epitope Tagging of Proteins

A major stumbling block for the localization of antigens using immunoelectron microscopy may be the lack of specific antibodies with strong affinity for the antigen to be localized. One novel solution is to work with commercially available, highly specific, monoclonal antibodies with proven affinity for a particular *epitope* or sequence of amino acids. Using molecular engineering techniques (Kolodziej and Young, 1991), one inserts the desired sequence of amino acids into the protein and then localizes the amino acid sequence using the available monoclonal antibodies. For example, you can incorporate the amino acids forming an epitope from the influenza hemagglutinin (HA) protein, a well-characterized protein for which monoclonal antibodies are available. When the cell produces the protein that one hopes to localize, the newly synthesized protein will contain the HA sequence. The monoclonal anti-HA antibodies will then locate the desired protein since it now contains the HA epitope (Figure 9.8). In one series of studies (Wente and Blobel, 1994), this "guilt-by-association" technique proved particularly useful for localizing proteins of the nuclear pore complex in the yeast *Saccharomyces cerevisiae*.

Ultrastructural Tags

Under the proper conditions, immunoglobulins and immunoglobulin fragments are directly visible under the electron microscope (Figure 9.9), but their use for most localization purposes requires an attached electron-dense substance because untagged antibody may be confused cell surface coatings. Labels, stains, or tags, as they are frequently called, allow the site of antigen localization to be readily visualized. The types of tags fall into three major categories, based on the nature of the tag. Some tags are organic molecules that possess *structured electron opacity,* others are *enzymes* whose reaction product can be detected after the addition of the substrate, and yet others are *heavy metals* that can be visualized directly.

Some of the first tags discovered were those that possessed structured electron opacity. *Ferritin,* the principal storage protein for iron in mammals, is obtained primarily from horse spleen. The ferritin molecule is about 10 to 12 nm in diameter and has a molecular weight of about 450,000. The core of the molecule is rich in iron, which is responsible for its electron opacity (Figure 9.10). Several methods are available for conjugation of ferritin to immunoglob-

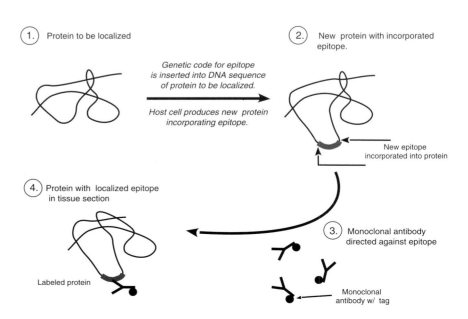

FIGURE 9.8 Illustration of steps involved in the epitope labeling procedure.

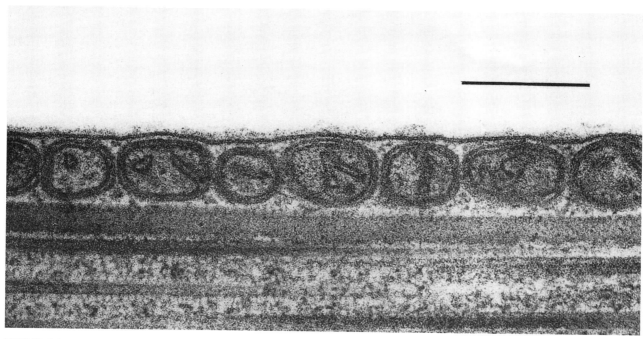

FIGURE 9.9 Micrographs showing IgG directly labeling a plasma membrane of a sperm tail. IgG appears as a fuzzy material seen on the plasma membrane. Bar = 0.25 μm.

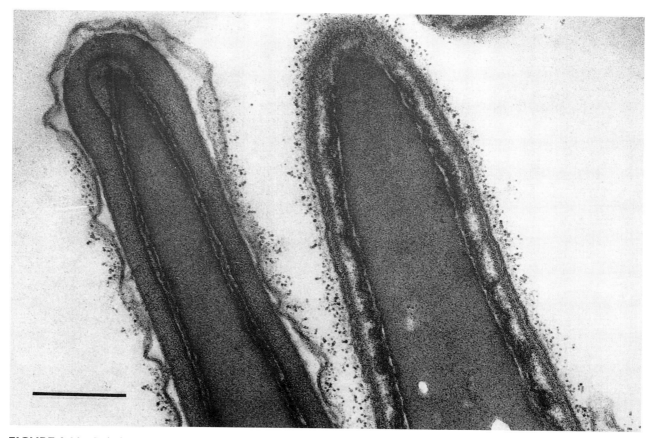

FIGURE 9.10 Labeling of the head region of two sperm with ferritin. An indirect technique was used to label cell surface antigens. Each ferritin molecule is about 10 nm diameter. Bar = 0.5 μm.

270   *Chapter 9*

ulins, although most ferritin-secondary antibody conjugates are usually procured commercially. Ferritin was used extensively in the early phase of immunocytochemistry; however, the development of better tags have made its use less popular. An example of ferritin labeling on a cell surface is shown in Figure 9.10. Other tags in the structured electron opacity category with ferritin are *hemocyanin* and intact *viruses*. The latter two tags are more often used in surface localization studies using the scanning electron microscope since they are quite large.

An enzyme tag is commonly attached to an immunoglobulin with the intent of using the catalytic properties of the enzyme to produce an insoluble reaction product visible in the electron microscope. *Horseradish peroxidase* (MW 40,000) is used almost exclusively in this category since its insoluble reaction product can be made electron dense with the addition of osmium. In the presence

of the hydrogen peroxidase substrate (H_2O_2), added 3-3' diaminobenzidine molecules serve as an electron donor to the reaction. The oxidation products of diaminobenzidine form an insoluble precipitate that appears colored red/brown at the light microscope level. For visualization of the precipitate at the electron microscope level, it is necessary to render it electron dense by chelation with osmium tetroxide where it appears as an extemely dense precipitate (Figure 9.11). Thus, the reaction can be visualized at both the light and electron microscope levels. The immunocytochemical reaction employing horseradish peroxidase is summarized in Figure 9.12.

An advantage of the enzyme method is that its reaction product is amplified by being continually deposited at the reaction site. For the same reason, the localization may be less precise than desired in relation to the antigen, since there is an indeterminate amount of reaction product present. A modifi-

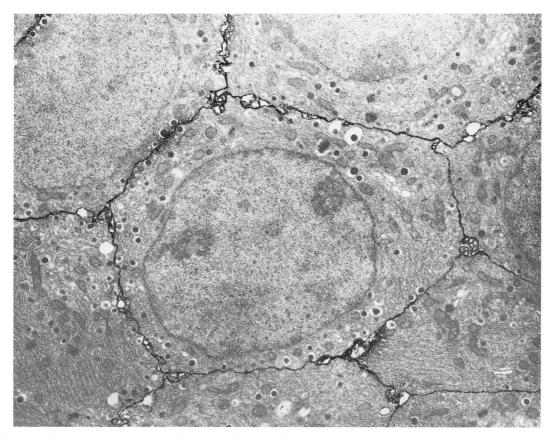

FIGURE 9.11   *Horseradish peroxidase* used in an indirect localization protocol for a plasma membrane protein in islet cells of the pancreas. The dense label is seen at the cell periphery. The tissue was not stained with lead or uranium, thus rendering the localization more prominent.
(Micrograph courtesy of O.K. Langley.)

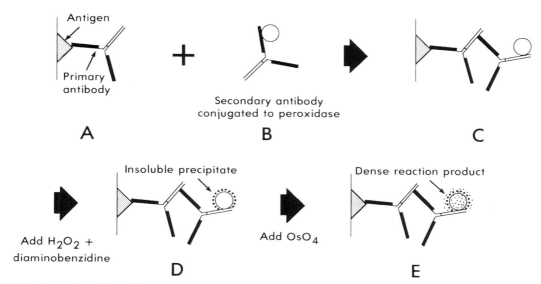

FIGURE 9.12 Steps in indirect labeling with peroxidase: (A) the antigen is attached to the primary antibody; (B) a secondary antibody from another species, directed against the IgG class of the primary antibody, is labeled with horseradish peroxidase and (C) allowed to bind. The addition of hydrogen peroxide and 3-3′ diaminobenzidine (D) causes an insoluble precipitate to form (E) which appears dense under the electron microscope after the addition of osmium.

cation of the enzyme method described above achieved popularity in the 1970s. The *PAP* (peroxidase-antiperoxidase; MW = 420,000) *technique*, as it is called, uses a soluble complex of peroxidase and antiperoxidase (Figure 9.13), which is available commercially. The PAP complex is essentially three enzyme molecules linked by two antibody molecules directed against peroxidase.

The PAP protocol, as shown in Figure 9.13, first calls for the binding of a primary antibody that has been made to react with the antigen. Next, one adds a secondary antibody, from another species, which is directed against immunoglobulins of the class of the primary antibody. The PAP complex is then added. The PAP complex binds to the secondary antibody because the antiperoxidase molecules were prepared in the same species as the primary antibody, but they exhibit different

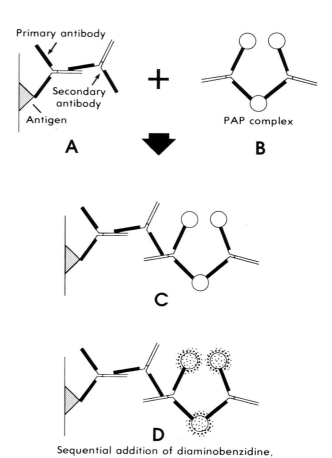

FIGURE 9.13 Steps in the peroxidase-antiperoxidase (PAP) technique. (A) The antigen has been sequentially exposed to the primary and secondary antibodies; (B) the PAP complex is added to the complex after excess primary and secondary antibodies have been washed away (C). The PAP complex reacts with the secondary antibody (since the PAP complex was produced in the same species as the primary antibody); (D) a dense reaction product is formed after osmium impregnation of the insoluble diaminobenzidine precipitate (see Figure 9.12).

determinants than the primary antibody does. Thus, the secondary antibody, being directed against all immunoglobulins of the first species, binds both the primary antibody and the PAP complex. With the addition of diaminobenzidine and later osmium, a characteristic electron-dense reaction product is visible under the electron microscope (Figure 9.14; for examples see Sternberger, 1979, and Moriarty and Garner, 1977). Gold or nickel grids must be used in this procedure to avoid reaction of copper grid bars with osmium. The PAP technique and the other aforementioned tags are now rarely used at the ultrastructural level and have been replaced by the more popular colloidal gold tags.

Although several heavy metals have been used occasionally as immune tags, the most common tag currently used is *colloidal gold* (Figure 9.15). Colloidal suspensions of gold may be prepared easily and readily tagged to immunoglobulins or protein A. Of particular note is that these discrete, highly electron-dense particles can be made in sizes from approximately 3 nm in diameter and upwards. Gold particles are easily detected on tissue sections because of their high electron scattering properties. Furthermore, they can be made small enough to allow clear visualization of surrounding tissue structure (Figure 9.16). Antigen binding sites are more easily quantifiable using colloidal gold tags than with other techniques because of the discrete properties of this tag. Many colloidal gold-immunoglobulin conjugates are available commercially, as is protein A-gold. Caution is in order since colloidal gold is not irreversibly bound to an antibody and may be released from its antibody by a change in pH.

The ease with which colloidal gold tags may be made or obtained commercially as well as their stability is largely the reason for the popularity of this label. Several current references employing colloidal gold are provided at the end of this chapter.

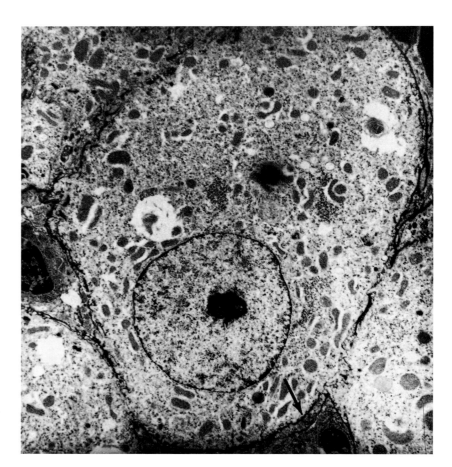

FIGURE 9.14 Use of the PAP technique to localize a surface antigen. Dense staining is seen along the cell surface (*arrow*).
(*Micrograph courtesy of A. Mayerhofer.*)

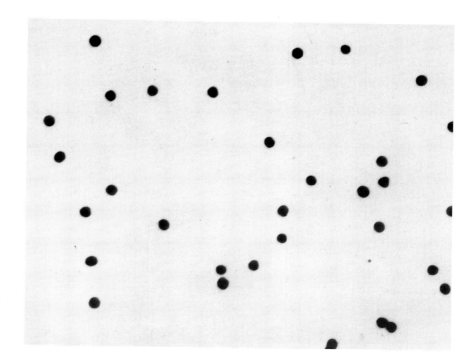

FIGURE 9.15 Colloidal gold. 100 nm gold particles stand out in sharp contrast with the background support film.

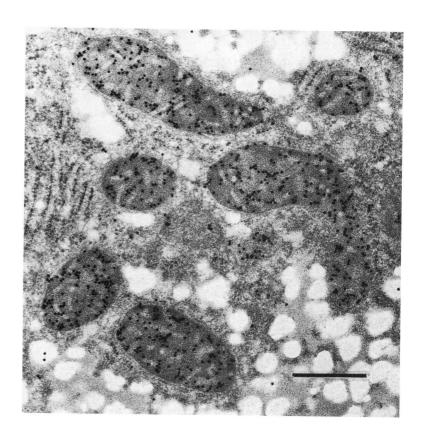

FIGURE 9.16 Localization using colloidal gold. Localization of carbamyl phosphate synthetase in rat liver tissue fixed with 1% glutaraldehyde and embedded in Lowicryl K4M. The protein A-gold tag was employed. Gold particles are seen overlying mitochondria. Bar = 0.5 μm. *(Micrograph courtesy of M. Bendayan.)*

General Considerations in Performing an Immunocytochemical Experiment

Rarely are any two experiments employing immunocytochemistry alike. There are many factors that govern the course and specific details of an experiment. Unfortunately, trial and error play a large role in the eventual outcome of an experiment, although the methodology section of published papers may not indicate all the "trials and tribulations" that the investigator has experienced to produce the final result. Some of the factors to consider in an experiment follow.

Immunohistochemistry/Immunofluorescence

A sound knowledge of light microscope immunohistochemistry serves as an important introduction to immunocytochemistry. As with most localization procedures, it is often best to try experiments at the *light microscope level* initially. In some cases, the tag may be the same as used at the electron microscope level, but most of the time FITC (fluorescein isothiocyanate) is employed for immunofluorescence. A rough idea of antigen localization can be achieved readily (in about 4 hours) using light microscopy. Controls may be employed at the light microscope level to establish the validity of the localization with generally similar methods being used for electron microscopy. Many of the conditions (fixation, etc.) that optimize localization may be determined with light microscopy. Several tags that are used for electron microscopy can also be visualized by light microscopy (e.g., horseradish peroxidase, colloidal gold).

Obtaining and Applying the Primary Antibody

It is important to obtain a primary antibody directed against the specific antigen to be localized. Although the antigen may be named similarly in different species, that does not imply that it is structurally identical in these species. An antibody may not cross-react with antigens from one species to the next when there are slight molecular or conformational dissimilarities in the antigen. For example, an antibody made against a specific blood antigen in one species may not react with the same named antigen from a different species.

Many primary antibodies are available commercially. They are usually expensive and provided in small quantities. Although commercial sources may indicate that the antibody meets certain specifications, it is important to verify this independently. Many investigators are generous in supplying high quality antibodies to other investigators at no cost. These are often of better quality than those commercially obtained.

Preparing one's own antibody from scratch may be a necessity, especially if one has more than a casual interest in localizing the antigen. This is time consuming and often costly. Polyclonal antibodies are made by immunizing a species, other than the one in which localization will be attempted, with either pure antigen or a mixture of antigens. This immunogen is the antigen that is to be localized. An adjuvant is frequently used to augment the immune response. The mouse, rabbit, and goat are commonly used species for producing antibodies. The purer the antigen injected, the more specific the response. Several booster shots may be required to increase the *titer* (concentration) of the antibody in the antisera. Because some animals may not respond at all, several animals must be injected to find one that is a good responder. It is important to use standard immunological procedures to determine the presence and specificity of the antibody (Harlow and Lane, 1988).

After obtaining blood from the animal, the serum fraction should be isolated. Serum may be used for localization purposes, or it may be further purified into an immunoglobulin fraction or even to IgG or IgG directed only against the antigen, for example. Purification of immunoglobulins is by standard immunological techniques (Harlow and Lane, 1988) The primary antibody is then diluted serially (e.g., 1:100, 1:200, 1:400, 1:800) for use in immunocytochemistry. If the postembedding technique (see below) is used, the primary antibody is often applied over a long period, from a few hours up to a day. This increases the possibility that all available antigenic sites will be bound by the primary antibody.

Obtaining the Secondary Antibody

Unlike the primary antibody, the secondary antibody is usually available commercially in large amounts and at reasonable cost. It is relatively simple to produce large amounts of a secondary antibody against IgG from one species (mouse, for example) in the goat, an animal in which large amounts of serum may be obtained. The secondary antibody must be chosen to react against both the immunoglobulins of the species (mouse) and the class (e.g., IgG or IgM, etc.) of the primary antibody employed. The secondary antibody is usually only briefly exposed to the tissue (1–2 hours).

Tissue Fixation

The overall goal of tissue fixation for immunocytochemistry is to preserve the tissue as close to its natural state as possible, while at the same time maintaining the ability of the antigen to react with the antibody. This is often much easier said than done, since the antigenic determinant is easily altered (denatured) or masked during fixation. As it turns out, for most antigens, glutaraldehyde must often be used in concentrations less than 1% because it is a strong denaturing agent. Osmium is rarely used prior to performing the labeling reaction. It is clear that antigens do not respond uniformly to fixation processes, necessitating experimentation with a variety of fixatives and fixative combinations. Formaldehyde, in concentrations as high as 4%, has been used in combination with low percentages (e.g., 0.5%) of glutaraldehyde to give good results for localization of many tissue antigens. Tissue contrast in postembedding procedures where osmium is not used is, not surprisingly, very low.

Preembedding or Postembedding Labeling

A major decision to be made is whether to perform the immunocytochemical applications prior to or after the embedding procedure (see Table 9.2). There are distinct advantages and disadvantages to both types of procedures.

Preembedding labeling or "staining," as it is sometimes erroneously called, is advantageous for *surface labeling* of live cells or cell fractions, although cell interior labeling is also conducted. Caution should be exercised because the labeling procedure itself may cause redistribution of the antigen on the cell surface. Mild fixation with 0.2% glutaraldehyde, for example, may stabilize antigens somewhat prior to labeling. Preembedding labeling is

TABLE 9.2 Two Major Types of Labeling Protocols

Steps in a Postembedding Protocol	Steps in a Preembedding Protocol
1. Primary fixation	1. Mild primary fixation, blocking and application of primary antibody
2. Dehydration, infiltration, sectioning, and embedding	
3. Etching of embedding media (optional)	*or*
4. Blocking	1. Application of primary antibody to live cells
5. Application of primary antibody	2. Washing
6. Washing	3. Blocking
7. Blocking	4. Application of secondary antibody with tag
8. Application of secondary antibody conjugated to a tag	5. Washing
9. Washing	6. Fixation (including osmium)
10. Postfixation with osmium	7. Dehydration, infiltration, embedding, sectioning
11. Washing	8. Staining with lead and uranyl acetate
12. Staining with lead and uranyl acetate	

useful when the label has easy access to the antigen (e.g., on the cell surface).

Because antibodies do not penetrate cell membranes, live cells may be broken open with *detergents* to reveal antigenic determinants within the interior of a live cell. Of course, ultrastructural preservation is severely compromised under such conditions. Sections, about 50 μm in thickness, made from solid tissue with a special vibrating or chopping microtome knife can be used to expose some cells to immunolabels. It is also possible to use *frozen ultrathin sections* (see Chapter 4) for immunocytochemistry. Fresh frozen sections are placed on grids and immunolabeled. Sections can then be fixed and stained.

The main advantage of preembedding labeling is that antigenicity is maintained during the labeling procedure. Tissue contrast is enhanced since the tissue may be osmicated, having already been labeled. Labeling with ferritin, colloidal gold, and peroxidase have been commonly used with this technique. An example of preembedding surface labeling is provided (Figure 9.10).

In the postembedding technique, the labeling procedure is carried out directly on tissue sections from embedded specimens. When using the postembedding technique, it is always presumed that antigenicity has been partially compromised by the fixation protocol. Tissue sections embedded in epoxy may be *etched* (partially eaten away) with materials such as hydrogen peroxide to expose hidden antigenic sites. A more recent mothodology for the extraction of epoxy has been published (Ris and Malecki, 1993).

The colloidal gold technique is especially suited for postembedding labeling. Tissue section contrast may be enhanced somewhat if osmication is performed on the thin section after labeling.

Embedding

In postembedding labeling procedures, the embedding process and the embedding medium itself may influence the antibody binding. Dehydration, infiltration, and embedding steps must be carried out under conditions that optimally preserve antigenicity. In postembedding procedures, embedding media (LR White, methylacrylate, JB-4, Lowicryl) with special characteristics (e.g., low temperature embedding) that will facilitate the preservation of antigenicity can be purchased.

Blocking Nonspecific Labeling

At several steps during a typical protocol, the tissue is exposed to normal, or nonimmune serum (i.e., serum from animals that have not been exposed to the antigen), or albumin for the purpose of *blocking* sites that may react nonspecifically with any antibody. Normal or nonimmune serum and albumin proteins are used to react with, for example, any residual nonreacted aldehyde sites. If this step had not been performed, these sites may have reacted with one of the antibodies and given false results. Because aldehydes such as glutaraldehyde are bifunctional agents [see Chapter 2], commonly reacting at one end, but remaining unreacted at the other end, they may bind IgG molecules nonspecifically.)

Staining

Lead and uranium staining are usually employed at the end of the protocol to enhance tissue contrast. In some instances, staining is omitted if the tissue contrast obscures the electron-dense tag employed.

Controls

Immunocytochemical experiments are virtually worthless without a battery of controls that should be conducted simultaneously with the localization attempt. There is a tendency for the novice to believe, initially, that any aggregation of label represents specific localization. Unfortunately, the controls will often reveal considerable localization artifact that might lead one to an erroneous conclusion if only experimentals (showing the same localization artifact) were examined. Localization "veterans" tend to be skeptical about their results until all the controls have verified that the localization seen is real. Several types of controls follow. Not all of the controls must be performed in every experiment.

Adsorption

The primary antibody is reacted with an excess of antigen, a process termed adsorption. After elimination of the antigen-antibody complex, the remaining solution, free of antibody (it is hoped), is exposed to the tissue. This control, although not always feasible due to sparse antigen supply, will show that the antigen and not some other substance is responsible for the localization seen. It is also possible to adsorb the primary antibody with homogenized tissue that contains the antigen. Essentially, the tissue antigen, which is in excess of the antibody, precipitates the antibody leaving the solution free of antibody. The adsorbed solution should not react with the antigen in localization protocols.

Use of Tag or Unlabeled Antibody

The tag and unlabeled antibody are used separately in place of the conjugated antibody. This establishes that the specific properties of the labeled antibody are responsible for the localization and the antigen or the tag alone are not responsible for the localization.

Omission of Primary or Secondary Antibodies

In theory, the labeling should *not* be successful if either the primary or subsequent antibodies are deleted. If labeling is seen under these conditions, then it is not the labeling intended as the result of the sequence of antibodies exposed to the antigen.

Use of Pre-Immune Sera

Collection of sera from animals prior to production of the primary antisera (so-called pre-immune sera) and its use in place of the primary antiserum or antibody will show if some component in the immune sera, other than the specific IgG, is responsible for antibody binding and the localization pattern recorded. If pre-immune sera cannot be obtained, then it is possible to employ normal serum from nonimmunized animals of the same species.

Dealing with Soluble Antigens

Soluble molecules are those freely capable of movement within the cell. For example, steroid hormones or soluble proteins are free to move about the cell within its aqueous environment. The processes of fixation, dehydration, etc., allow the further movement and/or diffusion of molecules away from their original site; thus, these processes are detrimental to localization of soluble molecules. *Ultrathin frozen sections* (see Chapter 4), *freeze-substitution*, and special embedding media have been used occasionally to localize soluble antigens.

Immunolocalization Using Cryoultramicrotomy

It is possible to use ultrathin frozen sections in immunocytochemical localizations (Tokuyasu, 1986, 1989). An example of a typical protocol follows. One must be aware, however, that proper dilutions of antibodies must be determined for each system.

After obtaining ultrathin frozen sections as described in Chapter 4, nickel grids containing the sections are placed onto droplets of ice-cold Tris buffered saline (TBS, 0.05 M, pH 7.4, 0.9% sodium chloride) containing 2% fish gelatin (to block non-specific binding of the primary antibody). After a number of grids have been obtained, the grids are

floated for various times on the following droplets of reagents: (a) TBS containing 0.1 M glycine for 10 minutes (to inactivate any residual aldehyde groups); (b) TBS for 30 seconds; (c) TBS containing a protein such as 10% goat or calf serum, 0.25% bovine serum albumin, 1% non-fat dry milk, or 0.1% horse serum (to block nonspecific binding) for 5 to 10 minutes; (d) a dilution of the primary antibody (generally 1/50 to 1/500) in TBS/protein for 15 to 30 minutes; (e) TBS/protein for four 5 minute changes; (f) secondary antibody labeled with gold (1/50 to 1/100 dilution) in TBS/protein for 15 to 30 minutes; (g) TBS/protein for four 5 minute changes; (h) distilled water, several quick rinses; (i) 2% uranyl acetate in distilled water for 3 to 5 minutes; (j) 0.1% aqueous uranyl acetate containing 2% methyl cellulose for 1 minute. After removing excess liquid, the grids should then be allowed to dry slowly in a moist chamber.

LR White Embedding Medium

As an alternative to cryoultramicrotomy, which requires a specialized apparatus, conventional ultramicrotomy may be used on tissues embedded in LR White acrylic embedding resin. In this system, specimens are lightly fixed in aldehyde alone and ultrathin sections obtained from the polymerized block. The sections are then placed onto nickel grids and floated on a series of droplets containing the reagents described for cryoultramicrotomy with the exception that the grids need to be stained only with 2% uranyl acetate. Methylcellulose is not needed to protect the sections from damage caused by drying since the tissues are surrounded by acrylic plastic.

Multiple Labeling Option

Colloidal gold markers can be purchased in uniform sizes. It is easy to distinguish 5 nm gold particles from 10 nm gold, or the latter from 20 nm gold. Thus one may localize two (or more) different antigens on a tissue section, one with a small gold tag and one with a larger gold tag. Figure 9.17 shows a double labeling micrograph where two different proteins are localized on a thin section of boar sperm.

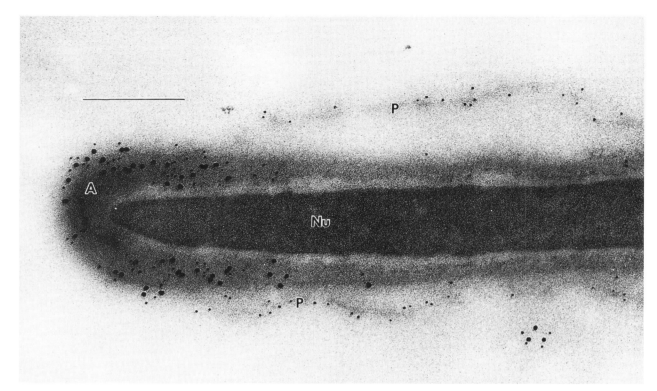

FIGURE 9.17   Localization of two proteins using gold tags of 5 nm and 10 nm diameter. Note the two antigens are present in two different sites. The larger particles localize an enzyme (proacrosin) in a boar sperm cell acrosome (A) while the smaller particles localize a protein (AP_z) found in the plasma membrane (P) as well as the acrosome. Bar = 0.5 μm.
(Courtesy of R. N. Peterson and K. Polakoski.)

Interpretation of Micrographs

If control experiments are adequately conducted, there should be little difficulty in interpreting micrographs. It is important to remember that, although localization using immunocytochemistry is very specific, the tag may lie a slight distance away from the antigen. Each IgG molecule is about 12 nm long. If the PAP method is employed, then the tag may be several times this distance in any direction from the antigen source. It is possible to quantify some labeling, especially colloidal gold (DeHarven, 1992; Enestrom, 1993).

REFERENCES

Classic Literature

Faulk, W. P., and G. M. Taylor. 1971. An immunocolloid method for the electron microscope. *Immunochemistry* 8:1081–83.

Graham, R. C., and M. J. Karnovsky. 1966. The early stages of adsorption of injected horseradish peroxidase in the proximal tubule of the mouse kidney: Ultrastructural cytochemistry by a new technique. *J Histochem Cytochem* 14:291–302.

Heitzman, H., and F. M. Richards. 1974. Use of the avidin-biotin complex for specific staining of biological membranes in electron microscopy. *Proc Natl Acad Sci USA* 71:3537–39.

Nakane, P. K., and G. B. Pierce. 1967. Enzyme-labelled antibodies for light and electron microscopic localization of tissue antigens. *J Cell Biol* 33:308–18.

Singer, S. J. 1959. Preparation of an electron-dense antibody conjugate. *Nature* 183:1523–25.

Sternberger, L. A., et al. 1965. Indirect immunouranium technique for staining embedded antigens in electron microscopy. *Exp Mol Pathol* 4:112–25.

Sternberger, L. A., et al. 1970. The unlabeled antibody enzyme method of immunohistochemistry. Preparation and properties of soluble antibody-antigen complex (horseradish peroxidase-antihorseradish peroxidase) and its use in identification of spirochetes. *J Histochem Cytochem* 18:315–33.

Methods References

Bullock, G. R., and P. Petrusz. 1982. *Techniques in immunocytochemistry.* London: Academic Press.

DeHarven, E. 1992. Immunogold labeling: Its efficiency for quantifying cell surface expressed antigens. *EMSA Bull* 22:45–47.

Eneström, S. 1993. Quantitative ultrastructural cytochemistry. *USA Microscopy and Analysis* (July) 27–29.

Griffiths, G., and H. Hoppeler. 1987. Quantitation in immunocytochemistry: Correlation of immunogold labeling to absolute number of membrane antigens. *J Histochem Cytochem* 34:1389–98.

Harlow, E., and D. Lane. 1988. *Antibodies: A laboratory manual.* New York: Cold Spring Harbor Laboratory.

Kolodziej, P. A., and R. A. Young. 1991. Epitope tagging and protein surveillance. In *Methods in Enzymology*, vol. 194. Academic Press, Inc., pp. 508–19.

Mollenhauer, H. H., and D. J. Morré. 1991. Golgi apparatus form and function. *J Electron Microsc Tech* 17:2–14.

Moriarty, G. C., and L. L. Garner. 1977. Immunocytochemical studies of cells in the rat adenohypophysis containing both ACTH and FSH. *Nature.* 265:356–58.

Polak, J. M., and I. M. Varndell. 1984. *Immunolabeling for electron microscopy.* Amsterdam, The Netherlands: Elsevier Science Publishers B.V.

Ris, H., and M. Malecki. 1993. High resolution field emission scanning electron microscope imaging of internal cell structures after Epon extraction from sections: A new approach to correlative ultrastructural and immunocytochemical studies: *J Struc Biol* 111:148–57.

Russell, L. D., R. N. Peterson, and T. A. Russell. 1982. Visualization of anti-sperm plasma membrane IgG and Fab as a method for localization of boar sperm membrane antigens. *J Histochem and Cytochem* 30:1217–27.

Silver, M. M., and S. A. Hearn. 1987. Post embedding immunoelectron microscopy using protein A-gold. *Ultrastruct Pathol* 11:693–703.

Sternberger, L. A. 1979. *Immunocytochemistry.* New York: John Wiley and Sons.

Tokuyasu, K. T. 1973. A technique for ultracryotomy of cell suspensions and tissues. *J Cell Biol* 57:551–65.

Verkleij, A. J., and J. L. M. Leunissen. 1989. *Immuno-gold labeling in cell biology.* Boca Raton, FL: CRC Press Inc., 368 pp.

Wente, S. R., and G. Blobel. 1994. NUP145 encodes a novel yeast glycine-leucine-phenylalanine-glycine (GLFG) nucleoporin required for nuclear envelope structure. *J Cell Biol* 125:955–69.

Recent Localization Reports

Eggli, P. S., and W. Graber. 1995. Association of hyaluronan with rat vascular endothelial and smooth muscle cells. *J Histochem Cytochem* 43:689–97.

Kleymann, G., C. K. Ostermeier, W. Heitmann, et al. 1995. Use of antibody fragments (Fv) in immunocytochemistry. *J Histochem Cytochem* 43:607–14.

Login, G. R., T-C. Ku, and A. M. Dvorak. 1995. Rapid primary microwave-aldehyde and microwave-osmium fixation: Improved detection of rat parotid acinar cell secretory granule α-amylase using a postembedding immuno-gold ultrastructural morphometric analysis. *J Histochem Cytochem* 43:515–23.

Thiry, M., and F. Puvion-Dutilleul. 1995. Differential distribution of single-stranded DNA, double-stranded DNA, and RNA in adenovirus-induced intranuclear regions of HeLa cells. *J Histochem Cytochem* 43:749–59.

Tokuyasu, K. T. 1986. Application of cryoultramicrotomy to immunocytochemistry. J. Microsc., 143:139–149.

Tokuyasu, K. T. 1989. Use of poly(vinylpyrrolidone) and poly(vinyl alcohol) for cryoultramicrotomy. Histochemical Journal 21:163–171.

Yi, J., Odile, M., C. Sassy-Prigaud, et al. 1995. Electron microscopic location of mRNA in the rat kidney. Improved postembedding in situ hybridization. *J Histochem Cytochem* 43:801–9.

Books

Beesley, J. E. 1989. *Colloidal gold: A new perspective for cytochemical marking.* RMS Handbook, No. 17. Oxford Science Publications, 64 pp.

Hayat, M. A., and B. T. Eaton. 1992. *Immuno-gold electron microscopy in virus diagnosis and research.* CRC Press Inc., Boca Raton, FL. 432 pp.

Hayat, M. A. 1989. *Colloidal gold: principles, methods and applications,* 2 vols. CRC Press Inc., Boca Raton, Oxford: Blackwell Scientific Publications. FL. 572 pp.

Lewis, P. R., and D. P. Knight. 1992. *Practical methods in electron microscopy,* vol. 14, *Cytochemical staining methods for electron microscopy.* A. M. Glauert, ed. Elsevier, Amsterdam. 321 pp.

Newman, G. R., and J. A. Hobot. 1993. *Resin microscopy and on-section immunocytochemistry.* Springer-Verlag, NY. 221 pp.

Ogawa, K., and T. Barka, 1992. *Electron microscopic cytochemistry and immunocytochemistry in biomedicine.* CRC Press Inc., Boca Raton, FL. 786 pp.

Polak, J., and S. Van Noorden. 1987. *An introduction to immunocytochemistry: Current techniques and problems,* rev. ed. RMS Handbook No. 11. Oxford Science Publications, 72 pp.

Roos, N., and A. J. Morgan. 1990. *Cryopreparation of thin biological specimens for electron microscopy.* RMS Handbook No. 21. Oxford Science Publications, 116 pp.

Chapter 10

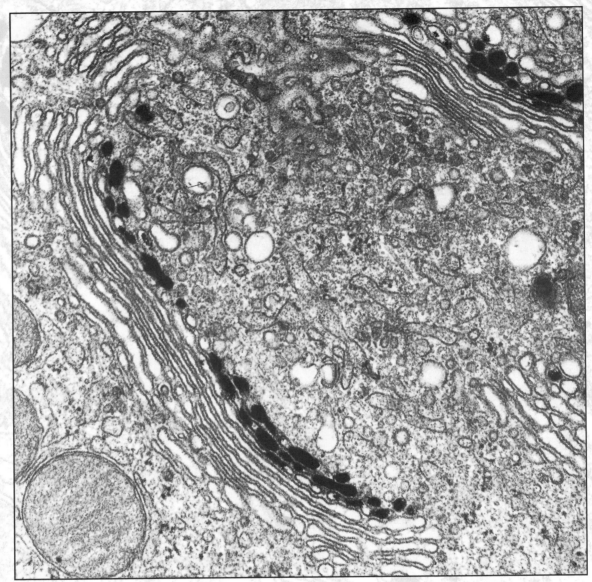

Courtesy of D. Friend.

Enzyme Cytochemistry

THE MODIFICATION OF ORGANIC MOLECULES in cells takes place via enzyme-catalyzed reactions. These reactions play a major role in regulating the metabolic pathways within cells. *Enzyme cytochemistry* utilizes the functional properties of enzymes to localize the site(s) of their activity. It is one of several localization modalities used in both light and electron microscopy. *Enzyme histochemistry* is a term applied to the localization of enzymes in tissues and cells and is generally undertaken at the light microscope level. Since the electron microscope affords increased resolution, the study of the localization of enzymes in subcellular compartments is usually undertaken with this tool. Most studies at the ultrastructural level are termed *enzyme cytochemistry*.

Enzyme cytochemistry has played a major role in helping us understand the structure and functions of cells. It has been an effective bridge between morphology and biochemistry and has helped elucidate pathways for uptake, processing, and discharge of various cell products. Since cytochemical reactions are associated with specific organelles, cytochemistry may be used with subcellular fractions to identify the organelle one wishes to study and to determine the purity of a subcellular fraction. Enzyme cytochemistry facilitates the determination of the purity of cultured cells as they grow and differentiate in tissue culture. One cell type may contain a marker enzyme that differentiates it from contaminating cell types. This technique is also important in pathology, where it can be used to distinguish cells by the specific enzymes they contain. Otherwise, cell types may be difficult to distinguish, leading to confusion in diagnosis and treatment.

A large proportion of cellular activity is mediated by enzymes. The cell may be viewed as a chemical machine containing enzymes that determine the types and rates of chemical reactions that occur. Metabolic reactions are numerous and their locations are varied. Generally, enzymatic reactions are compartmentalized within the cell. Enzymes may be bound to a cellular constituent, such as the mitochondrion or the plasma membrane, or may catalyze reactions within the soluble fraction of the cell. To understand cellular activity, it is important to know the site of cellular reactions.

Basis of Enzyme Cytochemistry

Enzymes speed up a reaction, but are not consumed by the reaction. The substance acted upon by the enzyme is the *substrate*. Enzyme cytochemistry targets enzymes by adding a large amount of substrate from an exogenous source to generate large amounts of the specific reaction product of interest. An exogenously added *trapping agent* is used to bind to and contain or trap the reaction product and make it directly visible under the electron microscope. Sometimes further reactions with the trapping agent are necessary to make it visible. The general scheme of enzyme cytochemical reactions may be summarized as follows:

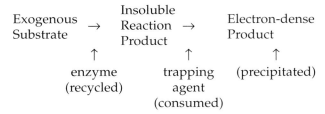

Enzyme cytochemistry is a natural outgrowth of enzyme histochemistry. To improve the resolution of localization, investigators turned to the tool of electron microscopy. The enzyme methods for light microscopy of the 1930s and 1940s (reviewed by Davenport, 1960; Barka and Anderson, 1963; Pearse, 1985) developed into techniques for electron microscopy in the 1950s, 1960s, and 1970s. The addition of new methodologies in recent years has made available a large reservoir of methods for localization of an even larger number of enzymes. Selected publications that had a major impact on the field are cited at the end of this chapter (see "Classics" in the "References" section). Although enzyme cytochemistry reached its peak in the 1970s, it remains a powerful tool for the electron microscopist. Papers employing enzyme localization at the electron microscope level may be found in *The Journal of Histochemistry and Cytochemistry* and various other journals cited in Chapter 1.

Requirements for Performing Enzyme Cytochemistry

As described above, the principle behind enzyme cytochemistry is inherently simple—a substrate and trapping agent are added to the biological material. The nonreacted components are later washed out and the tissue prepared for electron microscopy using more or less standard methods. Difficulties with cytochemical procedures arise when the conditions for the experiment are not optimal. In fact, it is rare that the conditions are optimal, sometimes making the technique technically demanding. There are at least five basic points to be considered if one is to perform enzyme cytochemistry successfully.

Preservation of Tissue Structure and Enzymatic Activity

Enzymes, being proteins, are vulnerable to fixatives that will cross-link them and cause denaturation. For most purposes, tissues must be fixed prior to performing cytochemistry. Thus, there is a delicate balance between preservation of tissue structure and preservation of enzymatic activity. Generally, percentages of glutaraldehyde from 0.2 to 2% are employed. Occasionally, formaldehyde is used in conjunction with glutaraldehyde to facilitate rapid fixation. The fixation protocol must be adjusted for each enzyme since not all enzymes are equally vulnerable. An advantage of fixation is that it breaks down membrane permeability to allow entrance of substrate that would not normally penetrate the cell.

Maximization of Reaction Conditions

Enzymatically catalyzed reactions proceed *in vivo* under carefully controlled conditions. They occur at *temperatures, pH,* and in the presence of *substrate concentrations* that are optimal for the production of the reaction product. Therefore, one normally would attempt to reproduce these conditions during the cytochemical procedure. Additionally, the *concentration of the trapping agent* should be optimized. Buffers used to maintain pH should be compatible with the reaction, and none of the chemicals used in tissue processing should interfere with or extract the localization products. The trapping agent should not diffuse from the initial reaction site. Considering all the factors that are important for the reaction to take place, it is sometimes surprising when cytochemical procedures work as planned.

Facilitation of Substrate Penetration

The ability of the substrate and/or trapping agent to diffuse through the tissue is frequently limited to less than 100μm. Therefore, solid tissues must be thinly sliced prior to incubation in the reaction medium. This presents a problem, since most fresh or weakly fixed tissue is often soft and therefore difficult to slice. Special instruments with sharp vibrating razor blades on simple microtomes, termed *vibrating microtomes,* have been developed to slice unfixed or weakly fixed tissue. Usually, tissue to be sliced is loosely embedded in a warmed agar mixture prior to sectioning. Agar, when cooled, provides a moderate amount of support for the vibrating knife to cut the tissue.

Use of Appropriate Controls

As in all scientific experiments, it is imperative that controls be performed. Usually controls are carried out concomitantly with the experimental localization. There have been many instances where controls have revealed that the localization was all or, in part, artifactual. Whenever possible, localization experiments are first carried out at the light microscope level for a preliminary determination of the adequacy of controls and to provide, as well, a tentative hint as to localization. If either the substrate or trapping agent are deleted, the reaction should not occur. Inhibitors or potentiators of the enzyme should modify the amount of reaction product present.

If it is possible to biochemically detect the presence of the enzymatic reaction products in cell fractions, then the electron microscopic localization in tissues at this same site is usually real.

Visualization of the Reaction Product

Tissue sections should always be viewed first with no staining reagent other than that provided by the enzymatic reaction product. In stained tissue, the reaction product may not be sharply contrasted against the tissue. On the other hand, if the reaction product is sufficiently electron dense it may appear prominent even after tissues have been stained. Stain contamination may be easily mistaken for reaction product.

Trapping Agents

Several types of trapping agents are used in enzyme cytochemical reactions. By far, the most common is the *insoluble heavy metal salt,* which is inherently electron dense. Lead especially, but also manganese, barium, cadmium, cobalt, and copper salts are favorites for this purpose. For example, the phosphate group released by a phosphatase is commonly trapped by lead to form insoluble lead phosphate.

Trapping agents may be *enzymes* themselves. In the protocol they react with other added substrates to produce an electron-dense precipitate. The most common of these is *peroxidase,* which may be attached to the reaction product. In the presence of hydrogen peroxide, diaminobenzidine, and then osmium, a dense reaction product is formed (see Chapter 9 for reaction of peroxidase).

Marker Enzymes

Cytochemists have found that enzymes tend to be *compartmentalized* such that each organelle may perform a specific task. Thus, an enzyme may serve as a *marker enzyme* for an organelle or subcellular compartment. When a subcellular fraction is isolated for biochemical studies, it is often useful to perform cytochemistry to determine if a specific cell fraction is contaminated with other cell fractions. For example, if biochemical studies on plasma membranes are contemplated, it is first important to know if the cell fraction is pure or contaminated with another organelle such as endoplasmic reticulum. Enzyme cytochemistry will also allow one to visualize the cytochemical reaction in a subcellular fraction and to compare it with the reaction in the intact cell.

It is not a hard and fast rule that an enzyme associated with, for example, the plasma membrane is indeed so in every cell type. However, cell biologists have examined many cell types and have found regular associations of cellular enzymes with specific compartments. The location of the marker enzymes listed in Table 10.1 is accurate for most mammalian cell types and have been confirmed by biochemical analyses of cell fractions. (See also Figures 10.1 and 10.2).

TABLE 10.1 Compartmentalization of Some Marker Enzymes in Mammalian Cells

Enzyme	General Location
acid phosphatase	Golgi, endoplasmic reticulum and lysosomes (GERL)
acetyl cholinesterase	plasma membrane
adenosine triphosphatase	plasma membrane
adenylate cyclase	plasma membrane
alkaline phosphatase	plasma membrane
catalase	peroxisome (microbody)
cytochrome oxidase	mitochondrion
5′ nucleotidase	plasma membrane
nucleoside diphosphatase	Golgi
peroxidase	peroxisome (microbody; Figure 10.1)
succinic dehydrogenase	mitochondrion
thiamine pyrophosphatase	Golgi (non-GERL; Figure 10.2)

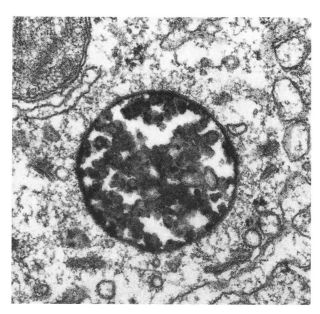

FIGURE 10.1 Peroxidase activity is represented by the dense material within this multivesicular body. *(Micrograph courtesy of D. Friend.)*

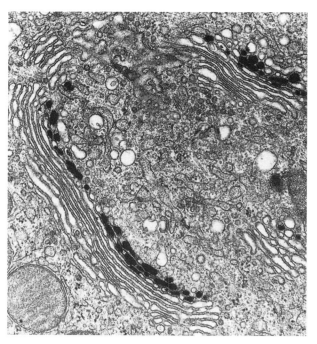

FIGURE 10.2 Thyamine pyrophosphate activity (dense deposit) is located in specific cisternae of this Golgi apparatus. *(Micrograph courtesy of D. Friend.)*

A Typical Protocol

In this section, a commonly used cytochemical method is summarized to give an example of how an enzyme cytochemical procedure is performed. For more details, consult the original reference sources.

The original electron microscopic method for localization of phosphatases, which are active at acid pH, was developed by Barka and Anderson (1962) for light microscopy. It was modified by Frank and Christensen (1968) for use at the ultrastructural level. The authors used this method to localize acid phosphatase in guinea pig Leydig cells. This localization report is reviewed briefly, emphasizing the steps of the protocol.

Fixation The testis was perfused with cacodylate-buffered glutaraldehyde for a period up to 2.5 hours. The concentration of glutaraldehyde (1.4%) was purposely kept very low to not destroy enzyme activity. After perfusion, the tissue was placed in fixative for up to 1 hour. After buffer washes, a vibrating microtome was used to make 50 to 100 μm sections to permeabilize the cells and effect substrate penetration.

Incubation Sections were incubated for 20 minutes in an incubation medium modified by Barka and Anderson (1962) from an earlier enzyme histochemistry procedure. The medium consisted of the *substrate,* sodium β-glycerol phosphate in tris-maleate buffer, and a *trapping agent,* lead nitrate. The rationale of the procedure is that phosphate ions will be liberated by the enzyme-catalyzed reaction and trapped by lead to form an insoluble lead phosphate precipitate.

Controls Controls consisted of deletion of the substrate from the incubation medium and use of sodium fluoride, an inhibitor of enzyme activity in the substrate. After incubation, the tissue was washed in acetate-veronol buffer containing sucrose.

Postfixation and Tissue Processing Osmium tetroxide (1%) was used as a postfixative. Some tissue was stained *en bloc* with uranyl acetate after postfixation, and other tissue was stained after dehydration, embedding, and sectioning. Note that osmium tetroxide would have destroyed enzyme activity had it been exposed to the tissue before the incubation.

Results Acid phosphatase activity of Leydig cells was seen in the inner cisternae of the Golgi apparatus, within lipofuscin pigment granules, and autophagic vacuoles.

Examples of Cytochemistry for Selected Enzymes

The method just described was employed by one of the authors of this text (L.D.R.) to localize acid phosphatase activity in seminiferous tubules of the testis. Photographic examples of acid phosphatase localization are illustrated (Figures 10.3 through 10.5).

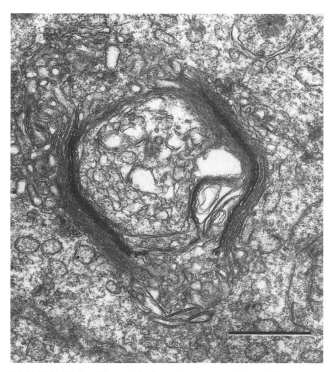

FIGURE 10.3 The acid phosphatase activity within certain of the Golgi cisternae of a germ cell appears very dense. Lead and uranium stained preparation. Bar = 1.0 μm.

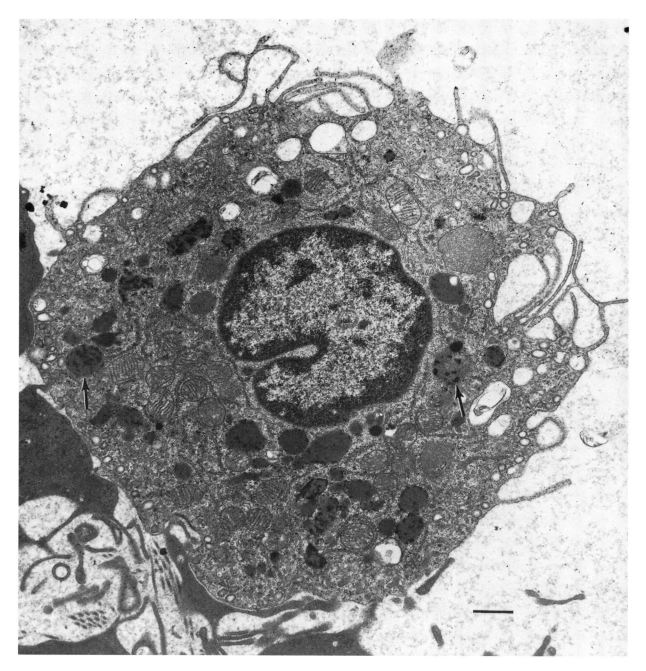

FIGURE 10.4 Acid phosphatase activity (*arrows*) within lysosomes of a macrophage. Numerous lysosomes are expected to be present in macrophages since these cells are active scavengers. Lead and uranium stained preparation. Bar = 1 μm.

Autoradiography/ Radioautography

AUTORADIOGRAPHY AND RADIOAUTOGRAPHY are interchangeable terms that refer to the use of radioactive isotopes and a radio-sensitive emulsion to trace cell processes or to localize substances within biological specimens. In brief, radioactive compounds are first administered to live organisms (Figure 11.1). Then the tissue is fixed, embedded, and sectioned, and the tissue section is brought into direct contact with a photographic emulsion. Over a period of weeks or months, as ionizing radiation is released from the radioactive substances, the radiation energy causes changes in the emulsion that are visualized after a photographic development process. In development, the photographic emulsion (usually silver bromide) is reduced to metallic silver in a way very similar to the development of emulsions used in the ordinary photographic process. The developed emulsion and tissue are viewed simultaneously. The result is an *autoradiograph*, a picture of both the tissue and the darkened spots on the photographic emulsion. The darkened spots or *grains* that appear on the tissue localize the general site of the radioactive emission. Autoradiography may be used either at the light or electron microscope levels (Figure 11.2).

Autoradiography has contributed to our knowledge of the functions of organelles. It has made a significant impact on our understanding of *dynamic* aspects of cells and tissues. There are numerous examples of how the intercellular pathways of compounds or drugs have been traced. In many instances, autoradiography has been key in our understanding of the site(s) of synthetic and metabolic pathways and has detailed secretory pathways in a very precise manner. Cell divisions have been mapped and quantified, and the fate of labeled cells has been traced throughout the organism. It has been an important localization technique, often used in circumstances where no other techniques were available. Receptors, vitamins, and a variety of other substances have been localized and their fate determined through autoradiography.

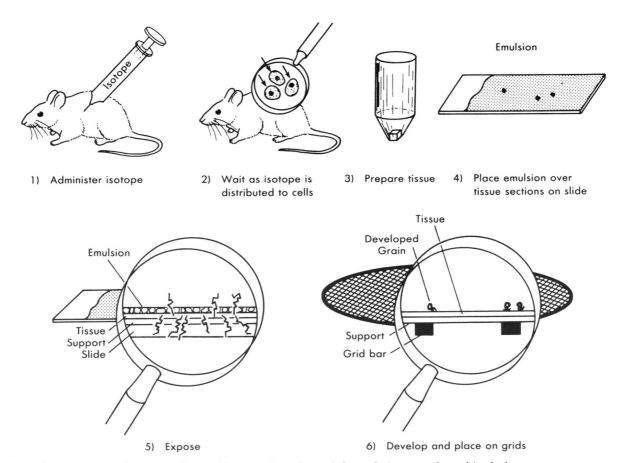

FIGURE 11.1   Steps in the autoradiographic procedure. Steps 4 through 6 are conducted in darkness.

proteins, as well as secretory material and lysosomal enzymes. The isotope ^{131}I is concentrated in the thyroid gland where it is incorporated into thyroglobulin. These are but a few examples of isotopes that are specific for biological processes.

It is also important to determine if the compound is accessible to intracellular sites. The experiment that is used at the end of this chapter, which illustrates how autoradiography is performed, shows how, in some circumstances, such a problem may be overcome. Also, labeled compounds may be given as a *pulse* (single exposure), or as a pulse followed by administration of the same, but nonradioactive, compound (*pulse-chase*), or as a *continuous infusion* or multiple infusions. The time that the labeled compound is available, in any significant quantity, to the tissue differs in each of the previous situations (pulse-chase, pulse, and continuous infusion from short to long availability), and the interpretation of the experiment in each of these circumstances may differ.

A little thought during the planning phase of an experiment may mean the difference between a successful and an unsuccessful experiment.

Planning the Experiment

Once the autoradiography technique has been chosen, it is important to plan the experiment well in advance. Radioactive compounds are costly and some have short half-lives, so it is desirable to make the first attempt a successful one.

CAUTION: Radioactive materials are extremely hazardous substances. At most institutions, licensing and/or certification procedures are required before an individual is allowed to handle them.

There are several considerations in planning the experiment. Specific activity of the radiolabel will be one factor in determining how much of the labeled compound is to be ordered. Too little radioactivity will not allow for practical exposure periods, and too much may cause damage to tissues. What is the half-life of the labeled compound? How much labeled compound should be used per administration? How many administrations of the compound? How many animals/plants and how many sampling intervals should be employed? All of these questions should be answered prior to beginning the experiment and especially prior to ordering the radiolabeled compound.

To determine sampling interval, it is important to estimate how long it will take the compound to reach the site(s) of interest. What route of administration will allow the labeled compound to reach the general site to be studied? Will the labeled compound be stabilized during fixation, or does its solubility pose problems with localization? Soluble compounds may be removed from the specimen during ordinary tissue preparation and may require specialized tissue preparative techniques such as freeze-drying (see Chapter 3).

Administration of the Radioactive Substance and Tissue Preparation

The current literature is often the best guide to the method and route of administration of radiolabeled substances. After administration, there is a variable period in which the labeled substance is taken up or incorporated into cells. If the experimental design is appropriate, the substance is specifically bound to the receptor or incorporated into the metabolic pathway of interest. At the sampling time (animal sacrifice), the tissue is fixed rapidly by methods that provide the best ultrastructural results. It is then dehydrated and eventually embedded in the medium of choice.

Light Microscope Autoradiography

Light microscope autoradiography, using thick (0.5 to 1.0 μm) sections, is usually undertaken initially. Thick sections contain more radioactivity and will expose more quickly. Since the exposure period for light microscopy is considerably less than for electron microscopy, valuable time will be saved if the thin (electron microscopic) sections are exposed simultaneously. The exposure time for electron microscopy may be estimated from the relative abundance of grains seen in light microscopy. About one week of light microscope exposure is equivalent to one month of exposure for electron microscope autoradiography. Developed grains at the light microscope level appear as spots (Figure 11.10). A preliminary determination of localization of grains may be made at the light microscope level. The pattern of grain distribution may be compared at various sacrifice intervals to determine which sacrifice intervals will be most interesting at the electron microscope level. Several light microscope slides may be made to check the progress of exposure.

Application of Emulsion

The most technically demanding step is the application of the emulsion to the thin sections. Part of the difficulty is that this operation must take place in a darkened room.

There are several methods for applying emulsion to tissue; none are easy for the novice. The thin section is placed on a grid that has been coated with a collodion support film. A partially gelled emulsion film is layered onto the grid using a wire loop that has been dipped in liquid emulsion so that a thin film of it is stretched across the loop. Alternatively, the grid with supporting film may be dipped in a liquid emulsion. These methods sometimes provide highly variable results often causing a nonuniform coat of emulsion to be layered onto the tissue.

More commonly used is a *stripping film* method in which a microscope slide is first coated with Collodion, thin sections placed on the coating, and then a thin layer of carbon evaporated onto the tissue. (The carbon is deposited to provide a better substrate for emulsion deposition and adhesion.) An emulsion layer is coated onto the slide by dipping the slide into warmed (liquified) emulsion. Special equipment may be used (Figure 11.5) to ensure that the emulsion is applied in a thin, uniform layer. Regardless of the technique used to apply emulsion, it is desirable to have a single molecular layer of closely packed silver bromide crystals uniformly covering the tissue section (Figure 11.6).

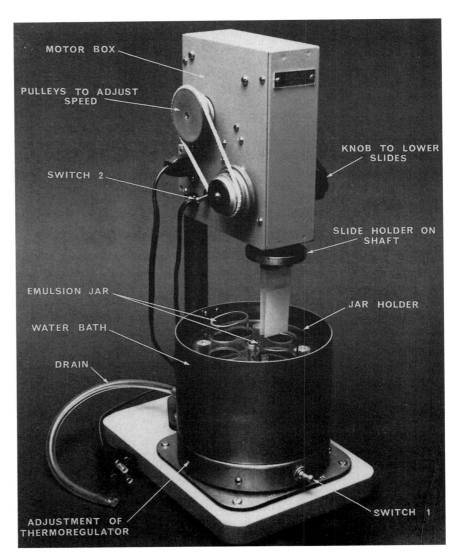

FIGURE 11.5 Semiautomatic coating device for autoradiography used to insure application of a monolayer of silver bromide emulsion.
(Courtesy of B. Kopriwa; taken from J Histochem Cytochem 14:923 with permission of the publisher.)

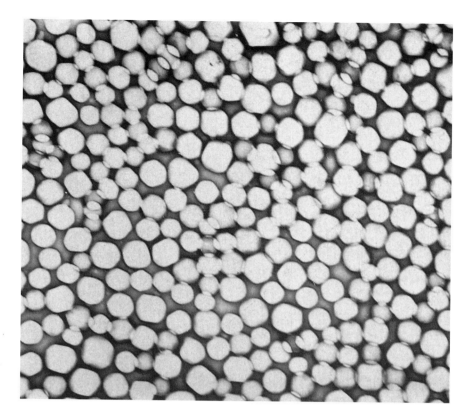

FIGURE 11.6 Ghost image of a monolayer of grains. In this figure the clear spaces represent the location of silver bromide crystals in the gelatin matrix. Crystals are absent in this micrograph due to their destruction by the electron beam.
(Micrographs courtesy of B. Kopriwa; taken from J Histochem Cytochem 14: 923 with permission of the publisher.)

Exposure of the Autoradiograph

After the emulsion is applied, the tissue is placed in a light-tight box where radioactive disintegrations occur creating latent images. The direction of radioactive emissions is random; only about 50% are thought to strike the emulsion, and even fewer strike silver bromide crystals to produce latent images. The half-life amount of compound administered, the amount and rate of uptake and turnover, as well as the specific activity all influence the exposure time. Exposure periods from one to ten months are common, although exposures may be shorter or longer. It is important that the autoradiograph show sufficient grains to provide information, yet not too many grains as to obscure tissue structure. Labeled compounds with half-lives under two weeks are difficult to work with given the constraint of rapid loss in radioactivity. [131]I is an exception if one is studying the thyroid, since the majority of injected iodine becomes rapidly concentrated in the thyroid.

Development of the Autoradiograph

Grids, or slides containing tissue, are successively transferred through developer, stop bath, fixer, and distilled water washes to produce grains from latent images and to remove unexposed grains. The fixation process removes the unexposed and undeveloped silver halide crystals but leaves the gelatin. After the fixer step, the lights may be turned on.

Staining of the Tissue

Tissue may be "prestained" or "poststained" or both. In prestaining, tissues are usually stained with uranyl acetate prior to the exposure phase. Poststaining with lead citrate (usually) takes place after development. It is important to obtain good staining that results in a contrasty image, since specimens supported with collodion and carbon coats will decrease the contrast.

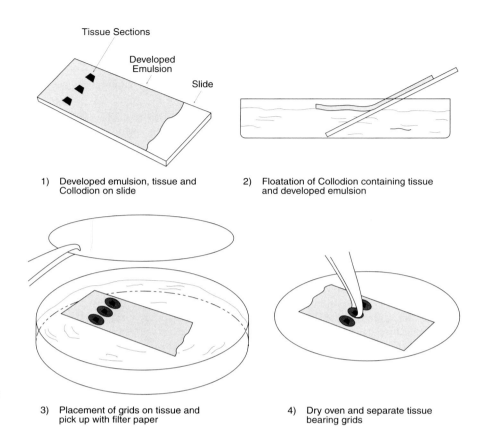

1) Developed emulsion, tissue and Collodion on slide

2) Floatation of Collodion containing tissue and developed emulsion

3) Placement of grids on tissue and pick up with filter paper

4) Dry oven and separate tissue bearing grids

FIGURE 11.7 Removal of exposed emulsion from the slide using water flotation and placement of grids on tissue sections.

Placement of Exposed Tissue on Grids

In the stripping film method, the Collodion coat containing the tissue section is floated free of the slide (Figure 11.7). Grids are placed on top of the tissue, and the Collodion holding the tissue and grids is lifted from the water from above. After drying, the excess Collodion is trimmed from the edges of the grid.

Interpretation of Autoradiographs

In transmission electron micrographs, developed grains appear as irregular tracks or single spots depending on the specific developing procedure utilized (see Figure 11.4). All autoradiographs show some nonspecific grains or background, but excessive background is an indication of aging of the emulsion, excessive heating of the emulsion, expo-

sure of the emulsion to a light leak, or spurious radioactive disintegrations.

The major obstacle to accurate interpretation of autoradiographs is the precise determination of the source of the disintegration. If all disintegrations were directed straight above, then the grains would directly overlay the source of emission. However, disintegrations are randomly directed and may travel some distance before striking a silver halide crystal. A circle with the source as the center, or *resolution circle*, represents the possible span of radioactive emissions (Figure 11.8).

Numerous methods have been developed to assess the probability that specific structures are labeled. The most commonly used isotope, tritium (^{3}H), has a range of less than 1 μm from the source. Some investigators draw a resolution circle around a grain that encompasses all of the possible sources of emission. It is then possible to calculate the probability that a given structure is the source of the emission. Others simply score the organelle that is under the grain. Yet others determine the grain density by counting the number of grains over a structure and dividing this by the area occupied by the structure.

The larger the cell structure acting as the source of radioactive emissions, the easier it is to identify the source (Figure 11.9). In some experiments, it is obvious to the eye that grains are exclusively associated with one organelle. One must resort to sophisticated methods of quantitation when the source of radioactivity is a small structural element or when there are several sources of emissions.

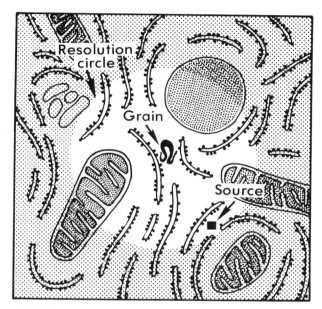

FIGURE 11.8   A resolution circle may be drawn with a grain as its center to indicate the probable boundaries of the location of the source. Any of the organelles within the resolution circle may be the source of radioactivity.

A Typical Experiment Employing Autoradiography

As an example of the application of autoradiography, the article by Bennett and O'Shaughnessy (1981) is used to show how information is gained through the technique of autoradiography.

The authors point out that *sialic acid* residues are common peripheral constituents of carbohydrate side chains of secretory and membrane proteins. Furthermore, sialic acid residues have been implicated in important cell functions such as determining the life span of circulating cells. The purpose of the experiment was to determine the subcellular site at which sialic acid residues are added to carbohydrate moieties within hepatocytes.

Sialic acid does not, itself, enter cells. Therefore, a chemically related radiolabeled precursor product, [^{3}H] N-acetylmannosamine, which readily enters cells and is converted by cells to sialic acid, was utilized. It was important first to determine that this precursor was specifically incorporated into protein as sialic acid. Biochemical studies verified this, allowing the interpretation that, after a period of time, the precursor substance that would appear as grains in autoradiographs would be converted to sialic acid.

Given this specificity, the authors proceeded with autoradiography at both the light and electron microscope levels. Animals were killed at varying intervals after injection of the radioactive sialic acid precursor. Light microscopic autoradiography (Figure 11.10) revealed abundant radioactive incorporation into hepatocytes and specific intracellular localization of

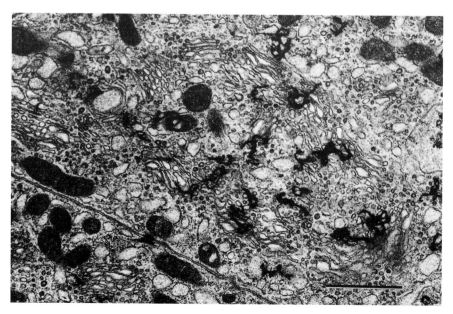

FIGURE 11.9   For large organelles such as this Golgi apparatus the source of grains is apparent. Bar = 1.0 μm.
(Micrograph courtesy of C. Flickinger; taken from Anat Rec 210:435 *with permission of the publisher.)*

grains within the cytoplasm of hepatocytes. With time, grain localization was seen at the cell surface.

To perform electron microscopic autoradiography, thin sections were placed on Collodion-coated glass slides; the sections were carbon coated by thermal evaporation (see Chapter 5) and stained with uranyl acetate. Slides were coated with emulsion (dipping technique) in a darkroom and allowed to expose in the dark for a period up to eight months. Slides were then developed and portions of the collodion support containing sections were circumscribed, placed on 300 mesh grids, and poststained with lead citrate.

At 10 minutes after injection, grains were located primarily over the Golgi region (Figure 11.11), specifically the trans face of the Golgi apparatus, but some were over secretory granules and other organelles. At longer intervals (1 to 24 hours) after the death of the animal, the proportion of grains over the Golgi apparatus decreased whereas the proportion over the cell surface at the hepatic sinusoidal region and within lysosomes increased (Figure 11.12). At three and nine days after injection of precursor, the location of grains was similar to the one-day interval, although few total grains were recorded.

The authors concluded that, because grains were concentrated in the Golgi apparatus, sialic acid is pri-

marily incorporated into glycoproteins at this site. More specifically, the incorporation of glycoproteins was in the trans face of the Golgi apparatus indicating that only a portion of the Golgi was actively incorporating sialic acid. Once formed, the glycoproteins migrated to secretory products and lysosomes and to the plasma membrane. That label was diminished at longer time intervals suggesting that there was continuous renewal of sialic-acid-bearing glycoproteins.

This journal article is typical of many good autoradiographic experiments found in the literature. Although, in this particular experiment, a dynamic synthetic process is traced using a labeled precursor, it should not be inferred that all autoradiographic experiments are designed similarly. Some are designed for localization purposes and do not involve multiple sacrifice intervals. The highlights of the article are only briefly summarized here; consequently, full treatment of the subject may be appreciated only by reading the article in its entirety.

Source: Bennett, G., and D. O'Shaughnessy. 1981. The site of incorporation of sialic acid residues into glycoproteins and the subsequent fate of these molecules in various rat and mouse cell types as shown by radioautography after injection of [3H]N-acetylmannosamine I. Observations in hepatocytes. *J Cell Biol* 88:1–15.

FIGURE 11.10 Light microscope autoradiographs of sialic acid ([3H] N-acetylmannosamine) incorporation into rat liver. (1) At 10 minutes after injection, grains are seen over what appears to be the Golgi region (*horizontal arrows*) and sinusoidal surface (*vertical arrows*). (2) At 30 minutes after injection, some grains are over the Golgi region and some are over the sinusoidal region. (3) and (4) At 4 hours and 24 hours, respectively, after injection most grains are over the sinusoidal surface of hepatocytes (*vertical arrows*) and lateral surfaces (*oblique arrow*). Few remain over the bile canalicular region (*horizontal arrow*).
(This and subsequent micrographs in this chapter courtesy of G. Bennett, taken from J. Cell Biol *88:1,1981 with permission of the publisher.)*

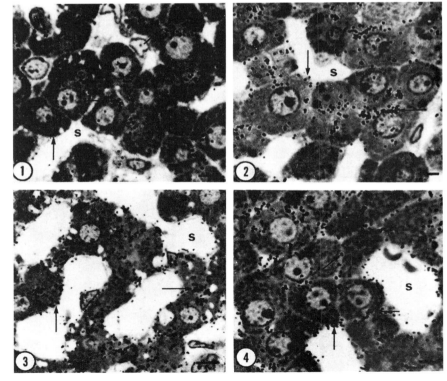

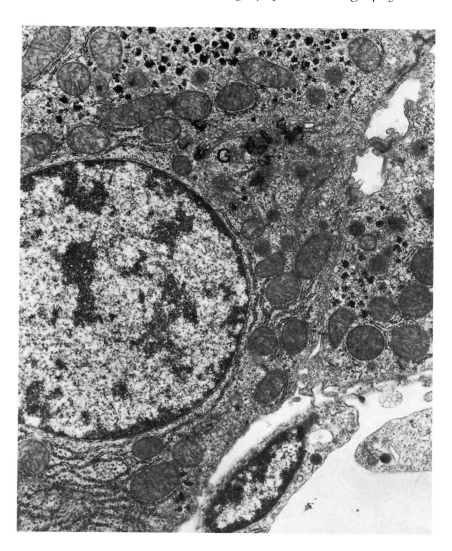

FIGURE 11.11 Ten minute autoradiograph of sialic acid ($[^3H]$ N-acetylmannosamine precursor) incorporation into a rat hepatocyte showing grains exclusively over the Golgi region (G).
(Courtesy of G. Bennett.)

In Situ Hybridization Using EM Autoradiography

In situ hybridization is a molecular biology technique. For introductory reading in molecular biology see Clark and Russell (1997). The in situ hybridization (ISH) technique employing radioactive tags utilizes specific radioactively labeled nucleic acids rather than antibodies to localize a complementary nucleic acid. The technique relies on the highly specific pairing of complementary strands of DNA/DNA, RNA/RNA, or DNA/RNA. In the reaction, one of the strands (the probe) is labeled so that after it pairs with its complementary, unlabeled strand (the target) the location of the tar-

get is revealed using either an autoradiographic or immunocytochemical technique. The procedure has proved particularly useful to localize the nucleic acids of replicating viruses, genes on chromosomes, messenger RNA and ribosomal RNA. In a typical localization, one would produce a probe (either DNA or RNA) containing a label such as radioactive sulfur (^{35}S), biotin or digoxigenin. Such probes may be produced using a variety of molecular techniques (nick translation, random primer extension, viral vectors, polymerase chain reaction) to force incorporation of the label into the nucleic acid probe. The labeled probe is then applied to cells that are fixed using only an aldehyde fixative (because osmium tetroxide destroys the reactivity). Several options for processing of the tissue are possible since the label may be applied to specimens that have been: (a) cut

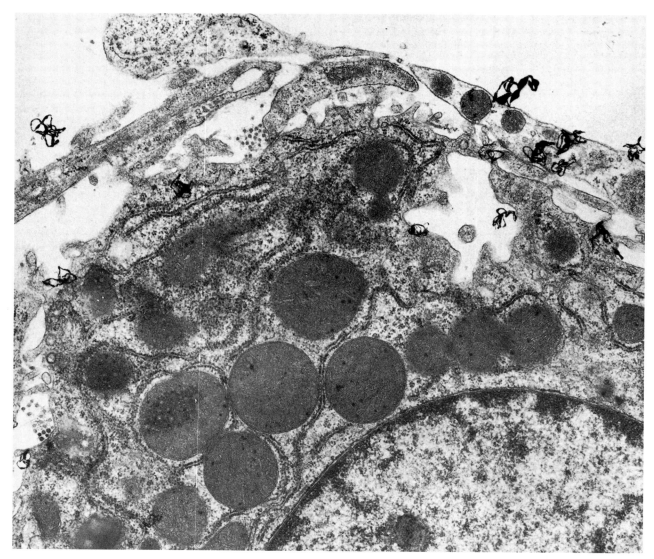

FIGURE 11.12   A twenty-four hour autoradiograph of sialic acid ([^{3}H] N-acetylmannosamine precursor) incorporation into a rat hepatocyte showing that most of the grains are over the sinusoidal surface of the cell. From the foregoing series of micrographs it was shown that the radioactive tracer injected has dynamic properties within the cell. *(Courtesy of G. Bennett.)*

using a vibratome into 50 to 100 μm thick sections or disrupted using mechanical or chemical treatment, (b) embedded in an appropriate resin such as LR White or Spurr's or (c) frozen sectioned. The probe is applied for 3 to 24 hours under conditions that facilitate pairing of probe and target (37°C, 0.6 M NaCl, 50% formamide) and the unpaired probe is removed by a series of washes. The probe/target hybrids are then revealed either autoradiographically (in the case of ^{35}S-labeled probes) or immunocytochemically using anti-biotin or anti-digoxigenin antibodies attached to colloidal gold or peroxidase. Le Gellec et al. (1992) published an elegant article

that localized growth hormone using various specimen treatments (vibratome-generated thick sections, ultrathin frozen sections, plastic-embedded sections). The probe was biotin-labeled and the target was revealed using either biotin-specific antibodies or avidin attached to either peroxidase or gold particles. This study determined that the optimal combination for ultrastructural preservation, sensitivity and resolution was obtained using aldehyde-fixed, plastic embedded specimens. For further reading on this technique, see the book edited by Morel (1993). A summary of the ISH technique is given in Figure 11.13.

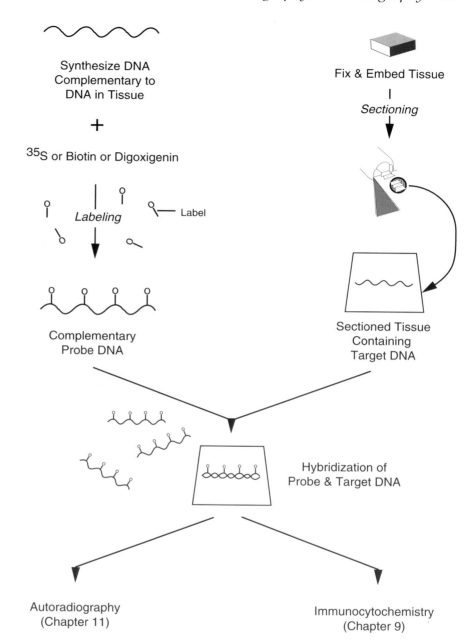

FIGURE 11.13 Summary of the *in situ* hybridization procedure.

REFERENCES

Classic and Historical References

Belanger, L. F., and C. P. Leblond. 1946. A method for locating radioactive elements in tissues by covering histological sections with photographic emulsions. *Endocrinology* 9:386–400.

Jamieson, J. D. and G. E. Palade. 1967. Intracellular transport of secretory proteins in pancreatic exocrine cell *J Cell Biol* 34:597–615.

Leblond, C. P. 1987. Radioautography: Role played by anatomists in the development and application of the technique. In *The American association of anatomists: Essays on the history of anatomy in America.* J. E. Pauly, ed., Baltimore Williams and Wilkins, pp. 89–103.

Liquier-Milward, J. 1956. Electron microscopy and radioautography as coupled techniques in tracer experiments. *Nature* (London) 177:619.

Revel, J. P., and E. D. Hay. 1961. Autoradiographic localization of DNA synthesis in a specific ultrastructural component of the interphase nucleus. *Exptl Cell Res* 25:474–97.

Technique References

Becker, W., and A. Bruce. 1985. Autoradiographic studies with fatty acids and some other lipids: A review. *Prog Lipid Res* 24:325–46.

Clark, D. P., and L. D. Russell. 1997. *Molecular biology made simple and fun.* Cache River Press, Vienna, IL. 470 pp.

Kopriwa, B. M. l973. A reliable, standardized method for ultrastructural electron microscopic radioautography. *Histochemie* 37:1–7.

Le Gellec, D., A. Trembleau, C. Pechoux, F. Gossard, and G. Morel. 1992. Ultrastructural nonradioactive in situ hybridization of GH mRNA in rat pituitary gland: Pre-embedding vs ultra-thin frozen sections vs post-embedding. *J Histochem Cytochem* 40:979–86.

Morel, G. (ed.). 1993. *Hybridization techniques for electron microscopy.* CRC Press, Inc., Boca Raton, FL. 360 pp. ISBN 0–8493–4414–X.

Roth, J. L., and W. E. Stumpf. 1969. *Autoradiography of diffusible substances.* New York: Academic Press.

Salpeter, M. M., and L. Bachmann. 1972. Autoradiography. In *Principles and techniques of electron microscopy,* M. A. Hayat, ed., New York: Van Nostrand Reinhold Co., pp. 219–78.

Salpeter, M. M., and F. A. McHenry. 1973. Electron microscope autoradiography. In *Advanced techniques in biological electron microscopy,* J. K. Koehler, ed., New York: Springer-Verlag, pp. 113–52.

Stumpf, W. E., and H. F. Solomon. 1995. *Autoradiography and correlative imaging.* San Diego: Academic Press, 597 pp.

Williams, M. A. 1977. *Autoradiography and immunocytochemistry,* A. M.Glauert, ed., North Holland, Amsterdam: Elsevier, pp. 77–197.

Studies Using Autoradiography

Bennett, G., and D. O'Shaughnessy. 1981. The site of incorporation of sialic acid residues into glycoproteins and the subsequent fate of these molecules in various rat and mouse cell types as shown by radioautography after injection of [3H] N-acetylmannosamine I. Observations in hepatocytes. *J Cell Biol* 88:1–15.

Bissionnette, R., et al. 1987. Radioautographic comparison of RNA synthesis patterns in epithelial cells of mouse pyloric antrum following 3H-uridine and 3H-orotic acid injections. *Am J Anat* 180:209–25.

Haddad, A., et al. 1987. Localization of glycoproteins in insulin secretory granules by ultrastructural autoradiography. *J Histochem Cytochem* 35:1059–62.

Mazariegos, M. R., et al. 1987. Radioautographic tracing of 3H proline in the endodermal cells of the parietal yolk sac as an indicator of the biogenesis of basement membrane components. *Am J Anat* 179:79–93.

Chapter 12

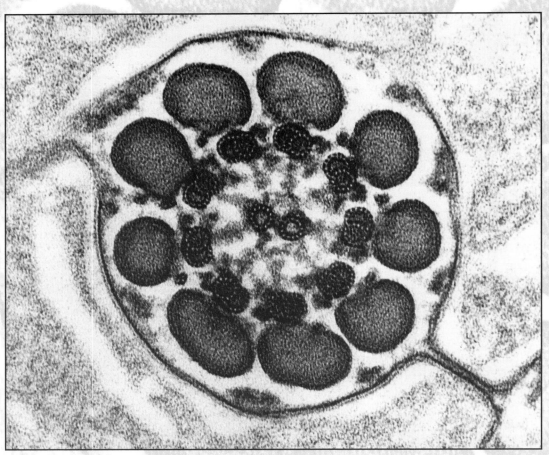

Micrograph Courtesy of R. Dallai.

Miscellaneous Localization and Enhancement Techniques

LOCALIZATION TECHNIQUES based on enzymatic, antigenic, and radioactive properties molecules have been covered previously (Chapters 9, 10, and 11). However, there are several other methods for localization of specific substances or cell structures that have proven useful for electron microscopy. These, in general, are based on some property of the particular molecule/structure that is to be localized.

Some localization techniques may be based on preferential staining of particular structures by what are called *cytochemical stains*. These stains are generally less specific than has been demonstrated for most other forms of localization. Even the routine stains used for electron microscopy, lead and uranium, are, to some degree, a means for localization of cellular constituents based on the chemical properties of these constituents (see Chapter 5). They are less specific than the cytochemical stains that will be considered in this chapter.

What follows is a list of selected localization and staining techniques to illustrate how methods have been developed to localize specific molecules or cell structures.

Actin

The natural affinity of the heads of the myosin molecule for actin causes myosin to bind to actin and provides the basis for a localization scheme for actin. Fragments (the heads) of myosin molecules are first isolated and purified (Ishikawa et al., 1969). They are then exposed to actin in cells that have been broken open, usually by detergents such as Triton X-100 or Nonidet NP 40. In the electron microscope, filaments containing actin appear "decorated" with an arrowhead pattern formed by the head region of myosin molecules (Figures 12.1 and 12.2).

FIGURE 12.1   Actin filaments that have been decorated with the heads (S-1 fragment) of the myosin molecule. The filaments are fuzzy, appearing to display uniformly oriented arrowheads at periodic intervals along the filament. The arrowheads are the myosin fragments. Bar = 0.25 μm.

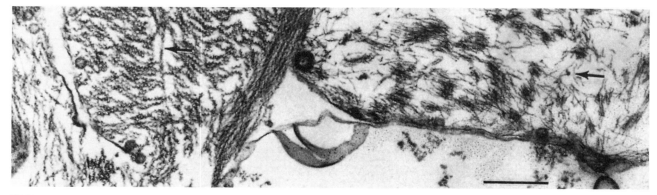

FIGURE 12.2   Actin filaments (*indicated by arrows*) of cell 1 appear decorated with the S-1 fragment of myosin as shown in Figure 12.1. Filaments in an adjoining cell to the right are not decorated indicating that they have not been exposed to myosin due to the impermeability of the cell. Bar = 0.25 μm.

TABLE 12.1 Specificities of Selected Plant Lectins for Oligosaccharide Residues

Lectin	Oligosaccharide Bound
Concanavalin A	α-D-glucose
Soybean lectin	α-galactose and N-acetylglucosamine
Wheat germ lectin	N-acetylglucosamine
Lotus seed lectin	fucose

Carbohydrates/Oligosaccharides

Using Lectins

Lectins are plant agglutinins from various species that have affinity for specific sugar sequences in oligosaccharide residues. They may be conjugated to a variety of ultrastructural tags and used for localization purposes. Over fifty lectins are available, and these possess a wide range of specificities. Selected lectins and their specificities are shown in Table 12.1.

Using Tannic Acid and Metals

The mechanism of the reaction of tannic acid and metals with carbohydrates is uncertain, but it may involve hydrogen bonding as described in a paper by Sannes et al., 1978.

Tannic acid placed in the fixative in a final concentration of about 1% often results in improved fixation for certain cell components. Generally, membrane structure is improved, revealing the trilaminar structure of membranes as well as membrane-associated carbohydrate moieties. In addition, Tubulin subunits of cytoplasmic microtubules as well as those of flagella and cilia (Figure 12.3) are visualized with this technique (Dallai and Afzelius, 1990).

Using a Modified PA-Schiff Reaction

Aldehyde groups are produced when periodic acid is used to oxidize carbohydrates. These are then reacted with alkaline silver solutions in which silver is deposited at the reaction site (Figures 12.4 and 12.5).

Alkaline bismuth stains certain polysaccharides incorporating the PA-Schiff reaction.

Using Colloidal Iron and Colloidal Thorium

Colloidal solutions of some metals will allow a clear demonstration of some carbohydrates. Acid mucopolysaccharide components will react with colloidal solutions of iron and thorium. A dense amorphous precipitate is seen at the site of carbohydrates.

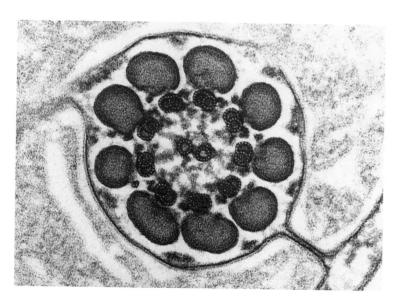

FIGURE 12.3 Tannic acid added to the fixative was used to demonstrate the microtubule (tubulin) subunits (*arrow*) in the microtubular components of human flagellum. Note that these are 13 protofilaments. (*Courtesy of R. Dallai.*)

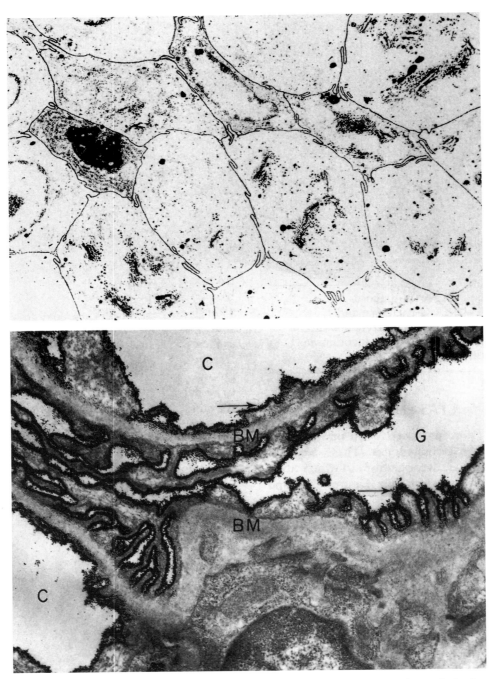

FIGURES 12.4 and 12.5   The PA-chromic acid-silver technique is used to localize complex carbohydrates on the intestinal epithelium (Figure 12.4, *top*) and in the kidney (Figure 12.5, *bottom*) in unstained sections. In the top figure, which is at low magnification, the glycocalyx is stained as are the Golgi apparatus and lysosomes and rough endoplasmic reticulum. A heavily stained Golgi of a goblet cell is seen at the left of the figure. In the bottom figure, which is at a moderate magnification, the glycocalyx of the capillary (C) of the glomerulus (G) of the kidney is heavily stained (*arrow*).

(Micrograph courtesy of C. P. Leblond taken from J Cell Biol, Vol. 40:395–414 and Vol. 32:27–53 respectively, and used with permission of the publisher.)

Golgi Complex/Multivesicular Body

It has long been recognized from light microscope studies that osmium will stain the Golgi apparatus. Using the method described by Friend (1969), osmium is reduced and deposited at specific saccules of the Golgi by a mechanism that remains unknown (Figure 12.6).

Glycogen/Membranes

This method utilizes a mixture of osmium tetroxide and potassium ferrocyanide (Karnovsky, 1971; Russell and Burguet, 1977). When the two solutions are mixed, the mixture appears dark brown as the osmium tetroxide is reduced by the potassium ferrocyanide. The mechanism of enhancement of glycogen and membranes is unknown, but is thought to be due to reduction of osmium by potas-

sium ferrocyanide (See Chapter 20, Figure 20-133). Membranes appear trilaminar (see Chapter 19, Figure 19-5), and glycogen stands out in sharp contrast to the cytoplasmic matrix. The cytoplasmic matrix is leached out during the postfixation procedure, which yields additional contrast to the overall tissue preparation. A major drawback is that ribosomes are poorly visualized. Overall, the technique often imparts a pleasing appearance to micrographs. For many, this technique is not only used to localize glycogen but as a routine fixation procedure for the examination of tissues.

Ions

Anions and cations are frequently employed as probes to localize ions of the opposite charge. The techniques are based on charge affinity of tissue ions (electrostatic forces) for the oppositely charged heavy metal.

One method employs pyroantimonate, which in the presence of cations selectively precipitates as the cation-antimonate complex (Figure 12.7).

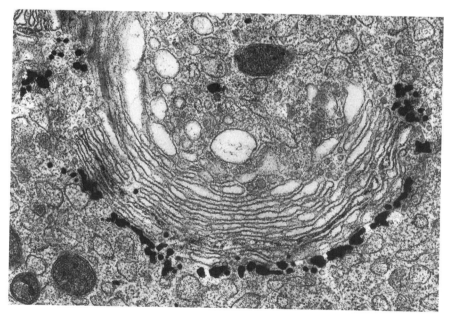

FIGURE 12.6 This Golgi apparatus impregnated with osmium indicates that the Golgi is not a chemically homogeneous organelle. Osmium is only impregnated on the cis-face of the Golgi.
(Micrograph courtesy of D. Friend.)

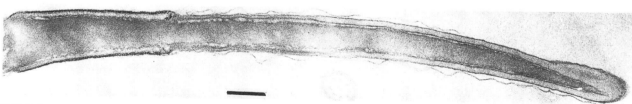

FIGURE 12.7 Localization of cations on a sperm plasma membrane with pyroantimonate. This unstained preparation shows that cations (*dark lines at the left*) are not evenly distributed over the cell surface. Bar = 1.0 μm.

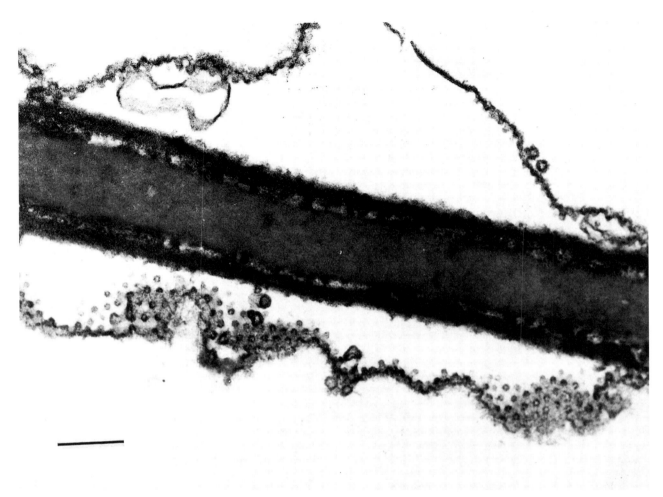

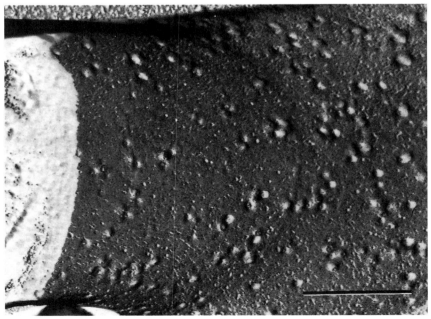

FIGURES 12.8 and 12.9   Filipin sterol membrane perturbations shown in thin section (Figure 12.8, *top*) and freeze fracture (Figure 12.9, *bottom*). In thin section, filipin sterol complexes within the membrane appear as small corrugations of the membrane (*arrowhead*) and in freeze fracture they appear as bumps and indentations (*arrows*) that are much larger than intramembranous particles. Bar = 0.25μm.

Nucleic Acids (DNA/RNA)

Techniques for staining nucleic acids, which are referenced at the end of this chapter, have several different bases for their ability to localize nucleic acids. Each paper should be consulted for details of the localization protocols as well as for the theoretical basis of staining.

Protein

The phosphotungstate technique is generally regarded as a rather nonspecific technique for protein localization. In the referenced technique (Silverman and Glick, 1969), sodium phosphotungstate most likely acts as an anionic stain for positively charged groups of proteins. The heavy tungsten molecule of phosphotungstate imparts the electron density to the protein molecules being localized.

Sterols

Two compounds, filipin (polyene antibiotic) and saponin (plant glycoside) are known to form a complex with certain sterols such as cholesterol. In doing so, they cause a perturbation of the membrane, which appears distinctive in thin sections and in freeze fracture preparations (Figures 12.8 and 12.9).

REFERENCES

General Reference

Lewis, P. R. and D. P. Knight. 1992. *Practical methods in electron microscopy, Vol. 14: Cytochemical staining methods for electron microscopy.* A. M. Glauert, ed. Amsterdam, The Netherlands: Elsevier/North-Holland Publishing Co.

Actin

Ishikawa, H., R. Bischoff, and H., Holtzer, 1969. Formation of arrowhead complexes with heavy meromyosin in a variety of cell types. *J Cell Biol* 43:312–28.

Russell, L. D. 1986. Characterization of filaments within the subacrosomal space of rat spermatids during spermiogenesis. *Tissue Cell* 18:887–98.

Carbohydrates/Oligosaccharides

Bernhard, W., and S. Avrameas. 1971. Ultrastructural visualization of cellular carbohydrate components by means of Concanavalin A. *Exptl Cell Res* 64:232–36.

Nicolson, G. L. 1978. Ultrastructural localization of lectin receptors. In *Advanced techniques in biological electron microscopy. Specific ultrastructural probes,* J. K. Koehler, ed., Heidelberg: Springer-Verlag, pp. 1–38.

Roth, J. 1983. Application of lectin-gold complexes for electron microscopic localization of glycoconjugates on thin sections. *J Histochem Cytochem* 31:987–99.

Using Tannic Acid and Metals

Dallai, R., and B. A. Afzelius. 1990. Microtubular density in insect spermatozoa: Results obtained with a new fixative. *J Biol* 103:164–79.

Sannes, P. L., et al. 1978. Tannic acid-metal salt sequences for light and electron microscopic localization of complex carbohydrates. *J Histochem Cytochem* 26:55–61.

Using a Modified PA-Schiff Reaction

Ainsworth, S. K., et al. 1972. Alkaline bismuth reagent for high resolution demonstration of periodate-reactive sites. *J Histochem Cytochem* 20:995–1005.

Rambourg, A., et al. 1969. Detection of complex carbohydrates in the Golgi apparatus of rat cells. *J Cell Biol* 40:395–414.

Using Colloidal Iron and Colloidal Thorium

Albersheim, P., et al. 1960. Stained pectin as seen in the electron microscope. *J Biophys Biochem Cytol* 8:501–6.

Rambourg, A., and C. P. Leblond. 1967. Electron microscopic observations on the carbohydrate-rich cell coat present on the surface of cells in the rat. *J Cell Biol* 32:27–53.

Golgi Complex/Multivesicular Body

Friend, D. S. 1969. Cytochemical staining of multivesicular body and Golgi vesicles. *J Cell Biol* 41:269–79.

Glycogen Membranes

Karnovsky, M. J. 1971. Use of ferrocyanide-reduced osmium tetroxide in electron microscopy. *J Cell Biol, Abstract* No. 284.

Russell, L. D., and S. Burguet. 1977. Ultrastructure of Leydig cells as revealed by secondary tissue treatment with a ferrocyanide-osmium mixture. *Tissue Cell* 9:751–66.

Ions

Gasic, G. J., et al. 1968. Positive and negative colloidal iron as cell surface stains. *Lab Invest* 18:63–71.

Happel, R. D., and J. A. V. Simpson. 1982. Distribution of mitochondrial calcium: Pyroantimonate precipitation and atomic absorption spectroscopy. *J Histochem Cytochem* 30:305–11.

Nucleic Acids (DNA/RNA)

Bendayan, M. 1981. Electron microscopical localization of nucleic acids by the use of enzyme-gold complexes. *J Histochem Cytochem* 29:531–41.

Bendayan, M., and E. Puvion. 1984. Ultrastructural localization of nucleic acids through several cytochemical techniques on osmium-fixed tissues: Comparative evaluation of the different labelings. *J Histochem Cytochem* 32:1185–91.

Maul, G. 1993. *Hybridization techniques for electron microscopy.* CRC Press, Boca Raton, FL. 360 pp.

Protein

Silverman, L., and D. Glick. 1969. The reactivity and staining of tissue protein with phosphotungstic acid. *J Cell Biol* 40:761–67.

Sterols

Elias, P. M., et al. 1979. Membrane sterol heterogeneity: Freeze-fracture detection with saponin and filipin. *J Histochem Cytochem* 27:1247–60.

Chapter 13

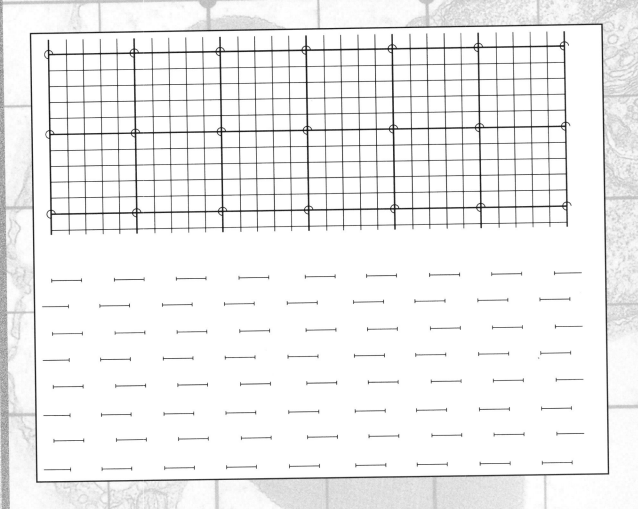

Quantitative Electron Microscopy

THE PURELY DESCRIPTIVE STUDIES that predominated at the beginning of biological electron microscopy have gradually evolved to include more experimental approaches and specialized techniques. One such area of specialization is the analysis of thin sections to obtain *quantitative information.*

As described in Chapter 15, quantative information about tissue composition can be gained using the analytical electron microscope. This chapter deals with quantification of structural components such as organelles.

During the period of descriptive morphology, vague terms such as *few, sparse, numerous, abundant, plentiful,* and *many* were frequently used as the only expression of quantity. Imprecise expressions of this type are no longer considered appropriate in situations where it is important to make a definitive statement about cell or tissue features. It is well known that micrographs may be selected arbitrarily by the investigator to portray a finding as being representative when, in fact, the micrograph shows atypical features. Unfortunately, this has a negative impact on the reputation of all microscopists. Additionally, Chapter 19 shows how tissue sections may give misleading information about the size and shape of structures. Thus, for a finding to be a credible one, it is important to use quantitative techniques, whenever possible, to assure the reader that what is being shown is representative of the entire sample.

In some experimental situations, it is important to express findings quantitatively, but in a *relative* way (e.g., "the volume of a particular structure was increased three-fold after treatment"). This statement is a quantitative measure and, as such, is much preferred to estimates that are subjective and based on a simple examination of tissue features. It provides a relative measure of volume and is useful when made in comparison to the volume of another structure. It also is necessary to express quantity in an *absolute* sense to correlate structural findings with physiological responses or biochemical measurements (e.g., "the volume of a particular structure was 210 μm^3. After a particular treatment of the animal, its volume increased three-fold to 630 μm^3."). Absolute volumes may either stand alone as a descriptive finding or be used to make comparisons in an experiment. Expression of structural features in quantitative terms, using either relative or absolute measures, has greatly enhanced the credibility of many morphological studies.

Quantitative measurements may be obtained directly from the microscope (see Chapter 15), from negatives using a densitometer or a measuring device, from micrographs using a ruler, or from serial thin section reconstruction. The obvious challenge is to obtain quantitative information about two- and three-dimensional biological structure from thin sections. Because of its extreme thinness in relation to most of the elements contained within it, the thin section is considered essentially a *two-dimensional structure* having length and width, but no thickness. However, biological structure is three-dimensional. Membranes, although three-dimensional in nature, may be considered as two-dimensional because their thickness is minor relative to the other dimensions. Figure 13.1 shows that a structure with three dimensions (e.g., expressed as μm^3) in life will reveal only two dimensions (expressed as μm^2) in section. In a similar way, a two-dimensional object (μm^2) will reveal only one dimension (μm) in all thin sections except in grazing sections. One-dimensional objects appear as points and are essentially dimensionless. (Table 13.1).

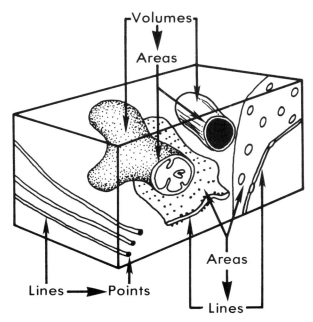

FIGURE 13.1 Three-dimensional images versus images in section as diagrammed on the cut face of a portion of a cell. On the section face, volumes are viewed as areas, areas are viewed as lines, and lines are viewed as points. The surface of a section shows N minus 1 dimensions as compared with the three-dimensional object.

TABLE 13.1 Appearance of Dimensionality in Nature Versus Electron Micrograph

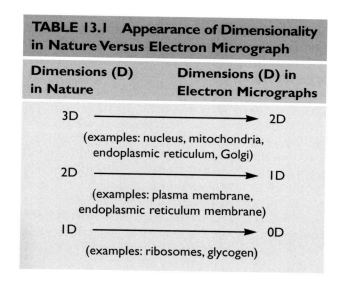

Dimensions (D) in Nature	Dimensions (D) in Electron Micrographs
3D ⟶ 2D	
(examples: nucleus, mitochondria, endoplasmic reticulum, Golgi)	
2D ⟶ 1D	
(examples: plasma membrane, endoplasmic reticulum membrane)	
1D ⟶ 0D	
(examples: ribosomes, glycogen)	

The major aspect of quantitative electron microscopy that deals with quantitative interpretations of thin sections is called *morphometry* or *stereology*. Technically, stereology is the geometric interpretation of sections, which are for most purposes two-dimensional images. Morphometry utilizes the principles of stereology to quantify structural components. In a practical sense, both terms apply to studies in which geometric and/or quantitative information may be derived from thin sections. Some investigators use the terms interchangeably. Stereology will be employed in the following discussion.

Because stereology is rarely performed the same way by two individuals, the strategy is to first learn the principles of stereology and then apply them selectively to a particular system. Usually bits and pieces are taken from the theory and the overall plan is accomplished in a "modular" manner. What follows in this chapter are descriptions of a variety of methodologies that may be utilized to construct an overall plan. A good understanding of stereology is gained through courses, practical sessions, or software tutorials. Currently, there is an excellent tutorial software package available through the University of Washington (see Bolender in the "Software" section of the References).

Basic stereological principles, as they apply to biological specimens, are largely borrowed from the field of geology. In geology, it is necessary to determine the composition of rocks with regional differences in their composition based on a view of a cut surface of the rock. The principle formulated by Delesse (1847), which directly relates fraction area (or area occupied by a structure) as viewed on the surface of a geologic specimen, to volume density (or the fraction of the volume of the structure as compared with the total volume) of the internal aspects of a geological specimen, is used heavily in biology.

Biological stereologists have expanded on the early geological principles to provide new and innovative ways to apply stereological principles. The *International Society for Stereology* has standardized the language of stereology and promotes this field of quantitative biology.

When to Use Stereology

The number of quantitative determinations that may be made on biological specimens is virtually unlimited. Stereological determinations are usually very time consuming, so it is important to decide first if stereology is the proper procedure, and then to determine the scope of the stereological determinations.

There are two major types of stereological analyses, descriptive and experimental. Descriptive stereology involves determinations for many tissue features. A *descriptive* study is used to quantitate many features of a structural component. For example, the initial stereological studies of lung or liver simply describe the various volumes and surface areas of the constituent cells and the organelles (Weibel et al., 1969; Bolender et al., 1978). These kinds of studies serve as a reference base for future studies that have an experimental or comparative basis. It also allows a comparison between structure and function (see the example of stereology at the end of the chapter).

Experimental stereology may be used to prove or disprove a particular point or may be used to determine any difference between control and treatment groups. The difference between descriptive and experimental is the breadth of the stereological determination. In experimental stereology, a very limited number of stereological determinations are usually indicated. In this instance, one sets out with a hypothesis about some quantitative tissue feature. There may be a debate, for example, of whether or not smooth endoplasmic reticulum increased after a particular treatment.

Although a study of this kind targets only smooth endoplasmic reticulum, a structure realized only by the electron microscope, one must usually use both light and electron microscopy to obtain the final answer. In a study to determine differences between control and treatment samples, there are many parameters potentially examined. Essentially, the entire gamut of tissue features must be quantitated to answer this question. The important point is to determine which parameters are likely to differ. In this case, before electron microscope studies are undertaken, it is usually beneficial to perform a subjective study at both the light and electron microscope levels. Sometimes preliminary (rough) quantitative studies are useful. Based on the findings, one is usually able to limit the stereological study to keep it from becoming unwieldy.

As mentioned previously, the need to use stereology is usually indicated from subjective observations in a descriptive study. Once the decision is made to proceed with stereology, then one usually conducts a light microscopic study followed by an electron microscopic study. An example of sequential stereology undertaken in this manner has been provided by Sinha Hikim et al. (descriptive study, 1988a; light microscope study, 1988b, electron microscope studies, 1989a and b).

General Scheme of Stereology

There are many ways to obtain quantitative information about biological specimens. Some are simple and some are exceedingly complex. There is no way that an introductory text can cover them all satisfactorily. Therefore, only basic principles will be discussed for the kinds of stereology of primary interest to biologists. If more information is required or if one is planning a stereological experiment, consult the references at the end of this chapter for more detailed information.

There is a standardized schematic or plan that applies to most stereological determinations (Figure 13.2). The *specimen* is prepared for electron microscopy and sections are made. *Micrographs* are taken and *test systems* (points, intersections, lines, etc.) applied to them in a systematic manner. From the *counts* obtained, formulas are applied to obtain three-dimensional information or *results*. *Statistical tests* are performed to analyze the quality of the data or to make comparisons between groups or formulate *conclusions*. Although, at a glance, this plan appears simple, there are numerous considerations at every step of the way.

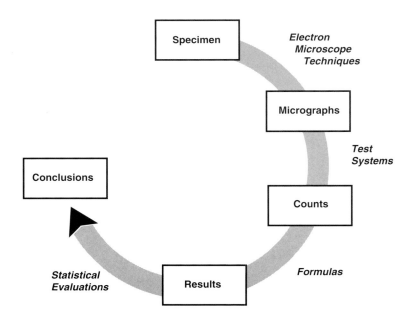

FIGURE 13.2 Scheme of stereology.

In planning a stereological experiment, it is generally best to work backwards—think about the procedure from end to beginning. Begin by determining what kind of structural features will be evaluated statistically. Determine what parameter (volume surface area, etc.) is desired. Determine what formulas will lead to a given parameter and, finally, what test systems will give the desired counts, and so forth.

Parameters Measured and Symbols Used in Stereology

What are some of the parameters obtainable through stereological techniques? Basically, these are *area, volume, surface area, length* and *number* although, less commonly, other parameters are determined. Each of these are abbreviated according to convention (Table 13.2). The kinds of counts used (points, intersections, etc.) in test systems are also abbreviated for convenient use in formulas. Subscripts identify how a particular value is expressed. For example, V, by itself, stands for volume (e. g., μm^3) and V_V stands for a volume of a structure per unit test volume or volume fraction ($\mu m^3/\mu m^3$ or expressed $\times$ 100 as a volume percentage; V_V %). For example, a nucleus might be said to have a volume (V) of 600 μm^3 or to occupy 21% (V_V%) of the entire cell. Sometimes additional subscripts are added to refer to a particular cell or organelle (e.g., V_{Vcell} means volume fraction of a cell).

Tissue Compartments or Spaces

Although it may seem obvious, it is always important to remember that a living organism occupies space. An organ system occupies part of the space or a *compartment* within the organism. Each organ can then be subdivided into a number of smaller compartments (e.g., tissues, tissue spaces, cells, organelles, etc.) based on structural properties seen with the naked eye or with the light and electron microscopes. The smaller compartments may be divided again into even smaller compartments based

TABLE 13.2 Convention for Abbreviation of Stereological Symbols

Symbol	Meaning	Expressed as
V	Volume of a structure	μm^3
V_V	Volume fraction or density	$\mu m^3/\mu m^3$
v	Mean volume of an individual element	μm^3
A	Area	μm^2
A_A	Area fraction or density	$\mu m^2/\mu m^2$
$\bar{A}$	Mean profile area (A_A/N_A)	μm^2
L	Length of a test line	μm
L_L	Linear fraction, i.e.,length of a line on a feature per unit test length	$\mu m/\mu m$
L_A	Length per unit test area	$\mu m/\mu m^2$
L_V	Length per unit test volume	$\mu m/\mu m^3$
S	Surface area	μm^2
S_V	Surface area per unit test volume	$\mu m^2/\mu m^3$
$\bar{S}$	Mean surface area	μm^2
S_V	Surface area to volume ratio	μm^{-1}
P	Number of points	self explanatory
P_P	Number of points on structure/total points(P_T)	self explanatory
N	Number of features	self explanatory
N_A	Number of profiles of a feature/unit test area	μm^{-2}
N_V	Number of profiles of a feature/unit test volume	μm^{-3}
Q	Number of transection points	self explanatory
I_L	Number of line intersections/line length	cm^{-1}
d	Profile diameter	μm
$\bar{d}$	Mean diameter of profiles	μm
$\bar{D}$	Mean diameter	μm
T	Section thickness	μm

on finer structural subdivisions, until the smallest compartments that can be resolved at the electron microscope level are identified (for an example of compartments in one organ, see Figure 13.3). Generally, stereological analyses focus on only one aspect of a particular organ or tissue, due to the *labor-intensive* nature of conducting even a limited stereological study. For every tissue compartment in Figure 13.3, one may determine volume, surface area, number, et cetera, making the possible data accumulation unwieldy. It is best to limit the information obtained to tissue spaces that are likely to prove most interesting and to expend one's efforts in performing a limited, but a quality, stereological study.

Since many compartments can be visualized in their entirety at the light microscope level, their evaluation is best conducted at that level. Large tissue spaces may be evaluated using low magnification light microscopy, whereas smaller ones require higher magnification light microscopy. The same principle applies to the electron microscope evaluation for measurements of cellular and subcellular compartments where stereologic evaluations may take place at two or more different magnification ranges. It is important to note that even limited stereological evaluations commonly use both light and electron microscopy in concert.

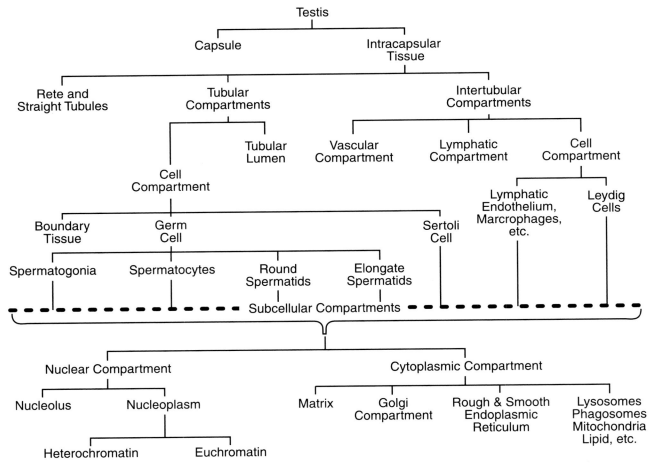

FIGURE 13.3 Compartments within an organ using the testis as an example. For each cell type listed immediately above the dotted line, the subcellular compartments are shown below the dotted line.

Test Systems

A variety of test systems may be overlaid on micrographs to obtain counts that will be substituted into formulas. Each of these test systems is to be used only under specified conditions. For now, the test systems will be described briefly and illustrated. Their full significance will not be appreciated until their use in obtaining counts is discussed later.

The most commonly used test system is the *point lattice* (ordered system of points), which is either directly overlaid on the micrograph as a transparency, or fitted directly into the eyepiece of a microscope. In practice, the same effect as a point lattice is usually obtained with a *square lattice*, where points are considered to be the intersection of any two lines. Square lattices may be designed with widely spaced bold lines indicating broad subdivisions and much finer lines for closely spaced subdivisions (*double lattice*; Figure 13.4). These systems are primarily used in volume density determinations. In performing volume density determinations, one might use only the intersections of broad lines for objects that, because they are either numerous or large, occupy a large percentage of a micrograph. All intersections (bold and fine) should be used for objects that occupy a small portion of the micrograph. A double lattice system thus allows simultaneous and efficient determinations of volume density of bigger objects with the bold grid and determinations of smaller or sparse structures with the finer grid.

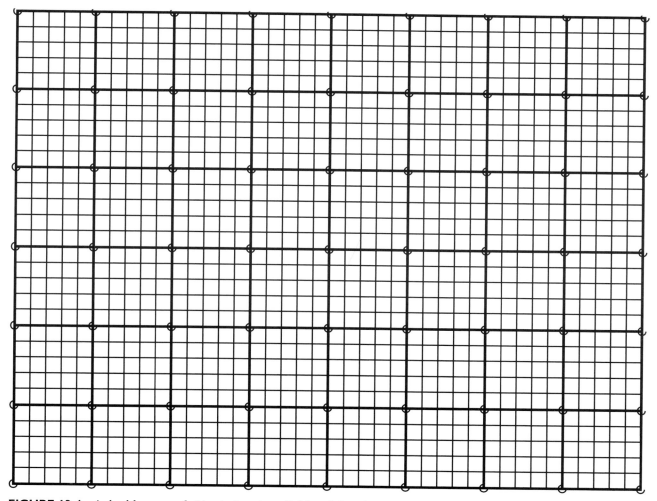

FIGURE 13.4 A double square lattice test system. Bold and fine divisions are shown.

FIGURE 13.5 Multipurpose test system.

The *parallel line test system* or a specially designed system of short lines known as a *multipurpose test system* (Figure 13.5) can be overlaid on micrographs to determine intersections of a test line with structures that appear to be single dimensional on a micrograph (e.g., a membrane). Both the length of the test line and the number of intersections that the test line makes with the structure of interest will be needed to obtain surface density and surface area. The multipurpose test system was designed to allow both volume density (the intersections at the line ends are used to count point hits) and surface area determinations (intersections of the line with pertinent surface contours on the micrograph structure) on a single test system. The *Mertz curvilinear test system,* based on similar principles, but employing curved lines, is used in certain circumstances where surface density and surface area determinations are needed (Figure 13.6). It is used for *anisotropic* structures, which are structures that vary their orientation depending on the plane of section. For example, a Mertz grid is used to determine smooth endoplasmic surface parameters when it is ordered in sheets within the cell, and the positioning of sheets is not random within the cell.

Basic Types of Determinations and Associated Formulas

Under the appropriate conditions (see "Assumptions and Conditions" later in the chapter), counts obtained from test systems may be entered into mathematically derived formulas that are used to obtain parameter estimates. The following parameter estimates described are volume, area, surface area, and number.

Volume Determinations

The point or square lattice test systems are placed over and affixed firmly to the micrograph. Points or intersections coincident with (overlaying) the particular structure (e.g., mitochondria, P_{mit}) of interest are counted. These are commonly termed "hits." Figure 13.7 illustrates a micrograph with a superimposed square lattice. The total number of hits or points overlying the cell cytoplasm containing the structure (P_{tot}) are also ascertained. To

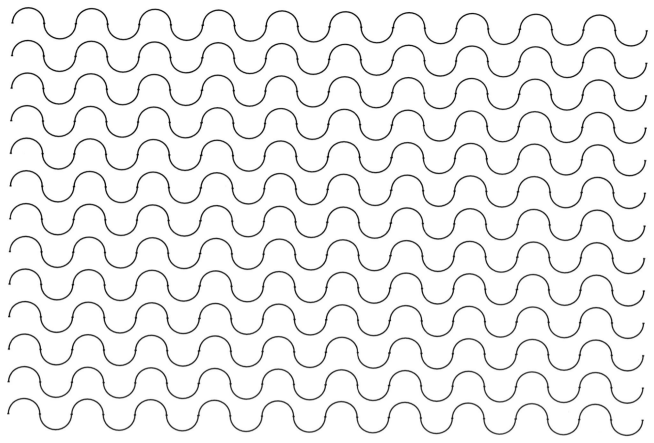

FIGURE 13.6 Mertz curvilinear test system.

determine the relative volume of the mitochondria or volume fraction (V_V), the following formula is applied:

Equation 13.1

$$V_V = \frac{P_{mit}}{P_{tot}}$$

The result obtained may be multiplied by 100 to express the data as a volume percentage ($V_V\%$).

Having obtained the volume fraction, it is next often desirable to know the absolute volume (V) of the structure in question. Before the absolute volume can be obtained, it is necessary to know the absolute volume (V) of the larger tissue space in which the smaller is contained. To continue with the example used above, mitochondria are contained within the cell cytoplasm so that the absolute mitochondrial volume per cell ($V_{mit./cell\ cyto}$) can be expressed as the product of the volume of the cell cytoplasm ($V_{cell\ cyto}$) and

the volume fraction of mitochondria (V_{Vmit}) as follows:

Equation 13.2a and b

a. $V = V_{V\ organelle\ of\ interest} \times$ reference volume

or

b. $V_{mit} = V_{Vmit} \times V_{cell\ cytoplasm}$

One can continue to extend the reasoning pattern set forth by finding the volume of the mitochondria in the particular cell cytoplasm to finding the volume of those mitochondria in the cell (nucleus and cytoplasm) and even the volume of mitochondria for that particular cell type in the organism. Equations 13.1 and 13.2 are, therefore, basic for most stereologic determinations.

At some point in the sequence of making determinations, it becomes important to obtain the *volume of a cell*, often one of the more difficult determinations in stereology. The approaches for accomplishing this are numerous and sometimes

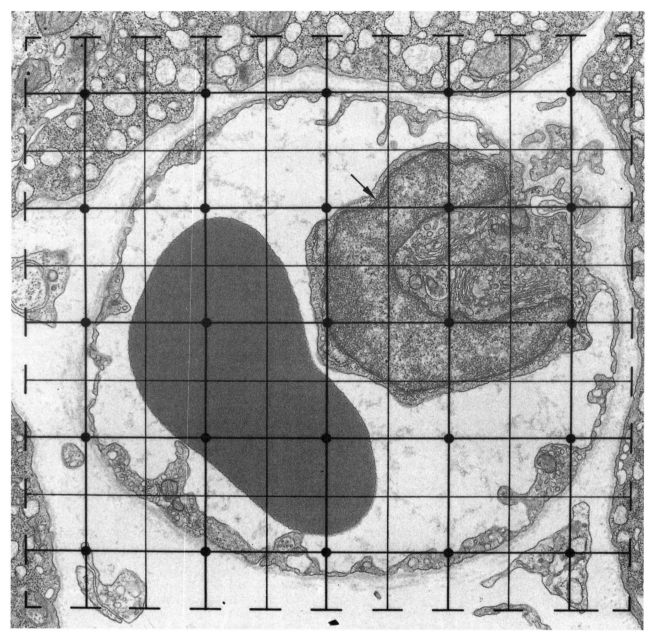

FIGURE 13.7 A micrograph overlayed with a double square lattice. *(Courtesy of R. P. Bolender with permission of the* Anatomical Record.*)*

difficult. All possess some degree of inherent error. A few approaches are described following.

Volume Determination Methods for Spherical Objects

If the cell or object (e.g., nucleus) in question is spherical and each cell of one type is of uniform size, then one simply need measure the diameter (or radius) of the cell and apply the appropriate formula:

Equations 13.3 and 13.4

$$V_{cell} = \frac{4}{3}\pi r^3 \text{ or } V_{cell} = \frac{1}{6}\pi d^3$$

(Cells, especially those in compact tissues, are usually not isolated such that their cell boundaries are clearly distinguishable by light microscopy, nor are they usually spherical. However, their nuclei are usually clearly distinguishable and frequently spherical. This is why most stereologic determinations are directed first at determining the volume of nuclei as a prelude to determining cell volume).

Equations 13.3 and 13.4 imply that one can determine the diameter of cells or nuclei from their appearance in sections. The real diameter (Figure 13.8), as measured in light microscope sections, or *profile diameter* (d) is obtained by selecting and measuring the largest profiles in thick sections (0.5–2 μm). This is only true if the largest profiles are selected from a much larger population since only a certain percentage of cells or nuclei will have been sectioned through their middle or through their real diameter. By systematically searching through serial sections, the largest profile diameter or real diameter may be selected. Also, by cutting extremely thick sections and staining them heavily, one is able to find the focal plane, which reveals the largest profile diameter. In the last technique, focusing up and down on the section should cause the spherical cell (or nucleus) to appear small, then appear progressively larger, reach a maximum diameter, and then progressively diminish in size. The maximum profile diameter is to be measured and used as the real diameter.

Most light microscopes can be equipped with a special measuring device in the eyepiece called an *ocular micrometer*, which first must be calibrated with another micrometer that fits on the stage of the microscope, the *stage micrometer*. The stage micrometer is an absolute scale that is divided into

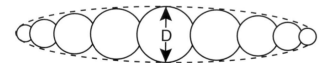

FIGURE 13.8 The real diameter (D) is measured from the largest profile diameters in a set of serial sections.

microns (or fractions of millimeters), whereas the ocular micrometer is a ruler with an arbitrary scale that must be calibrated for each magnification. The tissue elements are measured with the aid of the ocular micrometer and then converted into absolute measurements using the stage micrometer (Figure 13.9).

When using spherical cells or nuclei, the option is also available to calculate a mean diameter ($\overline{D}$) from a population where the mean of profiles diameter ($\overline{d}$) is known. The formula is as follows:

Equation 13.5

$$\overline{D} = 4/\pi \times \overline{d}$$

For this method to be accurate, it is important that all profile diameters in a randomly selected area be measured. In a strict sense, this formula applies only to profiles from a population of uniform size.

Mean cell volume (V_{cell}) in systems with spherical nuclei can be determined by using both the numerical density ($N_{V\,nuc}$) measurements of the nucleus (assuming cells are mononucleate) obtained, for example, by the Floderus equation (Equation 13.12), and the volume density measurements of the cell (V_{Vcell}) obtained by point counting. The relationship used to obtain cell volume is as follows:

Equation 13.6

$$V_{cell} = \frac{V_{Vcell}}{N_{Vnuc}}$$

The cell volume can also be obtained from the nuclear volume, if the volume fraction of the nucleus (N_{Vnuc}) within the cell is known. Any of the previously mentioned methods to calculate nuclear volume (V_{nuc}) can be employed. To determine the cell volume (V_{cell}), the following expression is used:

Equation 13.7

$$V_{cell} = \frac{V_{nuc}}{N_{Vnuc}}$$

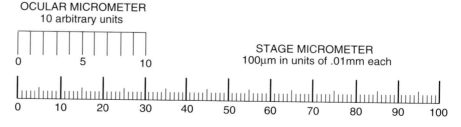

FIGURE 13.9 Diagram of ocular and stage micrometers.

OCULAR MICROMETER
10 arbitrary units

STAGE MICROMETER
100μm in units of .01mm each

at the microscope magnification employed
10 arbitrary units on the occular micrometer match up with 0.3 mm,
therefore one arbitrary eyepiece unit = .03mm or 30μm

The volume fraction of the nucleus can also be obtained by point counting, usually at the light microscope level. V_{Vnuc} is expressed as the points over the nucleus divided by the total points over the nucleus and cytoplasm. Such a determination is one of the most common in stereology.

Volume Determinations for Nonspherical Objects

Unfortunately for the stereologist, most cells are not spherical and/or they vary in size. Many nuclear types are also nonspherical. Determining the volume of nonspherical structures necessitates using different approaches than for spherical objects to obtain mean cell or nuclear volume determinations. If the cells are of regular geometric shapes and their dimensions known, specific geometrical formulas may be applied to calculate volume after measuring some parameter at the light microscope level. Individual size differences are handled by giving a mean size accompanied by some measure of variability such as standard error or standard deviation.

A time-consuming, but relatively accurate method for obtaining object size is to *reconstruct an object from serial sections* of known thickness. The areas of individual profiles are summed to obtain the total area. The volume (μm^3) is then the product of the summed areas (μm^2) of the profiles of interest and the section thicknesses (μm). Figure 13.10 is a summary diagram of this method. Thick sections ($\sim$0.5 μm) are often used to make serial reconstructions and areas determined at the light microscope level by one of the methods which follow.

Although microtomes have various gauges indicating the thickness of the sections produced at a particular setting, the scales on the microtome may not be an accurate measure of section thickness. Many investigators will reembed thick (0.5 to 2.0 μm) sec-

tions in epoxy and make thin sections perpendicular to them to measure section thickness in the electron microscope. Figure 13.11 shows such a section, which is approximately 0.9 μm in thickness.

The thickness of paraffin sections may be obtained readily by high power light microscopy. If the bottom and top portions of the sections are brought into focus separately, the distance traveled on the fine focus knob in going from one to another is directly proportional to the section thickness. The measurement equivalent to the gauge reading may be determined from the microscope manual. The accuracy of this method is sometimes questioned.

Various methods for the determination of the thickness of ultrathin sections have been reviewed by De Groot (1988). The thickness of ultrathin sections, as determined by their interference color, (see Chapter 4), may only be approximated. A simple technique has been developed to determine the thickness of ultrathin sections, but only in situations where an elevated fold appears in the tissue section (Figures 13.12 and 13.13). At high magnification, this fold has an electron dense line which is the middle space of the fold that is occupied by stain. The section thickness is roughly 1/2 the measured width of the fold. If one is making serial sections of spherical objects, the mean section thickness may be determined by finding out how many sections are necessary to traverse the sphere of a given diameter. Determine the diameter of the sphere from the largest profile section of the sphere and divide it by the number of sections that traversed the sphere to obtain section thickness.

Recent developments in light microscopy have opened the door to making certain volume determinations possibly a routine and relatively easy task. With the new *confocal microscopes* (see Figure 13.16), it is possible to examine fresh or fixed cells and to "section" them optically with lasers. An optical section refers to a narrow in-focus plane of a much thicker specimen. A computer interfaced with the confocal microscope will determine the areas of the sections and compute the volumes based on the sum of the thicknesses. Although relatively expensive at present, approaching the cost of a transmission electron microscope, it is expected that confocal microscopes will soon facilitate the difficult problem of volume determinations, especially when cells and nuclei are irregularly shaped.

$$V_N = \Sigma \text{Areas (A)} \times \text{section thickness (t)}$$

FIGURE 13.10 Tracings of the profiles of a serially sectioned nucleus that is of an irregular shape. The summed areas (ΣA) of all profiles ($1 + 2 \ldots + 9$) times the section thickness (t) equals the nuclear volume (V_n).

Area Determinations

Rarely do investigators wish to determine *area* parameters (A or A_A) as an end point in their measurements. They do, however, use area density (A_A)

FIGURE 13.11   Reembedded thick section measuring 0.9 *μm* in width (thickness). Bar = 0.25 *μm*.

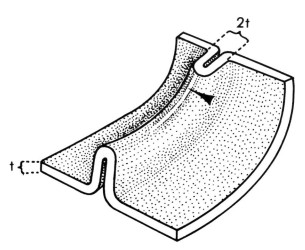

FIGURE 13.12   Diagram of a small fold in a tissue section of the kind utilized to make section thickness determinations. The thickness of the fold is approximately twice as thick as the section thickness. (See also Figure 13.13).

measurements from micrographs and equate them with volume density or volume fraction ($A_A = V_V$) if certain conditions are met, such as random sectioning and sampling from compartments of the same

size (see "Assumptions and Conditions" later in this chapter). Volume density is then used in a *relative* way to make comparisons between samples. For example, area density (A_A) measurements of lipid droplets can be compared to determine if there is a difference in the amount of lipid in one sample as compared to another. There are many ways to determine area density and, in general, they are some of the least difficult stereological procedures. Micrographs are two-dimensional images from which area density measurements may be made directly.

Area density (A_A) is converted to area (A) using the following formula:

Equation 13.8

$$A = A_A \times A_{ref\ vol}$$

where $A_{ref\ vol}$ is a known area, such as the area of a cell profile. To calculate area (A) from area density measurements, one must take into account the magnification of the micrograph (Equation 13.9).

Equation 13.9

$$\text{Area (A)} = \frac{\text{Area (A) as measured on the micrograph}}{\text{magnification}^2}$$

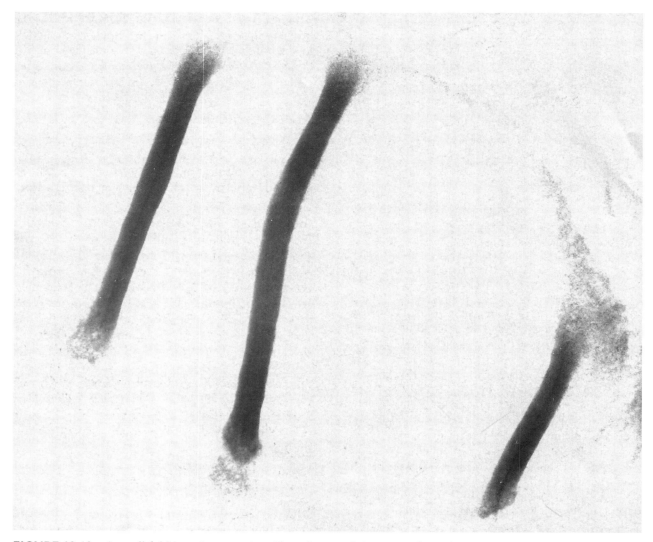

FIGURE 13.13 A small fold in a tissue section. The micrograph is purposely underexposed to show the internal detail of the fold. There is a slight space through the middle of the fold occupied by stain.

Area at the magnification of the micrograph may be determined with a simple measuring device called a *planimeter* or a slightly more sophisticated instrument called a *digitizer*. Either of these instruments will give the area if the perimeter is circled. For a relative measure, some investigators have actually cut out images from micrographs and weighed them on a sensitive balance (*cut-and-weigh method*). The additive weights for one sample may be used to make direct comparisons with those from another sample if the total test area of the two are similar. To obtain area (at the micrograph magnification) from A_A using the cut-and-weigh method, a known area of photographic paper (1 cm²) is weighed and used as a reference [A = sample wt × reference wt(1 cm²)].

Areas may be determined from microscope projections (i.e., *camera lucida*) drawings, although the area obtained must be divided by the magnification[2] (Equation 13.9) to determine the actual area (unmagnified). A_A measurements can be made utilizing a *point counting test system*. As already mentioned, A_A and V_V are equivalent given that sampling is random.

Surface Density and Surface-to-Volume Determinations

The *parallel line test system* is used for surface density determinations. The general principle employed is that there is a relationship between the number of times that test lines intersect a one-dimensional structure (line) in a micrograph and the surface area of a two-dimensional structure. Of course, in the micrograph, the two-dimensional structure in the tissue was sectioned to appear in the micrograph as a line (see Figure 13.1) . The formula used is:

Equation 13.10

$$S_V = 2 \times I_L$$

where S_V is the surface area per unit volume ($\mu m^2 / \mu m^3$) and I_L is the density of intersections on a test line length (i.e., the number of intersections of the linear probes, or test line, with the surface contour of a given profile divided by total length of the test lines).

A *multipurpose test system* (Figure 13.5) is used to determine surface-to-volume ratio. Generally, this test system is used to measure organelle surface area per cell volume. This system has lines that can be used to determine intersections with the structure of interest as well as end of line intersections that may be used as points for volume density determinations. The surface-to-volume ratio (S/V) is expressed as cm^2/cm^3 and is obtained as follows:

Equation 13.11

$$S/V = \frac{4 \times I}{P_t \times Z}$$

where I is the number of intersections. P_t is the number of points at the ends of the lines and Z is the length of the line expressed in relationship to the magnification of the micrograph (i.e., the distance in micrometers on the micrograph equivalent to the line length is measured).

Numerical Density Determinations

The number of structures may be expressed per organ (N), per area (N_A in cm^2) or per volume (N_V in cm^3). Once the *numerical density* (N_V) is obtained, it is easy to determine the number of structures per organ by multiplying the numerical density (which is the number of a given structure in a unit volume of an organ) by the organ volume. Let us first examine how numerical density can be obtained.

The *Floderus equation* is the most commonly used means to determine numerical density of spherical cells. This equation is frequently employed at the light microscope level to provide data essential for the logical extension of the study to the ultrastructural level (see "A Typical Published Report" later in this chapter). The formula is

based on counts of nuclei as seen in section (profile counts).

Equation 13.12

$$N_V = \frac{N_A}{(T + d - 2h)}$$

A calibrated device fitted into the ocular of a microscope is used to determine the number of nuclei per unit area (N_A). The average nuclear profile diameter ($\bar{d}$) and section thickness (T) are determined by one of the methods described above. The height (h) of the *cap* (which is the height of the smallest observed nuclear profile in the section) is placed in the formula. Cap size serves as a correction for the small polar regions of a nucleus, which become lost to the eye since they are grazed during sectioning and therefore are only present in part of one section. A cap is shown in Figure 13.14. There is no practical way to determine cap size. Operationally, many investigators simply assume cap size to be about 1/3 of the section thickness or 1/10 of the diameter of the spherical object.

Sometimes the investigator knows both the mean nuclear volume (v_N) and the total volume of a population of cells (V_{Vnuc}). It is then relatively simple to calculate the numerical density:

Equation 13.13

$$N_V = \frac{V_{Vnuc}}{v_N}$$

Given the test systems and work necessary to obtain the counts to enter into formulas, it is easy to

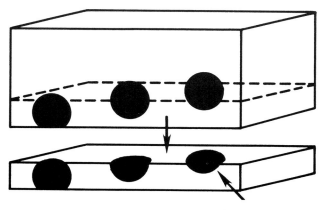

Polar Region or Cap

FIGURE 13.14 The polar region or *cap* of the nucleus of a spherical object in a section removed from the block face.

see why stereology is often considered tedious. After considerable effort and calculations the final results are obtained; however, having data that is objective in nature is usually worthwhile. Only after one working with the preceding equations will one really understand stereologic schemes.

Assumptions and Conditions

Stereological methods are statistical in nature, so it follows that the investigator can inadvertently prejudice the data through selection and bias. If one wants to determine, for example, the volume density of a structure, it is important that the particular structure neither be selected for nor avoided. Stereological formulas were derived with the idea that structures will be encountered randomly with a histological or a thin section. *Randomness* in selecting animals, tissue blocks, sectioning planes, and areas for morphometry will assure that the likelihood of encountering an object or a surface is proportional to its occurrence in the tissue section. When performing stereological determinations, the added work to assure randomness is well worth the effort. For example, if there is any doubt about whether or not the sampling is being done randomly, one can partition the available tissue, assign each portion a number, and then draw a number(s) from a hat. Being very careful to ensure randomness will instill confidence in the final outcome.

Many tissues exhibit an ordered pattern of their elements. The choice of a sampling method becomes particularly important in these systems so that a particular element is neither selected nor avoided. Since it is difficult to generalize broadly about procedures for all possible conditions, it is best to examine the literature and determine how sampling has been conducted for a particular tissue.

Multiple Stage Sampling

As already mentioned, stereology at the electron microscope level is often performed initially at the light microscope level. This is usually a requirement when absolute, rather than relative, measurements are required at the electron microscope level. For example, the volume fraction or relative area (A_A) of mitochondria may be obtained rather readily by point counting in both normal and treat-

ed animals at the ultrastructural level. Suppose an investigator recorded a 50% decrease in V_V of mitochondria. If this investigator based a conclusion solely on the initial volume density determinations, he or she might have erroneously suggested that there are fewer mitochondria. This data indicates only the area density of mitochondria, but not necessarily whether the cells containing the mitochondria have grown, shrunk, multiplied, etc., to produce this effect. It is very possible that this same treatment resulted in a 50% decrease in the volume density (V_V) of mitochondria, but that the total volume (V) of a population of mitochondria has remained unchanged (Figure 13.15). This could occur if the cell increased in size. There are other possibilities as well, one of which would be a change in size of mitochondria. This example illustrates that volume density data could not, by itself, provide this type of information. For most tissues, it is only possible to determine both relative and absolute parameters such as V_V and V if light and electron microscopy are employed together in a *multiple stage sampling protocol,* which is a long term meaning the use of both light and electron microscopy or microscopy at different magnifications.

To obtain an absolute volume, the volume of the largest tissue compartment must be determined at the onset of the morphometric study. This volume serves as a reference for all of the volume determinations made for smaller compartments (Figure 13.3). The largest tissue space

 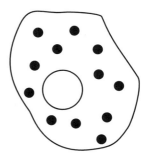

CONTROL   TREATMENT

FIGURE 13.15 In the two cells shown, V_V measurements would have recorded a volume density for the solid structure in the control as being twice that seen after treatment. Measurement of the cell size at the light microscope level using multiple stage sampling indicates that the volume (V) of the structure has not changed on a per cell basis; the cell has simply grown.

(usually an organ) is weighed and/or its volume determined by submersion. The volume of the fluid displaced by submersion is equal to the volume of the submerged object. Often it is inadvisable to use water to submerse biologicals, since it is nonphysiologic and may have deleterious effects on tissue structure. Isotonic solutions are preferable.

A more accurate way to determine volume is to measure fluid displacement by a weighing method. The weight difference (W) of an isotonic solution before and during submersion of the tissue can be used to determine the tissue compartment volume (V) if the specific gravity (sp gr) of the fluid is known. The relationship is as follows:

Equation 13.14

$$V = \frac{W}{sp\ gr}$$

Processing and sectioning will usually cause some change (usually a decrease) in a tissue's original volume, such that another determination of volume may be necessary and a correction factor employed in the calculations.

After tissue processing and sectioning, the volume fraction (V_V) of the compartment of interest is determined by point counting. The volume (V) of that smaller compartment is the product of the volume fraction of the smaller compartment and the volume of the larger tissue compartment (or organ) in which it is contained (Figure 13.3).

Equation 13.15

$$V_{smaller\ comp} = V_{V\ smaller\ comp} \times V_{larger\ comp}$$

Once the volume of the larger compartment is known, it is evident that one may tackle progressively smaller and smaller compartments in a systematic manner to obtain absolute volume data. Equation 13.15 is similar to Equation 13.2 and is one of the most commonly utilized formulas in stereology.

Light microscopy is useful to obtain point counting information from the larger tissue compartments. As far as analysis of smaller compartments, one cannot usually advance beyond nuclear and cytoplasmic compartments (using $\times$ 1,000 magnification) to obtain useful point counting information simply due to the limited resolution of the light microscope. Electron microscopy is then used beyond this point. Two or three steps (magnification levels) of electron microscopy may be necessary to determine volume fractions and surface areas for both large and small organelles.

How Much Data from Test Systems Is Needed?

When performing point counting or using the line intersection method, one must consider how many points need to be counted or intersections measured to obtain adequate results. As a general rule, as data is added, the standard error lowers to reach a value that changes little with the addition of more data. Thus, the standard error first will lower rapidly, and then begin to level off as more data are added.

Most stereologists accept that a standard error ($e_A{}^2$) of less than 10% of the mean is adequate. In such circumstances, the number of points (P) to be counted, where x equals the investigator's preliminary estimate of the percentage of the sample occupied by the structure of interest, is given in Equation 13.16. To achieve a standard error of 10% of the mean with the tissue component estimated to occupy about 5% of the sample, it is necessary to record about 1,900 points. The equation for calculation of points is as follows:

Equation 13.16

$$P = \frac{1 - x}{E_A{}^2 \cdot x} = \frac{1 - 0.05}{(0.1)^2 \times 0.05} = 1,900$$

Of course, the actual standard error should be calculated after the actual data are obtained.

Computer-Assisted Stereology

An advantage of employing stereological techniques is that they are not particularly expensive. They may be performed by hand and calculations made on a hand calculator. Computers and software, if purchased specifically for stereology, are a major expense. They will speed up certain aspects of the project such as data acquisition, storage, tabulation, and computations using stereological formulas and also statistical evaluation of the data. *Automatic image analyzers* take information from two-dimensional images such as videotapes and then record the counts more accurately than simple point counting by eye. Computers readily assess areas and particle numbers. Stereological formulas may then be applied by directing the computer to do so. Automated image analyzers have limited

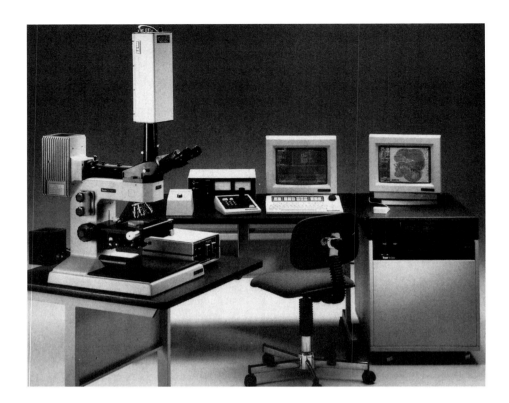

FIGURE 13.16 A modern-image analysis system. This system utilizes a confocal microscope.

abilities to recognize micrograph images, so it is first important to determine if this equipment will aid in the specific project under consideration. Both computer-assisted methods and automatic image analyzers are recommended when stereology will be used on a routine basis (Figure 13.16).

A Typical Published Report

As an example of how stereology is performed in an experimental setting, the published article by Mori and Christensen (1980) follows, in brief. To be fully appreciated, the article must be read in its entirety. The methods section is especially helpful in describing how the stereology was performed.

Rationale for the Study. This study is a good example of descriptive stereology. Besides providing baseline data on the Leydig cell population in the rat testis, the authors wished to correlate the cellular and subcellular structure of the *Leydig cell* with its major function of *steriod production* (testosterone). Since testosterone production rates are known and the particular cell type and the organelles involved in this process are well known, the authors targeted the Lyedig cell and particularly its mitochondria and smooth endoplasmic reticulum for morphometric determinations.

Methods. Figure 13.3 shows the space or compartment occupied by the Leydig cell and its organelles. The *organelles* are within the *Leydig cell* compartment and the Leydig cell is within the *intertubular compartment*. The intertubalar compartment is considered to be within a *parenchymal compartment* within the *testis*. With this information, it was possible to plan a stereological experiment. The basic plan of the methods employed were as follows.

Tissue Preparation. Specific gravity of the tissue was determined by floating the testis in sucrose solutions of known concentration and then dividing the weight difference (weight after addition of the testis − weight before addition of the testis) by the specific gravity of the fluid. Tissues were fixed in glutaraldehyde and the osmolarity controlled to prevent excessive tissue shrinkage. After fixation, the tissue was sliced and alternating slices used for light microscopy (methacrylate embedding) and electron microscopy (epoxy embedding). Measurements were made during tissue processing to determine extent of shrinkage. For electron microscopy, ferrocyanide was added to the osmium fixative to enhance membrane contrast since many membrane compartments would be measured. Methacrylate sections were cut at 2 μm for light microscopy and epoxy-embedded thin sections used showed silver to pale gold interference colors.

Morphometric Procedures at the Light Microscope Level. The volume of the parenchyma was determined by subtracting the volume of the capsule (connective tissue covering of the the testis) from the total volume (parenchyma and capsule). The capsule volume was estimated as the product of the thickness of the capsule and the surface area of the testis. A point counting system, employing a square lattice, was used for low magnification microscopy to obtain the volume density (V_v) of the interstitial cell compartment. At higher magnification, the volume density of the Leydig cell compartment was also determined by point counting. Leydig cell nuclear volume density was measured in the same manner. The numerical density of Leydig cells was determined by the Floderus equation. (Strictly speaking, this formula is for spherical nuclei, although most Leydig cell nuclei are not spherical.) The mean volume of the average Leydig cell was obtained by dividing the volume density (V_V) by the numerical density (N_V). The number of Leydig cells in the testis was found as the product of the numerical density and the testis weight.

Stereologic Determination at the Electron Microscopic Level. Since Leydig cells occupied only a small portion (2.7%) of the tissue, sampling throughout the tissue could not be performed with random electron micrographs. Instead, Leydig cells were photographed, taking care not to duplicate micrographs of the same cell. (This sampling procedure was not a good example of random sampling, but it is difficult to sample randomly when such a small percentage of volume of a tissue has the cell type of interest.) A two-stage sampling system was utilized to perform analysis on Leydig cells at the ultrastructural level. Micrographs printed at a final magnification of ×10,800 were used to determine the volume density of organelles and to determine the surface density of the plasma membrane. The surface density of organelles was determined at high magnification (×72,000). Point counting with a double lattice grid was used for volume density measurements, and line intersections on a multipurpose grid were used to determine surface density. The appropriate formula for each was employed (for V_V see Equation 13.1; for S_V see Equation 13.11). The authors used a complicated scheme to approximate a determination of numerical density of mitochondria (see the actual publication). There is an excellent discussion of actual and possible sources of error in the procedures used, which appears in the methods section. In some cases, correction factors were derived to be applied to raw data.

Results and Significance. The number of published parameters obtained for Leydig cells in this descriptive study were numerous and too extensive to reiterate; thus, only a few will be listed here. The number (N) of Leydig cells in one testis was 22 million although this population of cells comprised only 2.7% of the testis volume (V_v%). The mean Leydig cell volume ($\bar{v}$) was 1,210 μm^3 and the mean surface area ($\bar{s}$) was 1,520 μm^2. The two sites of steriodogenic enzymes, the mitochondria and smooth endoplasmic reticulum, were examined in detail. The surface area (S_{cell}) of smooth endoplasmic reticulum, was about 10,500 μm^2/cell or expressed on a per organ basis about 2,300 μm^2/cm^3 of testicular tissue. The mitochondrial inner membrane demonstrated a surface area of 2,920 μm^2 and there was 644 cm^2 of inner mitochondrial membrane/cm^3 (S/V) of testis. When correlated with published endocrine data, the average Leydig cell secretes about 0.44 pg of testosterone per day. Each square centimeter of smooth endoplasmic reticulum and inner mitochondrial membrane produces 4.2 ng and 15 ng testosterone per day, respectively. This study was the first to correlate cell and organelle structural parameters with functional parameters. As a descriptive morphometric study, it established baseline values for future studies on Leydig cell structure and function.

REFERENCES

Technical Articles and References

Baak, J. and Oort J. 1983. *A manual of morphometry in diagnostic pathology.* Springer-Verlag, NY.

Bolender, R. P. 1981. Stereology: applications to pharmacology. *Ann Rev Pharmacol Toxicol* 21:549–73.

De Groot, D. G. 1988. Comparison of methods for the estimation of the thickness of ultrathin tissue sections. *J. Microsc* 151:23–42.

Elias, H., and D. M. Hyde. 1980. An elementary introduction to stereology (quantitative microscopy). *Am J Anat* 159:412–46.

Elias, H., et al. 1971. Stereology: applications to biological research. *Physiol Rev* 51:158–200.

Loud, A. V. 1968. A quantitative stereological description of the ultrastructure of normal rat liver parenchymal cells. *J Cell Biol* 37:27–46.

Mayhew, T. M., and F. H. White. 1980. Ultrastructural morphometry of isolated cells: methods, models and applications. *Pathol Res Pract* 166:239–59.

Ross, J. 1986. *Practical stereology.* Plenum Press, NY.

Toth, R. 1982. An introduction to morphometric cytology and its application to botanical research. *Am J Bot* 69:1694–1706.

Weibel, E. R. 1969. Stereological principles for morphometry in electron microscopic cytology. *Int Rev Cytol* 26:235–302.

Weibel, E. R. 1973. Stereological techniques for electron microscopic morphometry. In *Principles and techniques of electron microscopy,* M. A. Hayat, ed., New York: Van Nostrand Reinhold Co.

Weibel, E. R. 1979. Stereological Methods. In *Practical methods for biological morphometry.* New York: Academic Press.

Williams, M. A. 1977. Quantitative methods in biology. In *Practical methods in electron microscopy,* A. M. Glauert, ed., Amsterdam: Elsevier North Holland Publishers, pp. 1–84.

Classic Articles

Bolender, R. P 1978. Correlation of morphometry and stereology with biochemical analysis of cell fractions. *Int Rev Cytol* 55:247–89.

Bolender, R. P., et al. 1978. Integrated stereological and biochemical studies on hepatocyte membranes. Membrane recovery in subcellular fractions. *J Cell Biol* 77: 565–83.

Delesse, A. 1847. Procede mechanique pour determines la composition de roches (extrait). *CR Acad Sci* (Paris) 25:544–60.

Eisenberg, B. R., et al. 1974. Stereologic analysis of mammalian skeletal muscle. *J Cell Biol* 60:732–54.

Floderus, S. 1944. Untersuchungen uber den bau der menschlicken hypophyse mit besonderer Berucksichtigung der quantitativen mikromorphologischen Verhailtnisse. *Acta Pathol Microbiol Scand Suppl* 53:1–276.

Loud, A. V., et al. 1965. Quantitative evaluation of cytoplasmic structures in electron micrographs. *Lab Invest* 14:996–1008.

Weibel, E. R., et al. 1969. Correlated morphometric and biochemical studies on the liver cell. *J Cell Biol* 42:68–91.

Recent Reports

Bowers, B., et al. 1981. Morphometric analysis of volumes and surface areas in membrane compartments during exocytosis in *Acanthamoeba. J Cell Biol* 88:509–15.

Buschmann, R. J. 1983. Morphology of the small intestinal enterocytes of the fasted rat and the effects of colchicine. *Cell Tissue Res* 231:289–99.

Mori, H., and A. K. Christensen. 1980. Morphometric analysis of Leydig cells. *J Cell Biol* 80:340–54.

Ryoo, J. W., and R. J. Buschmann. 1983. A morphometric analysis of the hypertrophy of experimental liver cirrhosis. *Virchow's Arch Pathol Anat* 400: 173–86.

Sinha Hikim, A. P., A. Bartke, and L. D. Russell. 1988a. The seasonal breeding hamster as a model to study structure-function relationships in the testis. *Tissue Cell* 20:63–78.

———. 1988b. Morphometric studies on hamster testes in gonadally active and inactive states: light microscope findings. *Biol Reprod* 39:1225–37.

———. 1989a. Correlative morphology and endocrinology of Sertoli cells in hamster testes in active and inactive states of spermatogenesis. *Enodcrinology* 125: 1829–43.

———. 1989b. Structure/function relationships in active and inactive hamster Leydig cells: a correlative morphometric and endocrine study. *Endocrinology* 125:1844–56.

Weber, J., et al. 1983. Three-dimensional reconstruction of a rat stage V Sertoli cell. II Morphometry of Sertoli-Sertoli and Sertoli-germ cell relationships. *Am J Anat* 167:163–79.

Wong, V., and L. D. Russell. 1983. Three-dimensional reconstruction of a rat stage V Sertoli cell. I Methods, basic configuration and dimensions. *Am J Anat* 167:133–61.

Software

Bolender, R. P. *QM 2000 current methods in quantitative morphology.* Contact the Health Sciences Center for Educational Resources distribution unit SB-56, University of Washington, Seattle, WA 98195. (Phone: 206–685–1186).

"NIH Image" http://rsb.info.nih.gov/nih-image/ (This software is downloadable free of charge from the World Wide Web and is excellent for stereological determinations.)

Chapter 14

Courtesy of Balzers AG.

Freeze Fracture Replication

FREEZE FRACTURE is a technique for the replication of fractured surfaces of frozen specimens for their examination in the transmission electron microscope. It is especially suited for studies of *membrane structure* and particularly the internal aspect of the lipid bilayer of the membrane. Freeze fracture provides three-dimensional information encompassing the general organization of tissues and cells to the macromolecular organization of the membrane. The technique has contributed much of what we know about the structure of cell *junctions* and has confirmed that many cell types, such as the epithelial cell or sperm cell, have organized regions of their plasma membrane which are functionally diverse.

To understand how freeze fracture derives its usefulness, it is important to recognize how a typical *membrane* is organized. The current view of membrane organization was proposed by Singer and Nicolson (1972). In this model, proteins exist in a sea of *phospholipid* molecules. The phospholipid of membranes is organized into a bilayer with the polar heads of the molecules facing and interacting with the polar aqueous environment of both the cell interior and cell exterior. The hydrophobic nonpolar carbon chains of each phospholipid layer are directed inward to face each other (Figure 14.1). Thus, the

physicochemical properties of the internal and external aspects of the lipid bilayer differ.

If a very sharp knife is used to cut a frozen specimen, the specimen acts like a brittle object and undergoes a fracture rather than being cut by the knife (Figure 14.2). In the frozen state, the hydrophobic membrane interior is most subject to fracture when external forces are exerted on the specimen. Since the energy required to split the lipid bilayer is less than is necessary to fracture ice or cytoplasm, the lipid bilayer is frequently split to reveal the *membrane interior* or *membrane faces.*

An example of how a simple cell, the red blood cell, might fracture after it had been frozen is illustrated in Figure 14.2. The ice surrounding the cell is fractured erratically; however, when the fracture reaches the plasma membrane, the lipid bilayer splits yielding two complementary fragments, each attached to the ice. After the fragments are separated from each other, it is possible to look down on each of them (if one is turned over) and view the split surfaces or fracture *faces.* Specific names have been given to each of the fracture faces. The portion of the bilayer associated with the exterior of the cell is termed the *E-face* or *extracellular face.* The portion associated with the interior of the cell is termed the

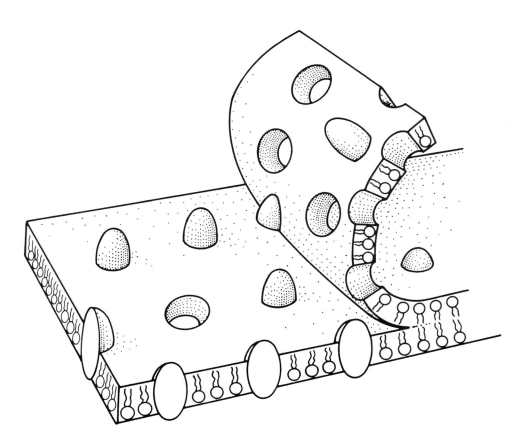

FIGURE 14.1 A lipid bilayer is split open to reveal its internal faces. After splitting, large proteins embedded in a phospholipid matrix remain with one of the two membrane halves.

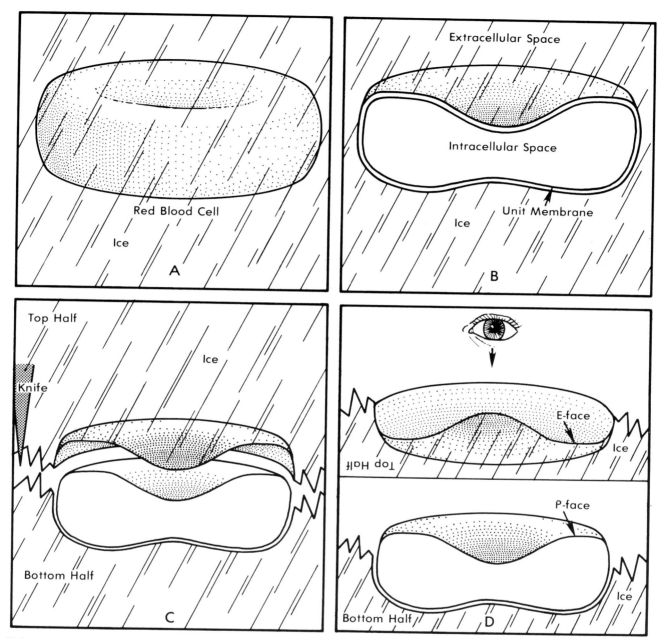

FIGURE 14.2 Fracture of a frozen red blood cell. (A) A cell is seen frozen in ice. (B) For ease of explanation it is depicted split in half from top to bottom allowing its lipid bilayer to be visualized during the subsequent fracturing process. The bilayer separates the intracellular space, containing frozen cytoplasm, and the extracellular space containing ice. Upon fracturing (C) with a knife the lipid bilayer, in the upper portion of the cell, is split down the middle. When both halves of the fractured surface are viewed from above after the top half is flipped over (D), the bottom fragment is seen to contain the face associated with the cytoplasm (protoplasm), or *P-face*, whereas the top fragment (now turned over) contains the face associated with the extracellular space, or the *E-face*.

P-face or *protoplasmic face*. (Figure 14.1).[1] Another way of thinking of this is to imagine poking a pin through a particular membrane face. Because the micrograph shows three dimesionality, one's eye is the judge of what appears closer to the viewer (convex) and is the structure first encountered by the pin. If the point of the pin exits the cell, then the face being examined is an E-face. On the other hand, a pin poked through the P-face or protoplasmic face would enter the cell interior or protoplasm.

---

[1]In the early freeze fracture literature, the P-face was designated the A-face and the E-face the B-face. The present designation allows a better association of the face with its nearest neighbor.

Assume the fracture traversed a red cell plasma membrane in the manner shown in Figure 14.3. Follow the fracture faces in each surface view to see how complementary regions of the same membrane are viewed and named.

To this point, only examples of fractured plasma membranes have been provided. Of course, other organelles are bounded with unit membranes, which, like the plasma membrane, may be given *P* or *E* designations. In membrane-bounded structures such as lysosomes, secretory granules, Golgi, and endoplasmic reticulum, the interior of the structure is considered extracellular space and the exterior is considered protoplasm, thus allowing the associated faces to be named E-face and P-face, respectively. The nucleus and mitochondrion are

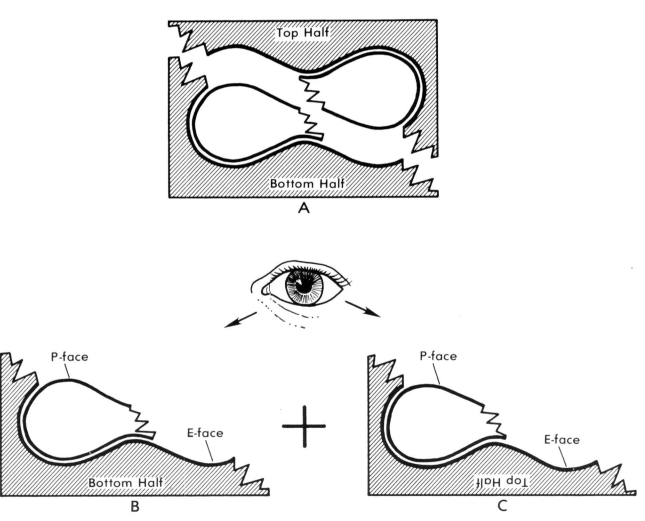

FIGURE 14.3 Fracture across a red blood cell. (A) The fracture describes a path through both the top and bottom surfaces of the cell. The bottom (B) and top (C) fracture fragments are displayed separately to show how membrane faces are designated. In (C), the fracture fragment is turned over so that the fracture surface can be viewed from above.

bounded by a double lipid bilayer. The internal aspect of the mitochondrium and nucleus is considered protoplasm and the space between each bounding membrane is considered extracellular space. Figure 14.4 shows how internal membrane faces are designated.

To provide an accurate interpretation of freeze fracture micrographs is a challenge. Even experienced investigators have mistakenly misnamed fracture faces in the scientific literature. Since the transmission electron microscope does not view the surface of fractured tissues, but rather examines a *replica* of the surface, some guidelines to interpretation of the final image are necessary. Interpretation of freeze fracture is aided by a knowledge of how replicas are produced from biological materials.

How to Produce a Replica

A replica is a thin *platinum-carbon cast* of the fractured tissue surface that is viewed in the transmission electron microscope. Areas of high metal density (those facing the platinum source) are dark

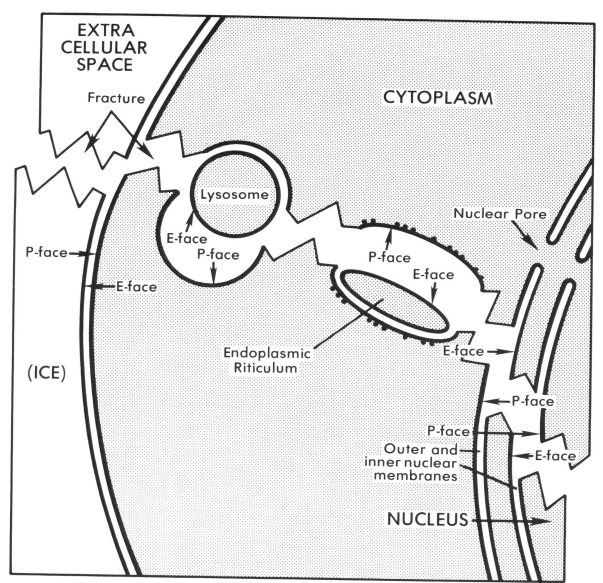

FIGURE 14.4 Fracture faces of intercellular components.

in appearance since the electrons are deflected by heavy metals. Areas of low metal density allow electrons to pass through the replica. The topological features of the biological specimen are conveyed by the replica. An image is formed on the fluorescent screen after penetration of the replica by electrons.

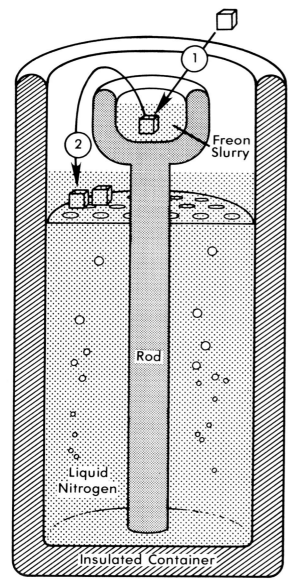

FIGURE 14.5 Freezing of a biological specimen for freeze fracture. The specimen is first (1) placed in a slurry of liquid Freon that has been cooled by a rod extending down into liquid nitrogen. Next (2), the specimen is rapidly transferred to liquid nitrogen where it remains frozen on a perforated tray until utilized in the freeze fracture apparatus.

The tissue (epithelia, cell suspensions, cell fractions, cell cultures, or solid tissue) is prepared for fracturing after mild fixation in 0.5% to 3% glutaraldehyde. Small pieces of tissue are infiltrated with 20% to 30% glycerol, a cryoprotectant that inhibits ice crystal formation by hydrogen bonding to water, interfering with its ability to crystallize. Due to the low thermal conductivity of most tissues, it is necessary to freeze tissues extremely rapidly in order to prevent ice crystal formation (Figure 14.5). Rapid freezing takes place in Freon that has been chilled to a slushy consistency at $-150°$ to $-155°C$ using liquid nitrogen. Although it is not as cold, freon is more effective in rapidly freezing the tissue than liquid nitrogen ($-196°C$) since its contact (wetting property) with the specimen optimizes heat transfer. The specimen is rapidly transferred into a liquid nitrogen storage vessel until needed for fracturing.

Replicas are produced in a freeze fracture apparatus of the kind diagrammed in Figure 14.6. A modern commercial instrument with the same basic features is shown in Figure 14.7. All of the activity related to the specimen takes place in a chamber at the top of the apparatus. The chamber contains a stage on which the specimen is mounted, a moveable knife (razor blade), and electrodes for vaporizing platinum/carbon and carbon onto the specimen. Vacuum lines feed into the chamber from below to allow the carbon and platinum evaporation processes to take place under high vacuum (10^{-6} to 10^{-7} Torr or 10^{-4} or 10^{-5} Pa) and also to prevent excessive water condensation on the specimen. Liquid nitrogen lines are used to cool both the stage and the knife. In the transfer of the specimen, fracturing, and replication process, tissues must remain solidly frozen. If they were to thaw in the transfer process, they would refreeze on the specimen stage, but this time more slowly with unwanted ice crystal formation.

Replicas are produced from frozen tissue in the following manner (Figure 14.8):

1. The frozen specimen is rapidly transferred to a precooled ($-150°C$) stage.
2. The chamber is closed and pumped to high vacuum.
3. The stage is warmed to $-100°C$ and the tissue fractured with a razor blade attached to a rotating arm.
4. Platinum/carbon are evaporated at an angle of about $45°$ relative to the surface of the specimen.

5. Carbon is evaporated onto the specimen from directly overhead to add strength to the replica.
6. The chamber is brought to room temperature and pressure.
7. The thawed specimen is gently placed in 5% sodium hypochlorite[2] to digest all organic

material from the replica. The tissue will bubble as it is digested from the replica. In some tissues, the replica may float free of the specimen.
8. The replica is moved to an additional chlorine bleach change and then through 2 or 3 distilled water washes.
9. The replica is picked up on a mesh grid and allowed to dry.
10. The replica is examined using a transmission electron microscope.

_____
[2]Grocery store chlorine bleach will suffice, although in some instances strong acids may be necessary to digest away organic matter.

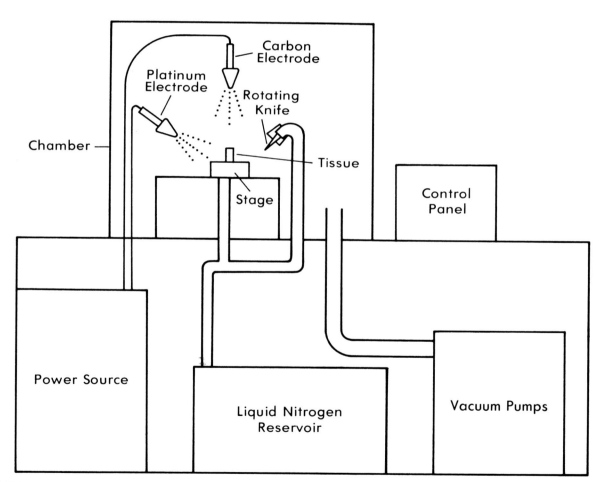

FIGURE 14.6 Diagrammatic representation of the essential features of a freeze fracture apparatus. Fracturing of tissue takes place in a *chamber*. The chamber is maintained at high vacuum by vacuum pumps. Both the stage and the knife are cooled by liquid nitrogen. Platinum and carbon electrodes have a power source below. All of these features are operated from the control panel.

FIGURE 14.7   Modern commercial freeze fracture apparatus. Fracturing occurs in the boxlike chamber to the left. The console at the right controls the temperature of the specimen and knife as well as the firing of the electrons. The console at the bottom maintains vacuum conditions.
(Photograph provided courtesy of the Balzers AG.)

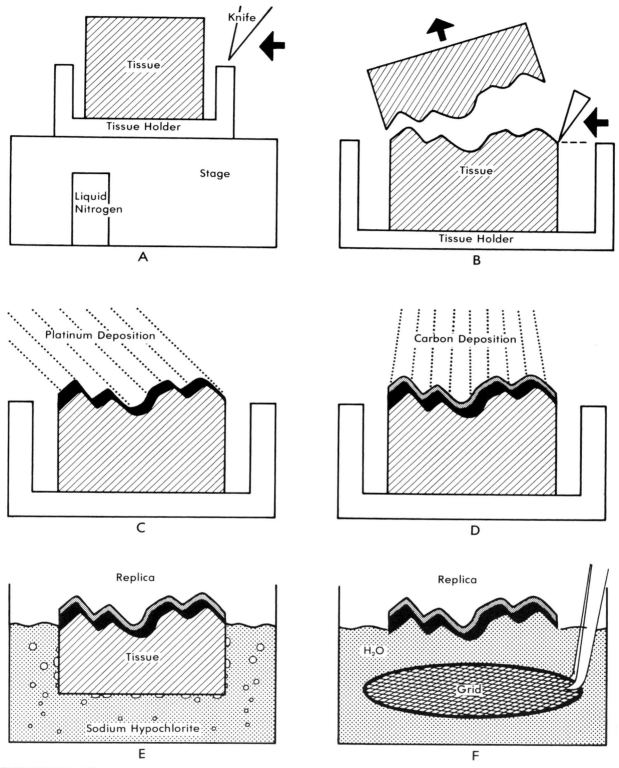

FIGURE 14.8 Diagram showing the steps in the fracturing and replication process. The frozen, immobilized tissue (A) is fractured (B) by a moving knife. Platinum (C) is deposited at a 45° angle and carbon is deposited (D) directly from above. The replica has a nonuniform layer of platinum and a relatively uniform carbon support layer. (E) The tissue is digested from the replica in sodium hypochlorite. (F) After water washes, the replica is picked up by a grid from below. Only steps A through D are conducted in a freeze fracture apparatus.

Interpretation of Freeze Fracture Replicas

A replica is a delicate, tissue-free, platinum carbon foil that displays the surface contours of the fractured tissue. Figure 14.9 is an electron micrograph that shows that platinum has been deposited differentially on the tissue surfaces that are perpendicular to, or closely aligned with, the direction of platinum deposition. A thicker deposition of platinum retards the penetration of the electron beam more so than does an area with sparse platinum. Some light areas may have no platinum and other darker areas may have platinum up to 5 nm in thickness. The variable and graded nature of platinum deposition imparts a three-dimensional appearance to freeze fracture micrographs, which mimics the actual contour of the fractured surface. At first glance, images appear like the lunar surface and display what appear to be craters, mountains, and valleys. The platinum particles that have been evaporated onto the specimen are about 2 nm in diameter. Resolution is limited by their large size.

In order to relate the replica image to actual tissue structure, certain guidelines have been developed, which, in most instances, will allow the correct membrane and membrane face to be identified. With experience, the investigator gains added confidence and the reasons for specific rules become clear. The guidelines for interpretation of freeze fracture replicas follow.

Finding a Membrane Face

Membrane faces are smoother areas than where ice alone or cytoplasm is fractured. Membrane faces contain *intramembranous particles (IMPs)*. IMPs are thought to be proteins embedded in the lipid bilayer, the three-dimensional structure of which is revealed upon fracture. These generally are the smallest distinct units (7 to 15 nm across or most frequently 8.5 nm) visualized in the bilayer (Figure 14.10). At low magnification, intramembranous particles may be difficult to resolve.

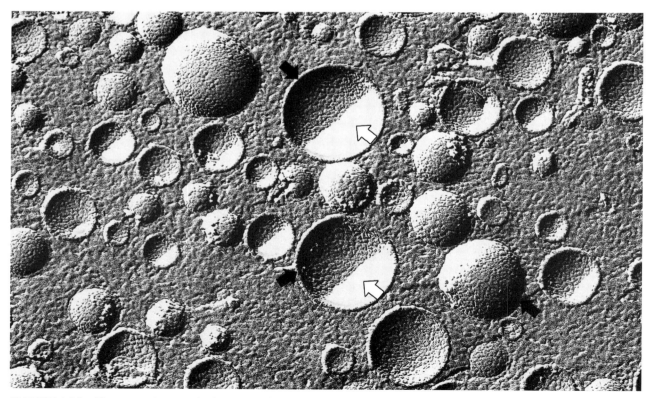

FIGURE 14.9 Electron micrograph showing differential deposition of platinum in membrane vesicles. The light areas have little platinum (open arrow) whereas the dark areas have a heavy deposition of platinum (dark arrow).

Orienting the Micrograph so that the Shading of IMPs Is from Below

First, a determination of the direction of platinum shadowing is made by observing which side of the IMP appears darkest. The micrograph is then oriented with the dark portion of the IMP facing downward. When the micrograph is oriented in this manner, what appears concave to the eye was indeed a concave feature of the replicated frozen tissue. The same is true of convex features. Figure 14.11 is an upside down view of Figure 14.9. Note how the concavities and convexities of the two micrographs are reversed when the micrographs are reversed in position.

Determining Which Membrane Is Being Viewed

The interpreter's basic knowledge of cells will help to determine what is being viewed. Cells, of course, are generally rounded or convex overall (Figure 14.12). Protoplasm (cytoplasm) intervenes between the cell membrane and the nucleus and contains numerous organelles and inclusions. This is also the topography represented by the replica. The plasma membrane has the features characteristic of the cell, such as microvilli, surface ruffles, junctions, etc., and can be so recognized. The nuclear membrane is identified by nuclear pores. Simple logic dictates that if the interior of the nucleus is being viewed, then to find the cell exterior it is first necessary to visualize sequentially the nuclear membrane, the cytoplasm, and the plasma membrane. Imagine that you are a minute creature able to walk from cell to cell, and imagine what logical steps must be taken to do so.

Determining Whether the Membrane Face Being Viewed Is a P-Face or an E-Face

When the surfaces of two cells are adjacent to each other and the plasma membrane faces of both cells are visualized, the E-face is always topographically on top of the P-face (Figure 14.13). This is logical since, to exit one cell, it is necessary to first go

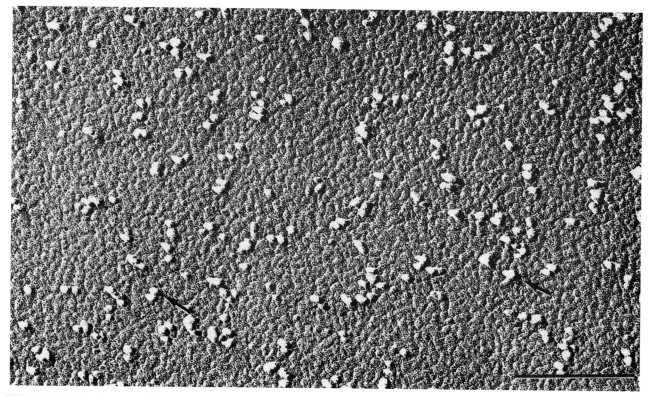

FIGURE 14.10 Intramembranous particles at high magnification (arrows). Bar = 0.25 μm.

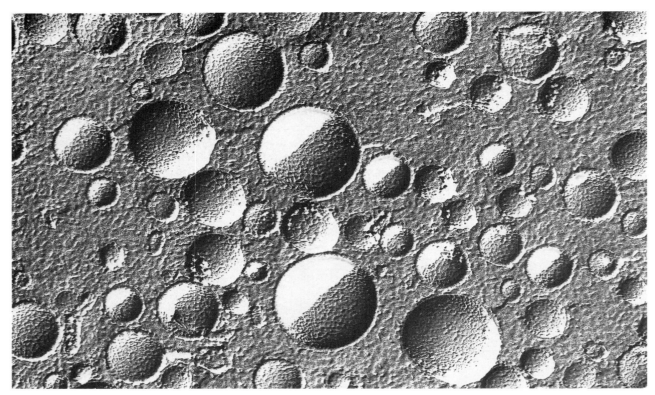

FIGURE 14.11 Upside down view of the freeze fracture micrograph shown in Figure 14.9. The topography of one micrograph appears exactly the opposite of the other micrograph, however only Figure 14.9 is oriented correctly.

through an E-face to enter the extracellular space. Then go through the P-face to enter the cytoplasm of the adjacent cell. This rule holds for the double lipid bilayer forming the nuclear and mitochondrial membranes. For example, if the fracture enters the nucleus from the cytoplasm, the E-face of the outer nuclear membrane will lie on top of the P-face of the inner nuclear membrane.

Generally, the P-face contains more IMPs than does the E-face (Figure 14.14).

P-face and E-face may be ascertained by determining the next deepest structure. If the next deepest structure is the extracellular space or its equivalent, then an E-face is being examined. If, on the other hand, a poke through a membrane face leads to protoplasm, a P-face is the face in question.

Figures 14.15 and 14.16 are provided as examples of how to interpret freeze fracture images. The legends to each of these figures allow the reader to follow a pattern of reasoning that might be used to interpret each micrograph. Further

examples of freeze fractured tissues are provided in Chapter 20.

Complementary Replicas

A technically difficult, but fascinating, procedure allows one to retrieve two replicas from the same tissue that will show both faces of the same membrane. Instead of fracturing the tissue by scraping the surface, as described above, the frozen tissue may be fractured by breaking it in half. The two surfaces that are replicated form *complementary replicas.* (Figures 14.17 and 14.18). The technique for producing complementary replicas is added proof that the lipid bilayer is split down the middle during tissue fracturing. It has also been shown using complementary replicas that, in some cases, there are complementary pits to match IMP particles on the opposing face. Finding areas where replica faces match is a supreme test of an individual's patience.

FIGURE 14.12 General view of a cell with freeze fracture. The cell has a rounded nucleus that is central in the figure and which is bounded by inner (INM) and outer nuclear membranes (ONM). The plasma membrane (PM) and nuclear envelope sandwich the cytoplasm. Profiles of various organelles (O) are seen sparsely scattered within the cytoplasm (C). Other cells adjoin this cell. Bar = 1 μm.

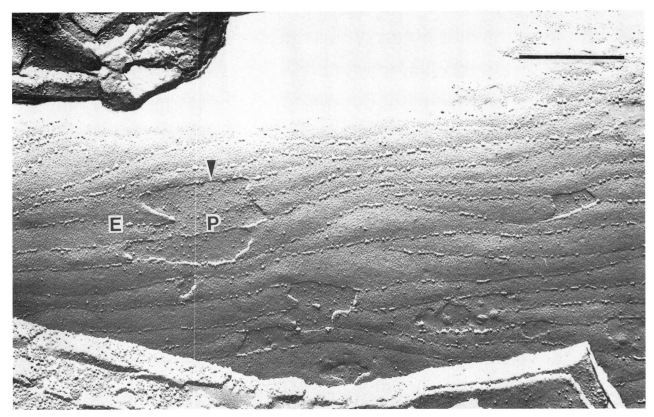

FIGURE 14.13 Membrane faces of adjacent cells. Where the fracture plane leaves one plasma membrane and jumps down to another, there is a distinct "step down" at this site (*arrowhead*). The membrane on top is the E-face of the cell that is more superficial and the membrane on the bottom is the P-face of the deeper cell. The rows of particles seen are sites of tight junctions. Bar = 1 μm.

Freeze Etching

This variation of freeze fracture is employed to examine actual membrane surfaces and not membrane faces as described previously. The term freeze etching is often used erroneously to describe the entire freeze fracture process, whereas it is only a modification of the freeze fracture process. To etch a specimen, the tissue is first cryoprotected, frozen, and fractured in the way already described. The knife is placed directly over the specimen, and the slightly cooler knife facilitates sublimation of the ice from the surface of the specimen onto the knife. Etching of ice from the surface occurs at 2 to 3 nm/sec at a temperature of $-100°$C. Removal of ice by sublimation allows the *true surface* of the membrane to be exposed (Figures 14.19 and 14.20). Herein lies the value of the technique, since the true

surface inevitably provides different information about membrane organization than either one of its faces.

Quick Freeze, Deep Etch, Rotary Shadow Techniques

This fascinating technique, exploited relatively recently, allows an examination of certain insoluble surface and intracellular structures. The three-dimensional detail of membranous structures and cytoskeletal elements is quite striking (Figure 14.21).

Tissue is frozen fresh or after mild glutaraldehyde fixation by a technique known as *quick freezing*. To do this, the tissue is slammed against a polished copper block, which itself is cooled with liquid helium. Amazingly, the tissue does not crush in this process, although the surface 10 to 15 μm is frozen extremely rapidly without ice crystal damage. Subsequently, the frozen tissue is fractured in a freeze fracture apparatus and *etched* for a period not to exceed 10 minutes. Platinum and carbon are deposited on the specimen as the specimen holder rotates, a process known as *rotary shadowing*. The result is a three-dimensional view of internal cell structures such as that shown in Figure 14.21.

Freeze Fracture Cytochemistry

More recently, techniques have been developed to localize substances using variations in the freeze fracture technique coupled with cytochemical techniques. These techniques are relatively simple, but their full potential has not been exploited. They include *label-fracture* (Pinto da Silva and Kan, 1984) and *fracture-label* (Pinto da Silva, et al., 1981). It is not within the scope of this text to describe these methods, and the reader is referred to the cited methods and reviews by Boonstra et al., 1991 and Severs, 1991. In addition, specific localization of lipid moieties has been achieved using freeze fracture techniques (Menco, 1986).

Problems and Artifacts

It is possible to produce many replicas and take pictures of them, all in one day—that is, if everything goes as planned. This rarely happens. Of the many problems that may be encountered, machine problems (especially the carbon and platinum guns), tissue problems, freezing problems, and replica problems contribute to making the technique technically demanding.

Poor freezing and accidental warming of tissue may lead to ice crystal formation in tissues, which causes deformation of structure and obscures tissue components (Figure 14.22). If tissues are not infiltrated adequately with glycerol, ice crystals will develop in all but the hardiest of cells (bacteria, yeasts, and spores). During the fracturing process, tissues may be torn from their tissue holder. Fortunately, most modern freeze fracture equipment allows several specimens to be fractured at one time. The platinum and carbon guns frequently

FIGURE 14.14 Relative abundance of intramembranous particles on the P- and E-faces.

FIGURE 14.15 Micrograph for self-interpretation. The key to figure interpretation is placed at the end of the chapter.

malfunction, leading to either too little or excessive deposition of metal on the specimen. More recently produced guns are more reliable.

By far, the most annoying problem is breaking and shattering of the replica during the final steps necessary to free the replica from the tissue and clean its surface. The replica is fragile and brittle and may fragment as the underlying tissue is dissolved. Transfer of the replica from solution to solution during the cleaning process frequently causes it

to fragment, a problem that is especially dismaying given the effort already expended to produce the replica.

Micrographs may show evidence of knife marks produced as the knife shaved, rather than fractured, the specimen (Figure 14.15). Extremely dense areas on the micrograph may indicate that tissue has been incompletely removed from the replica during the hypochlorite washes (Figure 14.23).

FIGURE 14.16 Micrograph for self-interpretation. The key to figure interpretation is placed at the end of the chapter.

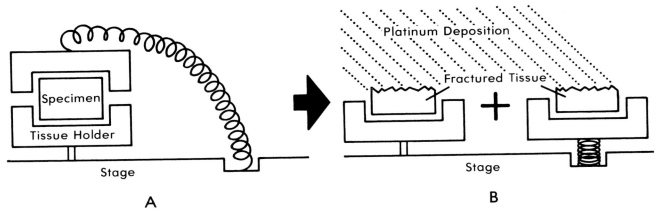

FIGURE 14.17 Production of complementary replicas. The tissue is held on the stage by a device, similar to that shown in (A). Fracturing occurs as the tissue is broken apart by placing tension on the specimen. Subsequently (B), the two matching surfaces are shaded with platinum and carbon to form complementary replicas.

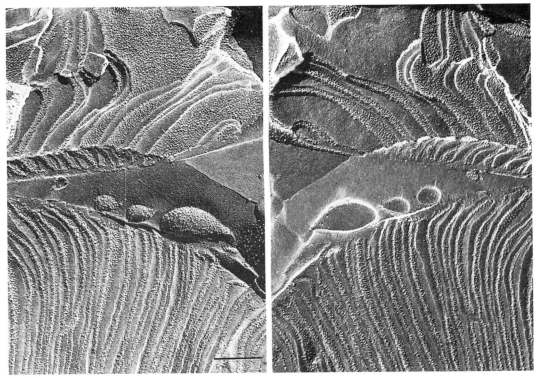

FIGURE 14.18 Complementary replicas. The two micrographs shown are membrane faces from guinea pig retina outer segments. As the replica was split, both portions of the split tissue were saved and replicated. Match each membrane component on the two replicas. Note differences in the relative abundance of intramembranous particles on complementary faces. Bar = 0.25 μm.
(Micrograph courtesy of R. L. Steere, Advanced Biotechnologies, Inc.)

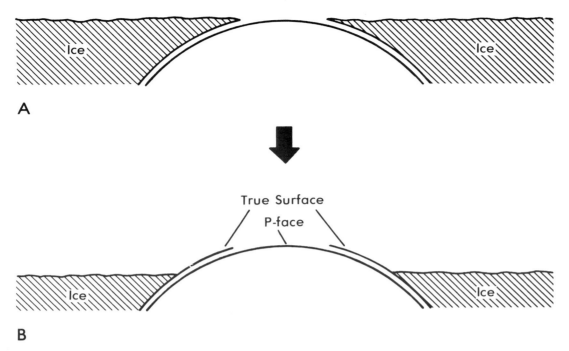

FIGURE 14.19   The process of etching. A typical fractured specimen (A) reveals the P-face of a membrane. After etching, (B) the ice level has receded revealing the *true surface* of the membrane.

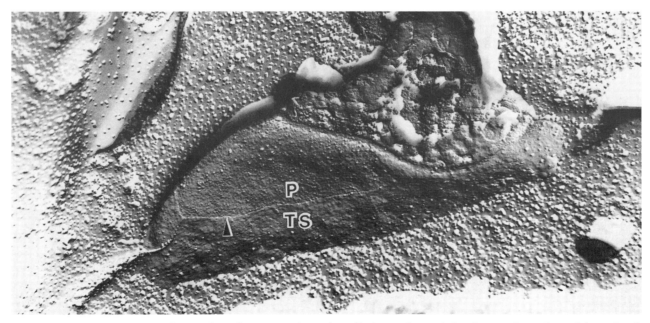

FIGURE 14.20   A micrograph revealing the true surface of a cell after etching. A thin line or "step down" (*arrowhead*) separates the P-face of a cell from the true surface.

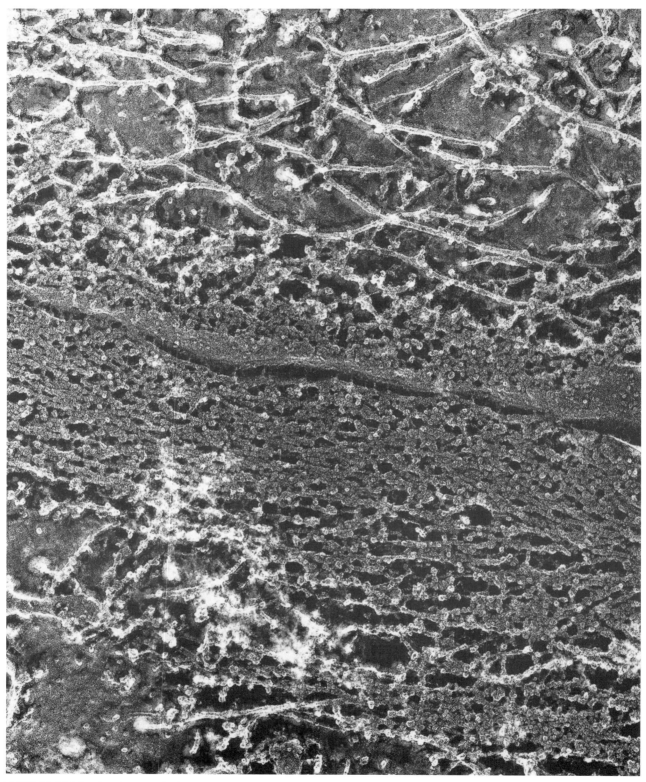

FIGURE 14.21   A micrograph showing an example of quick freezing, deep etching, and rotary shadowing. Two adjacent cells depict filaments. At the top of the micrograph intermediate filaments (10 nm) are seen whereas in the center one-third of the smaller (6 nm) actin filaments are visualized.
(Micrograph courtesy of R. Kelly.)

FIGURE 14.22 An example of numerous ice crystals that have formed on improperly frozen tissue.

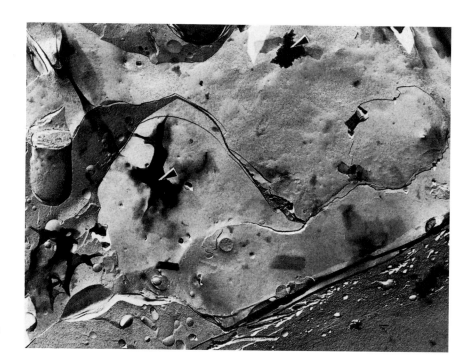

FIGURE 14.23 Undigested tissue and contamination appear dense on this freeze fracture replica.

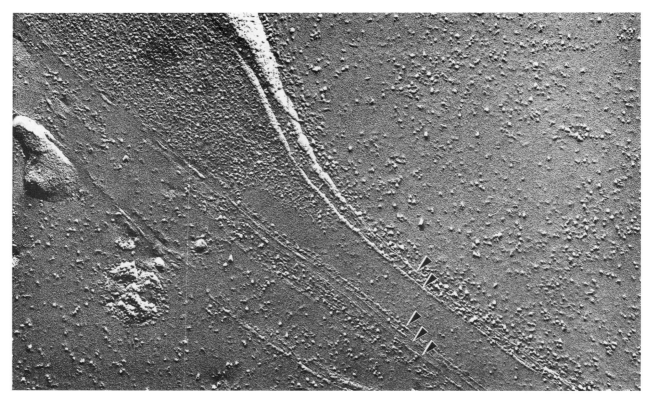

FIGURE 14.24   A freeze fracture micrograph showing extensive cross-fracturing of cell membranes (*arrowheads*).

FIGURE 14.25   Background artifact on a membrane face. The background condensation gives a cobblestone appearance (*arrow*) to the membrane face. Few membranous particles are seen.

Membrane faces should occupy about 20 to 60% of the image in replicas of solid tissues. If there are large numbers of *cross-fractures* of membranous elements, it is an indication that too high a concentration of glutaraldehyde was utilized or the fixation process was too long (Figure 14.24).

Replicated membrane faces may show a fine cobblestone effect, giving the appearance of closely packed, but extremely small, intramembranous particles (Figure 14.25). The exact cause of this defect is not known, but is thought to be due to vapor condensation on the specimen. It is important to begin the replication process as soon as possible after fracturing to prevent this artifact.

Analytical Electron Microscopy

IN THE EARLY YEARS following refinement of the technology, electron microscopes were used not only to reveal morphological information but also to map the distribution of various macromolecular entities in the tissues under investigation. Localization techniques (autoradiography, cytochemistry, immunocytochemistry) were used to reveal the architecture of the eukaryotic cell and to clarify the mechanisms of such physiological processes as protein synthesis, arrangement of DNA in the chromosome, recognition of antigenic sites by antibodies, and the structure of macromolecules such as enzymes. Electron microscopes are capable of providing other data as well.

When a high energy beam of electrons interacts with a specimen, the atoms of the specimen may cause the electrons to decelerate. The kinetic energy of the electron is then converted into other forms of energy and the lower energy electron will follow a trajectory that is different from what it would have followed had it not interacted with the specimen. The nature and the spectrum of the energy liberated, as well as the images formed by the new trajectories of the electron, can be captured by various detectors

attached to the electron microscope. Such detectors may reveal the atomic composition of the area struck by the beam of electrons and, under some circumstances, the quantity of the elements present.

Electron microscopes that are used to identify or characterize the chemical nature of components found in biological tissues are called *analytical electron microscopes.* A few applications of analytical electron microscopy might include: the localization of ions and electrolytes in various parts of the cell; a study of the changes in the distribution of ions during various physiological processes (growth, secretion, cell division, death); the identification of an unknown crystalline inclusion in a cell (asbestos fiber in a lung cell); investigating the pathways followed in the incorporation of toxic ions into the cell and in the detoxification process; and the confirmation of an enzymatic reaction product as a lead or osmium precipitate.

Interaction of an Electron Beam with a Specimen

Several different emanations or signals may be generated as a result of the electron beam striking a specimen. As illustrated in Figure 15.1, some electrons may pass through a suitably thin specimen with the loss of some energy. These are termed *inelastically scattered, transmitted electrons* and may be used in the conventional transmission electron microscope to reveal morphological information about a thin specimen. They also may be separated into various energy levels in an electron energy loss spectrometer for determination of elemental composition.

Other electrons may lose little or no energy upon interaction with the specimen. Such *elastically scattered* electrons may pass through the specimen as transmitted electrons or may be deflected back in the direction of the beam as *backscattered electrons.* Backscattered electrons may be detected using special detectors in scanning and transmission electron microscopes. As described in Chapter 7, such detectors may be used to discriminate areas of different atomic numbered elements: higher atomic numbered elements give off more backscattered electrons and appear brighter than lower numbered elements.

Secondary electrons, with energies typically under 50 eV, are a type of inelastically scattered electron that may be collected and imaged using secondary electron detectors in scanning or transmission electron microscopes. They are used pri-

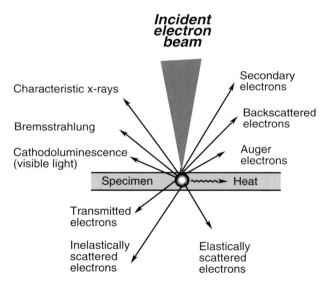

FIGURE 15.1 When an electron beam strikes a specimen, some of the kinetic energy is converted into various types of X rays, visible light, and heat. Some electrons may be transmitted through the specimen with the loss of some energy (inelastically scattered) or no loss of energy (elastically scattered). Other electrons may be given off from the top of the specimen as high energy (backscattered) electrons or lower energy (auger, secondary) electrons.

(Courtesy of Kevex Instruments, Inc.)

marily to reveal topographical features of a specimen in a scanning electron microscope, as described in Chapter 7.

Auger electrons are special types of low energy electrons that carry information about the chemical nature (atomic composition) of the specimen. Auger spectroscopy involves the collection and analysis of the spectrum of energy levels of these reflected electrons to give elemental composition from the upper atomic layers of the specimen. It is a powerful tool in the materials sciences for studying the distribution of the lighter numbered atomic elements on the surface of specimens. Since it has limited application in the biological sciences, and since a specialized instrument (scanning auger electron spectrometer) is needed in this analysis, it will not be discussed further in this textbook.

Cathodoluminescence results when the energy of the impinging electrons is converted into visible light. Certain types of compounds are capable of cathodoluminescence and may be detected using special detectors, as described in Chapter 7. Since relatively few naturally occurring biological macromolecules are cathodoluminescent, and since the resolution currently obtained is similar to the light microscope, the detection of such signals is only occasionally done.

Two types of signals that are commonly used in analytical studies are *characteristic X rays* and certain of the inelastically scattered transmitted electrons. These signals are detected using either energy or wavelength dispersive *X-ray analyzers* or *electron energy loss spectrometers,* respectively. Transmitted electrons may also be used to give compositional information when a transmission electron microscope is used in the *diffraction* mode. Since these are the three most widely used analytical techniques, they will be discussed separately in this chapter.

Microscopes Used for Detecting Analytical Signals

It is possible to attach a number of different detectors to electron microscopes. For instance, the scanning electron microscope may be fitted with secondary, backscattered, and transmitted electron detectors as well as detectors for X rays and cathodoluminescence. Besides these signal detectors, the transmission electron microscope may be equipped with electron energy loss spectrometers of various designs. Some accessories require little

modification to the standard microscope, while others may require additional lenses and beam control electronics for optimal performance.

Obviously, the most versatile analytical instrument would be one that combined the features of a scanning and a transmission electron microscope. Such an instrument became available in the mid-1970s. Termed a *scanning transmission electron microscope,* or STEM, the instrument is able to generate and precisely position a very fine probe of high energy electrons and to systematically scan the fine probe (as in the SEM) over a thin specimen while still being able to obtain diffraction patterns (as in the TEM). The miniaturization of the various detectors, as well as the small probe sizes generated, made possible the detection of the various signals from quite small areas of the specimen and thereby increased the resolution of the analytical techniques. Figure 15.2 is a photograph of an analytical electron microscope. A schematic diagram of the column of an analytical STEM is shown in Figure 15.3.

A number of design features of the STEM will be readily recognized. An illuminating system consisting of an electron gun and three condenser lenses is used to initiate and demagnify the fine electron probe. Double deflection coils are employed to generate the scanning raster (or to position the spot over the proper location on the specimen). A number of detectors (secondary, backscattered, X-ray) are positioned in close proximity to the specimen to enhance the sensitivity of detection. Several projector lenses (intermediate, P1 and P2) follow the objective lens, and an electron energy loss spectrometer may be positioned underneath the viewing screen and camera.

Because of the expense involved, it is unusual for such microscopes to be equipped with all of the analytical detectors. Instead, the most common configuration is a STEM equipped with an X-ray analyzer. Of course, all analytical electron microscopes using a TEM column have a variety of diffraction capabilities since this is a lens function that does not require specialized detectors. Table 15.1 summarizes the instrumentation necessary and results obtainable using various analytical techniques. Most STEM instruments are TEMs that have had an appropriate lens installed so that they may function in either the TEM, SEM, or STEM modes. There are also dedicated instruments, such as the Vacuum Generators HB-series STEM units, that are capable of extremely high resolutions in the STEM mode. However, they are less versatile for biologists since they are not readily operated in any of the TEM modes and are very expensive.

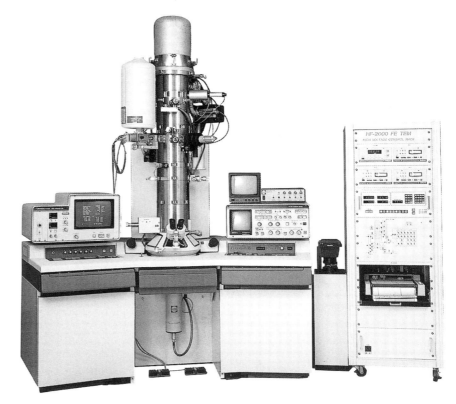

FIGURE 15.2 An analytical, cold field emission, 200 kV scanning transmission electron microscope equipped with an energy dispersive X-ray detector, secondary electron detector, and slow scan digital camera (under viewing port). The large console on the right houses electronics associated with the high voltage control and vacuum system. *(Courtesy of Hitachi Scientific Instruments.)*

TABLE 15.1   Summary of Main Features of Various Analytical Procedures

Procedure	Energy Dispersive X-ray Spectroscopy (EDS)	Wavelength Dispersive X-ray Spectroscopy (WDS)	Electron Energy Loss Spectroscopy (EELS)	Electron Diffraction
Microscope System Needed	TEM or STEM or SEM	TEM or STEM or SEM	TEM or STEM	TEM or STEM
What Identified	Elements with atomic numbers greater than 6 (11, normally)	Elements with atomic numbers greater than 3	Elements with atomic numbers greater than 3	Chemical identity of crystal
Quantitative	Yes	Yes	Yes	No
Smallest Area Analyzed	10 nm	10–100 nm	0.3–0.4 nm	1–10 nm
Detection Limit	10^{-13} gm	10^{-10} gm	10^{-19} gm	Not applicable
Specimen Type	Thin, ultrathin, bulk*	Thin, ultrathin, bulk	Ultrathin	Ultrathin

*Thin = 0.1–2 μm, ultrathin = 50–100 nm, bulk = intact tissue (chunk).

requires larger and more energetic electron probes, (2) able to detect only one element at a time, (3) less efficient than the energy dispersive detectors, (4) somewhat more expensive, and (5) unable to achieve high spatial resolution of elements due to large probe sizes required. On the positive side, the

WDS detectors: (1) offer ten times better capability than EDS to discriminate closely spaced X-ray energy peaks, (2) are better suited to light element analysis, (3) are better suited for trace element detection, and (4) do not require liquid nitrogen cooling of the detector.

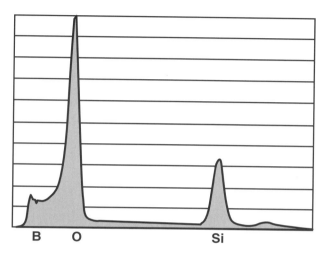

FIGURE 15.10A Characteristic X-ray analysis of glass containing 17% B_2O_3. Note the presence of boron, oxygen, and silicon peaks on the spectrum.

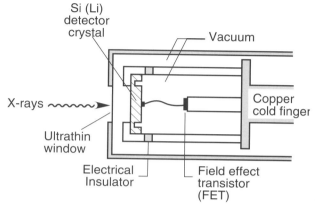

FIGURE 15.11A An energy dispersive X-ray detector. Diagram showing the basic components making up the detector.

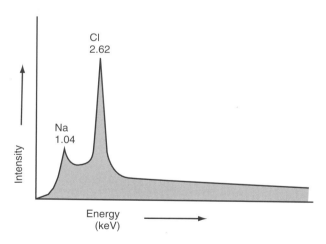

FIGURE 15.10B Absorption of the weaker Na X rays by the window of the energy dispersive detector makes quantitation difficult. Other factors that make quantitation difficult result from the fact that elements with high atomic numbers, Z, give off more X rays than lighter elements and due to the absorption, A, of the X rays by atoms of the sample and to a phenomenon termed secondary fluorescence, F, which results when an absorbed X ray gives rise to secondary X rays. For accurate quantitation, the ZAF correction factors must be entered into the quantitative equation using computers.

FIGURE 15.11B The detector is housed inside of the stainless steel rod and chilled from the large vessel containing liquid nitrogen. *(Figures 15.11A and 15.11B courtesy of Kevex Instruments, Inc.)*

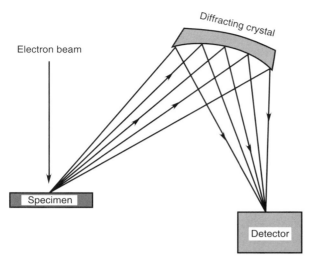

FIGURE 15.12A Principle of operation of wavelength dispersive X-ray (WDS) spectrometer. X rays generated by the specimen are "reflected" by the crystalline lattice into the detector. Very specific wavelengths of X ray are reflected by different lattice spacings so different crystals are needed in the spectrometer to cover various energies of X rays.

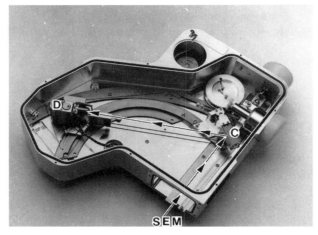

FIGURE 15.12B Cutaway view of WDS spectrometer showing location of crystal (C) and detector (D). Arrow shows direction X rays travel from specimen chamber to detector.
(Courtesy of Microspec Corporation.)

Information Obtainable Using X-Ray Analysis

The electronics of the energy dispersive (EDS) and wavelength dispersive (WDS) detectors are involved in the acquisition, sorting, and display of data as a spectrum of energies. Computers greatly expedite this process and assist in the interpretation of the spectrum. Figure 15.13 compares the output of a sample that was analyzed by EDS and WDS. The sharp separation of closely placed X-ray energies is evident in the largest peak displayed in the EDS spectrum. Here the two elements barium and titanium have not been resolved since their characteristic Xrays are so close in energy. On the other hand, WDS clearly resolved the two elements. In spite of this major advantage, WDS is seldom used in biological studies due to the excessive beam currents needed, which result in damage to most biological specimens.

Data may be presented in several different ways from both types of detectors. In the *spot analysis* mode, a fine probe of electrons is focused on a single area of interest and a spectrum is generated, as shown in Figure 15.14. When this information is presented for publication, one first takes a micrograph of the area that was analyzed and places a pointer on the actual spot that was probed. It must be realized, however, that X rays come from larger areas than indicated by

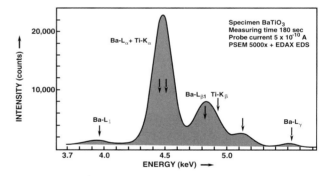

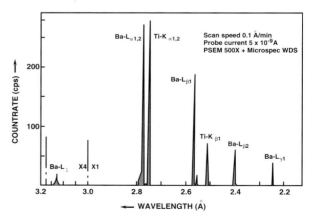

FIGURE 15.13 Spectral output from same sample analyzed by energy dispersive X-ray procedure (*top*) versus wavelength dispersive procedure (*bottom*). Note that WDS is able to resolve the two closely placed peaks that were summed together by EDS.
(Courtesy of D. B Williams.)

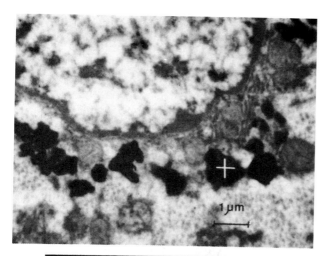

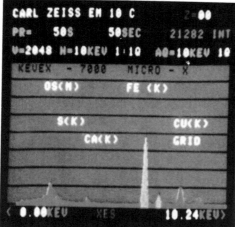

FIGURE 15.14 *(Top panel)* Energy dispersive X-ray analysis was conducted by focusing the probe on the spot indicated in the electron micrograph. This liver cell demonstrates hemochromatosis. *(Bottom panel)* The dense bodies are rich in iron as indicated in the X-ray spectrum taken from the marked spot.
(Courtesy of Dr. G. Schwalbach and Carl Zeiss, Inc.)

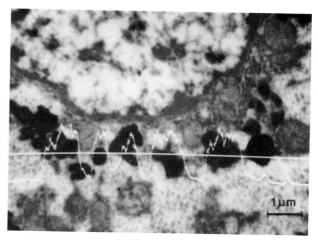

FIGURE 15.15 X-ray analysis was accomplished on the same liver cell shown in Figure 15.14 by scanning a straight line across the specimen as shown. As the specimen was analyzed for iron along the line scanned, the quantity of iron was indicated by the level of deflection of the peaks.
(Courtesy of Dr. G. Schwalbach and Carl Zeiss, Inc.)

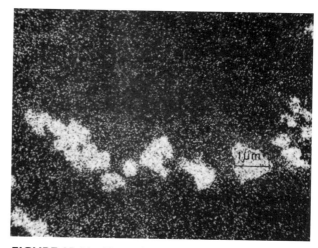

FIGURE 15.16 X-ray dot map showing distribution of iron in same liver cell shown in Figure 15.14.
(Courtesy of Dr. G. Schwalbach and Carl Zeiss, Inc.)

the spot size (see Figure 15.17) and this must always be taken into account when localizations are conducted. The spectrum may then be displayed in a separate photograph, or it may be superimposed over the electron micrograph of the specimen.

In the *line scan analysis* mode, the electron probe is moved in a straight line across the specimen (pausing for 100 seconds or so at each point to generate enough X rays), and the amount of a specified element is superimposed as a line graph over the micrograph (Figure 15.15).

In a *dot map*, the beam is moved across a large area of the specimen, pausing for a fixed amount of time at each point to generate X rays. This may take many hours, depending on the area scanned, so

computer control of the beam is very important to facilitate this process. Figure 15.16 is a dot map showing the distribution of iron in the same liver cell shown in Figure 15.14. Whenever a particular element (iron, in this case) is found in the specimen, this is indicated by a bright spot. Such data becomes quite informative when the dots are superimposed over an actual electron micrograph. With modern energy dispersive X-ray analysis systems, it is possible to simultaneously map many different elements by assigning various colors to the elements (e.g., sodium may be displayed as red areas, while phosphorus may be shown in green, etc.).

Specimen Preparation for X-Ray Microanalysis

Two of the goals of specimen preparation for X-ray analysis are to retain the elements of interest in their normal locations in the cell and to preserve the ultrastructure to the extent that it will be recognizable. These two goals are conflicting, because the fixatives, dehydrants, and embedments used in conventional specimen preparation procedures displace or completely remove diffusible ions. Obviously, the use of unfixed tissue presents an equally difficult problem of interpretating where in the cell the element is actually being localized.

For optimal results, the specimen should be thin, smooth, electrically and thermally conductive, and with discrete inclusions or compartments in the cell containing high concentrations of ions. Ideally the compartments should be surrounded by areas of lower atomic numbered elements or water so that interfering background X rays would not be present. These conditions are far removed from the actual situation in cytological material, so that X-ray microanalysis in biological systems has yet to fulfill its promise. Nonetheless, some useful information may be obtained under certain circumstances and with appropriate preparatory techniques (Morgan, 1985). Several categories of specimens may be analyzed.

Bulk Samples

It is possible to examine bulk samples (thick slices, intact tissues, pieces of fractured specimens) in the SEM or STEM instrument operating in the SEM mode. Such specimens are mounted onto aluminum or carbon stubs using carbon paint and coated with a conducting layer of thermally evaporated carbon. When using cold stages that maintain the specimen in the frozen state while under observation, it is possible to examine quick-frozen, uncoated specimens directly in the SEM or STEM without any drying procedures. Unfortunately, the irregular surfaces of the specimen restrict quantitative microanalysis, while the deep penetration of the probe into the specimen limits the spatial resolution to 4 to 8 μm (Figure 15.17). In addition, the probe may melt locally certain areas of the specimen, leading to redistribution of diffusible ions.

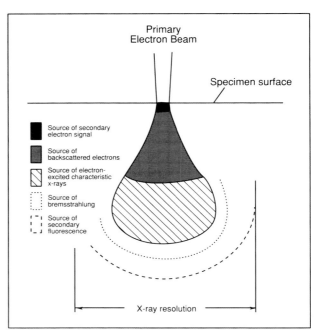

FIGURE 15.17 Diagram showing the depth and relative size of areas from which various signals may emanate. The enlargement of the signal source effectively diminishes resolution. Note that X-ray signals have the poorest resolution. Thin sections of specimens do not suffer from this problem since the probe does not spread out to this extent.
(Courtesy of Kevex Instruments, Inc.)

The Challenge: Quantitative Analysis of Bulk Specimens

Quantitative microanalysis in bulk samples is a laborious, exacting undertaking that requires the use of a standard specimen (containing known amounts of the elements to be analyzed) that closely approximates the properties of the sample to be analyzed. In addition, one must apply correction factors that take into account differences in mean atomic number between specimen and the standard, the absorption of some of the X rays by the detector, and a correction for extraneous X rays generated by other X rays in the specimen (*X-ray fluorescence* phenomenon) (Figure 15.18).

One method of correction for the various variables in bulk biological specimens is the *ZAF correction method* (Philibert and Tixier, 1968), as summarized in the following equation:

$$C_i = (ZAF)_i I_i / I_{(i)}$$

where C_i is the amount of the element i present; $I_{(i)}$ is the intensity of the X rays from a standard composed only of element i; I_i is the X-ray intensity measured in the specimen under analysis; and ZAF refers to the three corrections applied for *atomic number*, *self-absorption*, and *fluorescence*, respectively.

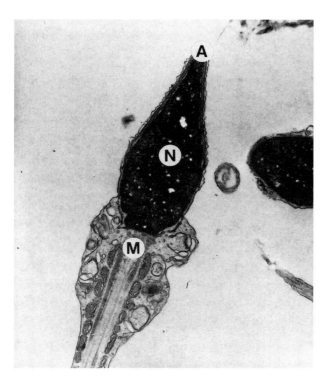

FIGURE 15.18 Section of human sperm cell showing areas where elemental X-ray analysis was conducted and quantitated in Figure 15.19.

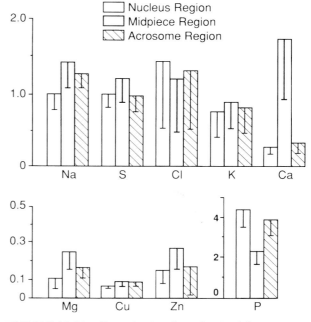

FIGURE 15.19 Quantitative data obtained from acrosome, midpiece, and nucleus of sperm cell similar to the one shown in Figure 15.18.
Adapted from Chandler, J.A. 1977. X-ray Microanalysis in the Electron Microscope. Practical Methods in Electron Microscopy, *Vol. 5, Pt. II, A. M. Glauert, ed., North-Holland/American Elsevier Publishing Co.)*

Single Cells, Isolated Organelles, Liquid Secretions, or Extracts

These types of specimen may be obtained from individual cells or cellular inclusions (particulate or fibrous) and deposited onto a carbon-coated grid and examined in a suitably equipped TEM or in a STEM operating in the transmitted mode. If the cells and constituents are thin enough, it is possible to obtain qualitative and semiquantitative data (Figure 15.19).

Sectioned Materials

The sections may range in thickness from 0.5 μm to less than 100 nm. They are particularly useful specimens for X-ray microanalysis since complicated correction equations (ZAF correction, for example) may not be necessary. Although thinner specimens are more desirable from a morphological and a quantitative standpoint, the levels of elements present may be too small to detect (see Table 15.1).

Some Precautions with Sectioned Materials for X-Ray Microanalysis

If it is necessary to fix tissues and embed them in plastic prior to sectioning, one should use glutaraldehyde alone (since osmium may mask some elements) and avoid buffers containing ions that are to be localized. The specimens should be embedded in an acrylic resin such as LR White because it has a lower background reading of certain elements (especially sulfur and chlorine) than do the epoxy resins.

Sections on the order of 100 nm may be cut and mounted on Formvar-coated grids. If copper is one of the elements to be analyzed, grids composed of noninterfering metals (carbon, nylon, aluminum, beryllium, titanium, gold, nickel) should be used. It is desirable to carbon-coat the sections to prevent the buildup of static charges and to stabilize the sections during the analytical procedure.

When it is necessary to localize diffusible ions, then the standard fixation, dehydration, and embedding process must be avoided and alternative techniques should be used. If the specimen is inherently very thin—such as some individual cells or cell fractions—it may be possible to deposit the material or to grow the cells directly on plastic and carbon-coated grids. Specimens may then be rapidly frozen and freeze-dried prior to examination, or one may use cold stages to maintain the specimen at 100°K. Obviously, ultrastructural details will be sacrificed, but it may be possible to recognize gross features such as mitochondria, membranous systems, vesicles, and so forth, in the dried or frozen specimens.

Cryoultramicrotomy (see Chapter 4) may be beneficial with thicker specimens when diffusible ions are to be localized. The best approach is to quickly freeze the unfixed specimen, cut thin sections, and mount them onto a coated grid. This must be done without using any liquids and while maintaining the frozen state of the sections.

If one is able to accomplish this task, then the frozen sections may be freeze-dried (usually while still in the cryochamber of the ultramicrotome) or a special transfer stage may be used to move the frozen-hydrated specimen onto a cold stage in the TEM (Figure 15.20). Freeze-drying of hydrated specimens must be undertaken with trepidation, since the removal of water will leave unsupported ions in such an unattached condition that displacement of the ions is likely. These procedures are quite difficult and require expensive and sophisticated accessories for the ultramicrotome as well as for the electron microscope.

Relatively few laboratories are able to accomplish microanalytical procedures on frozen hydrated specimens. Nonetheless, this combination of procedures is probably the only way that one may quantitate low concentrations of diffusible ions in biological specimens. For a discussion of the various correction methods needed when attempting quantitative analysis on frozen hydrated specimens see Warley (1997) and Hall (1986).

Electron Energy Loss Spectroscopy (EELS)

This technique is used to detect and differentiate various energy levels of electrons that have been transmitted through a thin specimen. As in EDS and WDS, the spectrum of electron energies is displayed and may be used to determine the elemental composition of the atoms in the specimen that caused the loss in energy of the beam electrons.

Differentiation of the various energy levels of electrons is carried out using an *electromagnetic spectrometer* placed either after the specimen or under the viewing screen of the TEM or STEM. As the beam electrons enter the electromagnetic field of the spectrometer, they are bent to various degrees and

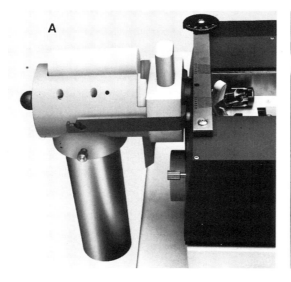
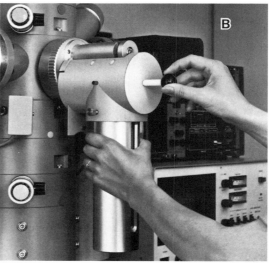

FIGURE 15.20 (A) Cryotransfer device attached to back of cryochamber of ultramicrotome. Frozen sections are transferred into the chamber and maintained at liquid nitrogen temperatures. (B) Frozen hydrated sections are moved into the electron microscope column for viewing. *(Courtesy of Gatan.)*

brought to focus some distance from the spectrometer. Lower energy electrons are deflected to a greater degree than are high energy electrons, so that the focal points of the various energy groups of electrons are physically separated. A movable plate with a narrow slit is positioned to permit electrons of a specific energy range to pass into a detector. The detector is similar to the scintillator/photomultiplier type used to detect secondary electrons in the SEM. Figure 15.21 is a schematic representation of a system commonly used in EELS.

A typical range of energy loss for 100 kV beam electrons is from 100 to 1000 eV. Modern spectrometers can detect energy losses in the 0 to 2000 eV range with resolutions of better than 5 eV, so that the spectrometer is able to accommodate and adequately resolve the spectrum of anticipated energies in most biological specimens. Since EELS detectors detect a primary event (loss of energy in transmitted electrons) rather than a secondary event (X-ray emission), EELS is 10 to 100 times more efficient than EDS. Spatial resolution in EELS (10 nm) is comparable to that obtainable in thin specimens analyzed by EDS methods (10 to 50 nm). Figure 15.22 shows some electron micrographs in which EELS is used to map the distribution of phosphorus in a mitochondrion and endoplasmic reticulum, as well as to reveal the location of iron, calcium, and oxygen in a section of a lung biopsy.

Quantitation in EELS is more straightforward than with X-ray techniques: ZAF corrections are not needed and comparable standards need not be run each time. In addition, despite some current limitations, it eventually should be possible to obtain true quantitation of the number of atoms/μm^2 present in a thin specimen.

Theoretically, EELS should be ideally suited for the detection of low atomic numbered elements that make up biological tissues. In fact, EELS is less frequently used than EDS and WDS for detection and quantitation. The principal obstacle has been the inability to produce thin enough specimens—several times thinner than the 60 to 80 nm slices obtained by ultramicrotomy—since thicker specimens increase the background levels due to multiple scattering events. This is the same situation as with chromatic aberration (Chapter 6) in thicker sections.

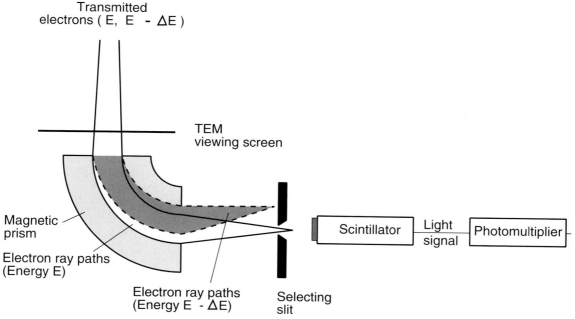

FIGURE 15.21 Schematic diagram of an electromagnetic spectrometer for electron energy loss investigations. Only two different energy levels of electrons are shown being focused in the plane of the selecting slit. Lower energy electrons *(dashed line)* are deflected to a greater extent than higher energy ones and thereby may be separated by the spectrometer. The separated electrons are then sampled by positioning the selecting slit in the proper location to allow the electrons to pass through the slit.
(Courtesy of D. B. Williams and Philips Electronics Instruments.)

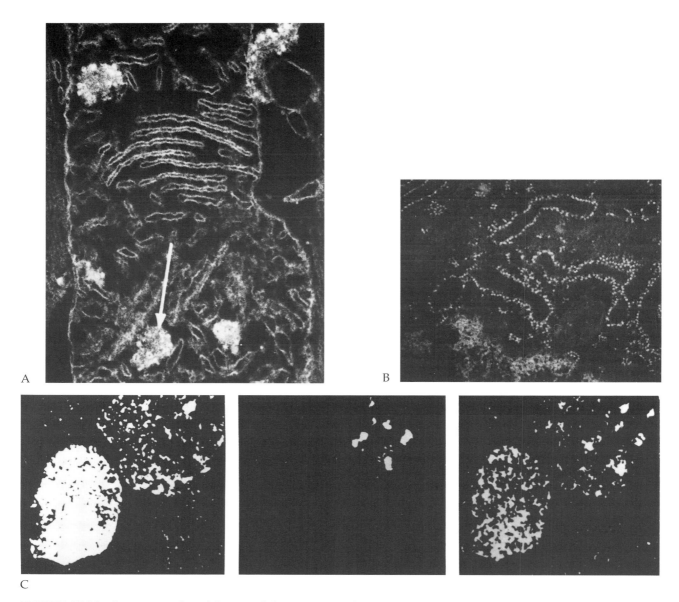

FIGURE 15.22 Some examples of the use of electron energy loss spectroscopy to detect particular elements in sectioned cells. (A) Portion of mitochondrion with areas rich in phosphorus appearing very bright.
(Courtesy of P. A. Schnabel.)
(B) Phosphorus localized in ribosomes along endoplasmic reticulum.
(Courtesy Carl Zeiss, Inc.)
(C) Elemental distribution of Fe, Ca, and O, respectively, in lung biopsy.
(Courtesy of C. H. W. Horne.) (All micrographs were provided by Carl Zeiss, Inc.)

Besides giving elemental composition and quantitation, the electron spectrometer may be used as an energy filter to enhance contrast and resolution in thicker specimens. In practice, an EEL spectrometer is used to filter out electrons of particular energies with the remainder being used to form the image. This would permit one to examine thicker specimens, since selected wavelength electrons (that give rise to chromatic aberration) may be removed (Figures 15.23 and 15.24). One commercial electron microscope manufactured by Zeiss has such capabilities (Figure 15.25).

A potentially exciting use of EELS lies in the imaging of single, heavy atoms on low atomic num-

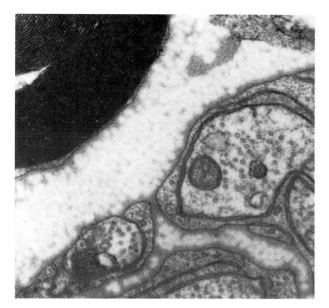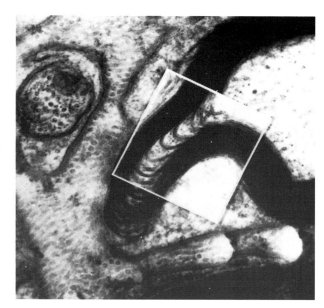

FIGURE 15.23 Use of an EEL spectrometer to diminish chromatic aberration in a thick section. Section on left is a conventional micrograph of a 70 nm ultrathin section of nerve tissue. Section on right is over seven times thicker and still usable since only a narrow wavelength of electrons was permitted to pass through the selecting slit. *(Courtesy of Carl Zeiss, Inc.)*

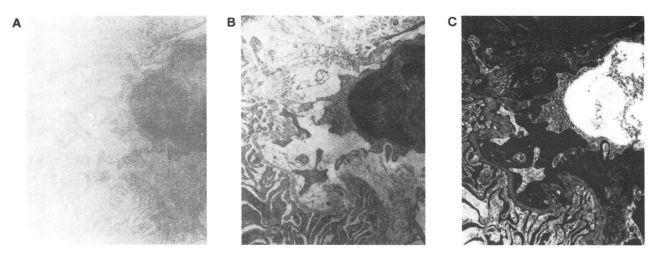

FIGURE 15.24 EEL spectrometers may be used to enhance contrast in unstained sections. (A) Unstained section viewed in conventional TEM mode. (B) Unstained section viewed in energy loss mode with spectrometer selecting only electrons that have not lost energy (zero energy loss). (C) In the spectroscopic mode, areas rich in phosphorus show up as bright areas indicating electrons that have lost 180 eV. *(Courtesy of Carl Zeiss, Inc.)*

bered substrates. For example, if one could prepare specific DNA or RNA probes that have been labeled with uranium atoms, one may be able to directly image a particular labeled gene along a strand of DNA in the chromosome.

In summary, EELS may be used to detect and quantitate the lower numbered atomic elements as well as to improve the imaging capabilities of a TEM or STEM. It is not as popular as X-ray analytical procedures in spite of its powerful capabilities primarily because most biological specimens are too thick and readily damaged by the beam during analysis.

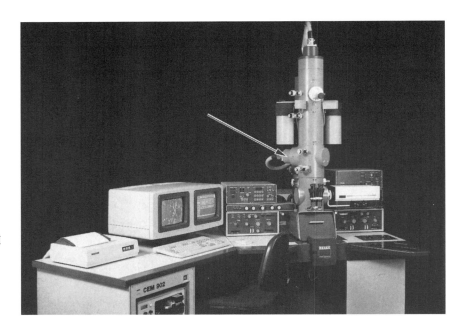

FIGURE 15.25 Commercial STEM equipped for electron energy loss spectroscopy, the Zeiss CEM 902. Arrow indicates location of spectrometer. *(Courtesy of Carl Zeiss, Inc.)*

Electron Diffraction

Even the most basic transmission electron microscope can generate a diffraction pattern from a specimen. This is because the diffraction pattern is always present in the back focal plane of the objective lens. Normally, most biologically oriented electron microscopists are interested in examining the *image* generated by the transmitted imaging electrons and therefore have no need for viewing the diffraction pattern. If one examines the ray diagram shown in Figure 15.26, it is apparent that the forward scattered diffracted electrons come to a focal point (this is the back focal plane of the objective lens) but are excluded by the objective aperture. As will be shown, one of the operational requirements to obtain diffraction patterns may involve removal of the objective aperture or the use of a larger aperture.

Although diffraction patterns are generated by all specimens, some patterns have more information about the nature of the specimen than do others. For instance, specimens with randomly or nonperiodically oriented atoms (the majority of biological specimens) generate a diffuse electron diffraction pattern that simply confirms that the atoms of the specimen are not arranged in a repeating or periodic manner (Figure 15.27). By contrast, whenever the specimen or parts of the specimen consist of molecules or atoms with a *repeating periodicity* (as in a crystalline lattice), then a diffraction pattern is

formed that may be useful in the identification of the crystal or molecule (Figure 15.28). Unlike the other analytical procedures of X-ray microanalysis and EELS, which identify and quantify amounts of individual elements present in an area, electron diffraction may give the spacing of the crystalline lattice and (since various crystals have unique lattice spacings and diffraction patterns) the chemical identity of the crystal. On the other hand, electron diffraction cannot be used to determine the quantity of a particular chemical that has been identified.

Formation of Diffraction Patterns

A crystalline object consists of identical atoms or molecules arranged as repeating units along certain planes. For example, imagine atoms or molecules arranged to form a single layer (like a sheet of paper). If one begins stacking additional layers upon the initial layer (a stack of paper sheets), one would generate an object with crystalline features. A model for such stacks of atoms or molecules is shown in Figure 15.29. Note that the crystal consists of the same basic repeating unit (atom or molecule) in various planes that are very precisely spaced relative to one another. Examples of crystalline specimens might include asbestos, sodium chloride, carotene, cholesterol stearate, ferritin, and myelin.

If a beam of electrons strikes a crystalline structure at an appropriate angle (so-called Bragg angle) the electrons will be diffracted or "reflected" from the lattice planes. The reflection follows Bragg's law of diffraction that is summarized in Equation 15.1.

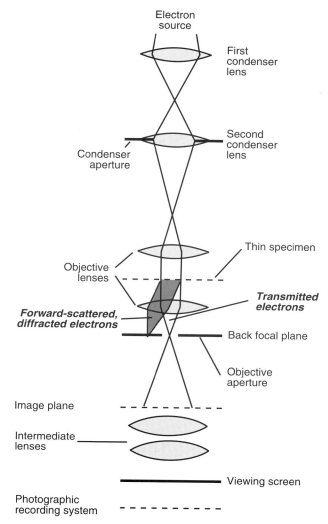

FIGURE 15.26 Schematic of lenses in a transmission electron microscope. Note the dashed line indicating one group of diffracted electrons that converge in the back focal plane of the objective lens.
(*Modified material courtesy of D. B. Williams and Philips Electronic Instruments.*)

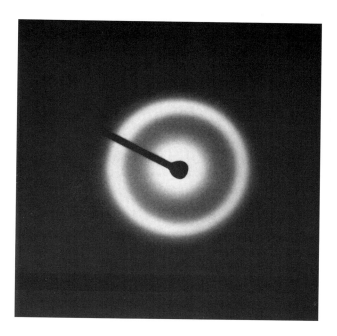

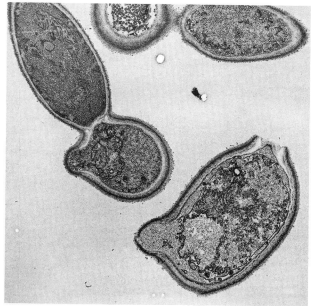

FIGURE 15.27 Electron diffraction pattern obtained from thin section shown on the bottom. The noncrystalline nature of the sectioned specimen generates a diffuse diffraction pattern.

Equation 15.1: Bragg's Equation

$$n\lambda = 2d \sin \theta$$

where n = an integer
λ = wavelength of the electron that is diffracted
d = crystalline lattice spacing
θ = angle of incidence and reflection of the electrons striking the crystal.

One will obtain a discrete diffraction pattern only when the incident electrons enter the crystalline lattice at the appropriate Bragg angle. With crystals, some of the electrons that enter the lattice at the proper angle will be reflected by the various lattice planes in the same direction and at the same angle to come to focus in the back focal plane. This generates the diffraction pattern. In the case of an amorphous specimen, the electron beam that enters the specimen is diffracted in multiple directions and at

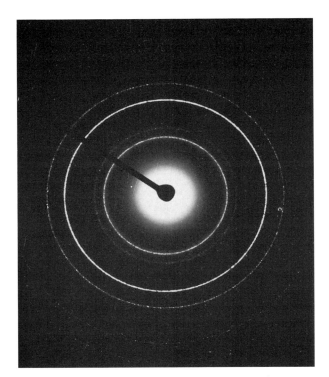

FIGURE 15.28   Electron diffraction pattern of a polycrystalline specimen. A thin film of gold was deposited onto a plastic film by evaporating the molten metal in a vacuum evaporator. The gold vapor settled onto the plastic and formed tiny crystals that are responsible for the pattern shown.
(Courtesy of B. DeNeve.)

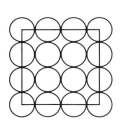 

FIGURE 15.29   (A) Array of atoms in a single plane. Each sphere equals one atom. (B) Stacks of planes generate the crystalline lattice. Here the crystal is viewed from one of the corners of the cube of atoms.

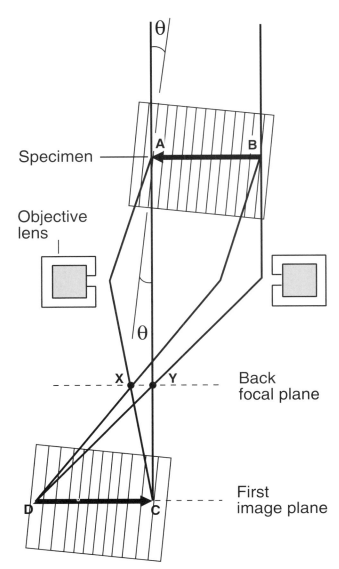

FIGURE 15.30   Diagram showing path that electrons take on striking a crystalline object. Some rays (C, D) converge in the image plane while others (X, Y) converge in the back focal plane to generate the diffraction pattern.

various angles so that the electrons are unable to converge into a discrete spot and form a diffuse ring pattern instead. With a crystalline specimen, in order to obtain the proper Bragg angle, it is necessary to orient the specimen very precisely by tilting

and rotating it relative to the electron beam until the diffraction pattern is obtained.

If one examines Figure 15.30, a number of phenomena may be observed. In this illustration, a crystalline object is struck by an electron beam so that some electrons pass through the specimen and are brought to a focus point (C or D) on an image plane (the first image plane). Note that the electrons that are brought together at the image points C and D are those electrons that were scattered from the same physical *location* in the specimen (A and B, respectively). Therefore in the standard imaging mode of the TEM, electrons that are scattered from the same point in the specimen come together at a common point in the image plane. These electrons impart information about the morphology of the specimen.

Again referring to Figure 15.30, one should notice another plane where electrons converge. It will be noted that in the back focal plane, some electrons converge at points X and Y. Careful examination of the ray diagram reveals that those electrons that were scattered in the same *direction* (rays that emerge from the specimen parallel to each other) converge along the back focal plane to form diffraction points. Therefore, in the diffraction mode those electrons that are scattered in the same *direction* (parallel to each other) converge at a common point in the back focal plane. These electron focal points when imaged on the viewing screen impart information about the atomic configuration and chemical identity of the specimen.

Single Crystal Versus Polycrystalline Specimens

A single crystal will generate a diffraction pattern consisting of spots (Figure 15.31, bottom), with the layout of the spots depending on the type of crystal lattice (14 different types exist) being illuminated and the orientation of the crystal to the beam. In practice, one photographs the diffraction pattern and, in a properly calibrated microscope, measures the distances and angles between the spots to determine the distance "d" between lattice planes. *Since the d spacings are unique for each crystal, one may look up the d spacings in a reference book (or computer system) and obtain an identification of the crystal.* If one has the capability of doing X-ray analysis, the identification of the unknown crystal will be greatly expedited since the possibilities will be limited to crystals composed only of those elements detected by the X-ray procedure.

In a polycrystalline specimen (Figure 15.31, top), many crystals are present all of which are generating spot patterns, so that the individual spots merge into rings surrounding a bright central spot (the undiffracted electrons). An example of a polycrystalline specimen would be an evaporated film of gold or aluminum where millions of tiny crystals of the metal have settled onto a plastic substrate such as Formvar. As in the previous example, the radius of the rings is related to lattice d spacings.

Determination of Spacings in a Crystalline Lattice

After one has recorded the spot or ring diffraction pattern, the negative is examined and distances are measured to be used in the following equation derived from the Bragg equation to calculate d spacing:

Equation 15.2: d Spacing in Crystalline Lattice

$$d = \frac{\lambda L}{R}$$

In this expression, R is the distance in millimeters from the central bright spot to one of the rings or spots, L is the *camera length* (distance in millimeters between specimen and photographic film), and lambda is the wavelength of the electron (based on the accelerating voltage: 50 kV = 0.00536 nm, 75 kV = 0.00433 nm, 100 kV = 0.00370 nm).

Equation 15.2 was derived from the Bragg equation (Equation 15.1) by substituting as follows. If one examines Figure 15.32, where R is the distance measured in millimeters (on the negative of the TEM diffraction pattern) from the central bright spot to the center of the diffracted spot and L is the camera length, then simple geometry tells us that:

$$R/L = \tan 2\theta$$

Furthermore, since the θ angles through which the electrons are diffracted are extremely small (about 1–2°), one may safely state the following:

$$\tan 2\theta = 2 \sin \theta$$

If one then recalls the Bragg equation,

$$\lambda = 2d \sin \theta,$$

upon following appropriate substitutions from the previous equations,

$$R/L = \lambda d,$$

one may obtain the final Equation 15.2,

$$d = \lambda L/R.$$

Because the critical three components in the equation are either known or may be measured from the electron micrograph of the diffraction pattern, it appears to be a simple process to determine the d spacings. As might be suspected, some background work must be done to confirm certain of these values.

The factor L, or camera length, is problematical since it may vary slightly every time a sample is examined. This may be due to variation in specimen positioning caused by bent grids, using a different specimen holder, a change in the vertical adjustment of the spec-

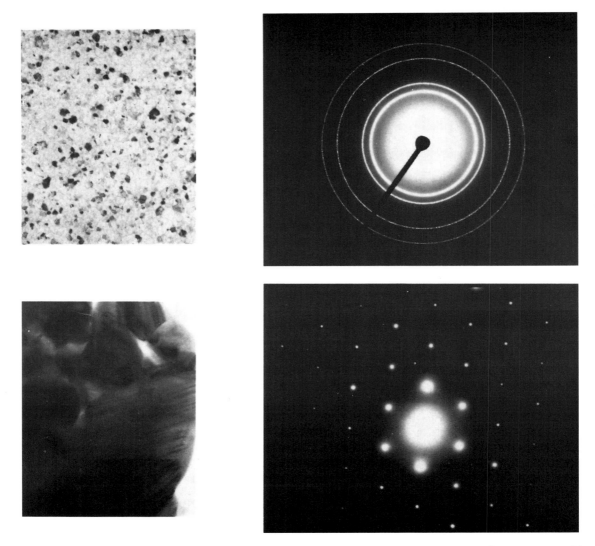

FIGURE 15.31   Comparison of diffraction pattern from a polycrystalline versus a monocrystalline specimen. The top left panel shows a standard TEM image of crystals magnified 4,500× and the top right panel shows the diffraction pattern obtained from this polycrystalline specimen. The lower left panel shows a TEM image of one of the crystals magnified at 90,000×. A diffraction pattern was obtained from the crystal and is shown in the bottom right panel. *(Courtesy of P. Tlomak.)*

imen holder in the stage (z axis control), and even cleaning of polepieces or readjustment of standard lens current settings. Consequently, for precise calculations, it is best to calibrate the camera length on a regular basis and to verify that the specimen is in the proper position (i.e., the eucentric position) following the specific TEM manufacturer's directions. Once the camera length is satisfactorily determined, it is multiplied by the wavelength of electron used and the result is now termed the *camera constant* (λL).

Determination of Camera Constant (λL)

1. Insert a standard specimen with known d spacings into the TEM and focus the image. Some standards include evaporated thin films of gold, aluminum, or thallous chloride. As an example, we will use gold as the standard (see Figure 15.28 and Table 15.2).

Electron
Beam

θ

Crystal

2θ

L

Photographic
film

R

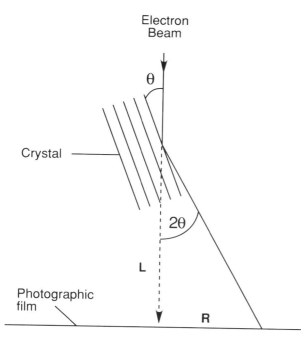

FIGURE 15.32 Diagram showing the relationship between camera length (L) and distance (R) between central beam spot and a diffraction spot. Using simple geometry it may be determined that R/L = tan 2 θ.

2. Verify that the specimen is in the eucentric position (check manufacturer's directions). In some microscopes this may involve tilting the specimen +/− 10° and observing the centration of the focused image on the screen. If the image shifts off center during the tilting, then the proper stage adjustments (x, y, z movements) are made until the image shift is minimal.
3. Remove the objective aperture, insert the diffraction aperture, and center it using mechanical controls.
4. Put the TEM in the diffraction mode (e.g., "DIFF" button).
5. Depending on the TEM model, obtain a focused diffraction pattern by varying the current to the proper lenses (intermediate or projector lenses).
6. Record the image of the diffraction pattern, as follows:
   a. adjust brightness with condenser lens 2 to barely see the faintest rings and refocus central spot with DIFF focus control,
   b. insert the central beam stopper to block the bright central spot,
   c. expose the negative for 20 to 30 seconds,
   d. remove the beam stop during the final 3 to 5 seconds of exposure.

TABLE 15.2 Gold Diffraction Standard

hkl	I	d	R	dR
111	100	2.355	____	____
200	52	2.039	____	____
220	32	1.442	____	____
311	36	1.230	____	____
222	12	1.1774	____	____
400	6	1.0196	____	____
331	23	0.9358	____	____
420	22	0.9120	____	____
422	23	0.8325	____	____
			Average	____

7. On a negative of the diffraction pattern:
   a. measure the diameter of each of the rings,
   b. divide the diameter by 2 to obtain the radius "R" in millimeters.
8. Look up the value for d_{hkl} for each of the rings (given in Å).

   Note 1: The designation d_{111} is a way of specifying a particular lattice plane (since many different planes are present in a crystal). It is important to be able to assign the d spacings to the proper set of lattice planes since the spacings of the various planes are different. The three subscripts in the expression d_{hkl} are termed the *Miller indices* and the process of assigning d spacings to the proper lattice plane is termed *indexing*. The process of indexing single crystals is complicated and requires a knowledge of crystallography. Details of this procedure are outlined in the references cited at the end of this chapter.

   Note 2: The rings represent various lattice planes in a polycrystalline specimen. In the gold specimen (Table 15.2) the rings are arranged in a particular order from the center outwards with the brightest ring, d_{111}, having a relative intensity "I" of 100%. Other rings are expressed in terms of relative brightness to d_{111} so that the d_{200} ring with I = 52 is nearly half as bright, etc.
9. Proceed to calculate camera constant for d_{111}:
   a. from Equation 15.2, we may derive the following expression for camera constant:

   $$\lambda L = dR$$

b. since we have already measured the radius R for the d_{111} ring from the negative, and since the actual d spacing for the d_{111} plane is known, simply multiply d by R to obtain the camera constant.

10. Repeat the calculation of camera constant for the remaining rings and fill in the values in Table 15.2 (make a photocopy of the page).

11. Average together all of the various camera constants (dR) in order to obtain a more accurate figure. This is the camera constant value to be used in subsequent determination of d spacings in unknown crystals. *Note: It is important to verify that the rings are symmetrical in order for the measurements to be accurate; therefore, several measurements must be made from various locations along the ring to verify the roundness of the rings.*

Once the camera constant has been accurately determined using the standard sample, one may then proceed to use one of the possible diffraction modes to determine the identity of an unknown crystal or group of crystals. After tilting and orienting the crystal to an appropriate zone axis to obtain the diffraction pattern, photographically record the diffraction pattern of the unknown crystal(s). One must now determine the crystal structure by matching the diffraction pattern of the unidentified crystal to one of the patterns of the 14 lattice systems (consult diffraction references given at end of chapter). After determining the d spacings from the negative, the pattern is indexed and the values are looked up in a reference book (*Mineral Powder Diffraction File*, for example) or by computer program (the *Mineral Powder Diffraction File* is now available on CD-ROM) to identify the unknown crystal. Energy dispersive X-ray analysis will greatly expedite this procedure by identifying the various elements making up the crystal.

Types of Diffraction Modes

Depending on the design of the TEM or STEM being used, it is possible to operate the microscope in up to six different types of diffraction modes. Only the two approaches that are primarily used in biological studies will be discussed in detail.

Selected Area Diffraction (SAD)

This is probably the most commonly used diffraction mode in TEM and STEM instruments. It can be used to generate diffraction patterns from crystalline materials with lattice spacings smaller than

2.5 to 4.0 nm. In practice, one positions a diffraction aperture of the appropriate size over the area from which the diffraction is to be conducted (thereby "selecting an area") and then adjusts the microscope to obtain the pattern (Figure 15.33). To obtain diffraction from areas as small as 0.5 μm involves the use of very small diffraction apertures (<5–10 μm in diameter) that contaminate too rapidly. In addition, spherical aberration in the objective lens causes inaccuracies in the selection of an area in the specimen. If apertures smaller than 5 μm are used, one is never certain what area of the specimen is actually generating the diffraction pattern. Most of the time, therefore, SAD is conducted from specimen areas in the 10 to 100 μm range. The general steps for accomplishing SAD are outlined in the next section (specific procedures will vary from one microscope to another, so the manufacturer's manual should be consulted).

Performing Selected Area Diffraction

1. Insert specimen, locate area of interest, and focus in normal imaging mode using objective lens controls. A side entry stage with capabilities for tilting (and possibly rotating) the specimen may be necessary to orient the specimen for single crystal diffraction studies. Verify that the stage is eucentric.

2. Remove objective aperture.

3. Press the "SA" button to put microscope in selected area diffraction mode. Select an appropriate magnification, center the object of interest on the phosphorescent screen, and focus the object of interest using the objective lens focus controls.

4. Insert a suitably sized diffraction aperture to just cover the object of interest and center the aperture over the object. Focus on the edge of the aperture until the edge is sharp using the "Diffraction Spot" focus control. Recheck focus of object of interest using objective lens focus control and spread the brightness control to illuminate the specimen so that it is just visible.

5. Switch the microscope into the diffraction mode by pressing the appropriate button (normally labeled "DIFF"). Select an appropriate camera length (0.2 or 0.4 m are normally used for viewing using the binoculars, while 0.8 m is used for recording the image). If the diffraction spot is not centered, use the intermediate alignment controls to correct the centration. Focus diffraction spot with appropriate control ("DIFF SPOT") to obtain the smallest, sharpest spot.

Intermediate and High Voltage Microscopy

CONVENTIONAL transmission electron microscopy employs accelerating voltages in the range of 25 to 125 kV (1 kV = 1,000 volts). Most commonly, 75 to 100 kV are used for routine work. Since most conventional microscopes have a limited voltage capability (up to 100 to 125 kV), they cannot be used for high voltage work. Instead, one would use a specially designed *high voltage electron microscope* (HVEM), which is essentially a standard transmission electron microscope with high voltage capabilities (Figure 16.1). Medium or *intermediate voltage electron microscopes* (IVEM) cover the voltage ranges from 200 up to 500 kV, while high voltage electron microscopes are used at accelerating voltages above 500 kV and well into the mega (million) voltage (3 MV) range.

There are three publicly funded centralized facilities in the United States where biological investigators may share usage of a high voltage electron microscope: the University of Colorado (Boulder), New York State Department of Health (Albany), and the University of California (Berkeley; two instruments). The microscope with the highest voltage capability is in Toulouse, France. This microscope has a three megavolt (3 MV) capability. High voltage microscopes are extremely expensive and usually require housing in a specially designed building. Intermediate voltage microscopes that also can be used at the normal 75 to 100 kV are more commonly seen as replacements for outdated conventional and high voltage microscopes. Presently, many of these instruments are in service in university and industrial settings.

Historical Perspective

High voltage microscopy was a natural outgrowth of the desire to examine whole cells and obtain three-dimensional information. In the early days of electron microscopy, Porter and colleagues (1945), in a now classic paper, examined whole cells, especially the thin edges of fibroblasts, at relatively low voltages. Although valuable information was obtained, many of the cells' features contained in the thicker portions of the preparation were impenetrable by the electron beam. For some time after the advent of good microtomes, investigators concentrated their efforts on obtaining information from the newly produced thin sections. The first high voltage microscopes were constructed in the late 1940s and the 1950s. After the high voltage microscope was operational,

Porter and colleagues returned to studying whole cells with this tool, this time with much more success (for examples, see Wolosewick and Porter, 1976; Porter and Tucker, 1981). Numerous other investigators have used the less than 50 high voltage electron microscopes scattered throughout the world. The Japanese, Americans, French, and English are especially active in biological applications of high voltage microscopy.

Advantages of High Voltage Microscopy

There are four main advantages to using voltages at or in excess of 100 kV. The first is the *increased resolution* obtainable at high voltages. High energy electrons possess a shorter wavelength than lower energy electrons. As was demonstrated in Chapter 6, the shorter the wavelength of the electron, the higher (better) the resolution capability. Figure 16.2 illustrates the resolution difference obtained by changing the accelerating voltages. Unfortunately, the kind of thicker specimens usually used in the high voltage microscope partially offset this increase in resolving ability since there is an accompanying increase in *chromatic aberration.* Chromatic aberration is caused by the energy loss of electrons as they collide with the thicker tissue section (see Chapter 6). It is evident when the longer wavelength (lower energy) electrons do not focus at the same spot as electrons with higher energy and shorter wavelengths.

The second advantage of using higher voltage is the *increased specimen penetrating capability* of high energy electrons (Figure 16.3). Sections 3 μm in thickness may be penetrated by the beam of a high voltage microscope to produce quality micrographs. In a practical sense, penetration is evidenced by the fraction of the beam coming from the specimen that passes through the objective aperture. It is possible to collect more electrons if a larger objective aperture is used; however, contrast is sacrificed (see Chapter 6). The penetration of biological materials is greater when a nonembedded specimen is critical point dried and contains no embedding media (see Chapter 3) than when specimens have been conventionally processed and embedded. Moreover, each specimen has its unique features that affect penetrating capability of the beam. As accelerating voltage increases, the penetrating power of the beam does not increase pro-

FIGURE 16.1 The National Institutes of Health-supported high voltage (1 MV) microscope at the University of Colorado in Boulder.
(Photograph courtesy of R. McIntosh.)

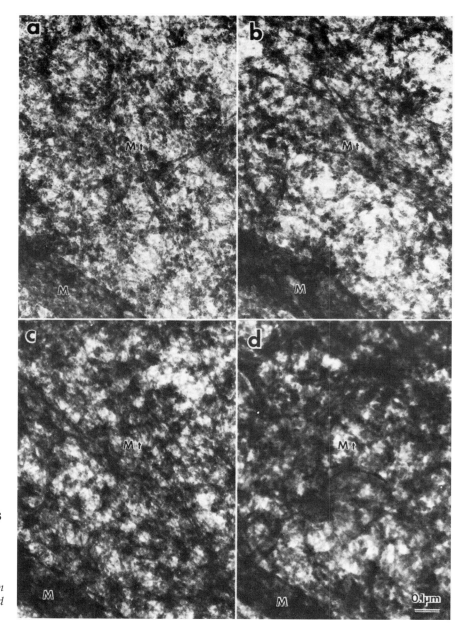

FIGURE 16.2 Resolution improvement in cytoplasmic structure at differing (A = 500kV, B = 200kV[uHR], C = 200kV, and D = 100kV) accelerating voltages. M = mitochondria; Mt = microtubules.
(Micrographs courtesy of K. Hama from J Electron Microsc 30:57–62. Used with permission from the publisher.)

portionally; a 3 MV microscope has only marginal advantage over a 1 MV microscope (Figure 16.4).

The third advantage of high voltage microscopy is the *increased depth of information* as compared with the lower voltage microscope. Ironically, the high voltage microscope has less depth of field because the shorter wavelength electrons actually decrease the depth of field (see Chapter 6). In spite of this, the information in the section is imaged at virtually all depths within the section,

providing important information about three dimensionality (Figure 16.5). Because of the great depth of information exhibited by high voltage microscopes, micrographs may be taken in pairs in which one image is tilted slightly (about 15° to 20°). Such images are called *stereo pairs* (Figure 16.6). Viewing structures from two different angles is the key to obtaining depth perception or three-dimensionality. Stereo pairs help resolve the visual confusion that results from superimposition of structures

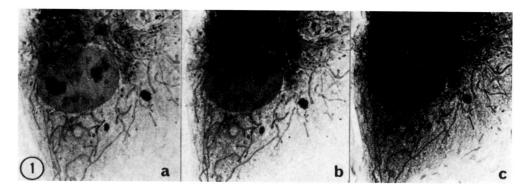

FIGURE 16.3   Series of micrographs taken at 1,000 (a), 200 (b), and 100 kV (c), demonstrating increased specimen resolution at the higher kVs.
(Micrographs courtesy of K. Hama from 38th Ann Proc Elec Micros Soc Amer, *pp. 802–5. Used with permission from the publisher.)*

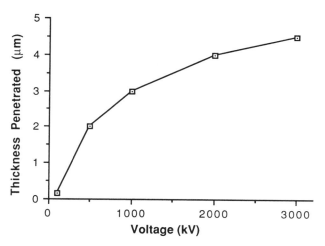

FIGURE 16.4   The relationship between accelerating voltage and the ability of the beam to penetrate specimens of various thickness.

in extremely thick (0.3 to 3.0 μm) sections. If one is wearing glasses designed for stereo viewing, the effect is an image that has a three-dimensional appearance.

The fourth advantage is that the *damage to the specimen at very high accelerating voltages is proportionably less* than at lower voltages due to less electron interaction with the specimen. Since the electrons are traveling substantially faster at the higher accelerating voltages, fewer electrons are deflected by the specimen. (Ironically, the thicker specimens often viewed by the high voltage microscope are more susceptible to radiation damage than are thinner specimens because there is simply more tissue and embedment present to interact with

electrons.) Taking advantage of both the increased penetrating capability and the overall reduction in specimen damage, investigators have made several attempts to view *living* or *wet specimens.* Only very limited information has been obtained from viewing live specimens. Many improvements are needed before the fine details of most living cells can be visualized.

One disadvantage of using higher accelerating voltages is that contrast is lessened. The decrease in contrast is partially offset by using stains or impregnation techniques that penetrate thicker sections and enhance contrast (Figure 16.7; Thiery and Rambourg, 1976).

Contributions of High Voltage Microscopy

Problems related to interpretations of three-dimensional images from two-dimensional electron micrographs are numerous (see Chapter 19). Short of extensive serial section reconstruction, high voltage microscopy has been the technique of choice to gain three-dimensional information. The complexities of the Golgi apparatus or the endoplasmic reticulum, the path taken by filaments and their relationships to other organelles have, among many other things, been visualized by high voltage microscopy. Complex relationships of one cell component to another have been realized. Serious attempts have been made at viewing living specimens, especially bacteria and viruses.

Most investigators agree that the high voltage work by Porter and colleagues (Wolosewick and

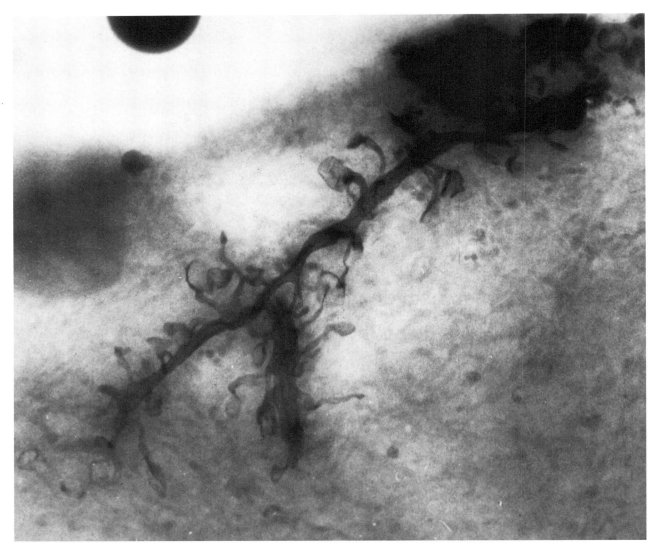

FIGURE 16.5 High voltage electron micrograph from a 3 μm thick plastic embedded specimen containing a dendrite stained by intracellular injection of horseradish peroxidase. The high resolution and penetration provided by the section thickness allow clear visualization of the fine dendritic appendages specialized for synaptic transmission. *(Micrograph courtesy of C. Wilson.)*

Porter, 1976; Porter and Tucker, 1981) is a major step in our understanding of how cells are organized. These fascinating papers present evidence that there is a highly organized and intricate ground substance in the cell that mediates, regulates, and directs transport within the cell. The system of filamentous elements composing the ground substance has been termed the *microtrabecular system or lattice* (Figure 16.8). It is suggested that the microtrabecular lattice organizes most cell components into a unified structure, the *cytoblast.* This work is an

important technical achievement advanced by high voltage microscopy, but most investigators feel that the hypotheses advanced by Porter and colleagues await further experimental testing.

The greatest advantages of high voltage microscopes have been achieved in the materials sciences. Microscopes currently available on the market are generally intermediate range voltage instruments, capable of imaging most biological specimen components that can be imaged by a high voltage microscope.

A

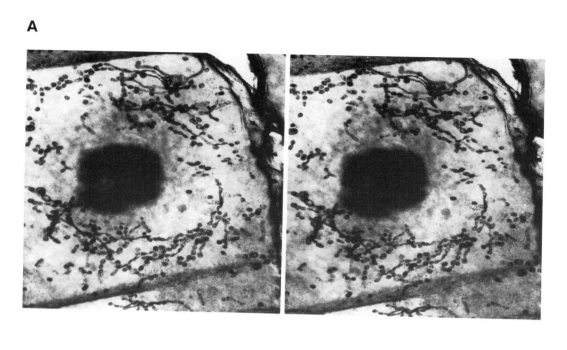

B

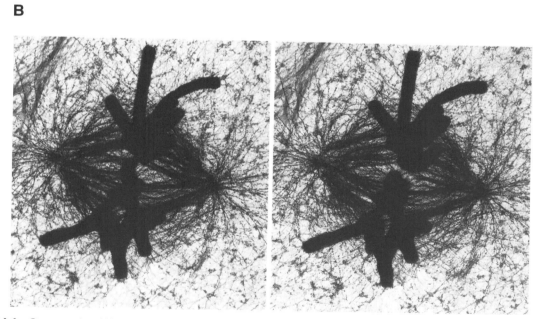

FIGURE 16.6 Stereo pairs. (A) High voltage micrograph of a 2 to 3 μm section of a pea root tip. (B) Isolated mammalian cell in mitosis whose microtubules have been darkened by the binding of 20 nm particles of colloidal gold. Glasses designed for stereo viewing should be worn to visualize three dimensions. *(Micrographs courtesy of P. Favard, N. Carosso, and R. McIntosh.)*

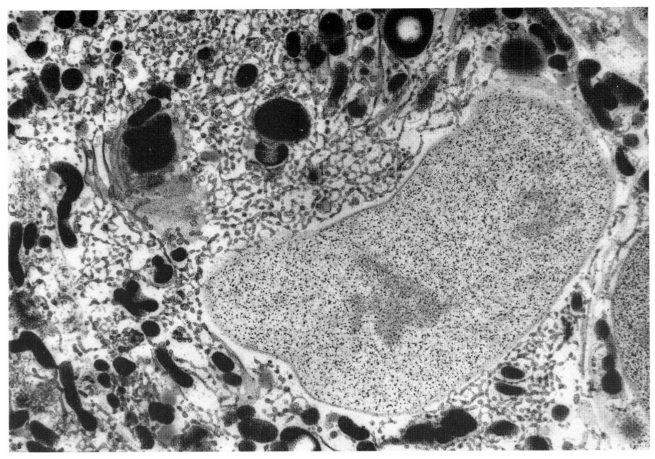

FIGURE 16.7 A 2 μm thick section impregnated with osmium and viewed using intermediate voltage conditions. Note how the thickness of the section allows the continuity of the endoplasmic reticulum (*arrow*) to be traced. Most of the dense structures in the cytoplasm are mitochondria.
(*Micrograph courtesy of L. Hermo.*)

FIGURE 16.8 The microtrabecular lattice (Mt) and mitochondria (M).
(*Micrograph courtesy of K. Hama from J Electron Microsc 30:57–62. Used with permission from the publisher.*)

REFERENCES

Dupouy, G. 1985. Megavolt electron microscopy. In *Advances in electronics and electron physics,* Suppl. 16, pp. 103–65.

Ellisman, M. H., B. E. Soto, and M. E. Martone. 1994. The merger of microscopy and advanced computing: a new frontier for the 21st century *Proc Elec Micros Soc Amer,* pp. 10–11.

Fujita, H. 1989. The research center for ultra-high voltage electron microscopy at Osaka University. *J Electron Microsc Tech* 12:201–18.

Hama, K. 1973. High voltage electron microscopy. In *Advanced techniques in biological electron microscopy,* J. K. Koehler, ed., New York: Springer-Verlag, pp. 275–97.

Hama, K., and F. Nagata. 1970. A stereoscopic observation of tracheal epithelium of mouse by means of the high voltage electron microscope. *J Cell Biol* 45:654–59.

Heath, J. P., and L. D. Peachy. 1989. Morphology of fibroplasts in collagen gels: A study using 400 KeV electron microscopy and computer graphics. *Cell, Motil and Cytoskel* 14:382–92.

Mazzone, H. M., et al. 1968. The high voltage electron microscope in virology. *Adv Virus Res* 30:43–82.

Parvin, B., D. Agarwal, D. Owen, M. A. O'Keefe, L. H. Westmacott, U. Dahmen, and R. Gronsky. 1995. A project for on-line remote control of a high voltage TEM. *Proc Elec Micros Soc Amer,* pp. 82–83.

Porter, K., and B. Tucker. 1981. The ground substance of living cells. *Sci Amer* 244:56–67.

Porter, K. R., et al. 1945. A study of tissue culture cells by electron microscopy. Methods and preliminary observations. *J Exptl Med* 97:727–50.

Ris, H. 1969. Use of the high voltage electron microscope for the study of thick biological specimens. *J Microscopie* 8:761–66.

Thiery, G., and A. Rambourg. 1976. A new staining technique for studying thick sections in the electron microscope. *J Microscopie Biol Cell* 26:103–6.

Wolosewick, J. J., and K. R. Porter. 1976. Stereo high-voltage electron microscopy of whole cells of the human diploid line, WI-38. *Am J Anat* 147:303–24.

Zaluzec, N. J. 1995. Tele-presence microscopy: an interactive multi-user environment for collaborative research using high speed networks and the Internet. *Proc Microscopy and Microanalysis,* pp. 14–15.

Chapter 17

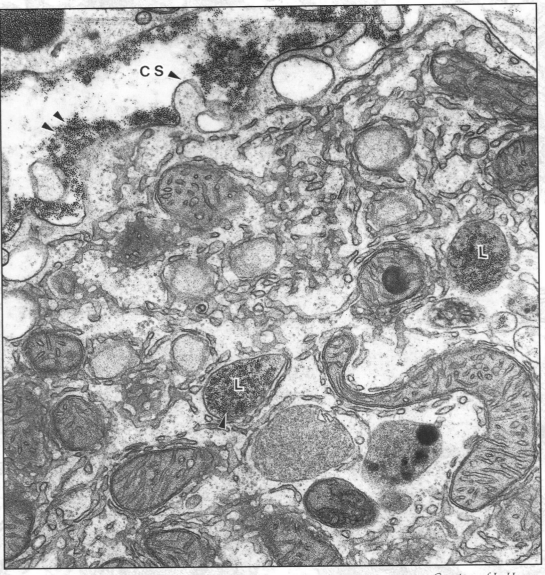

Courtesy of L. Hermo.

Tracers

Tracers are exogenously administered substances that, when visualized in electron microscope preparations, provide valuable information about cell compartments, junctional elements, and cell surfaces. Tracers are employed to delineate *extracellular spaces* and the limits to extracellular spaces. By highlighting extracellular spaces, trac-

ers can be used to determine the site(s) of *physiologic barriers*. A tracer may be used to determine the *permeability* of the vascular system, and specifically, the permeability of the vascular endothelium. Tracers may be used to *follow the paths of molecules* in a physiological system (Figure 17.1). For example, an electron-dense marker such as fer-

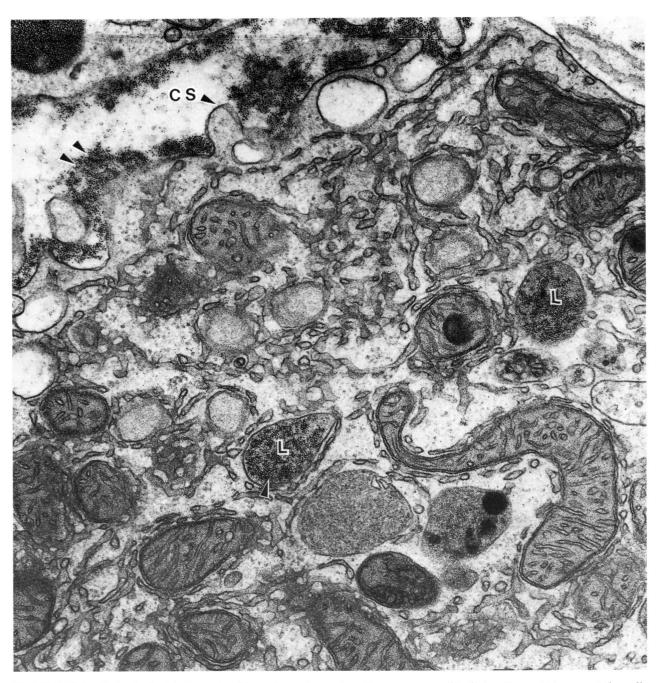

FIGURE 17.1 Cationic ferritin (*arrowheads*) used as a tracer. Ferritin was exposed to living tissue. It is seen at the cell surface (CS) and has been internalized and incorporated into lysosomes (L). *(Micrograph courtesy of L. Hermo.)*

ritin is frequently used to trace endocytic events and the fate of endocytosed materials. Since some tracers travel within the extracellular space, the extent of their excursion is often indicated by their binding to cell surfaces.

Tracers have been used to distinguish types of cell *junctions* (see Figures 20.5 through 20.21, Chapter 20) by virtue of their ability to delineate/highlight the extracellular space in the vicinity of the junction and to allow measurement of the intercellular space. For example, a tight or occluding junction excludes tracer, whereas a gap junction will permit a 2 to 4 nm wide deposition of tracer within the intercellular space (Figure 20.16). The width of the tracer between membranes forming the junction can thus be used to differentiate tight from gap junctions. A number of other junctional types are delineated and/or categorized by the use of tracers.

Some Specific Tracers in Use

Cationic and Native Ferritin

Cationic ferritin carries a positive charge and thus will bind to negatively charged moieties on cell surfaces. The internalization of bound substances, such as ferritin, is termed *adsorptive endocytosis,* a general term referring to uptake of specific molecules. One type of adsorptive endocytosis is termed *receptor-mediated endocytosis,* which requires receptor-ligand affinity for uptake. For many cells that show an endocytic process, vesicles form at the cell surface and pinch off within the cell. These contain the bound tracer. The cationic ferritin is successively shuttled to multivesicular bodies and then lysosomes. Figure 17.1 shows an example of cationic ferritin used as a tracer.

Not being a charged molecule, *native ferritin* is taken up, or endocytosed, by some cells in a similar manner as other molecules may be taken up in a nonspecific manner. This process has been termed *fluid-phase endocytosis*. In many cells, native ferritin is taken up during fluid-phase endocytosis by invaginations in the cell surface like pinocytotic vesicles that later pinch off inside the cell. Tracer is next usually transferred to multivesicular bodies and then to lysosomes.

Lanthanum

Lanthanum, with an atomic number of 57, produces contrast under the electron microscope. Lanthanum is a trivalent cation and will bind to negatively charged ions on the cell surface, especially negatively charged glycoprotein moieties.

As the pH of a lanthanum hydroxide solution increases, it forms a colloidal suspension. A colloidal lanthanum suspension is most commonly used as a tracer rather than a stain. Lanthanum will follow an intercellular route until it is blocked by a specific structure such as a tight junction (Figures 17.2 and 17.3). Lanthanum may circumvent tight junctions at certain points if they are not continuous around the cell. In doing so, lanthanum will define the site of the tight junction, which itself should appear free of lanthanum (Figure 17.3). By using lanthanum in this manner, it is possible to trace the pathway of molecules as they move through an epithelium and thus define the permeability of the epithelium. Junctions such as gap junctions and desmosomes will allow passage of lanthanum around them except where the intercellular space is sealed. Lanthanum will allow a clear definition of the width of the intercellular space and aid in the characterization of the junctional type.

Lanthanum cannot be used as a quantitative measure of permeability. It cannot be assumed that because lanthanum is blocked at a particular site that *all* substances are blocked under physiologic circumstances. Lanthanum grossly defines where a structural blockage exists within epithelia and whether or not an epithelium, after a treatment, has had a major change in its permeability to this substance.

Horseradish Peroxidase

Horseradish peroxidase is an enzyme with a molecular weight of about 40,000. It is used in much the same way as lanthanum to outline the intercellular space and locate the site of permeability barriers. It differs, however, from lanthanum in that it is not inherently electron dense. In the presence of diaminobenzidine, H_2O_2, and later osmium, the site of deposition of horseradish peroxidase is revealed as an amorphous electron-dense deposit (see Chapter 9). This enzyme marker is amplified by the enzymatic reaction catalyzed by peroxidase, and thus the osmicated reaction product presents as a very electron-dense marker (Figure 17.4).

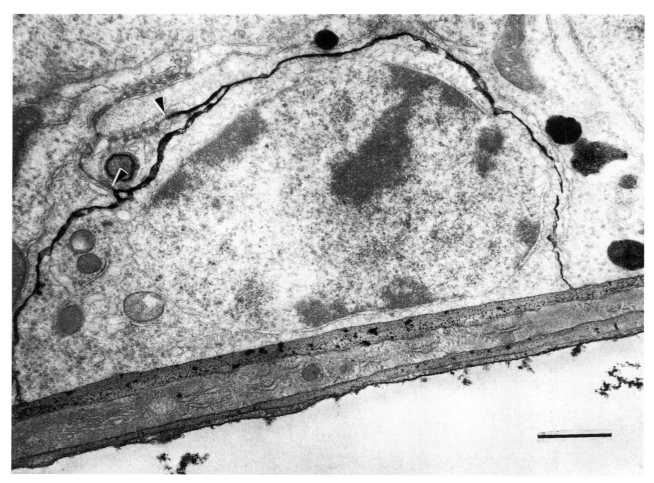

FIGURE 17.2 Lanthanum used as a tracer. Tracer was introduced into the vascular system along with the fixative. The tracer highlighted the borders of a cell, but was stopped from passing into the tissue by tight junctions at the site indicated by the arrowhead. Tissue was not stained conventionally using lead and uranium salts.

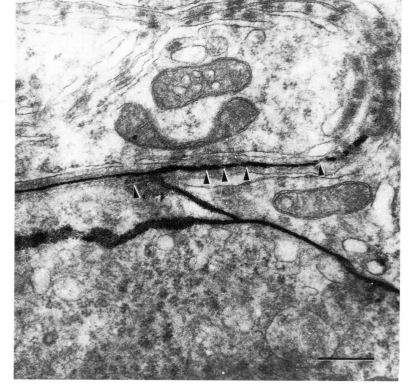

FIGURE 17.3 In a similar, but higher magnification micrograph to Figure 17.2, lanthanum mixed with the fixative was able to bypass some tight junctions, as evidenced by their presence as fine strands in negative relief (*arrowheads*). Lanthanum was, however, prevented from penetrating deeply within the epithelium by other tight junctions.

410

Image Processing and Image Analysis by Computer

THE GOAL OF ELECTRON MICROSCOPY is to produce an image containing information about the specimen under investigation. During the developmental stages of electron microscopy and cell biology, visual information alone was sufficient since the ultrastructure of various cells had yet to be determined. Once a reference library of images was established, advanced studies began to derive numerical information from images. Researchers asked such questions as: Are the volumes occupied by the organelles in quiescent cells different from cells undergoing growth or degenerative processes? How do the numbers and volumes of organelles vary as a result of a particular drug treatment? What effect does a new cancer regime have on the distribution and numbers of malignant cells in a tumor? What effect does a synthetic plant hormone have on cell wall growth?

Quantitative questions are best answered by analyzing the images using numerical measurements and applying a statistical analysis to the data. Indeed, the quantitative aspects of ultrastructural research are considered the best way to show differences between two samples. The processing of the images, measurement and analysis of the data, as well as generation of scientific reports is so greatly facilitated by the use of personal computers that it is unusual not to find computers used in most EM facilities. Typical steps in a computerized image analysis study are the same as those followed for a traditional stereological study (see Figure 13.2) and include: (a) obtaining and preparing the cells for study by electron microscopy, (b) capturing electron microscope images in a computer, (c) processing of digital images to enhance overall gray scale and to select features of interest, (d) computerized measurement of features of interest in the cells, (e) statistical evaluation of the numerical data, and (f) production of a report with appropriate conclusions (Figure 18.1).

In earlier chapters, we explained how to prepare specimens for examination in the electron microscope as well as how to record and process images using conventional photographic techniques. In this chapter, we explain how to obtain and process *digital* images and how to analyze the information using computers.

Capturing the Image: Conventional Versus Digital Methods

Conventional electron microscopes record images on a photographic film that requires development and printing onto photographic paper in an appropriately equipped darkroom. This generally takes at least several hours and generates noxious fumes and waste chemicals that must be disposed of in an environmentally conscious manner. In order to save time, instant films can be used to photograph the viewing screens of the SEM and occasionally the TEM. Such films can produce both a high quality negative and a print in the 4" × 5" size range. Unfortunately, waste chemicals and contaminated water are still generated because it is necessary to clear the negative of developing gel using a concentrated sulfite solution followed by a water rinse. If one wishes larger sized prints, it is necessary to enlarge the negatives in a darkroom.

To speed up the production of images for analysis and avoid the generation of chemical wastes, microscopists may acquire digital images from electron microscopes using electronic devices and cameras attached to personal computers. Such devices capture images nearly instantaneously without requiring any chemicals. The digital images can be stored on a variety of magnetic and optical media, retrieved by computer, processed and analyzed using various software programs, and printed out on high resolution or conventional laser printers at a very reasonable cost.

If you examine a magnified portion of a digital image, it becomes apparent that it is a mosaic composed of thousands of picture elements, or *pixels*, arranged in a grid of ordered rows and columns. During image capture, or *digitization*, each pixel is assigned a shade of gray or color, depending on the mode (gray scale or color) selected by the operator.

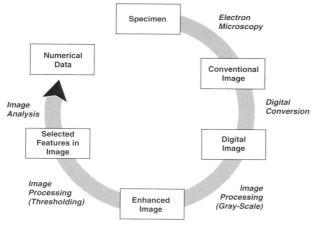

FIGURE 18.1   Steps involved in conducting a computerized image analysis of an electron micrograph.

For the conventional TEM, the process of capturing a digital image is more complicated because no electronic signals are available for the image capture card. A simple way to capture the TEM image is to mount a video camera so that it focuses on the viewing screen. Unfortunately, this system is awkward because the camera protrudes from the front of the microscope and the tilted viewing screen may distort the image. A better solution is to utilize one of the auxiliary ports of the TEM to mount the video camera (most TEMs have a port designed to accept a 35 mm film camera). A special high-resolution, retractable, phosphorescent viewing screen is then inserted opposite the camera, which sees the viewing screen through a glass window in front of the camera lens (Figure 18.5).

Although several varieties of cameras can be used, cameras equipped with an electronic chip containing arrays of light-sensitive, charge-coupled devices (CCD) are more prevalent since their sensitivity characteristics are closer to that of film. The chip of a standard CCD camera contains hundreds of thousands of tiny photosensitive detectors each about the size of a red blood cell (8 μm) and arranged in columns and rows. (More sensitive chips contain detectors 2 to 3 times larger than a red blood cell.) This array of detectors can vary in size from 512^2 to over $2,048^2$ units with prices rising to tens of thousands of dollars for cameras with larger high-quality chips. The resolution of the camera increases with the size of the chip and the number of photosensitive detectors. If one wishes to enlarge digital images electronically to the 5″ × 5″ to 6″ × 6″ range, then a minimum of a $1,024^2$ chip is recommended. Such chips will generate digital images that contain over one million pixels in an area approximately two-thirds of an inch. When the image is enlarged 12 times to obtain an 8″ × 8″ image, we have 128 pixels per inch. The enlarged digital images produced by such cameras are of a reasonable quality for study prints if not enlarged more than 5 to 6 times. For publication quality prints or images that must be enlarged significantly, however, most microscopists still prefer conventional photographic media.

It is possible to digitize conventional electron micrographs using a flat bed scanner or digitizer. The scanner should be set to acquire gray scale images with at least 256 levels of gray and at resolutions of 72 to 600 pixels per inch (ppi). Smaller contact prints should be digitized at higher resolutions while 8″ × 10″ prints can be digitized at lower resolutions depending on the final size at which the digital image will be used. If one digitizes an 8″ × 10″ print at 600 ppi and uses a final

digital print sized 4″ × 5″, this yields an image with resolution equivalent to 1,200 ppi. On the other hand, a 4″ × 5″ photograph scanned at a resolution of 600 ppi will have a final resolution of 300 ppi when enlarged 2 times. One must keep in mind that the file size of the acquired image will increase with the size of the photograph scanned, the resolution (in pixels), and the number of gray levels desired.

Microscopists often express disappointment with digital images when compared to conventional photographic media primarily due to lack of resolution in digital images. An important operating principle when working with digital images is to *always record the image at the proper magnification in the microscope* rather than attempting to enlarge the digital image in the computer beyond the useful range (5 to 6 times). You should plan the microscopy session so that the full magnification range needed for the study will be recorded. Likewise, you should be wary of the use of interpolative software methods in an attempt to "improve" the resolution of an image. Such computational manipulation of the pixels does not improve resolution although it will make the pixels less prominent by inserting a larger number of smaller pixels. The use of interpolation may be likened to *empty magnification* (see Chapter 6).

Pixels versus Dots Per Inch

Resolutions may be expressed as dots or pixels per inch, depending on the capture device and printer being discussed. The term *pixel* means that the information point will be stored as a shade of gray (the numbers of shades being variable as described previously). The term *dot* refers to a black or white point that lacks shades of gray. The resolution of conventional laser printers, for example, is often expressed in dots per inch (dpi) because they are not capable of printing in shades of gray. Instead such printers simulate shading by a process called *dithering*, or varying the size and placement of the dots, as for example, one might see in this book. Some printers, such as the dye sublimation printer, are capable of printing the points, or pixels, in as much as 256 shades of gray. Resolution of such printers, therefore, is more accurately expressed as pixels per inch (ppi).

A digital image from the electron microscope may either be displayed on a computer monitor for immediate processing and analysis or it may be stored on various recording media for use at a later time. The major types of storage media for digital

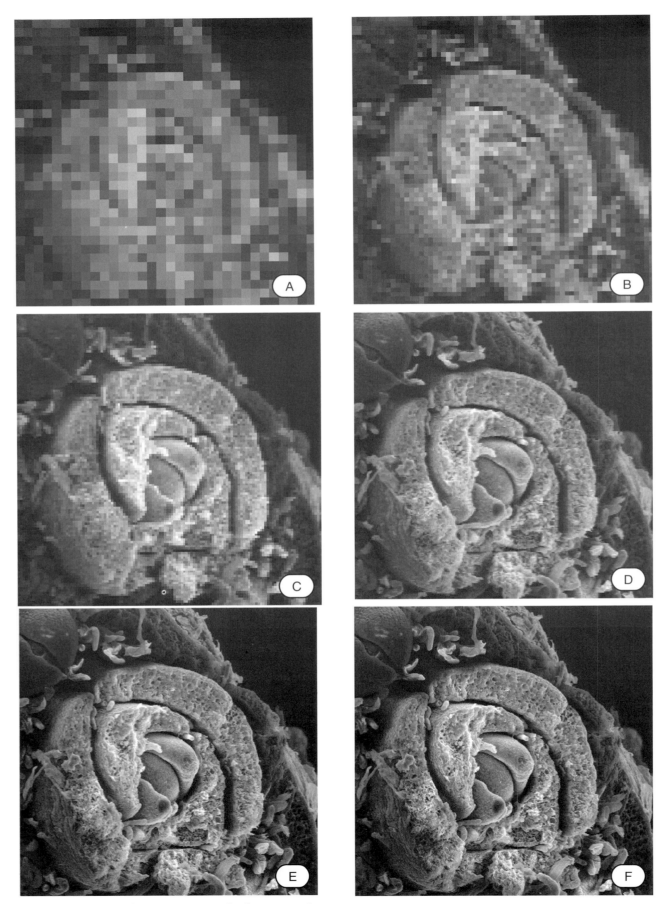

FIGURE 18.3 (See figure caption on the facing page.)

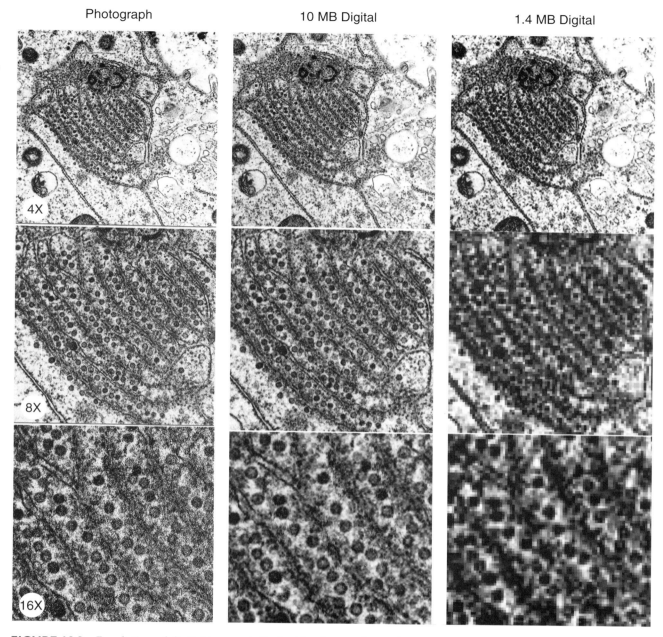

| Photograph | 10 MB Digital | 1.4 MB Digital |

FIGURE 18.3 Resolution of digital images is limited by the numbers of pixels in the image.
(*A–F, on the facing page*) Scanning electron microscope image showing the effect of variations in the number of pixels making up a digital image. Numbers of pixels range from 32^2, 64^2, 128^2, 256^2, 512^2, 1024^2, respectively. Image was digitized directly in SEM.

(*Above*) comparison of the effects of enlargement on images produced by conventional technology in a darkroom versus digital procedures. Left column shows images produced from a TEM negative in darkroom and enlarged at 4×, 8×, and 16× (from top to bottom, respectively). These images were produced using a point-source enlarger, Polycontrast II RC paper with an F5 filter and 105 and 80 mm lenses. Middle column shows digitally produced images obtained by scanning the TEM negative at a resolution of 2,000 pixels per inch (ppi) to generate a 10 megabyte file that was subsequently digitally enlarged 4×–16×. The quality of the image is marginally acceptable even at 16×. Right column shows the same image that was digitized at 300 ppi to generate a 1.4 megabyte file. Enlargements beyond 4× were unacceptable in this case.

421

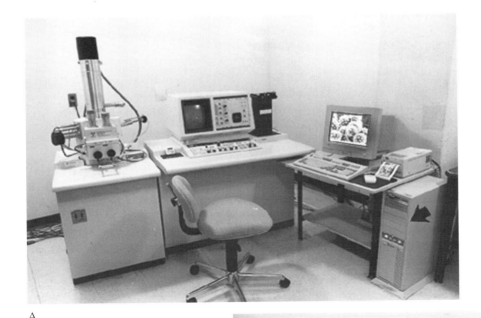

A

B

FIGURE 18.4   Conventional scanning electron microscopes may be modified to capture images digitally. (A) Shows an analog SEM adapted in this manner. To the right of the SEM is the image capture system that has been internally interfaced with an analog to digital converter. Images are fed into a personal computer (*arrow*) equipped with a high resolution capture card and are displayed on a monitor, printed out using inexpensive thermal printers, and saved for further processing. (B) Close-up of a digital image capture system showing an image on a display monitor. The software provided by the manufacturer permits one to make measurements, alter contrast and pseudocolor images, archive images with key words to facilitate searches, attach notes and descriptions to the image files, and print out the images on a variety of printers such as the thermal printer shown (*arrow*). This system is manufactured by SEMICAPS of Santa Clara, California.

images include larger hard disks using magnetic media (fixed or removable hard drives) as well as optical or magneto-optical disks that range in storage capacity from 100 megabytes up to several gigabytes (a gigabyte equals 1,000 megabytes).

Considering that a good quality gray scale image may occupy over a megabyte of space, floppy diskettes are generally not used to store digital images because they are extremely slow and can store only one image.

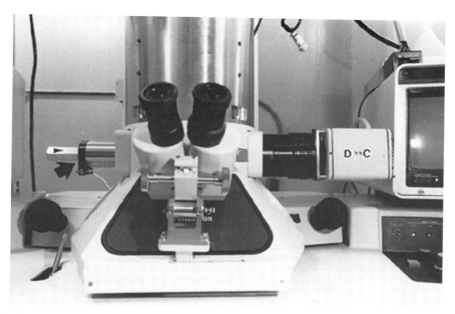

FIGURE 18.5 Digital camera attached to port of conventional TEM. A small display screen (*arrow*) similar to the phosphorescent viewing screen is inserted pneumatically into the beam and the image captured by a digital camera (DC) opposite the screen. The camera is outside the vacuum and views the screen through a glass port. The digital signal is sent to a personal computer similar to the one shown in Figure 18.4.

Image Processing

After a digital image has been obtained from an electron microscope, it is possible to enhance the overall quality of the image and to select features of interest for image analysis. Images that are electronically manipulated in this way are said to undergo *image processing*. After the image has been processed, it may be printed out for morphological scrutiny (e.g., a study print) or it may be numerically analyzed (image analysis), as discussed subsequently in this chapter. Specialized software will be needed for image processing and for image analysis. Although no single program will perform all possible functions, a good image analysis program such as *NIH Image* for the Macintosh or *Image-Pro Plus* (Media Cybernetics) for IBM-compatible computers will perform many of the basic processing and most of the analytical regimes. *Adobe Photoshop,* an image processing program, will accomplish a good bit of processing and analysis on both the IBM and Macintosh platforms. Such programs range in price from free (*NIH Image*) to several thousand dollars.

Among the more commonly used software with image processing capabilities you should be able to control contrast, brightness and gamma, burn and dodge, remove noise or objectionable backgrounds, homogenize illumination, sharpen or deblur, remove or enhance repeating structures, and highlight or select features for image analysis. These capabilities are described as follows.

Controlling Contrast, Brightness, and Gamma

For someone experienced with conventional darkroom procedures for controlling contrast, these capabilities are quite remarkable. Most imaging programs permit one to independently control brightness, contrast, and gamma either by means of entering numerical values into a table or sliding indicators along a scale (Figure 18.6A, B) or by changing the slope of an intensity plot of the gray levels (Figure 18.7A, B) on a logarithmic scale. The image usually changes as one makes the adjustments so that one is able to see immediately the effects of the changes. Nearly all image processing programs display a histogram of the dynamic range and contrast in a digital image (Figure 18.8A, B) that may be useful to detect deficiencies in the tonal range. In the histogram, the x-axis shows the index value (gray level) while the y-axis shows the number of pixels in the image that have this gray value.

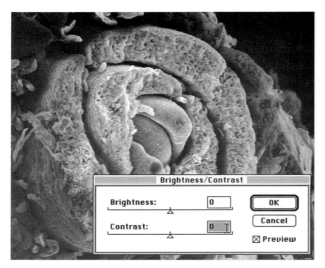

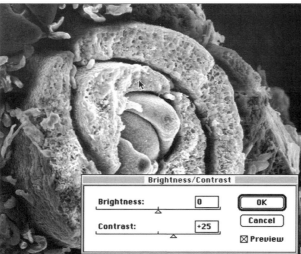

FIGURE 18.6 Adjustment of contrast using slider control. (A) Image from the scanning electron microscope (SEM) without any adjustment. (B) SEM image after increase in contrast.

In Figure 18.8A, a very flat image, the histogram is compressed and indicates a lack of highlight and shadow. By comparison, Figure 18.8B shows an image with increased contrast. The histogram of this image shows an expanded tonal scale with good highlights and shadows.

Gamma refers to a nonlinear mathematical correction that is applied to digital images. Gamma is most effectively used to enhance contrast in the dark areas of an image. This is possible because a nonlinear correction is applied to the original index values. For example, in Figure 18.9 (dashed line) we see that an index value of 125 (a middle gray tone) will be converted to a gray value of 20 (a considerably brighter tone) if we use the upper gamma cor-

rection curve. Study Figure 18.10 A–C which shows the effects of applying different gamma values to a digital image. Gamma values greater than 1 result in more dynamic range (contrast) in the dark areas, values of 1 produce no change, and values less than 1 decrease contrast in dark areas. Gamma is the single, most important processing tool that should be adjusted initially—even before brightness and contrast. *As a general rule: use gamma correction far more than brightness-contrast.*

Burning-In and Dodging

Some processing programs such as Photoshop (Adobe Systems, Inc.) permit one to burn-in and dodge in light or dark areas, respectively, on the digital image much like one would expect with conventional photographic media (Chapter 8). But, as is the case with conventional media, excessively under- or overexposed areas that lack true information cannot be improved by this procedure. The advantage to digital burning-in and dodging is that one may work gradually on selected areas of the image bringing up details until the optimum effect is achieved—always with the capability to reverse mistakes and to continue the improvement process. Figure 18.11A, B demonstrates the use of digital dodging to lighten overly dense areas. You must exercise some caution when using this procedure because selectively lightening or darkening areas may lead to misinterpretation of the information. For instance, if some cell types are naturally more dense or take up more stain than adjoining cells, this may serve as a means for differentiating cells (tumor cells, macrophages, bacterial or viral inclusions, etc.) so that the density should not be altered in that case.

Removing Noise

Noise appears as random speckles throughout the digital image. It may be removed or minimized in several ways. Some digital capture devices have the capability to *frame average* whereby multiple copies of the image (frames) are accumulated a specified number of times and the frames averaged. Random noise is thereby removed in the averaging process while signal is intensified. Figure 18.12 shows the results of using frame averaging on an image to remove electronic noise. *Recursive filtering* is similar to frame averaging with the exception that the program computes an exponentially weighted average of the last image so that the last image is more dominant. In both cases, noise will be filtered out after the images are averaged or recur so many times.

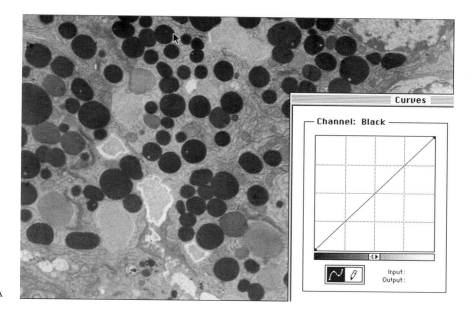

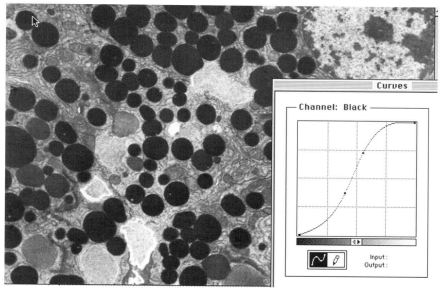

FIGURE 18.7 Alternate method of changing brightness and contrast in an image involves changing the plot of the gray levels, increasing or decreasing various grays. (A) Transmission electron microscope image of mammalian cells showing relatively low contrast. (B) After redrawing the curve of gray levels, contrast is improved significantly.

Other methods to remove noise (Figure 18.13A) involve applying special filters to the pixels so that out of place, excessively bright or dark pixels are adjusted to be similar to the neighboring pixels. Most image processing and analysis programs have versions of these filters that may go by different names. *Smoothing* is one method for removing noise whereby a pixel is replaced by the average of itself and its neighbors. The microscopist specifies the extent of the smoothing by selecting the number of pixels to be averaged surrounding a central pixel. As the name implies, smoothing reduces noise but at the expense of resolution. An example of a smoothing filter called a *low pass filter* (Figure 18.13B) can be used to minimize noise in an image by blurring sharp edges. This type of filter replaces the central pixel with the mean value of its neighbors. A *median filter* is useful to remove random lines (Figure 18.13D) or "shot noise" (Figure 18.13E) resulting from individual pixels occurring as black and white speckles (like buckshot) peppering the image. In this procedure, a specified number of pixels are compared (e.g., a 3″ × 3″ square containing 9 pixels) and the median value is used for the central pixel.

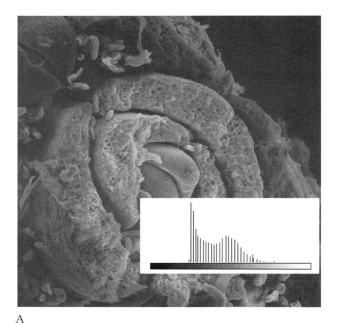
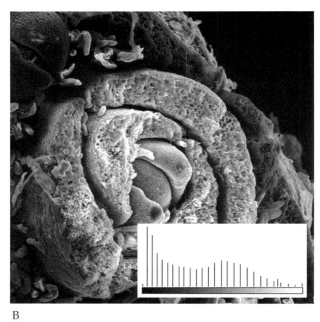

A   B

FIGURE 18.8 Histogram of gray levels in scanning electron microscope image of plant specimen. (A) A low contrast image shows a compressed histogram lacking highlights and shadows. (B) After adjusting contrast the histogram is expanded and shows highlights and shadows.

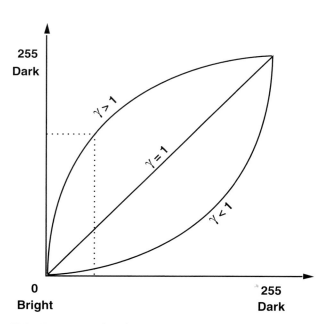

FIGURE 18.9 Plot showing conversion of input signal of 125 (on y-axis) into a gray value of 20 (on x-axis) when a gamma curve greater than 1 is selected. Compare Figure 18.10A–C for the result of such a conversion.

Filtering operations are carried out on a small group of pixels at a time. This small pixel group is also called a *kernel* and can consist of a matrix of pixels ranging in size from 3 pixels × 3 pixels, 5 pixels × 5 pixels, 7 pixels × 7 pixels, and so forth as selected by the analyst. When one changes the kernel size, this will affect the degree of change of the filtering process. Some experimentation is in order to study the effect of varying the kernel size.

The median filter can be applied many times until the noise is reduced to an acceptable level (Figure 18.13D). Fine details will be lost as the number of applications increase, but the overall boundaries of objects will be preserved resulting in an effect similar to posterization. In another type of filter, termed a *top hat filter,* any bright pixels that stand out (or stick up like the crown of a top hat) from the surrounding neighbors (e.g., the brim of the hat) are replaced. In the opposite case, where dark pixels constitute the noise, a *rolling ball filter* will replace the dark pixels with values similar to neighboring ones.

Besides the burning, dodging, and filters that may be applied to images that show areas of uneven illumination or brightness, one may use combinations of various neighborhood filters

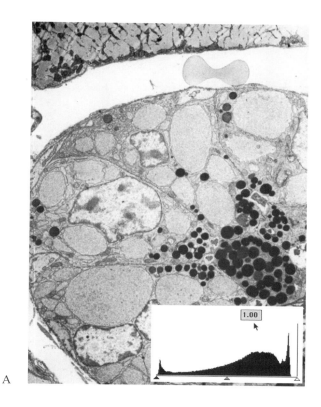

A

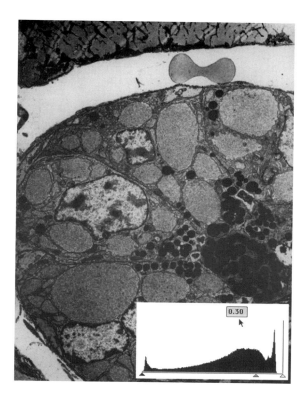

B

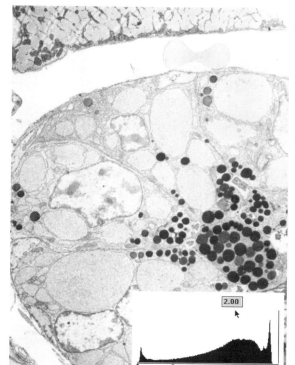

C

FIGURE 18.10 Effects of applying gamma correction to a digital image. (A) Transmission electron microscope image with a gamma value of 1. This is the uncorrected image as captured in the microscope. (B) When a gamma value of 0.30 is applied, note the overall loss of contrast in the shadows. (C) When a gamma value of 2.0 is applied, the shadows show an increase in contrast and an overall lightening so that one is able to discern more information in the shadow areas of the image.

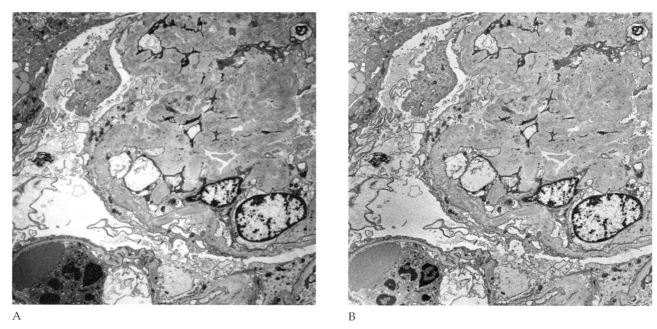

A   B

FIGURE 18.11   Ultrathin section in transmission electron microscope captured on digital camera. (A) Original image
shows dark areas in top and bottom left corners. (B) After digital dodging, the darkened areas show more detail and
the overall image is of the same brightness.
(Courtesy of B. Armbruster, Gatan Instruments.)

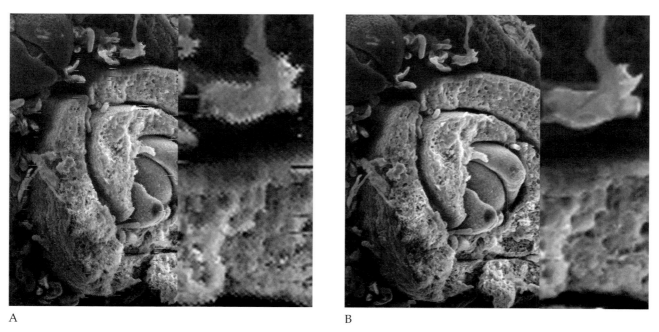

A   B

FIGURE 18.12   Scanning electron microscope image showing the effect of frame averaging. (A) Image shows a single
frame captured digitally from the SEM. Notice electronic noise present as random lines or streaks. (B) Image represents
15 frames that were averaged together to remove electronic noise. Right portion of both images is enlarged 3× to show
details.

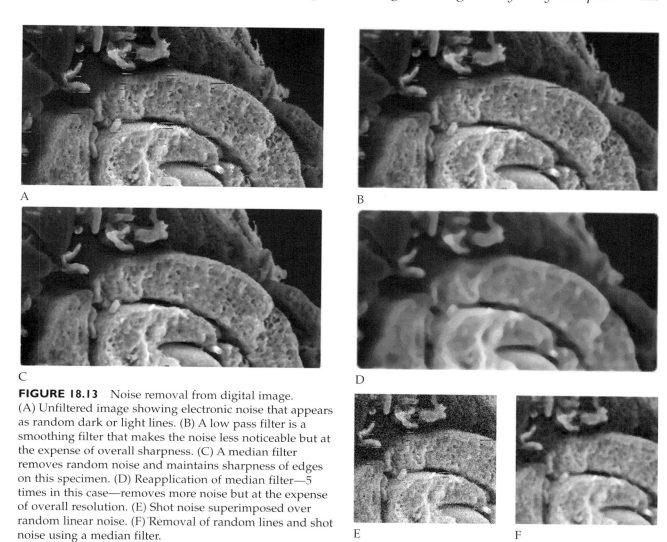

FIGURE 18.13   Noise removal from digital image. (A) Unfiltered image showing electronic noise that appears as random dark or light lines. (B) A low pass filter is a smoothing filter that makes the noise less noticeable but at the expense of overall sharpness. (C) A median filter removes random noise and maintains sharpness of edges on this specimen. (D) Reapplication of median filter—5 times in this case—removes more noise but at the expense of overall resolution. (E) Shot noise superimposed over random linear noise. (F) Removal of random lines and shot noise using a median filter.

followed by smoothing and subtraction from the original image to obtain excellent results. This procedure requires much practice and an appropriate specimen. For details, consult *The Image Processing Handbook* (1994).

Background Removal

It is possible to remove consistently appearing noise or artifacts such as uneven illumination, dirt, or lint by subtracting the artifact out of the image. To accomplish this, an image is acquired in the usual way and a second image is acquired but without the specimen present so that it contains only the artifacts. If the artifactual image is subtracted from the first image, then only true data will remain. This procedure is very important when one uses CCD cameras in the TEM to capture images from a phosphorescent screen because the granularity of the screen or perhaps dirt or lint fibers may be present in all of the images unless they are subtracted (Figure 18.14).

Sharpening

You can increase the apparent sharpness of a digital image in several ways. Many programs contain a *sharpening filter* that one simply selects to achieve the sharpening. This uses the *unsharp masking filter* technique and works by combining a somewhat blurry or unsharp version of an image with the original image (Figure 18.15). As a result, contrast is enhanced in high contrast areas (edges between light and dark areas) while low contrast areas are largely unaffected. This will result in a sharper appearing image but at the expense of smoothness because the pixels making up the image along edges now appear more prominently. *High pass*

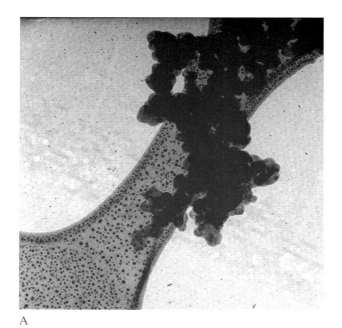
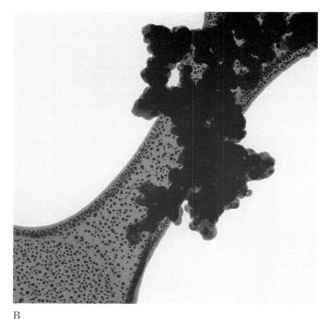

A   B

FIGURE 18.14 Background removal. (A) Unprocessed digital image of carbon particles on holey film viewed in the transmission electron microscope. Note the presence of dirt and random bright spots on viewing screen. (B) After subtraction of image of plain viewing screen, noise and bright spots are removed and only specimen is imaged. (*Courtesy of B. Armbruster, Gatan Instuments.*)

filters(Figure 18.16) replace central pixels with significantly brighter pixels as compared to the neighboring pixels resulting in higher contrast and apparent increase in sharpness. If one wishes to exercise more control over the process, it is possible to use the high pass filter to increase contrast at local boundaries and then to add the filtered image to the original image. With appropriate images, this will result in a sharpened image with a reasonable overall gray scale (Figure 18.17).

Other filters that may increase sharpness by selectively enhancing contrast along edges include the *Laplacian filter* that enhances all edges and the *Sobel filter* that enhances the principal edges in an image.

Fast Fourier Transforms

The conventional images that most microscopists deal with exist in what is termed the *spatial domain.* It is in the spatial domain that we previously demonstrated how to effect the contrast, sharpness, noise, and background removal. Another, mathematically intensive way of dealing with images for more profound processing is to numerically transform the spatial domain (gray scale) information into the so-called *frequency domain* by applying the

mathematical equations developed by Joseph Fourier. This converts the image into a display of frequencies displayed as an energy spectrum. The energy spectrum is plotted as an array of bright points symmetrically arranged around a bright central area that represents the average intensity across the image. The farther one moves away from the center, the higher the frequency of occurrence of a particular intensity. The degree of brightness of each point indicates the amplitude while the angular direction from the center shows the direction of the waveform.

The Fourier transform may be accomplished using lenses, in which case this is termed an *optical transform.* An example of an optical transform is an electron diffraction pattern. Another way of achieving the conversion without resorting to lenses is to use a computer to apply the Fourier equations to a digitized image. When the computer carries out these calculations, the conventional image is converted into a diffractogram. Once this conversion takes place, it is possible to apply intensive control over the processing of the image.

For example, if we perform a Fourier transform (usually termed a "fast" Fourier transform or FFT due to the accelerated algorithm used) on an image of a cell, we will obtain results similar to the

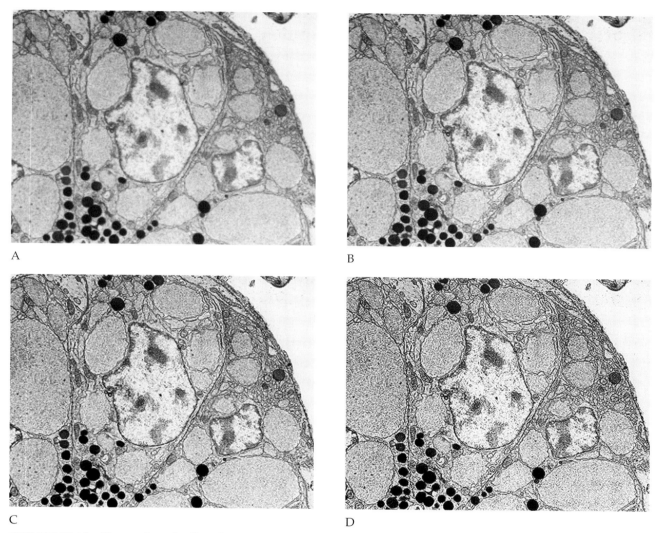

A

B

C

D

FIGURE 18.15 Sharpening of a digital image using unsharp masking technique. (A) Original digital image of an ultrathin section captured from the transmission electron microscope. (B) Sharpening of original image after a single sharpening pass. (C) After two sharpening passes, the sharpening becomes excessive. (D) By a third pass, overall details become obscured.

diffractogram shown in Figure 15.27. Notice the central, circular cloud that is brighter and denser as one approaches the center. On the other hand, if we transform an image containing a periodic structure such as a crystal, the FFT (or diffractogram) will reveal a pattern-like arrangement of spots representing the repeating lattice lines of the crystal (see Figure 15.31, lower right and Figure 18.18). With a digital image, it is possible to edit the FFT on the viewing screen and remove stray noise around the spots (Figure 18.19A). This will emphasize the repeating structure and exclude nonrepeating information, thereby increasing the sharpness and contrast of periodic structures

when one reconstructs the image by performing an inverse FFT (Figure 18.19B and Figure 18.20). Alternatively, removing the spots, will remove the periodic information of the crystal upon performing an inverse FFT. An example in which this might prove useful would be to de-emphasize an artifact with periodicity such as knife chatter marks in a section (Figure 18.21). Because knife chatter marks are not straight and regular, some marks will still remain. FFT would also be useful to remove periodic structure such as muscle fibers to see detail that would otherwise be hidden, or to remove electronic noise or vibration from an image.

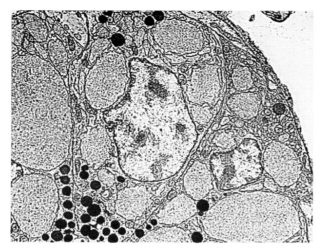

FIGURE 18.16 A high pass filter may be used to increase contrast along edges resulting in an apparent increase in contrast.

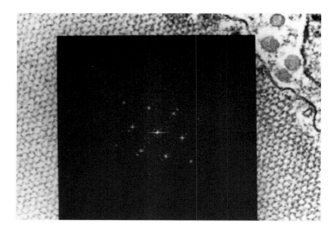

FIGURE 18.18 Fourier transform of Figure 20.142. Energy spectrum shows repeating structures evidenced as bright points. Nonrepeating information is shown as randomly occurring spots.

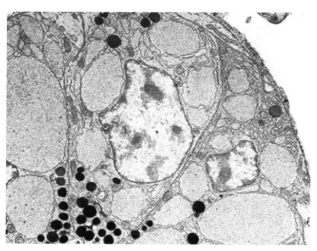

FIGURE 18.17 By adding images as in Figures 18.15A (unfiltered image) and 18.16 (high pass filter) one obtains an image with good sharpness and tonal range as shown in this figure.

Besides enhancing or de-emphasizing periodic structures, the FFT can be used to detect certain conditions or problems in a specimen that may not be readily apparent in the spatial domain, or gray scale image. Among the conditions that the FFT may be useful to detect are: (a) optimum focus as well as amount of defocus of an image, (b) presence and direction of astigmatism, (c) detection of drift and removal of blurring due to movement, (d) misalignment of apertures, and (e) determination of helical structures.

Look-Up Tables, Thresholding, Binary Images, and Pseudocoloring

As the digitized image signal passes to and from the computer's random access memory, or RAM, it passes through a component of RAM that specifies the palette or range of grays that are to be used. This palette is usually termed a look-up table (LUT) because the input signal is compared against data in a stored table to determine the value of the final output signal. In Figure 18.22A, for example, we see some input signals that pass through the LUT and are converted to different output values, in this case resulting in a reversal (or inversion) of the image contrast thereby yielding a negative from a positive image. *Segmentation* and *thresholding* are ways of adjusting the LUT so that we are able to select features of interest within an image by specifying the range of gray values that make up these features. For instance, in Figure 18.22B, a LUT was set up that converted all gray levels below a threshold value of 100 to zero (black) but left all of the other gray levels unchanged. In this case, the feature of interest (which had grays ranging from 0 to 100) would now be converted to a very prominent black. A type of thresholding, termed segmentation or contouring, involves specifying upper and lower limits that are to be selected. In this way one is able to select the gray levels making up the feature of interest. An example of a LUT for segmentation is shown in Figure 18.22C and the results of such a segmentation are shown in Figure 18.26B.

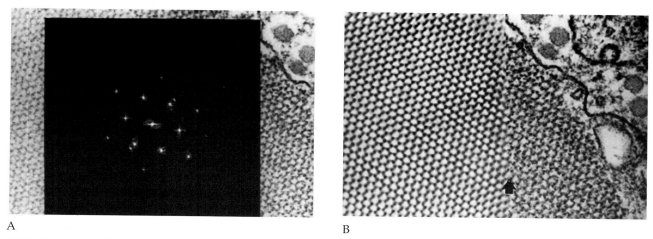

A   B

FIGURE 18.19 Editing of the FFT followed by image reconstruction by means of an inverse FFT results in an image showing an enhancement of periodic structures. (A) Edited version of the FFT shown in Figure 18.18. Nonrepeating information has been removed leaving behind only bright spots that represent repeating structures. (B) Inverse FFT showing the results of editing the spectrum in (A). Notice how the repeating structures are more evident. Arrow indicates the boundary between edited/inverse transformed and original information on right.

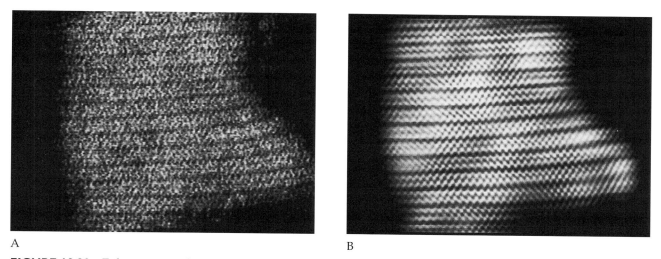

A   B

FIGURE 18.20 Enhancement of repeating structure by means of Fourier transform editing. (A) Original image of negatively stained ATPase enzyme suggesting repeating structure. (B) Image following Fourier transformation, enhancement of repeating spots, removal of nonrepeating information and inverse transformation to obtain an improved image showing repeating subunit structures of the enzyme macromolecules.
(Specimen courtesy of P. M. D. Hardwicke.)

Binary images are those that have been thresholded so that features are either black or white. Binarization is an important way of selecting the features for further image analysis (Figure 18.23). For instance, a common procedure in image analysis involves counting particles or objects present in a specimen. Under ideal conditions one first makes the features of interest binary and then instructs the program to do the counting. Unfortunately, most specimens are less than ideal and the binary image often must be further processed to achieve accurate

counts. A common problem involves touching or connected particles that are seen as singlets rather than discrete units. Several filters may be used to separate particles connected by bridges or projections. The process of *erosion* will remove pixels from the perimeter of objects much like the removal of layers in an onion. The process may be repeated as often as necessary until the erosion has separated the objects and an accurate count may be made. One cannot use erosion alone, however, if one wishes to know the area of the objects because we have

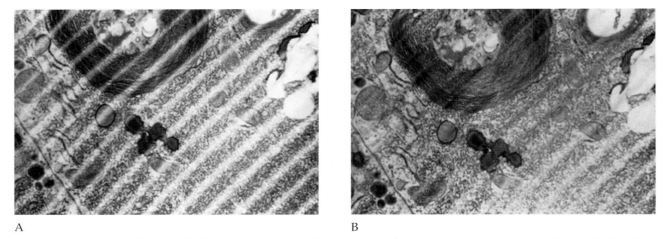

A   B

FIGURE 18.21   Partial removal of repeating structure (chatter marks) by using Fourier transform editing. (A) Original image. (B) Image following Fourier transformation, editing of repeating spots, and reverse transformation to obtain an improved image.

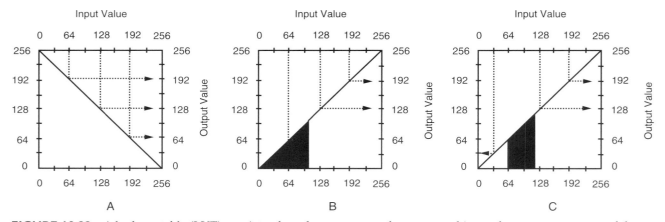

A   B   C

FIGURE 18.22   A look-up table (LUT) consists of a palette or range of grays stored in random access memory of the computer. By selecting a particular LUT one is able to effect changes in brightness and contrast or to eliminate certain parts of the gray scale. (A) This LUT will result in inversion of the image and convert a positive to a negative or vice versa. (B) This LUT will convert all gray levels below a value of 100 to zero (black) but leave all other gray levels unchanged. (C) This LUT will define upper and lower limits (64 and 120, respectively) to select the features of interest.

eroded some of the particle. An *opening filter* is needed in this situation. This filter first applies an erosion to separate the particles and then a dilation to return the particles back to the proper area (Figure 18.24). A closing filter, on the other hand, performs a dilation followed by an erosion. This is often used to remove or close small, dark holes in a bright area.

Pseudocoloring is another capability of LUTs. In this case instead of converting one gray level into another, a particular gray level is converted into a color specified by the analyst. In some instances, pseudocoloring is useful to highlight certain features that may not be evident using a gray scale.

Since the human eye can only discriminate 20 to 30 shades of gray but is able to differentiate hundreds of colors, an image composed of numerous gray tones might best be presented using pseudocoloring. Although it is possible to obtain dramatic renditions of biological specimens using pseudocoloring, one must take care not to subvert the overall gestalt or real contents of the image. Pseudocoloring works best with flat specimens having defined contours. Segmentation, thresholding, and pseudocoloring are important processing steps prior to image analysis because the features of interest to be quantified or analyzed are selected by these processes.

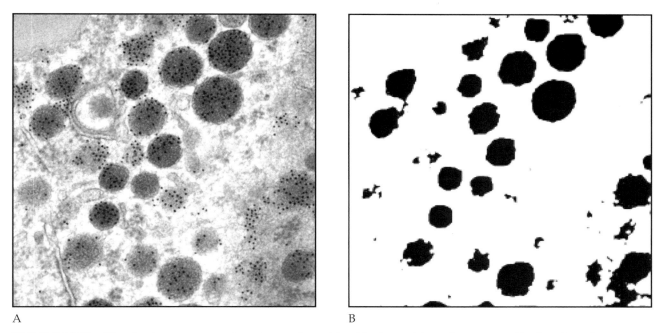

A   B

FIGURE 18.23   Binarization converts selected grays into either black or white to better quantify the selected tones. (A) Original image showing labeling of secretory granules with colloidal gold. (B) Binarization of the original image results in a conversion of the secretory granules into black areas that may then be quantified by image analysis. *(Courtesy of B. Armbruster, Gatan, Inc.)*

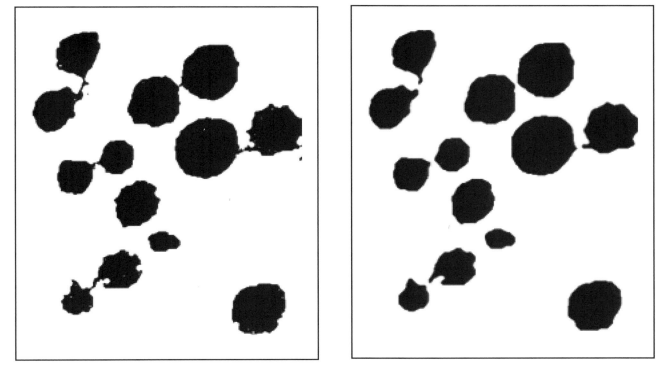

FIGURE 18.24   To count or quantify the area of touching particles (*left*) it is necessary first to separate them by using an erosion procedure followed by a dilation procedure to obtain the results shown (*right*).

Image Averaging and Computer Enhancement

There is often a need to provide an idealized "model" of images that are poorly resolved at very high magnifications. In the past this was accomplished photographically to some degree by rotation of symmetrical images (Markham et al., 1963). In some biological materials, images are of periodic or symmetrical structures with features that show up poorly against the background. The computer can be used to strengthen images by comparing features in a number of similar images to determine what is and what is not a typical feature of a micrograph (Afzelius et al., 1990). A number of like images are required for this purpose to determine an average. Once the image has been captured in a computer, the like images can be manipulated (rotated, magnified, demagnified, modified to similar elipticity) to overlay one another. Computer software will also remove known distortions from single images. The computer is directed to *average* several images to provide a single image with more sharpness than any of the original micrographs. The example shown in Figure 18.25 shows how the com-

puter averaging method is used to visually sharpen structures that are less well visualized in an electron micrograph.

Final Display of Digital Images for Publication and Presentation

After one has processed and adjusted the digital images to maximize the information, one ultimately will wish to display the image either in a publication or a poster presentation. Some journals will accept digital images on disk provided that they are in the proper format and of sufficient resolution. A generally accepted format is the TIFF format that can be used by Macintosh and IBM-compatible systems. If one uses a high-quality software program to generate and manipulate the image, it is best to use this program to produce the TIFF file because there may be various features that are better rendered by the native program. Some images may be quite large and take up an entire floppy diskette, so it is best to inquire with the

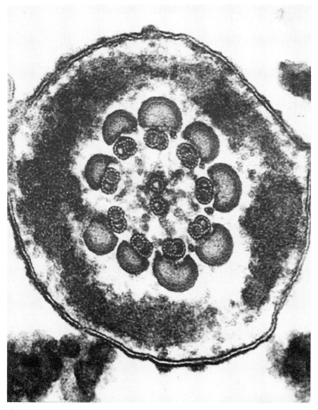

A

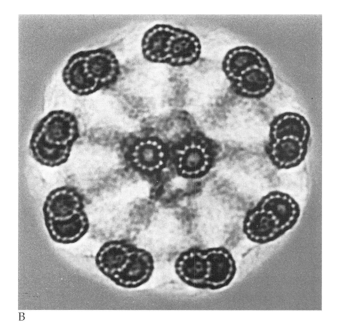

B

FIGURE 18.25 Computer averaging of several micrographs will improve the clarity of detail. (A) Original micrograph. (B) Computer-averaged micrograph
(Courtesy of R. Dallai.)

publisher if they are able to accept diskettes or larger format removable disks (Syquest, Iomega Zip or Jaz disks, magneto-optical or CD ROM) and the format (IBM or Macintosh) that they require. Modern publishers will be able to accommodate a wide range of formats and it may even be possible to submit the publication as an electronic document.

For hard copies of the digital images, one has many different printer options. The least expensive option is an ink jet (or bubble jet) printer that sprays tiny droplets of ink onto a specially coated paper. One printer in particular, the Epson Stylus, produces a remarkably high quality print with a resolution over 700 dots per inch (dpi) that is certainly suitable for posters and even some reports. This printer costs around $400 and is compatible with most computers. Laser printers will produce reasonable quality study prints but even at 1200 dpi, the quality is not as good as some of the ink jet printers. Dye sublimation printers produce the highest quality print that approaches that of photographs. They range in price from under $2,000 to over $10,000—the difference being speed more than print quality. The lower cost printers take 5 to 10 or more minutes to produce a print that the higher cost units produce in less than 1 minute. Unless one intends to produce many prints, the lower cost unit should suffice. At the top of the line, the best quality output is from high resolution printers and typesetters that are used to produce output for books and professional journals. Such devices start at several thousand dollars and can easily cost tens of thousands of dollars; however, they offer high resolution at reasonable speeds. Digital images may also be input into slide makers, devices for generating high-quality slides on 35 mm film for projection. Slide makers will generate transparencies with resolutions of 4,000 to 8,000 lines and may even be used to generate a photographic negative.

Electronic Image Processing and Image Analysis for Electron Microscopy

The acquisition of an electronic image is the necessary prerequisite for analyzing that image electronically. Computer analysis offers the advantage that the analysis of the image may take place without ever having printed that image, a saving of both time and materials. Furthermore, in most instances the computer analysis may be more accurate than the manual means. (If the image is to be analyzed by conventional means, then a print can be made from the computer by a high resolution printer and the task accomplished manually.)

Image analysis by computer has developed for many fields of study, simultaneously. Images from satellites are often analyzed for one particular feature or another. Much of the important early work in image analysis was driven by the space program and took place at such locations as the Jet Propulsion Laboratory in California. In relation to electron microscopy, image analysis usually leads to a quantitative expression of some parameter (length, area, volume, etc.). Thus, the general purpose of image analysis is to perform certain tasks electronically that are a part of the scheme of stereology as presented in Chapter 13 (see Figure 13.2).

What follows is a description of some of the basic capabilities of image analysis software. Each software package is unique and employs somewhat different terminology. Although some software may have been designed specifically for analysis of biological samples, most have similar features. Thus what follows are common measurements obtainable with most software packages.

Particle Counts

The objective is to count the number of particles of a particular type. Image manipulation is minimal if the particles to be counted do not make contact with each other and are black against a white background. If particles are of reasonably similar sizes the computer will allow you to specify (in pixels) the minimum and maximum size of particles that will be counted. In some software packages the computer numbers, outlines, or removes particles as they are counted.

If particles are the only discernible feature in the micrograph then the *binary* feature of the software will create only a black on white image. If this is not possible then the thresholding function may allow the investigator to bracket the gray level of the particles for binary imaging. In some instances, overlapping gray levels in the object to be counted and other areas in the micrograph force one to manually select the objects to be counted. Overall this does not result in time saving. Assuming particles are of a sufficiently different tonal value than the background, one can threshold or specify a range of gray levels that will be counted.

When thresholding or binary imaging of the selected objects is obtained, the *count* function will provide nearly instantaneous particle counts. This procedure is most valuable when there are a large number of particles to be counted in the field of view or when there are a large number of fields to count. Otherwise, the task may be performed manually in just as rapid a manner.

Area

The area of desired objects can be measured electronically to give a volume density (V_V) or volume occupancy percentage ($V_V\%$). The computer-stored electron microscope image is placed on the screen and manipulated as described in the section on "Particle Counts" to give a binary image. The area of the blackened region is determined using the *area* feature of the program. If the thresholding function does not isolate the structure to be measured even after image processing, then one must manually outline the areas to be determined. No matter which way the areas are highlighted, the computer will give a V_V based on the area of the desired structure to the area of the micrograph or a predetermined boxed in reference area.

The *area* function of the computer is the most time-saving function because area measurements made using manual digitizers are more time consuming and probably less accurate. *Area* is a commonly desired measurement that may stand on its own or may be used to determine the volume of a particular structure when the reference volume (usually cell or cytoplasmic volume) is known (Equation 18.1).

Equation 18.1:

$$V_{structure} = \frac{V_V \text{ of structure}}{\text{of interest}} \times \text{reference volume}$$

An alternate and more time-consuming electronic method to obtain V_V or $V_V\%$ information is to overlay a point lattice grid on the micrograph. Some software features contain a variety of lattices that may be accessed. The investigator then manually records points as being over the structure of interest or as part of total points and employs the following formula (Equation 18.2). It is generally far more efficient and less error prone to count grid points to determine area fraction and obtain V_v.

Equation 18.2:

$$V_V = \frac{\text{points over structure of interest}}{\text{total points}}$$

Because area and volume determinations are among the most common measurements, we have illustrated this process (Figures 18.26 and 18.27). It was our objective to determine the volume density of secretory granules that are the dense bodies (Figure 18.26A). This image was digitized directly from the microscope and exported to an image analysis software system. The thresholding tool was used to determine the range of gray that would produce a binary image shown in Figure 18.26B. The program then determined the blackened area of the image as compared with the total area of the micrograph. The result was that 40.8% of the microcraph was composed of secretory granules yielding a V_V of 0.408.

In yet another example of an area determination, it was the desire of the investigator to determine the area of the swollen endoplasmic reticulum in Figure 18.27A compared with the area of the glandular acinus outlined 18.27B containing the endoplasmic reticulum. The gray level of the endoplasmic reticulum and its contents did not allow for thresholding, because the grays of other structural components of the micrographs overlapped with the grays of the endoplasmic reticulum. Thus no useful binary image could be easily produced. The *trace* tool was utilized to outline each individual saccule of endoplasmic reticulum and to determine the area of each. Their cumulative area was obtained and divided by the area outlined in the micrograph to obtain the V_V (.357) (volume occupancy) or $V_V\%$ (35.7).

Mean Particle Diameter

The diameter of particles or structures that are not round is dependent on where measurements are taken. If you average a large number of diameter measurements taken through the centroid of the feature of interest you obtain the *mean particle diameter*. First use the tools of thresholding to produce a black-and-white image, which selects and contrasts the particles of choice. The computer is then directed to determine the mean particle diameter.

Grid Overlays for Surface Density Measurements

A commonly used stereologic measurement for surface area is the line-intersection method. For those computers that have multipurpose grids or Mertz grids (see Figure 13.6) these grids can be brought up to overlay micrographs. The investigator then uses the grids to count both points over and line intersections with the desired structure to obtain the surface density (S_V) or surface density percentage ($S_V\%$)

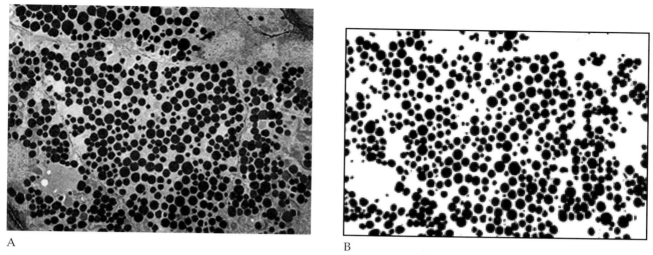

FIGURE 18.26   Thresholding of a gray scale image. (A) Micrograph of secretory gland cells of the mouse reproductive tract in which one hopes to determine the volume density of secretory granules (SG) in the micrograph field. (B) Binary image of (A) thresholded to show only secretory granules.

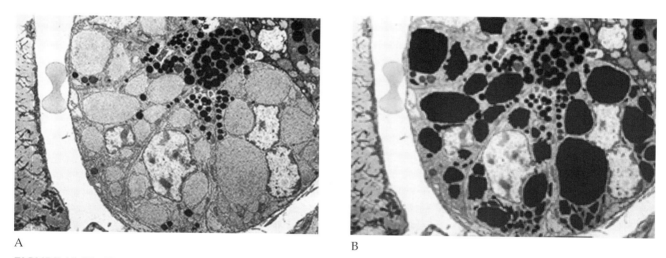

FIGURE 18.27   Determination of area occupied by specific organelle. (A) Micrograph of secretory gland cells of the mouse in which one wishes to determine the volume density of the swollen endoplasmic reticulum (ER). An attempt to threshold the ER was not successful because gray levels overlapped those of other parts of the cell. (B) To determine the area of the ER, individual secretory granules were outlined manually (made binary). The areas of the blackened granules were then determined as well as the much larger area of the cell encircled by the line.

measurements. Formulas (Chapter 13) are used to calculate surface density parameters. Remember from Chapter 13 that the product of the surface density and the reference volume yields surface area (S).

Length

The length of items that show one dimension on electron micrographs, but are in reality two-dimensional (i.e., a surface), can be determined in digital micrographs if the line on the screen is unbroken and clear. By one method, you must select the line of choice with a tool on the computer toolbox. The computer will trace the line with the *length* function and will calculate the length of the line. In the case of biological micrographs, many cuts are not perpendicular to membranes, yielding fuzzy lines that are not traceable by the computer. To overcome this problem, software will allow you to make a freehand line over the line of interest and direct the computer to *measure* this line or you may *skeletonize* the blurred line down to 1 to 2 pixels in width. On the other hand, if the length to be measured represents a perimeter of an object, then the *perimeter function* can easily measure the perimeter of the sectioned profile.

Determination of Angles

Software packages may have a tool to measure angles created by images that have been thresholded. The computer will commonly perform this function on angular structures or a special *angle tool* may be used to measure angles.

Center of a Mass (Centroid)

The center of a mass or weighted mean center of all coordinates can be determined using a function operating on a selected thresholded (binary) image.

Image Analysis Software

Most of the features already discussed are available through a free (public domain) software package obtainable from the National Institutes of Health. This package known as *NIH Image* was developed for the Macintosh system and written by Wayne Rasband. It be downloaded from the Internet using the address http://rsb.info.nih.gov/nih-image. Part of the software package is a detailed manual for its use. *NIH Image* is an ideal beginning program because it offers excellent support from a community of users, runs on an inexpensive and common computer, and is a low-risk way to acquire basic familiarity with the possibilities of image processing and analysis.

Three-Dimensional Imaging

Slices or image profiles that have been taken serially by light or electron microscopy can be aligned and stacked on one another so that one may view a transparent stack of images. This gives the investigator a good concept of the three-dimensional configuration of the structure. In the process of stacking images in the computer you must utilize alignment techniques to ensure that one image is properly stacked on another. The software allows the investigator to position each profile relative to another by moving in the x- or y-axis as well as to rotate and/or invert the image.

Most commonly, software is dedicated to the reconstruction of planar images into three-dimensional images. Like the stacking procedure employed earlier, the three-dimensional image is constructed by aligning images with respect to one another and requesting that the computer reconstruct a final three-dimensional image using the knowledge of the section thickness. When the reconstructed image is completed, the three-dimensional creation often appears impressive. It may be rotated to see all aspects of the reconstructed structure. The final reconstruction may contain internal features that were input as the surface of the structure was input, so that internal components of the reconstructed structure are viewed in one color while the surface is viewed in another (Royer and Kinnamon, 1988).

Practicality and Utility of Computer Image Analysis: A Caveat

The speed and variety of tasks make the computer a desirable tool to analyze micrographs and imported images. Repetitive analysis of images using the same or similar protocol puts the computer's capabilities to best use. However, with all the available technology, computer software is not always readily adaptable for the analysis of biological materials. We can see enemy tanks from satellite images and match fingerprints in forensic medicine, but the computer is not yet able to recognize biological structure (e.g., filled vacuoles versus nuclei) nearly as well as the human eye. Thus in determining whether or not to purchase hardware and software for image analysis one must carefully consider what is to be done and if the computer is a practical solution to answering the problem at hand. There are instances of individuals spending tens or hundreds of thousands of dollars with the expectation that hardware and software will solve problems that are, in reality, more readily achieved by manual stereology (see Chapter 13). There are just as many examples of sophisticated computers remaining idle while waiting for individuals knowledgeable in software use.

TelePresence Microscopy

To make electron microscopes accessible to more users, software is being finalized that will enable researchers to observe images and control many of the functions of the microscopes over a wide area network (Ellisman et al., 1994; Parvin et al., 1995). At the present time, researchers send specimens to a suitably equipped resource center and establish an electronic link with the microscope after the specimen has been inserted. Researchers are able to observe the images on their computer monitors and control many of the commonly used adjustments of the microscope (specimen movement, illumination adjustment, magnification changes, capture of images, etc.). At present there is a time delay (due to transmission of the information and feedback to the operator) before selected changes take effect; however, as electronic connections improve the controls will approach the feel of real-time operations. Zaluzec (1995) has coined the term TelePresence microscopy to mean that an investigator will be able to sit at his personal computer and control an electron microscope located hundreds or even thousands of miles away with the impression of actually being present at the control console.

REFERENCES

Afzelius, B., et al. 1990. Microtubules and their protofilaments in the flagellum of an insect spermatozoan. *J Cell Sci* 95:207–17.

Castleman, K. R. 1979. *Digital image processing.* Prentice Hall, Englewood Cliffs, NJ.

Cookson, J. 1994. Three-dimensional reconstruction in microscopy. *Proc Royal Microsc Soc* 29(1):3–10.

Doherty, E. R., and J. Astola. 1994. *An introduction to nonlinear image processing.* SPIE, Bellingham, WA.

Ellisman, M. H., B. E. Soto, and M. E. Martone. 1994. *Proc Microsc Soc Amer,* 64–65.

Gonzalez, R. C. and P. Wintz. 1987. *Digital image processing,* 2d ed. Addison-Wesley, Reading, MA.

Gunderson, H. J. G., et al. 1988. Some new, simple and efficient stereological methods and their use in pathological research and diagnosis. *Acta Pathol Microbiol Immunol Scand* 96:857.

Inoué, S. 1986. *Video microscopy.* New York: Plenum Press.

Jain, A. K. 1989. *Fundamentals of digital image processing.* Prentice Hall, London.

Markham, R., S. Frey, and G. J. Hills. 1963. Methods for the enhancement of image details and accentuation of structure in electron microscopy. *Virol* 20:88–102.

Misell, D. L. 1978. Image analysis, enhancement and interpretation, vol. 7. In *Practical methods in electron microscopy,* A. M. Glauert, ed., North Holland Publishing Co., Amsterdam.

Parvin, B., D. Agarwal, D. Owen, M. A. O'Keefe, L. H. Westmacott, U. Dahmen, and R. Gronsky. 1995. A project for on-line remote control of a high voltage TEM. *Proc Microsc Soc Amer,* 82–83.

Prewitt, J. M. S., and M. L. Mendelsohn. 1966. The analysis of cell images. *Ann NY Acad Sci* 128:1035–53.

Rich, J., and S. Bozek. 1994. *Photoshop in black and white.* Peachpit Press, Inc., Berkeley, CA.

Royer, S. M., and J. C. Kinnamon. 1988. Ultrastructure of mouse foliate taste buds: Synaptic and nonsynaptic interactions between taste bud cells and nerve fibers. *J Comp Neurol* 270:11–24.

Russ, J. C. 1990. *Computer assisted microscopy.* New York: Plenum Press.

Russ, J. C. 1994. *The image processing handbook,* 2d ed. CRC Press, Inc., Boca Raton, FL.

Shotton, D. 1993. *Electronic light microscopy.* New York: Wiley-Liss.

Zaluzec, N. J. 1995. Tele-presence microscopy: An interactive multi-user environment for collaborative research using high speed networks and the Internet. *Proc Microsc and Microanal* 1995, 14–15.

Software

Photoshop—Adobe Systems, Inc., 1585 Charleston Road, Mountain View, CA 94039.

NIH Image—Wayne Rasband, National Institutes of Health, Research Services Branch, National Institutes of Mental Health, Bethesda, MD 20892.

Image-Pro Plus—Media Cybernetics, 8484 Georgia Ave., Silver Spring, MD 20910.

IP Lab—Signal Analytics, 440 Maple Avenue East, Suite 201, Vienna, VA 22180.

T3D-Fortner Software LLC, 100 Carpenter Drive, Sterling, VA 20164-4464.

Chapter 19

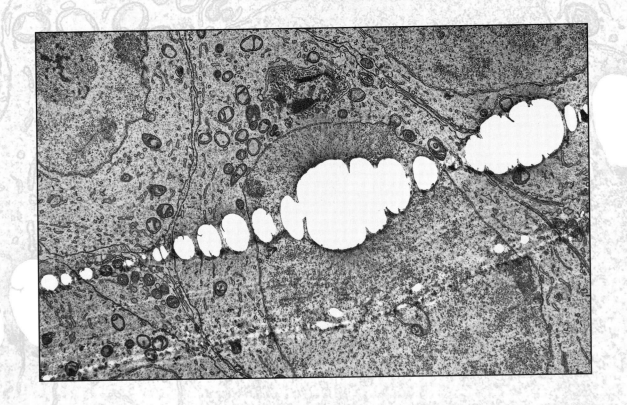

membranes of the Golgi stack due to the size and distribution of silver grains on the photographic paper and the resolving power of the unaided human eye (0.2 mm). The space in each bilayer is resolved only at the higher magnifications (see the section "Membranes" to follow).

To assist the human eye as it views micrographs, one must artificially increase its resolving power through enlargement of high quality negatives. A fine grain negative, as is used in electron microscopy, may be enlarged photographically as much as 10 times its original size to obtain the maximal information from it. To do this, the negative must be sharply in focus. For most electron microscope negatives, a 2 to 3 times enlargement will allow one to gain the great majority of information. Enlarging standard size negatives ($3\frac{1}{4}'' \times 4''$) about 2.6 times gives a final print magnification of approximately $8'' \times 10''$.

It is important to distinguish high resolution light micrographs from electron micrographs. Figure 19.3 shows great resolution considering it is magnified over 2,000 times. An image of this type is the best the light microscope can produce as evidenced by the ability to visualize numerous organelles and inclusions. Yet, the features of the micrograph are blurred. Electron micrographs at a comparable magnification would have sharp features, demonstrating the high resolution afforded by the instrument (refer back to Fig. 4.9).

Membranes

Membranes are a commonly encountered feature in most electron micrographs. Staining of the lipid bilayer (by osmium and heavy metals) is observed primarily over the polar regions of phospholipid molecules causing them to appear dense, whereas the center of the membrane appears translucent (*trilaminar profile of a lipid bilayer*; Figure 19.4). The way membranes are sectioned leads to one of the most common causes of micrograph misinterpretation. Figure 19.5 shows a micrograph that depicts several unit membranes, each of which in some area

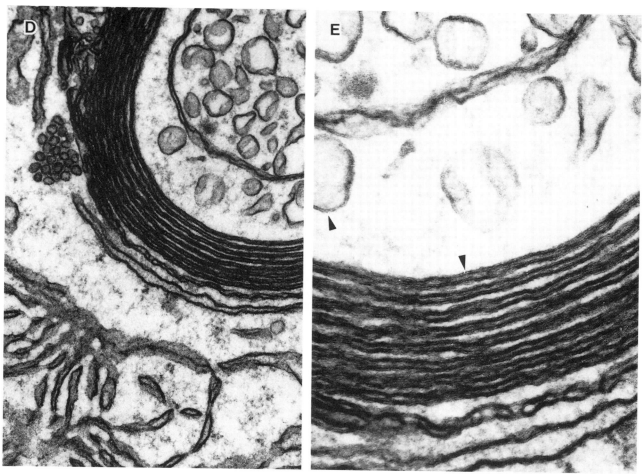

FIGURE 19.2 Continued

shows the typical trilaminar profile of a lipid bilayer. The microtome may enter the membrane bilayer at an infinite number of angles. Our brain expects to see membranes sectioned perpendicularly, where they are resolved as distinct "railroad tracks." How-

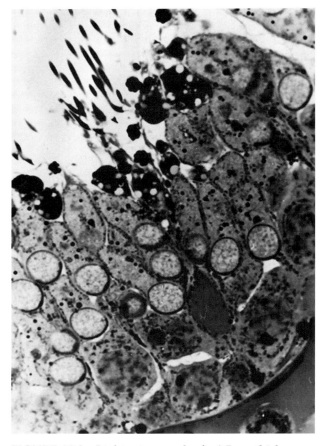

FIGURE 19.3   Light micrograph of a 0.5 μm thick section stained with toluidine blue.

ever, there are other possible ways that membranes may appear in sections. Their appearance depends on the angle at which the sectioning plane encounters the membrane and whether or not the membrane curves within the section (Figure 19.6). Keep in mind that the electron beam must pass through the entire thickness of a section and that any structural elements in the path of the beam will scatter electrons from the viewing screen. When viewing micrographs, membranes may appear to be broken or missing in certain areas when, in reality, they are traveling parallel or nearly parallel to the plane of section. Look for a slight fuzziness in the region of the missing membrane, which indicates the presence of an obliquely sectioned membrane (Figure 19.6). Usually, membranes that are really broken show sharp discontinuities (see Figure 19.19).

The less than ideal sectioning of membranes may be overcome to some extent by tilting the section on the specimen stage (see Chapter 2). In this way, the plane of the membrane bilayer may be placed parallel to the path of the electrons. This will overcome the fuzzy appearance of obliquely sectioned membranes. Figure 19.7 shows the effect tilting a specimen has on the appearance of membranes of a Golgi apparatus.

Shape, Kinds, and Number of Structures

It is important to remember that transmission micrographs are taken from three-dimensional material that has been finely sectioned in a single plane. It is almost impossible to ascertain from thin sections the *shape* of a structure that is larger (and most structures are) than the thickness of the section (Figure 19.8). Because

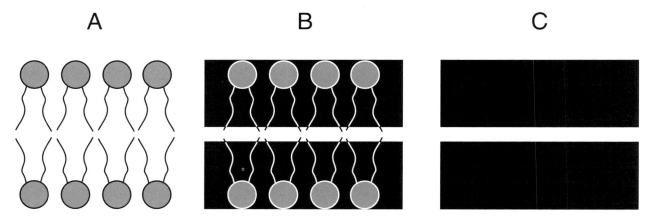

FIGURE 19.4   The phospholipid molecules of a membrane as shown in (A) are stained over their polar region and, to a large extent, over their nonpolar tails as shown in (B), giving rise to the trilaminar appearance of membranes in electron micrographs as shown in (C).

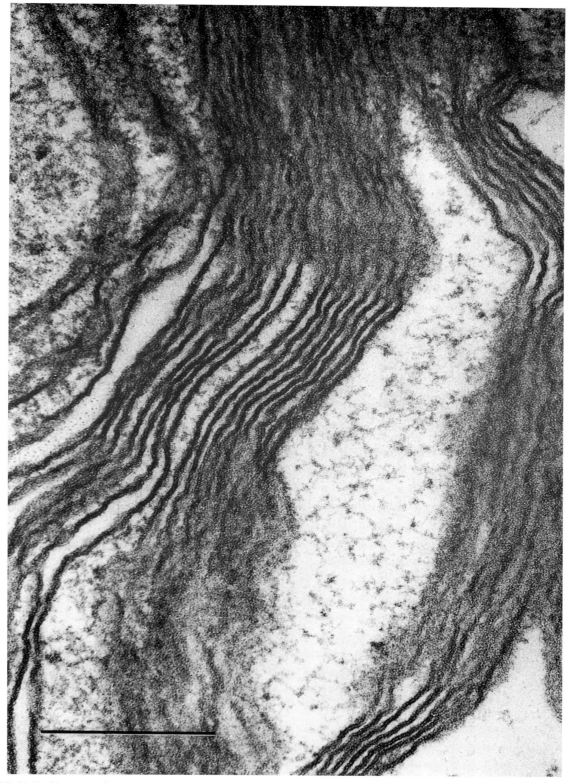

FIGURE 19.5 Unit membranes sectioned at various planes due to their change in orientation within thickness of the section. The membranes that are sectioned perpendicularly appear distinct and trilaminar, whereas, when those same membranes are sectioned obliquely and/or in the plane of the membrane (*en face*), they appear fuzzy. Membranes that change their orientation with respect to the plane of sectioning, may, at times, appear discontinuous. Bar = 0.25 μm.

FIGURE 19.6 Drawing showing the various planes in which membranes may be sectioned and how their corresponding images would appear in electron micrographs.

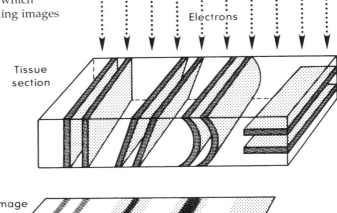

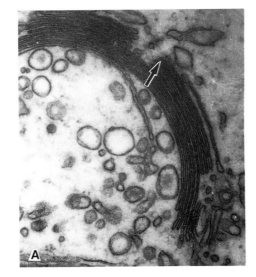

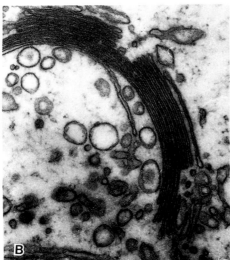

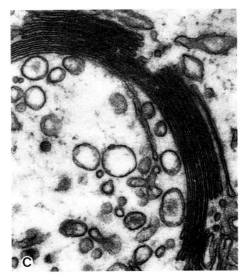

FIGURE 19.7 The effect of specimen tilting on the appearance of membranous elements of the Golgi apparatus. (A) The specimen that is in the normal position (0° angle) perpendicular to the electron beam shows regions where the membrane appears fuzzy (*arrow*). (B) After tilting the specimen 6° in one direction, the membranes in this region appear much sharper. (C) Tilting the specimen to a 12° angle makes them again appear fuzzy. However, in other regions they appear much sharper. The membrane bilayer is clearest when the membrane bilayer is in the plane of the electron beam. (See Figure 19.6.)

something appears rounded in a two-dimensional micrograph does not necessarily mean that the entire object is spherical in three dimensions. The shape and/or configuration of membranous organelles may be very deceptive. For example, rarely are sheets of smooth endoplasmic reticulum sectioned to reveal the configuration of the sheet as a whole. Fortuitous

sections in the plane of the sheet, or *en face* views, show the elaborate interconnecting pattern of a portion of the Golgi apparatus, a pattern that could not easily be imagined from most thin sectioned areas (Figure 19.9). Some complex structures may have very different appearances depending on the plane of section. For example, skeletal or striated muscle has

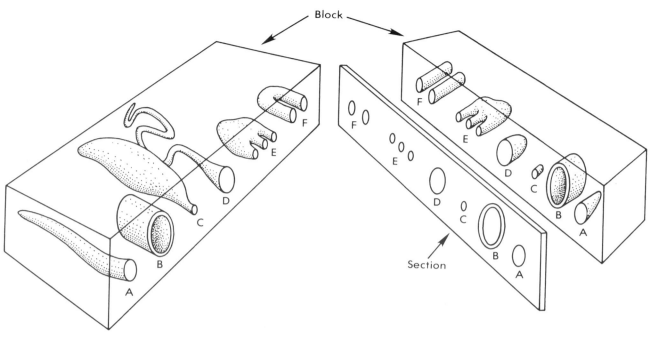

FIGURE 19.8 Sectioned profiles and their three-dimensional form in a tissue block: (A) A sphere seen in section is really an elongate structure; (B) a ring-shaped structure in section is really tubular in three dimension; (C) a structure which appears small in section is predominantly of large diameter; (D) a structure which appears large is predominantly of small diameter; (E) three structures in section are part of a larger single structure; (F) what appears as isolated sructures are part of a single convoluted structure.

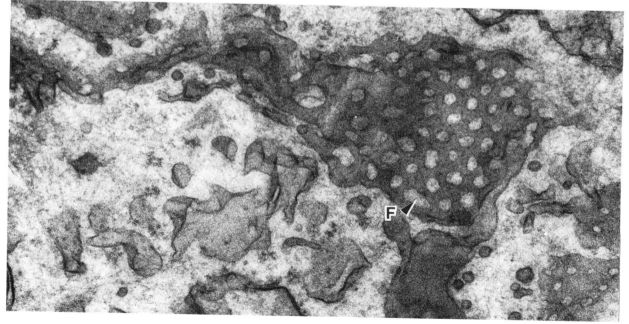

FIGURE 19.9 A cisternae of the Golgi apparatus captured within the plane of the section (*en face* view) shows the fenestrations (F) of the membranous saccule.

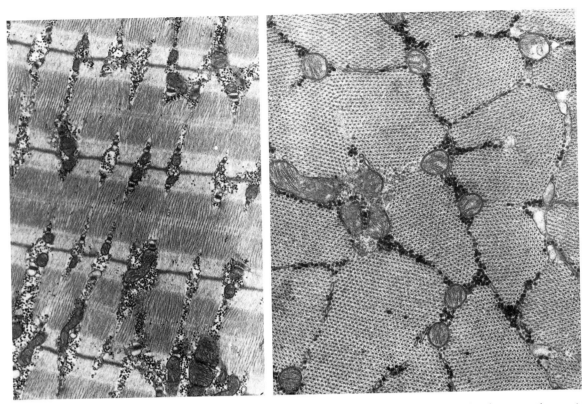

FIGURE 19.10 Striated muscle in longitudinal (A) and cross section (B) may not appear to be the same tissue at first glance. However, both figures are needed to show that the actin and myosin are filamentous in nature as opposed to sheet-like as might be inferred from (A) or simply particulate as might be inferred from (B).

a very different appearance from cross-sectioned striated muscle. Compare Figure 19.10A with B.

One method used to determine the three-dimensional structure or shape of an object or cell is to section it serially, trace the profile of the object or cell, and assemble a model. This process known as *three-dimensional reconstruction,* allows one to view the entire structure as it exists in nature. The techniques to perform this task are demanding because all or nearly all sections of the area of interest must be collected and photographed. Figure 19.11 shows a cell reconstructed from micrographs taken semiserially. Cells may be assembled manually or the profile drawing can be entered into a computer for computer modeling. A computer can rotate the reconstructed model and manipulate it to allow visualization of almost any aspect of its surface or internal aspect (McLean and Prothero, 1991; Aferson et al., 1991). Several software packages are available for this purpose. Information about complex three dimensional structures can also be obtained through use of high or intermediate voltage microscopy (see Chapter 16) or use of stereo pairs.

Identification of the *kinds* of structures in sections is often made difficult by the plane of section.

Figure 19.12 depicts what, at first sight, appears to be microvilli protruding into an internal cavity within a cell. Consider that the internal cavity may not actually be within the cell, but may instead be an indentation (invagination) into the cell. The site where the extracellular space joins the apparent internal cavity is not visualized in the plane of section; only the "internal cavity" is evident. The view that the cell is invaginated from the extracellular space is supported by the similarity of the extracellular material with the contents of the sectioned area. The nonuniform contour of the cell surface (microvilli) both outside the cell and also in the "internal cavity" indicates that the surface may have an invagination that is sectioned to make it appear intracellular when it is not.

Use of sectioned material may also lead to false interpretation of the *number* of structures encountered. If, for example, two profiles of a structure are seen within a section (similar to Figure 19.8F), can it be concluded that the cell has two structures that are separate entities? Not necessarily, because the structure depicted may be U-shaped. Only the limbs of the U may have been captured by the section. Long, thin objects may appear very numerous

FIGURE 19.11 Reconstructed Sertoli cell from the testis of a rat. A Plexiglass model was made from over 100 thin sections.
(From V. Wong and L. D. Russell. 1983. Am J Anat 167:143–61. Used with permission of Wiley-Liss.)

whereas, in reality, they are few in number but configured tortuously to give this false impression.

When structures are frequently encountered in sections and, moreover, are sectioned in many planes, it is possible to gain confidence in one's interpretations about the kinds, numbers, and shapes of structures (see Figure 19.13). Otherwise it may be necessary to make these determinations with serial thin sections or thick 1 to 2 μm sections examined by high or intermediate voltage electron microscopy, bulk examination of fractured specimens by scanning electron microscopy, or even more specialized techniques.

Fixation Artifacts

Fixation, in itself, is an artifactual process that distorts tissue structure from its living state. Cells are never completely free of fixation artifact. Even optimal fixation protocols for a particular tissue tend to emphasize certain features at the expense of others. For example, some fixation protocols emphasize membranes, but often the background matrix of the cell is virtually absent. Figure 19.14 shows the results obtained from four different fixation protocols employed for the same tissue. Naturally, investigators will publish their best micrographs, thus emphasizing the areas of better fixation. The goal of the investigator is to minimize fixation artifact to the point that the tissue has as close a resemblance to its living state as possible. It is important for the investigator to recognize what constitutes acceptable fixation and what does not. Furthermore, to avoid the repetition of errors, different kinds of fixation artifacts should be recognized.

Fixation artifacts are of many different kinds. One type or a whole cadre of fixation artifacts may be evident in one micrograph. The artifacts described here, while not a complete cataloguing of artifacts that may be encountered, are some of the more common.

The method by which tissues are exposed to primary fixative may result in artifacts. *Immersion fixation,* for most tissues, is less desirable than *vascular perfusion fixation* (see Chapter 2). The animal's own vascular system is ideal to disseminate fixative to tissue as compared with the relatively slow penetration of fixative in tissue fragments that have been mechanically disturbed by excision. For example, by comparing Figure 19.15A and B, which are micrographs of the same cell type from the same species fixed by immersion and perfusion, respectively, it is revealed that the smooth endoplasmic reticulum may lose its anastomosing tubular appearance and may become vesicular after immersion fixation. There are numerous other reasons to employ perfusion fixation whenever possible; however, the size of the animal being perfused often determines the accessibility to the vascular system. Thus, many animals cannot be perfused because they or their vessels are too small. Other large animals cannot be perfused because their vascular system would require an enormous volume of fixative and such an endeavor would not be practical in a laboratory setting.

Postfixation with osmium is important to obtain quality electron micrographs. The rate of penetration of osmium is very slow (less than 0.5 mm in 1 to 2 hours depending on the compactness of the tissue). Consequently, since osmium frequently does not reach the center of tissue blocks, the result is

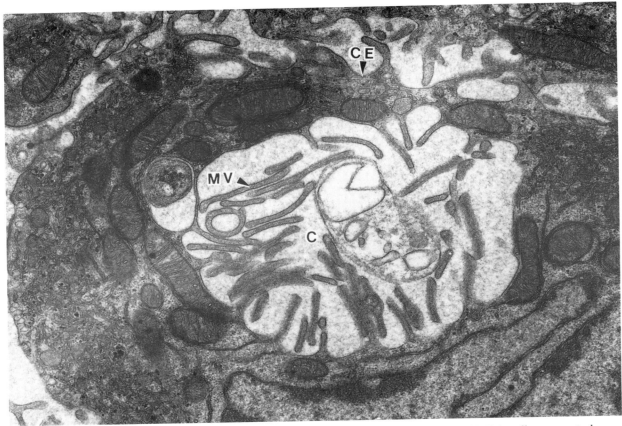

FIGURE 19.12 Structure appearing to be within a cell when it is actually outside the cell. This cell appears to have microvilli (MV) protruding from the cell interior into an internal cell cavity (C) as well as from the cell exterior (CE). In reality, all of the microvilli of the cell protrude from the cell exterior; however, an inpocketing of the cell gives the false impression that microvilli protrude from the cell interior.

poor fixation of numerous elements, most notably membranes. Figure 19.16 shows an example of inadequate osmium penetration. Note that it is difficult to discern cell outlines due to poor membrane fixation. For reasons that are unknown, osmium may penetrate some cells whereas other cells are only fixed on their plasma membrane (Figure 19.17).

Keeping tissue slices under 1 mm in any one axis will ensure rapid penetration of osmium. If large blocks are used, there may be a temptation to over-fix tissues. Over-fixation will also result in poor ultrastructural visualization of membrane components or a leaching out of structural components of the cell. Ironically, the appearance of over-fixed micrographs is similar to those that have been under-fixed.

A common sign of poor fixation is swelling of mitochondria. Under conditions of poor fixation, the mitochondrial cristae often become peripherally positioned. The central region of the mitochondrion

has a rarified matrix and/or swollen appearance (Figure 19.18), and for this reason is frequently referred to as "blown."

Membranes throughout the cell may also appear broken or discontinuous in certain areas (Figure 19.19). The membranes of the nuclear envelope may separate from each other causing a widening of the perinuclear cisternae (Figure 19.20). Some investigators add calcium chloride (1 to 2 mM) to the fixative to improve membrane preservation and to prevent mitochondrial swelling and swelling in the perinuclear cisternae of the nuclear envelope. However, in some cases, this may adversely affect the preservation of microtubules, especially in plant cells.

Fixation artifacts are frequently associated with the *osmotic strength* of the fixative. Exaggeration of the intercellular space is often the result of hyperosmolarity of the fixative. This can usually be remedied by decreasing the osmotic strength of the

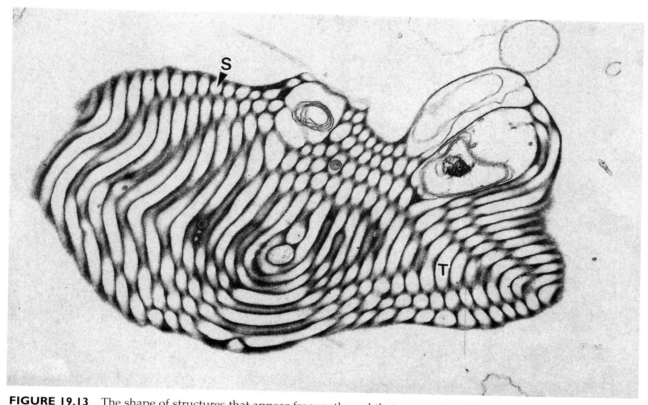

FIGURE 19.13 The shape of structures that appear frequently and that are sectioned from varying angles may be deduced by studying the various profiles seen. Although some profiles suggest that the intercellular structures shown are spheres (S), others suggest that this is an erroneous interpretation. It is likely that the plane of sectioning crossed structures that were tubular (T) in form and made them appear rounded. *(Micrograph courtesy of R. Sprando.)*

buffering system used with the fixative. Shrinkage due to tissue fixation may be obvious in tissue sectioning. In other instances, tissue spaces resulting from shrinkage may be present (Figure 19.21). Hypoosmotic solutions frequently cause swelling of cells and/or organelles.

Temperature, pH, duration of fixation, nature of the fixative, and *the nature of the buffer* are also important factors to consider when trying to prevent artifacts (see Glauert, 1978, for examples of the effects of buffers and fixatives). Because there is no standard way to fix all tissues, it is important to begin by following the past scientific literature that is applicable to the specific tissue under consideration. Sometimes there is no general agreement on the fixative conditions, and one must modify the published methodology to obtain the desired results. Mollenhauer (1993) discusses these and other artifacts caused by dehydration and embedding of tissue.

Dehydration, Infiltration, and Embedding Artifacts

Dehydration, infiltration, and embedding displace the water matrix of the specimen with an embedding material that is eventually polymerized into a hard plastic. Therefore, most artifacts are due to inadequate removal of either the water or the fluids used in the dehydration process. This usually results in *holes* in the tissue (Figure 19.22). A rare hole may be acceptable, but large numbers of holes like those illustrated should be cause for re-examination of the techniques used in embedding.

Holes may also be produced in tissues in zones of sharp transition from soft embedding material to a much harder tissue structure. This problem is usually solved by using a harder embedding resin with a hardness similar to the structure in question. Another solution is to use a low viscosity resin such as LR White, which penetrates hard tissues more readily than many

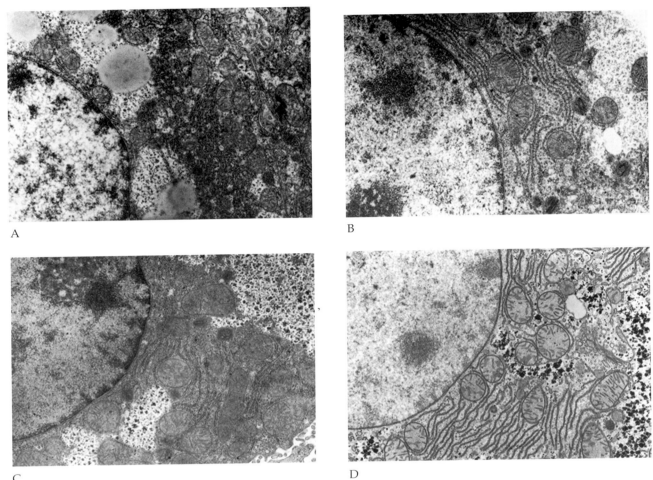

A

B

C

D

FIGURE 19.14   These four micrographs are examples of different fixation protocols. In (A), the tissue was fixed in glutaraldehyde by immersion, and postfixed in osmium tetroxide. In (B), perfusion fixation was utilized to administer glutaraldehyde. The tissue was postfixed in osmium tetroxide. Note that the rough endoplasmic reticulum is less disorganized in the perfused versus the unperfused tissue. There is also less clumping of cytosolic and nuclear constituents. Fewer clear spaces are seen in the mitochondria of (B) than are seen in (A). Overall, the tissue in (B) appears more pleasing to the eye. In (C), the tissue was perfused with Karnovsky's fixative (glutaraldehyde) followed by osmium postfixation. Note the overall homogenous appearance of the cytoplasmic and nuclear matrix. Ribosomes are less prominent and the Golgi glycogen zone begins to show evidence of glycogen particles. In (D), the primary fixation was with glutaraldehyde followed by an osmium:ferrocyanide postfixation. Note that membranes and glycogen stand out sharply. However, one does not observe ribosomes clearly and the contents of the nucleus (heterochromatin) are less pronounced. All tissues were processed subsequent to fixation at the same time by the same methodology.

other embedding media. Using a sharp diamond knife will usually reduce the severity of this problem.

Sectioning Artifacts

This category includes numerous artifacts associated with sectioning that are frequently bothersome to the investigator. The knife, whether it be glass or diamond, may have lost its sharpness or be defective or dirty in certain areas. This leads to *knife marks* *or scrape marks* on the section, which appear as lines or tears perpendicular to the knife edge (Figure 19.23). The beam may enlarge small holes formed by knife marks or a dirty knife (Figure 19.23).

An artifact called *chatter* is produced when vibrations in the block or knife edge occur as the result of improper sectioning speed or knife angle (Figure 19.24). Chatter rarely results from vibrations in the building in which the microtome is located. Chatter appears as parallel lines, or alternating

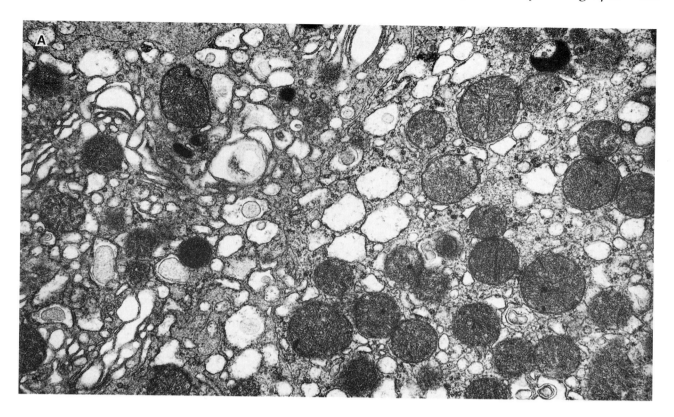

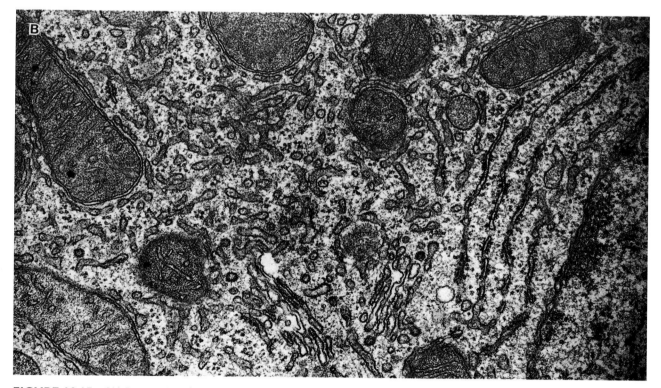

FIGURE 19.15 (A) Immersion-fixed Leydig cell and (B) perfusion-fixed Leydig cell showing the variable appearance of smooth endoplasmic reticulum. Note the apparent difference in density of the two cell types. (Rat tissue.)

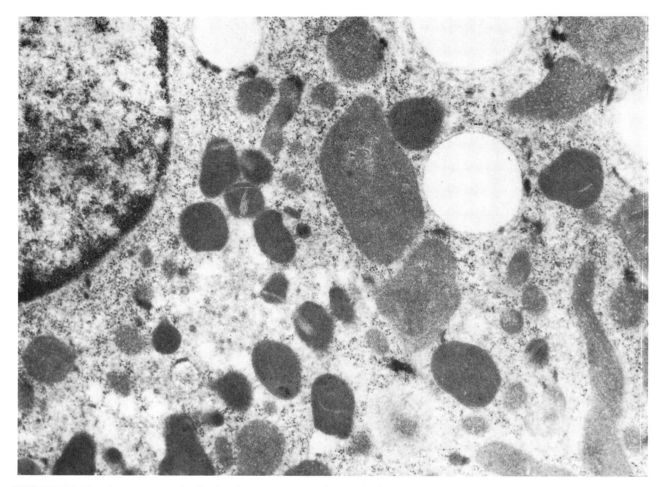

FIGURE 19.16   Appearance of cells that have not been adequately fixed with osmium. Note that membranes are not visualized.

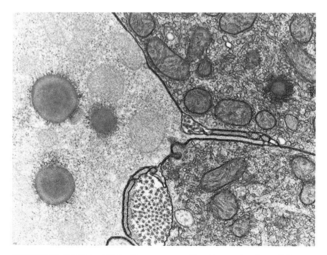

FIGURE 19.17   This tissue was fixed routinely with glutaraldehyde and osmium but the osmium only stained the cell surface of one cell and did not penetrate deeper. The cell in the left half of the figure remains unosmicated.

thick and thin areas, parallel to the cutting edge of the knife. An easy way to distinguish chatter from knife marks is that chatter marks are quite regular and always spaced evenly from each other, whereas knife marks may or may not be repetitive, but are rarely evenly spaced. In situations where both knife marks and chatter are seen, knife marks always occur perpendicular to chatter (Figure 19.25). If chatter is caused by sectioning, changing the knife angle and/or sectioning speed will usually remedy chatter marks, but a different knife or clean cutting edge may be needed to remedy knife marks or scrapes.

Compression is produced when very soft compressible materials are sectioned. The pressure exerted by the knife may compress the section such that its vertical dimension is less than that of the block face from which it was taken. Thus, tissue structure is distorted in one dimension by as much as 30% (Figure 19.26). Some stretching of the section

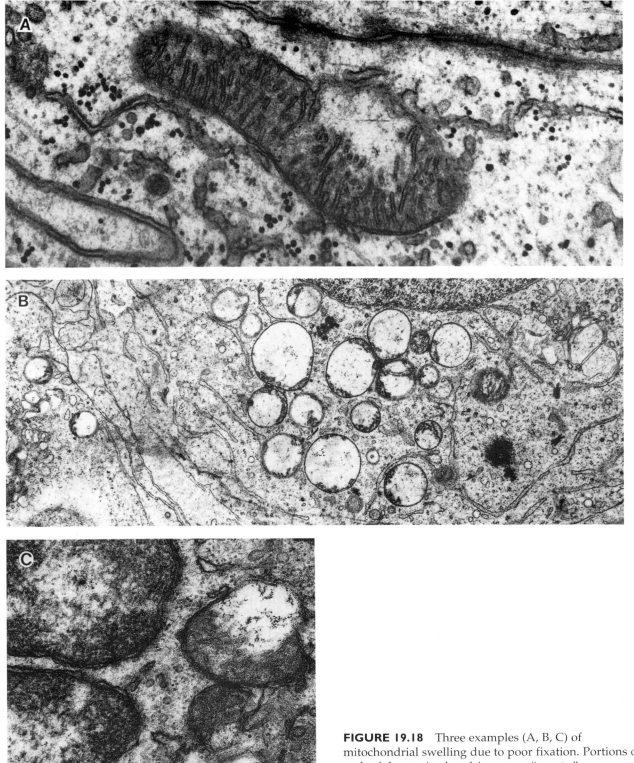

FIGURE 19.18 Three examples (A, B, C) of mitochondrial swelling due to poor fixation. Portions of each of these mitochondria appear "empty."

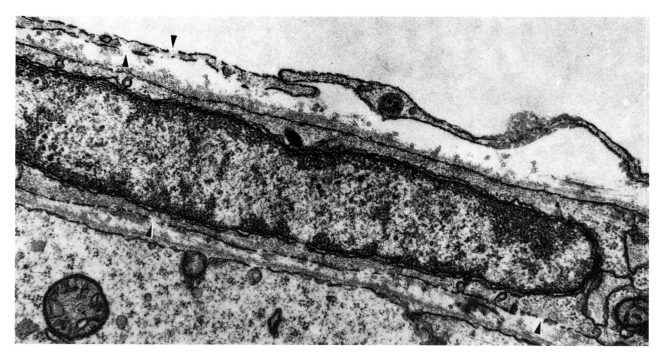

FIGURE 19.19 Membrane discontinuities (*arrowheads*) in this micrograph are due to poor fixation.

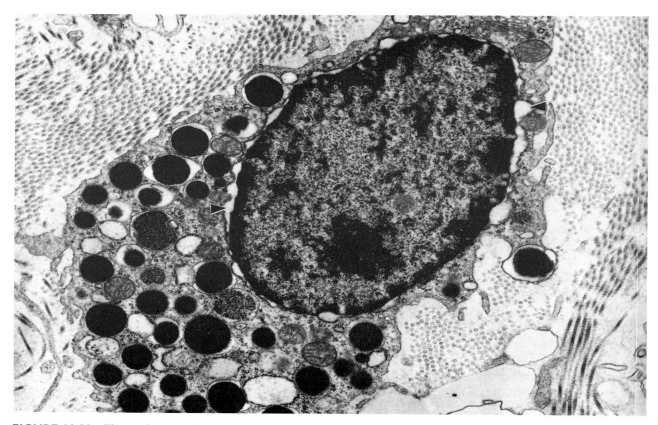

FIGURE 19.20 The nuclear envelope of this mast cell shows fixation artifact as evidenced by swelling of the perinuclear space (*arrowheads*).

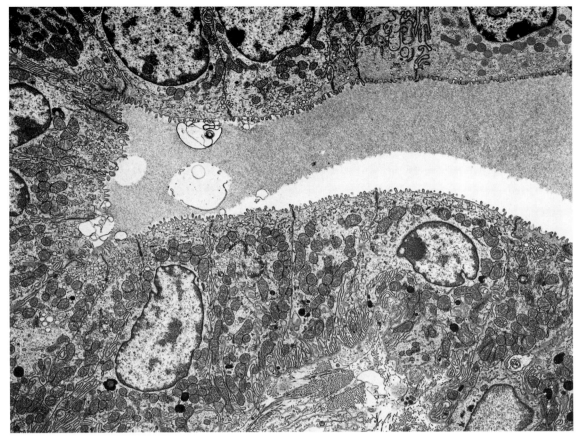

FIGURE 19.21 Tissue shrinkage has occurred in this epithelial lined duct system as evidenced by the clear space in the lumen of the duct. The clear space has formed from shrinkage of the epithelium away from the proteinaceous contents of the lumen. Shrinkage probably occurred during dehydration of the tissue. (Urethral gland of the mouse.)

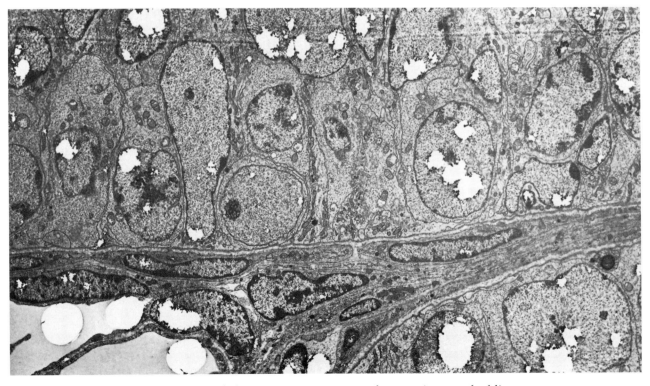

FIGURE 19.22 Holes in tissue due to failure to remove water or solvents prior to embedding.

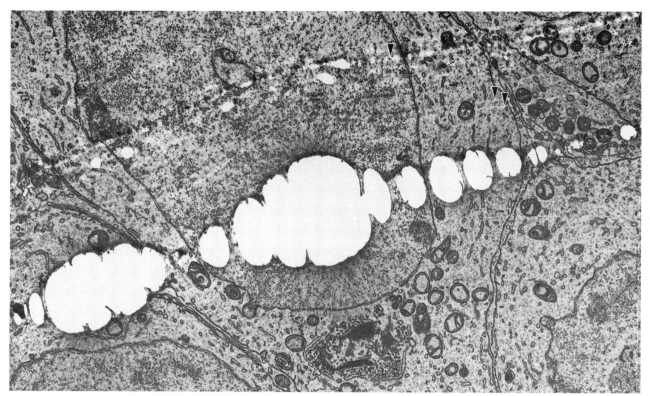

FIGURE 19.23 Knife marks (*arrowheads*) have scratched and caused irregularities in section thickness. The holes produced by other knife marks are enlarged by the heating of the electron beam.

by warming or by exposure to vapors of organic solvents such as xylene or chloroform may restore the compressed surface to approximately its normal size and shape.

While not a sectioning artifact, per se, *section thickness* is important in determining the overall appearance of the tissue. Section thickness must be tailored to the specific needs of the investigator. As a general rule, studies that require low print magnifications (<6,000) ideally employ relatively thick sections, such as those displaying silver-gold or gold interference colors (80 to 120 nm thick). By using thicker sections at low magnification, it is possible to enhance contrast, and thus image quality, through the increased superimposition of dense tissue structures against nontissue elements. Studies at intermediate magnifications (print magnifications of 15,000 to 40,000) may require sections displaying silver interference colors (60 to 70 nm). For high resolution, high magnification (>60,000 print magnification) studies, it is desirable to have gray or silver-gray sections (40 to 60 nm). At the higher magnifications, resolution of biological materials is enhanced by decreased superimposition of structures. A practical rule for

evaluating section thickness at moderately high and high magnifications is to look for the "railroad track" appearance of unit membranes. The better the railroad track is resolved, the thinner the sections (compare Figure 19.27A and B). Very thin sections tend to break or distort easily under the electron beam. An example of "thick" and "thin" sections adjacent to each other on the same grid is provided in Figure 19.28. When viewing tissue it is not uncommon to see that section thickness has changed, Figure 19.29.

If the boat used to pick up sections is not clean or the block is dirty prior to sectioning, the sections will undoubtedly show *boat contamination*. The appearances of this contamination are as varied as the nature of the contaminants themselves. The example provided in Figure 19.30 shows boat contamination produced by epoxy fragments remaining on the side of the block after it was ground with a rotating drill. If stale water is used bacteria may be present (Figure 19.31). Washing the block under a fast stream of water prior to sectioning may have prevented this annoying artifact. Figure 19.32 shows *tissue folds*, produced either during sectioning or as the tissue was picked up on the grid. Tis-

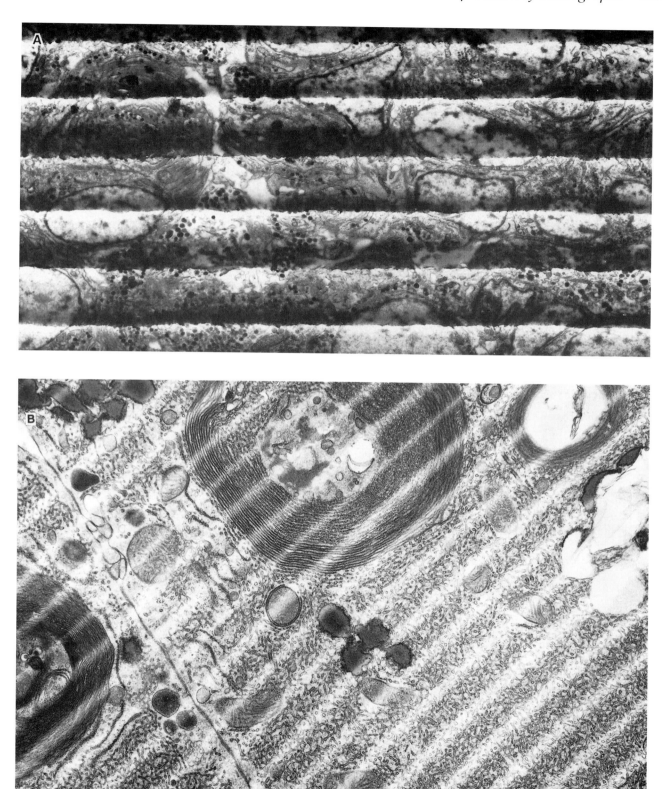

FIGURE 19.24 Two types of chatter, each showing alternating thick and thin areas. In (A) the chatter is more severe. *(Micrograph courtesy of students wishing to remain anonymous.)*

FIGURE 19.25   Chatter with knife marks seen in a section of epoxy. Note that chatter (evenly spaced) and knife marks (unevenly spaced) are perpendicular to each other.

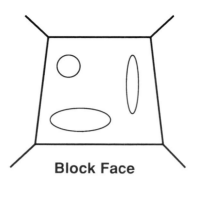

Block Face

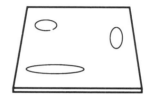

Section With Compression

FIGURE 19.26   Diagram of the shortening of one dimension of a section as compared with the block face due to compression. Note how tissue structures are compressed in the axis perpendicular to the knife face.

sue folds are easy to identify because the result is a malalignment of the image caused by part of the tissue still being within the fold. Another type of fold may mimic a tissue fold, however, in this instance, the image is not malaligned by the fold. This is a *fold in the support film* that occurred prior to sectioning when the film was placed on the grid (Figure 19.33).

If grids are not properly cleaned prior to retrieval of sections, the grids themselves may be a source of contamination often seen as an "oil slick" on the water surface. Cleaning the eyelash probe with the fingers or using oily forceps or troughs are common sources of oil problems.

For additional information on sectioning artifacts, see Chapter 4 on ultramicrotomy.

Staining Artifacts

Without a doubt, the most common and bothersome artifacts are those related to the staining of tissue. When other staining problems have been corrected, staining artifacts or *stain contamination* seem to occur again and again. Unfortunately, stain contamination seems to gravitate to tissue

sections at the site considered most desirable for photography! Meticulous detail must be paid to the staining procedure; even then there is no guarantee of success.

Stain artifacts may take a variety of forms on electron micrographs. *Lead precipitate* may appear as fine grains ("pepper"), crystals (Figure 19.34), or as dense spherical particles (Figure 19.35). *Uranyl acetate* contamination appears as dense amorphous aggregates or blotches of variable size (Figure 19.34A, B; Figure 19.35A). It is always prudent to use double-distilled water in making stains, as properties of the water may cause heavy metal precipitation on tissue sections. The water used to make lead stains should be boiled before the stain is made to remove the CO_2 that will precipitate the lead as lead carbonate. The importance of washing tissues numerous times after each stain application cannot be over emphasized. Staining artifacts (Figure 19.36) may be regionalized suggesting that during drying on the grid they may be concentrated at a particular site. Most stain contamination may be removed from the surface of the section using 0.5 N hydrochloric acid.

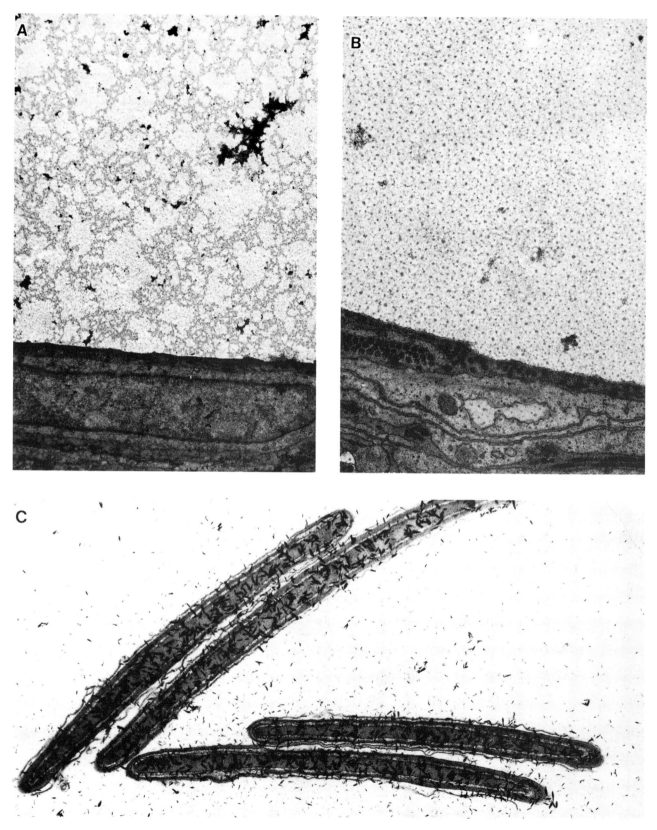

FIGURE 19.34 Various forms of lead contamination. (A) and (B) also depict uranyl acetate contamination (large irregular densities).

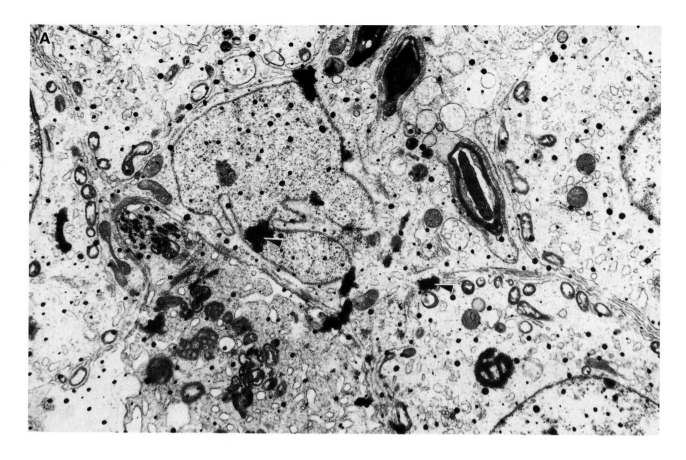

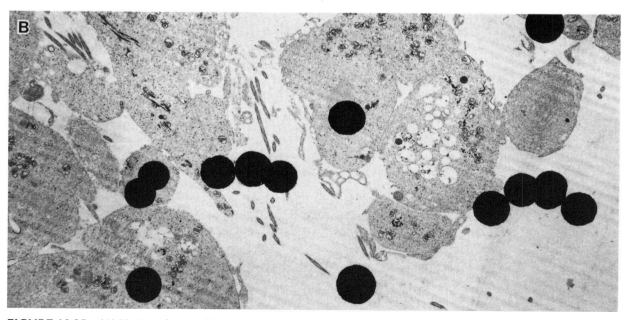

FIGURE 19.35 (A) Various forms of lead (spherical particles) and uranyl acetate (irregular densities; *arrowheads*) contamination. In (B) there is mild chatter artifact.

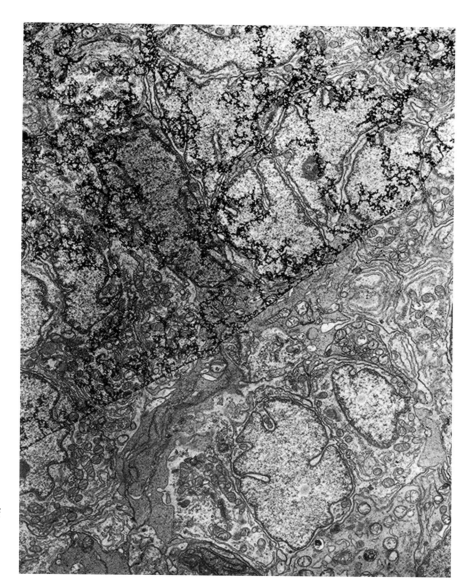

FIGURE 19.36 Regionalization of artifacts. Uranyl acetate contamination and contamination due to impure water is seen only over a certain region of the tissue.

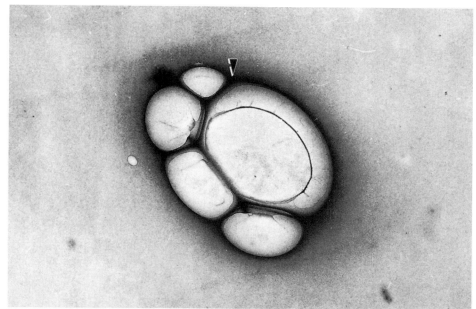

FIGURE 19.37 Stain contamination (*arrowhead*) is seen between holes in the support film and an epoxy section.

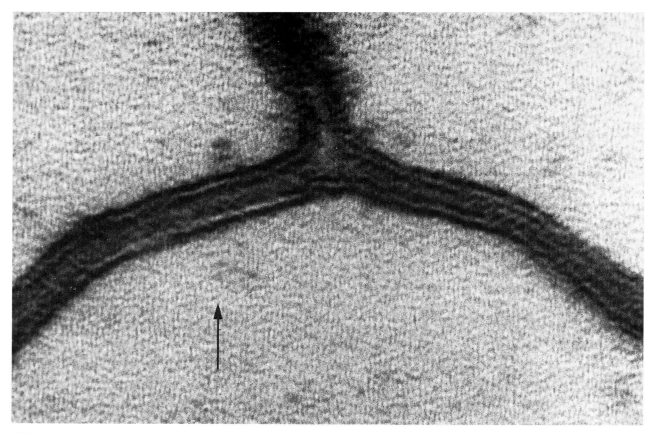

FIGURE 19.38 Astigmatism. Note there appear to be fine streaks on the micrograph in parallel with the direction indicated by the arrow.

FIGURE 19.39 Artifacts on a tissue section (mounted on a hexagonal grid) that are viewed on a scan setting of the microscope by turning off the objective lens and removing the objective aperture. Areas in which the semicondensed beam has focused on the section (prior to removal of the objective aperture) and changed the properties of the section appear lighter (*arrowheads*) than areas not exposed to the electron beam.

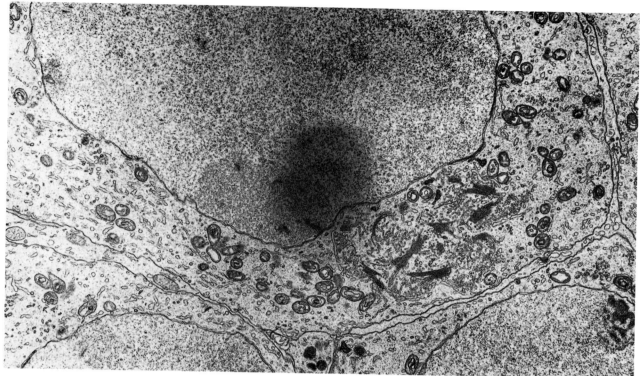

FIGURE 19.40 Hydrocarbon contamination (*center*) produced from focusing the beam too long on a single spot on the specimen.

Contamination within the microscope may interact with a focused beam to produce a rounded, dark spot on the specimen termed *beam contamination*. Usually hydrocarbons interacting with and being broken down by the electron beam are the source of the problem (Figure 19.40). Avoid concentrating the beam on any one area for a prolonged period. Modern microscopes are equipped with anticontamination devices or "cold fingers" (see Chapter 6), which condense contamination on chilled surfaces thereby preventing condensation onto the tissue section.

Photographic Artifacts

Ideally, the darkroom is more often used to eliminate, rather than to produce, tissue artifacts. Under normal circumstances, the contrast of tissue sections is usually less than desirable. In the process of making a negative and then a print from the same negative, one can usually remedy the low contrast of the tissue sections. Doing this requires using a grade of photographic paper designed to yield high contrast prints. Concentrated developing solutions, long low-light exposures, and a point-source enlarger can all be used to impart additional contrast to the final photographic print. Dodging of exposures can be used to counteract the uneven density of the negative. Dust or lint on a negative (see Figure 19.28) is readily remedied using a hand held gas dispenser or a fine brush. Artifacts due to improper photographic techniques are covered in Chapter 8.

Artifacts associated with photography usually involve procedural errors rather than actual artifacts inherent in the negative. They are overcome with experience. Over- or underdevelopment, improper exposure, improper fixation during the photographic process, unsuitable contrast, and scratching of the negative are common errors frequently made in developing and printing negatives and micrographs (see Chapter 8).

Interpreting Dynamic Processes from Static Images

When one wishes to follow events within biological structures, cautious interpretation of electron micrographs is recommended. Living tissue is in a dynamic state. Cell components are in constant movement and flux; growth and degradation processes are taking place. Ingestion and elimination of substances, cell division, and a variety of other processes characterize cellular activity. On the other hand, tissue prepared for electron microscopy is static because living processes have been purposefully halted in order to examine the tissue.

To conclude that some event is taking place when micrographs only *suggest* that it is happening is to extend the data beyond what is logically interpretable from the micrograph. Usually, several logical explanations are possible for how a given static image may participate in a dynamic process, which further limits one's ability to speculate. There are numerous examples in the literature of interpretations that have overextended the available data! The trend in analyzing dynamic processes is to use techniques that allow less room for error in interpretation, such as autoradiography (see Chapter 11) or specific labels. Tissues may be prepared at intervals to trace events from a particular starting point. Information can be quantitated at specific time points (see Chapter 13). In addition, other types of data (e.g., biochemical) may be used to support the interpretation. *It is questionable whether a conclusion about dynamic events in biological materials can be considered as fact if it is based strictly on descriptive information from micrographs.*

Estimation of Micrograph Magnification

The available range of electron microscope magnifications is great, extending from several hundred to over one million times. It is difficult, at first, to appreciate differences in this broad (thousandfold) range of magnifications.

One must realize that the magnification figures provided for micrographs may vary by ± 10% and still be within the manufacturer's specifications for the microscope. When accuracy is important, the microscope must be calibrated (see Chapter 6).

Modern electron microscopes print the negative magnification directly on the negative and/or display it on the scope panel. These features may not always be available to the individual. With experience, the investigator can provide an educated estimate of magnification. The terms *low, medium,* and *high magnification* are relative terms and should be used in a context that is defined. For our purposes, we have arbitrarily defined low magnification as a final image under ×6,000; medium magnification as being between ×15,000 and ×40,000; and high magnification as being above ×60,000. Gaps are intentionally left between the magnification ranges to indicate areas where they merge (e.g., medium low or medium high magnification).

If, for example, one is looking at an 8″ × 10″ micrograph of mammalian liver tissue, low magnification should reveal several cells within the micrograph. At medium magnifications, a single cell at most should be seen, or more likely a portion of a cell. At the lower range of high magnifications, one may see a few organelles. At the higher range, it is possible to visualize macromolecules.

Microscope magnification, as obtained from the settings on the microscope, is, of course, the magnification of the negative and not the print. To obtain the latter, measure the distance between two objects that are widely spaced on the negative, and measure the same distance between the two objects on the micrograph. The final magnification (mag) of the enlarged image is determined from the print and the negative (neg) as follows:

$$\text{final magnification} = \frac{\text{measurement from print}}{\text{measurement from negative}} \times \text{negative magnification}$$

There are a number of possible errors (machine and human) that may lead to slight deviations of the calculated magnification obtained from the microscope digital readout as compared with the actual magnification of a printed micrograph. It is generally assumed that these errors exist, so the final expression of magnification is usually rounded off to the nearest 100 times. For example, ×37,500 is a more appropriate expression of magnification than is ×37,687. Sometimes when magnification is not critical to the interpretation, the final figure is rounded off to the nearest 500 or 1,000 times.

These days the magnification of a negative is obtained from a gauge or digital display on the microscope. If this information is not printed on the micrograph, a relatively accurate way to estimate micrograph magnifications is to use the knowledge of the size of cell structures to calculate magnification. Resolved unit membranes in their unmagnified state are 7 to 9 nm across, or 8 nm on the average. In micrographs, their size is directly proportional to the magnification. (Of course, the lipid bilayer is only resolved well at moderately high or high magnifications.) If, for example, a membrane (8 nm across) is measured at 1.2 mm using a fine ruler, it has been magnified about ×150,000 according to the following formula:

$$\text{magnification} = \frac{\text{measured size}}{\text{known organelle size}} \quad \text{or}$$

$$\frac{1.2 \times 10^{-3}\,\text{m}}{8 \times 10^{-9}\,\text{m}} = 1.5 \times 10^5 \text{ or } 150,000$$

Membranes are not the only structures of relatively constant size that may be measured in micrographs. Keep in mind that calculations of this type provide only a "ball park" estimation of micrograph magnification. Other organelles that may be measured to obtain a rough idea of magnification include:

cilia	0.2 μm across
centriole	0.15 μm across
glycogen particles	30 nm
intermediate filaments	10–12 nm
microfilaments (actin)	6–7 nm across
microtubules	22–25 nm across
ribosomes (mammalian)	20 nm

REFERENCES

Aferson, J., et al. 1991. A microcomputer based system for three-dimensional reconstructions from tomographic or histologic sections. *Anal Quant Cytol Histol* 13:80–88.

Crang, R. F. E. and K. L. Klomparens. 1988. *Artifacts in biological electron microscopy.* New York: Plenum Press.

Glauert, A. M. 1978. *Fixation, dehydration and embedding of biological specimens.* Amsterdam, The Netherlands: North Holland Publishing Co.

McLean, M., and J. W. Prothero. 1991. Three-dimensional reconstruction from serial sections. V. Calibration of dimensional changes incurred during tissue preparation and data processing. *Anal Quant Cytol Histol* 13:269–78.

Mollenhauer, H. H. 1993. Artifacts caused by dehydration and epoxy embedding in transmission electron microscopy. *Microsc Res Tech* 26:496–512.

Wong, V., and L. D. Russell. 1983. Three-dimensional reconstruction of a rat Stage V Sertoli cell: I. Methods, basic configuration and dimensions. *Am J Anat* 167:143–61.

Chapter 20

Survey of Biological Ultrastructure

IT IS IMPORTANT TO UNDERSTAND the theoretical aspects of tissue preparation and image formation in the electron microscope. It is equally important, however, to be capable of *applying* this knowledge to obtain a better understanding of biological systems. This chapter introduces some of the more common features of biological ultrastructure. Other more extensive treatises and/or atlases of cell ultrastructure have been published and serve as excellent sources of illustrative and descriptive information. In addition, major textbooks of histology and cell biology are also good reference sources (see "References" at the end of this chapter).

What follows is a brief but systematic introductory description of, and most importantly, a *guide* to the identification of basic biological ultrastructure in animal and plant cells. Table 20.1 shows one way

---

BRIEF DEFINITIONS OF FLOWCHART ITEMS

1. **Plasma membrane.** Unit membrane (trilaminar) enclosing cellular contents
2. **Junction.** Structural relationship between adjacent cells or between a cell and the extracellular matrix having a particular function
3. **Cytoplasm.** Cell contents excluding the nucleus
4. **Nucleus.** Double unit membrane-bound compartment containing the chromosomes and nucleolus
5. **Endocytic apparatus.** Surface membrane involved in the uptake of extracellular material
6. **Glycocalyx.** Mat of fine filaments (2.5–.5 nm) extending from the external aspect of the plasma membrane for up to 0.5 μm
7. **Surface specialization.** Projection and/or invaginations of the plasma membrane and underlying cytoplasm
8. **Cytoskeleton.** Network of filamentous elements contained within the cytoplasm
9. **Inclusion.** Nonmembranous mass or body within the cytoplasm that generally serve as storage sites
10. **Organelle.** Any morphologically and functionally distinct membrane-bound compartment or structure characterized by biochemical activity
11. **Cytosol.** Space containing soluble elements not occupied by organelles, filaments, inclusions or other vesicles
12. **Nucleolus.** Electron-dense mass composed of closely packed fibers and granules; involved in RNA synthesis
13. **Nuclear envelope.** Double membrane enclosing the nuclear material; contains numerous pores (9–26 nm) having a diaphragm
14. **Basement membrane.** Layer (~100 nm thick), underlying primarily epithelial cells, composed of type IV collagen between plasma membrane and larger collagen fibers of the extracellular matrix
15. **Cell wall.** Rigid layer (30–100 nm) of thick amorphous material surrounding plant cells
16. **Type I collagen.** Fibrils of the extracellular matrix that exhibit cross striations every 67 nm and tend to form larger bundles several micrometers in diameter that run at right angles to one another
17. **Coated pit/vesicle.** Invaginations of the plasma membrane coated on their cytosolic surface with a densely packed material and involved in receptor-mediated uptake; a coated pit that has been internalized is called a coated vesicle
18. **Micropinocytotic vesicle.** Short invaginations of the cell surface 50–100 nm across
19. **Pinocytotic vesicle.** Large (>100 nm), irregular indentations of the cell surface
20. **Zonula adherens.** Membranes of adjacent cells separated by 24 nm and show a subsurface density binding actin filaments
21. **Macula adherens.** Membranes of adjacent cells that are parallel and separated by 25 nm at a region of subsurface density that is attached to intermediate filaments
22. **Glycogen.** Spherical particles (30 nm) which may occur singly or in the form of rosettes up to 0.1 μm; storage form of glucose plus enzymes; sometimes considered an organelle
23. **Lipid.** Homogenous bodies of variable size and osmiophility; often extracted by dehydration, readily reveal knife marks/chatter
24. **Crystalloid.** Geometric mass with an inherent lattice structure
25. **Ribosome.** 20 nm bipartite masses with fuzzy exterior when compared with glycogen; involved in protein synthesis
26. **Other.** Secretory granules, pigments, and cell specific products that are peculiar to the cell type
27. **Golgi complex.** Stack of up to ten perinuclear saccular elements; the complex usually displays a concave and convex aspect
28. **Chloroplast.** A plant specific organelle involved in photosynthesis possessing a double membrane, stacked membranous discs called grana, and sometimes electron translucent areas containing starch
29. **Vacuole.** (Plant specific) large, electron translucent or flocculent area within plants
30. **Lysosomal apparatus.** Machinery for intracellular digestion
31. **Endoplasmic reticulum.** Continuous, branching network of membrane sacs
32. **Nucleoplasm.** Contents of the nucleus within the nuclear membrane, but not including the nucleolus
33. **Hemidesmosone.** Cell–extracellular matrix relationship characterized by a subsurface plaque associated with intermediate filaments; functions as an adhering junction
34. **Gap.** Relationship between adjoining cells in which a 2 nm gap is present giving a septalaminar appearance; functions as a communicating junction
35. **Zonula occludens.** Local fusion of membranes giving a pentalaminar appearance and obliterating the intercellular space; functions as a permeability barrier
36. **Plasmodesmata.** A cytoplasmic connection (60 nm in diameter) between adjacent plant cells passing through the intervening cell wall
37. **Mitochondria.** Double membranous ellipse with extensive infolding of inner membrane into cristae
38. **Peroxisomes.** 0.2–0.4 nm in diameter elements some of which contain a dense crystalline core; can only be defined with certainty by immunocytochemistry or biochemical analysis
39. **Nuclear lamina.** Thin, dense layer of intermediate filaments underlying the nuclear membrane
40. **Chromatin.** Complex of DNA, histones, and nonhistone proteins within the nucleus
41. **Centrosome.** Region near nucleus from which microtubules emanate
42. **Granular (rough) endoplasmic reticulum.** Membrane saccules that appear beaded due to the presence of attached ribosomes
43. **Agranular or smooth endoplasmic reticulum.** Segments of the endoplasmic reticulum lacking attached ribosomes
44. **Flagella.** Long (150 + μm) cell projections with a 9 + 2 microtubule core (axoneme) usually associated with accessory structures
45. **Cilia.** Long (up to 15 μm), thin projection of cell surface containing a 9 + 2 microtubule axoneme core
46. **Lamellipodia.** Broad, sheetlike extensions of the cell surface containing actin filaments
47. **Filopodia.** Needlelike projections (up to 50 μm), longer than microspikes, which contain an actin core
48. **Microtubules.** Hollow, narrow (24 nm) cylinders serving a structural role or used as tracks for intracellular movements
49. **Actin filaments.** Filament 5.9 nm in diameter often appearing indistinct, often occurring in bundles, and often subplasmalemmal
50. **Intermediate filaments.** Filamentous elements ~10 nm in diameter often associated with desmosomes and hemidesmosomes
51. **Pericentriolar matrix.** Darkly staining amorphous material surrounding two centrioles
52. **Centrioles.** Cylinders (0.2 μm wide) composed of 9 microtubule triplets oriented in a pinwheel fashion
53. **Microspikes.** Long (up to 10 μm), needlelike (~0.1 μm wide) projection containing an actin core seen primarily in cultured cells
54. **Stereocilia.** Thin projection (up to 10 μm in length) from the cell surface containing an actin core
55. **Lysosomes.** Elements of variable size scattered irregularly throughout the cell, usually electron dense, and usually possessing a heterogeneous interior; can only be differentiated with certainty by enzyme cytochemistry or biochemical analysis
56. **Autophagosome.** Membrane-bound structure containing heterogeneous material ingested from the same cell
57. **Phagocytic vesicle.** Vesicle containing heterogeneous material ingested for degradation within the cell
58. **Multivesicular body.** Membrane-bound structure containing multiple smaller membrane-bound vesicles; involved in intercellular digestion
59. **Residual bodies.** Membrane-bound structures containing remnants of undigested material

TABLE 20.I Flowchart for Identification of Selected Tissue Elements Based Primarily on Fine Structural Characteristics*

(Number refers to definitions on opposite page.)

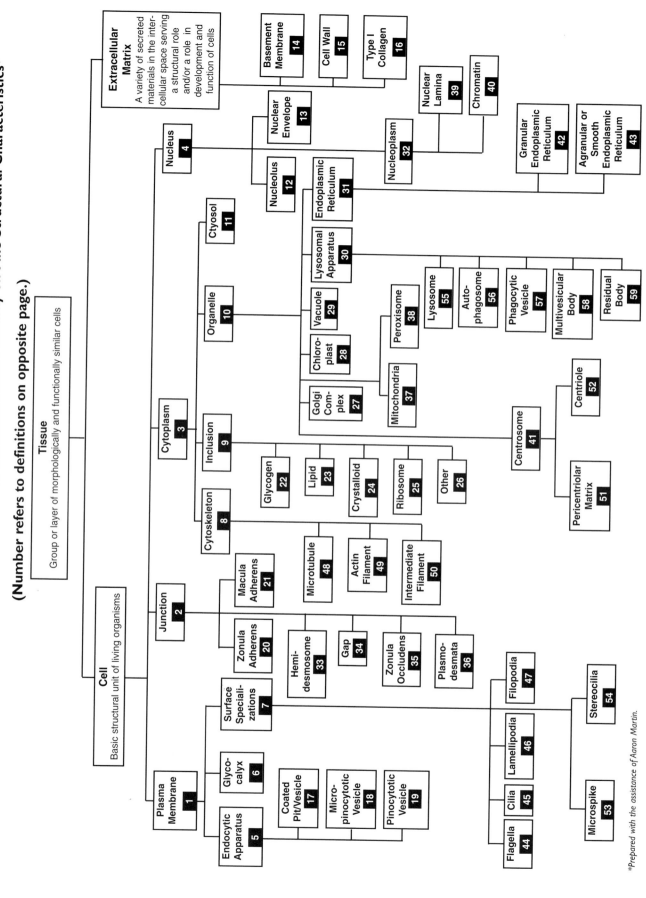

*Prepared with the assistance of Aaron Martin.

479

in which cells and cell components may be organized and classified. On the page opposite the flow-chart, brief definitions of cell components are provided. Some functional information is also included for each organelle or structure. It is important to associate the structure of each cell component with its function. Finally, key references are included within each section to help the reader obtain more in-depth information about a particular structure or organelle.

The Cell Surface

In animal tissues, the cell surface is the plasmalemma or plasma membrane. In plant tissues, the surface is the cell wall, a structure covered later under "Special Features of Plant Tissues." The plasmalemma interacts with the extracellular environment and regulates *transport* for both large and small molecules into and out of the cell. The cell membrane participates in ingestion or elimination of large bodies of material as well as small molecules. The cell surface forms *junctions* with other cellular and noncellular elements that participate in attachment, cell-to-cell communication, and regulation of permeability between cells. *Receptors* along the cell surface serve as recognition sites for binding various macromolecules. The binding event may then signal the cell interior to perform specific tasks or inform the cell about the identity of the other cell. The cell surface has configurational specializations that function to increase the overall surface area of the cell or to provide for motility of the entire cell or impart motility to one of the protruding cell processes.

The Lipid Bilayer of the Plasmalemma

At first glance, the cell surface may appear to be extremely simple, but a more detailed structural analysis reveals its extreme complexity. At low and medium magnifications, the cell membrane appears as an electron-dense line. At high magnification and with the appropriate fixation and staining protocols and sectioning perpendicular to the membrane, it, like all biological membranes, appears as a *bilayer*. Membrane structure is emphasized by osmium and electron-dense stains, which are preferentially deposited over the polar heads of the phospholipid molecules (Figure 19.4). This 7 to 11 nm thick membrane is termed a *trilaminar membrane* because of the three visible lines one sees. There are two electron-dense lines separated by one electron-translucent line (Figure 20.1). Because the lines are parallel, it is

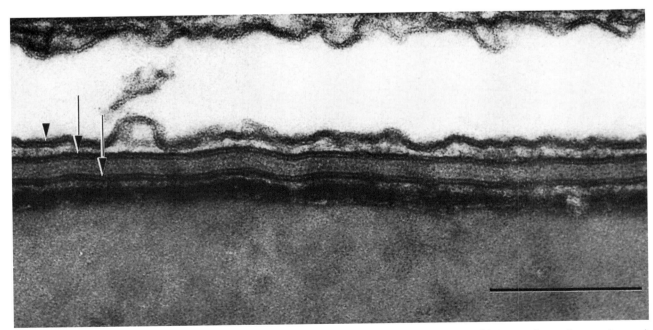

FIGURE 20.1 The cell surface membrane (*arrowhead*) or plasma membrane and the cell's internal membranes (*arrows*) are shown at high magnification. The plasma membrane and the other membranes show a distinct trilaminar or "railroad track" appearance. (Boar sperm.) Bar = 0.5 μm.

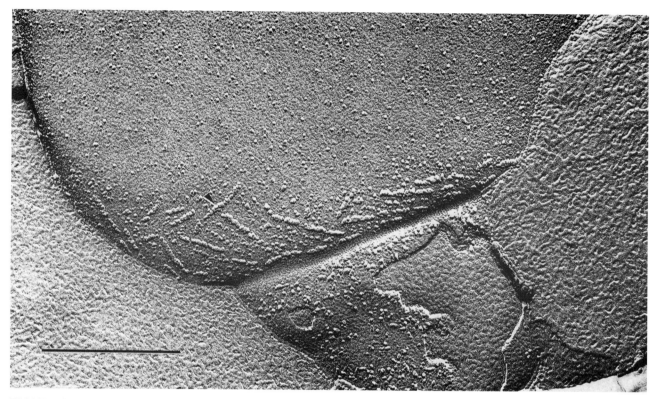

FIGURE 20.2 Intramembranous particles (IMPs) of various sizes are seen in a freeze fractured plasmalemma. The pattern and concentration of IMPs appear randomly distributed in some regions and in other regions discrete rows of particles (*arrowhead*) are seen. (Boar sperm.) Bar = 0.5 μm.

often called a "railroad track." Chapter 19 describes how membranes appear when they are sectioned in planes other than perpendicular to the plane of the membrane.

The freeze fracture technique splits the lipid bilayer so that the two membrane halves, or faces, are visualized from their internal aspect (see Chapter 14). Molecules that span the width of the bilayer, or transmembrane proteins, appear as intramembranous particles (IMPs) in the freeze fracture micrograph (Figure 20.2). The term *intrinsic* proteins is used to designate proteins that span the lipid bilayer. *Extrinsic* are those proteins that are bound only to the external surface of the plasma membrane.

The membrane bilayer (Figure 20.3) provides a lipid interface between the aqueous environment of the cell exterior and that of the cell interior. It forms a framework to contain or bind molecules such as proteins. According to the *fluid mosaic model* of membrane structure (Singer and Nicolson, 1972),

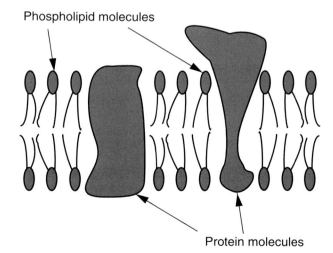

FIGURE 20.3 Biological membranes are composed of phospholipid molecules with their polar (hydrophilic) heads facing outward (*top and bottom of figure*) and their hydrophobic tails facing each other and are also composed of protein molecules embedded and protruding from the membrane.

the membrane proteins are capable of movement in the semiliquid phospholipid bilayer. Plasma membrane molecules serve a variety of transport, receptor, and recognition functions.

The Glycocalyx

Virtually every cell plasma membrane contains an exterior that appears fuzzy under the electron microscope. In some cells it is barely visualized and special stains are required to demonstrate its presence, but in other cells it forms a prominent fuzzy coat that extends many times the thickness of the membrane. Variously referred to as the *glycocalyx* or *cell coat*, it is composed of polysaccharides, especially those containing negatively charged sialic acid residues that are bound to the cell surface and are part of intrinsic proteins of the bilayer. Cell surface specializations that extend into a lumen frequently possess a well-developed glycocalyx (Figure 20.4).

The carbohydrates of the glycocalyx are important in cell-to-cell recognition and adhesion. Since the glycocalyx is the outermost aspect of the cell, it is assumed to function as a barrier to some substances. The net negative charge of the glycocalyx regulates, to some degree, the kinds of charged molecules that can approach the cell surface.

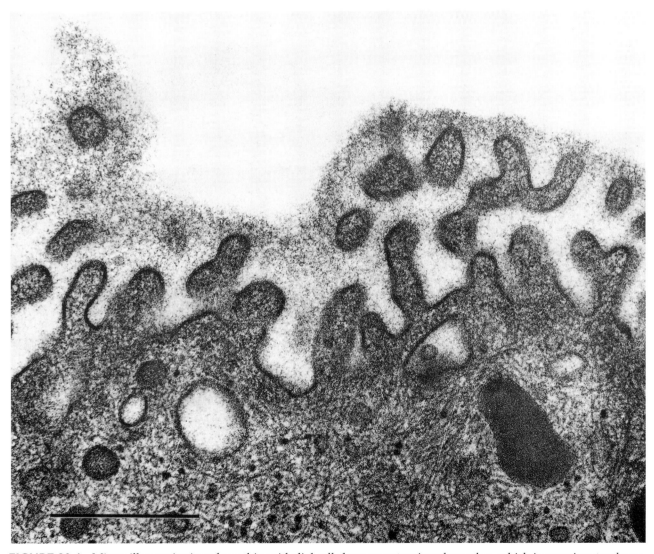

FIGURE 20.4 Microvillus projections from this epithelial cell show an extensive glycocalyx, which is prominent only on the lumenal surface of the cell. The glycocalyx appears as a fuzzy substance at the top of the micrograph and on close inspection reveals a fine filamentous texture. (Sloughed epithelial cell in the human male reproductive tract.) Bar = 0.5 μm.

Cell Junctions

Cells may join to other cells or may be attached to connective tissue elements by specializations of their cell surface. A great number of variations of junctions have been described in biological materials (Table 20.2). Only the most common of these will be illustrated. (The plasmodesmata of plant cells are considered under "Special Features of Plant Tissues.")

Tight or occluding junctions, also classified as *zonula occludens,* are linear fusions of membrane between adjoining epithelial cells that usually extend around the entire circumference of an epithelial cell. Most commonly, tight junctions are

TABLE 20.2   Summary of Some Junctional Types and Properties

Name(s)	Main Features	Common Location	Function
Tight or Occluding Junction, Zonula Occludens	Fusion of cell membranes giving pentalaminar appearance to junction Associated with actin Freeze fracture shows rows of particles and pits on the corresponding membrane face	Apex of epithelial cells as part of a junctional complex	Permeability barrier
Intermediate Junction, Zonula Adherens	Subsurface density with actin filaments converging on the junction	Apex of epithelial cells as part of a junctional complex	Intracellular adhesion Transfer forces from actin filaments in adjacent cells
Desmosome, Macula Adherens	Subsurface density Widened intercellular space Intermediate dense line Site of convergence of tonofilaments (intermediate filaments containing keratin)	Apex of epithelial cells as part of a junctional complex	Intracellular adhesion; transfer of forces from cell to cell
Hemidesmosomes	Appearance of a half desmosome	Between base of an epithelial cell and the basal lamina	Cell-to-extracellular matrix attachment
Gap Junctions, Nexus	Pentalaminar junction 2 nm intercellular space Hexagonally packed particles or associated pits in freeze fracture	No specific sites in most cases	Regulated intercellular communication of substances under 1,200 molecular weight
Continuous Junction, Septate Junction	Bars or septa crossing the intercellular space Associated with intercellular actin filaments In freeze fracture they appear as rows of particles and pits on the complementary face	Apical aspect of epithelial cells	Invertebrate form of tight junction Intercellular attachment device

sectioned such that they appear as a punctate fusion of the two plasma membranes, thus obliterating the intercellular space. At the junctional site, the resolved membranes give a five-layered or *pentalaminar* appearance (Figure 20.5). Some

TIGHT JUNCTION

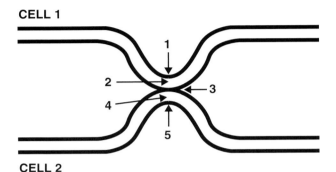

FIGURE 20.5 Diagram showing the numbering of layers of a pentalaminar tight junction between two cells.

protocols (Figures 20.6 and 20.7) for fixing tissue show the junctional particles in negative relief forming the fusion site between two plasma membranes.

In freeze fracture replicas, occluding junctional particles are seen as linear rows of particles within the lipid bilayer. Particles generally predominate on one membrane face (Figure 20.8) and complementary pits are apparent on the other face. Most epithelia contain multiple rows of junctions that either run parallel to each other or anastomose (join) with other rows to some degree. It is not known whether the fusing elements of the tight junction are protein or lipid. Recently, a protein "occludin" had been isolated that may be the elusive sealing element (Furuse et al., 1994).

Tight junctions have a major role in the regulation of paracellular transport of materials. Epithelia effectively regulate the environment of more deeply placed cells by excluding many substances from traveling along their exterior. Tracers (see Chapter

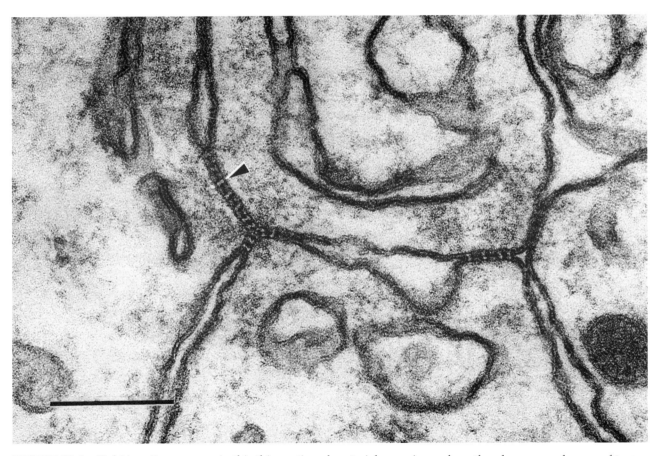

FIGURE 20.6 Tight junctions appear in this thin sectioned material as regions where the plasma membranes of two adjoining cells converge and join. At fusion sites, junctional particles are evident in negative relief as paired translucencies in the bilayer (*arrowhead*). (Sertoli cell junction from a rat.) Bar = 0.25 μm.

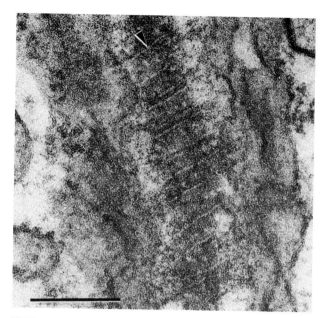

FIGURE 20.7 A grazing, or *en face,* section of tight junctions shows junctional particles in negative relief as seen in Figure 20.6. Because of the grazing plane of section, the membranes forming the junction appear fuzzy and the junctional particles collectively appear as linear arrays (*arrowhead*). (Sertoli cell junction.) Bar = 0.25 μm.

17) have been used to show the exclusion properties of tight junctions.

The *septate or continuous junction (zonula continua)* is the invertebrate counterpart of the tight junction. When viewed in thin section, the intercellular space (17 to 19 nm wide) does not narrow, but remains uniform and is traversed by numerous septa (Figure 20.9). In freeze fracture preparations septate junctions appear as particle-rich ridges on one face and corresponding pits on complementary fracture face (Figure 20.10).

Some continuous junctions exclude tracers, others do not. Continuous junctions appear to have adhering properties and some appear to regulate paracellular transport. They share some of the proteins of adherens junctions.

Adherens junctions contain a variety of morphological types, two of which will be described. The *intermediate junction* or *zonula adherens* forms a continuous "belt" around the cell. At the junctional site, the intercellular space remains constant at about 25 nm (giving the appearance of rigidity) and may display a vague intercellular line running in parallel with the junctional plasma membranes. The cytoplasmic surface of the cell has a density that often receives actin

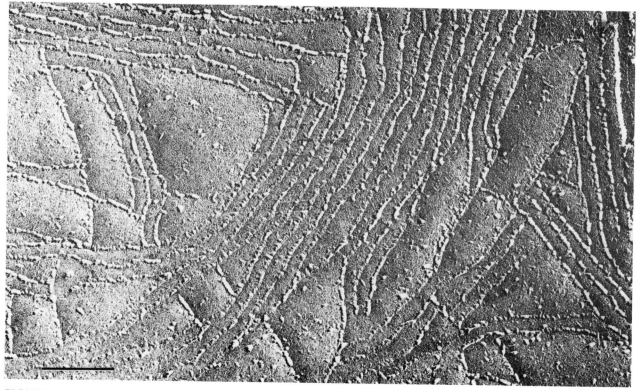

FIGURE 20.8 Tight junctions appear in freeze fracture replicas as extensive rows of intramembranous particles. Some tight junctional rows anastomose. (Sertoli cell junction.) Bar = 0.25 μm.

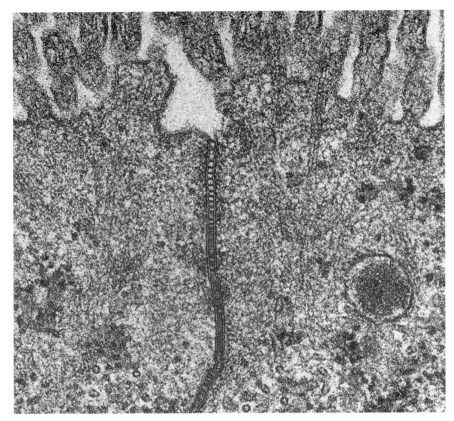

FIGURE 20.9 Continuous junction in a thin section of an invertebrate tissue. Numerous septa bridge the intercellular space between epithelial cells.
(Micrograph courtesy of D. Friend.)

filaments from areas far removed from the junctional site. It has an appearance similar to the desmosome described on pages 487–488.

The *desmosome* or *macula adherens* forms plaque-like structures that, in thin section, have an appearance similar to the zonula adherens, but the desmosome contains a definite intermediate dense line that lies within the intercellular space and parallels the cell surface. Intermediate filaments, which are composed of keratin (*tonofilaments*), make hairpin turns at the junctional density (Figures 20.11 and 20.12). Freeze fracture of desmosomes reveals that they are characterized as aggregations of intramembranous particles of unequal size (Figure 20.13).

Hemidesmosomes or *half desmosomes* are cell surface specializations found at the interface of some epithelial cells and the underlying connective tissue acellular material such as a basal lamina (Figure 20.14). Their appearance at the cell surface is one

whereby the junction resembles a desmosome joined to a basal lamina.

Adherens junctions, as the name implies, are important in cell-to-cell adhesion. They impart an overall cohesiveness to tissues, resisting forces that tend to pull cells apart. The function of hemidesmosomes is to anchor cells to the basal lamina.

Gap junctions or *nexus* and tight junctions were originally thought to be a single junctional type, but advances in both resolution capability, tissue preparation, and their visualization in freeze fracture have allowed them to be distinguished from one another. Gap junctions demonstrate a seven-layered (septalaminar) appearance, which suggests that the opposing membranes of the cell come close to each other, but do not fuse (Figure 20.15). The intercellular space appears to be reduced to 2 to 4 nm. In many instances, the thinness of the section and the orientation are

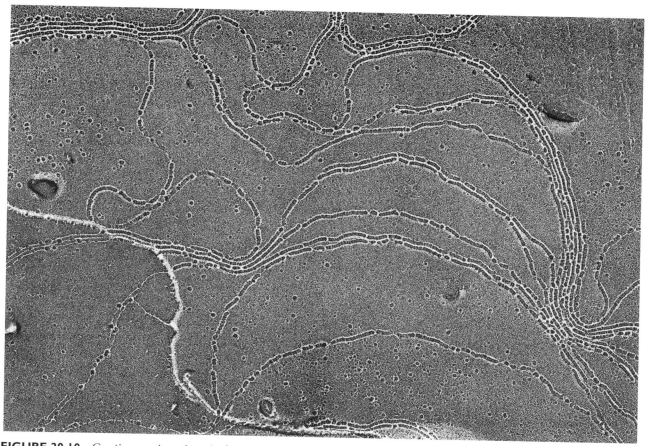

FIGURE 20.10 Continuous junctions in freeze fracture. This freeze fracture micrograph of a larva moth intestine has been rotary shadowed (see Chapter 14). On one face, particle rows predominate. On the other face, pits are in evidence. *(Micrograph of a moth larva courtesy of D. Friend.)*

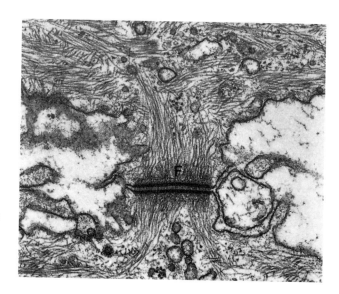

FIGURE 20.11 This desmosome junction shows the characteristic subsurface densities and the tonofilaments (f) that appear to insert into the densities. Tonofilaments actually make a sharp hairpin turn on reaching the junctional density.
(Micrograph of a newt desmosome courtesy of D. Kelly.)

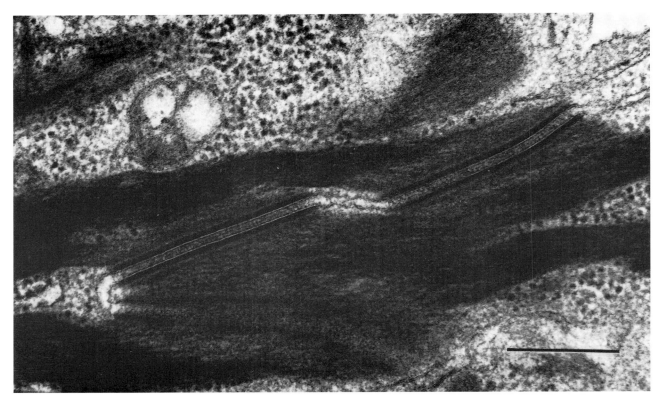

FIGURE 20.12 Features similar to that depicted in Figure 20.11 are depicted in this micrograph of a desmosome although the tonofilaments are more tightly packed. Bar = 1.0 μm.
(Micrograph courtesy of D. Kelly.)

such that the intercellular space is resolved (Figure 20.16).

Gap junction particles (*connexins*) may be seen in appropriately stained *en face* thin sections (Figure 20.17). Adjacent connexins in opposing cells join to form connexins that are known to span the lipid bilayers of both cells although this feature is not evident in a thin section view.

Gap junction particles are best seen in freeze fracture where they are most commonly found as plaque-like aggregations of hundreds of packed particles. Complementary pits are present on the opposite membrane face. Particles, measuring about 8 nm, are arranged hexagonally on one membrane face and are known to span the lipid bilayer of both membranes (Figures 20.18 and 20.19).

Gap junctions are regarded as sites of *intercellular communication*. Connexins contain pores that are thought to transport molecules of up to 1,200 molecular weight from one cell to another.

Junctional complexes join virtually all epithelial cells at their lateral surface near the apical aspect of the cell (Figures 20.20 and 20.21). Usually, these are present where the lateral surface of the epithelial cell meets a lumen such as seen in the gastrointestinal tract. The tight junction is the component closest to the lumen, followed by the intermediate junction, and then the desmosome, which is most basally positioned. Desmosomes may not be visualized in sectioned materials as part of the junctional complex since they are only periodic structures.

Cell Surface Specializations

Cells surfaces are variously configured to accomplish specific functions. Underlying cytoskeletal structures are responsible for the nonrounded appearance of many cells. These will be mentioned here, but described more fully in subsequent sections. Cell surface specializations are so numerous that it is impractical to describe all of them. A few more common forms will be illustrated. The scanning electron microscope and the freeze fracture technique are especially suited to visualize many of the three-dimensional aspects of complex cell surfaces. Cul-

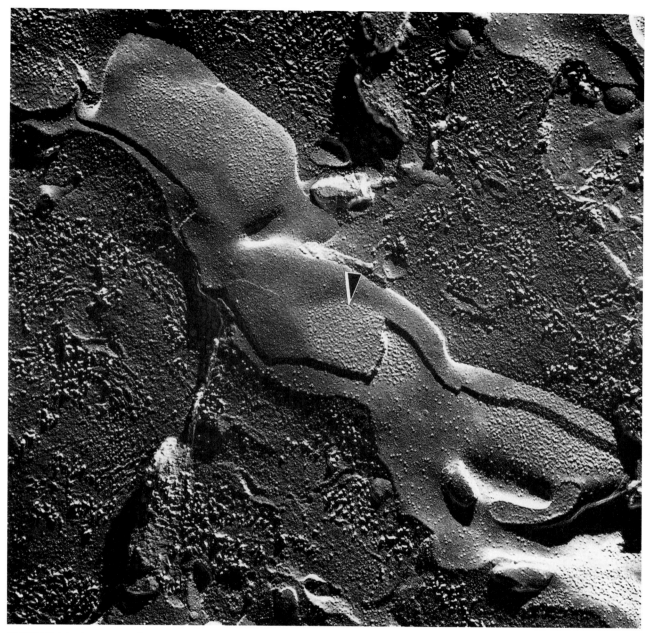

FIGURE 20.13 Freeze fracture image showing several desmosomes between two cells. The desmosome image (*arrowhead*) is recognized by the rounded aggregations of intramembranous particles of nonuniform size. (*Micrograph courtesy of D. Kelly.*)

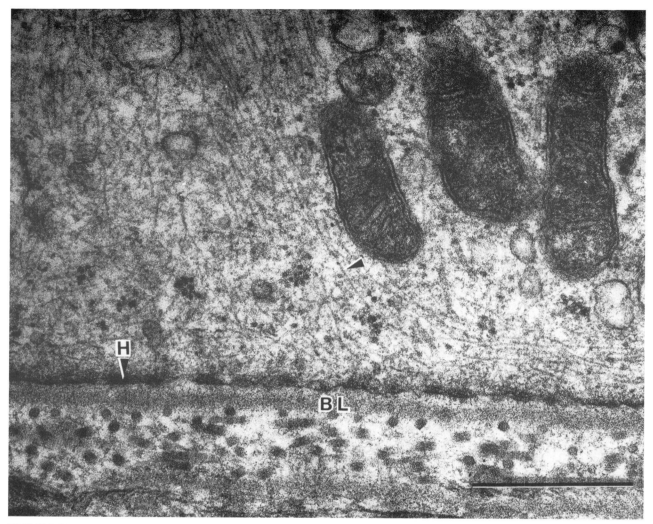

FIGURE 20.14   Transmission electron micrograph of a hemidesmosome (H). Numerous intermediate filaments (*isolated arrowhead*) converge to the cell surface in the region that it impacts the extracellular matrix of the basal lamina (BL). Bar = 0.5 μm.

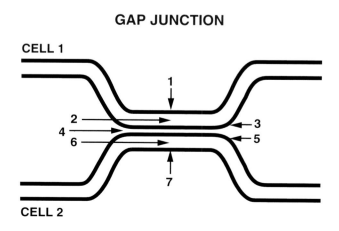

FIGURE 20.15   Drawing of a gap junction showing its seven-layered appearance.

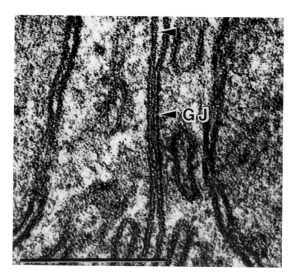

FIGURE 20.16   The intercellular space (*arrowhead*) at gap junction sites (GJ), narrows to the point that it is barely resolvable. Seven layers can be detected. (Micrograph of a liver gap junction.) Bar = 0.25 μm.

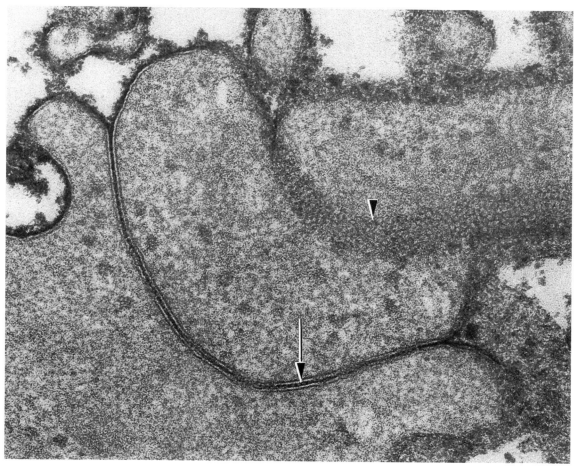

FIGURE 20.17   Gap junction particles in an *en face* thin section appear negatively stained (*arrowhead*) in this micrograph. In a cross-sectioned gap junction a plasma membrane-bound tracer (*arrow*) has been utilized to penetrate a 2 to 4 nm gap in the junction.
(Micrograph courtesy of D. Friend.)

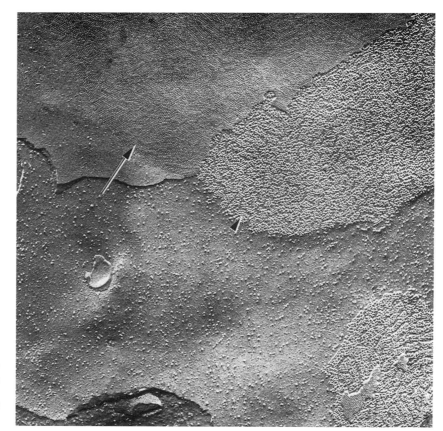

FIGURE 20.18 Gap junction in a freeze fracture preparation of chick skin. Closely packed gap junction particles (*arrowhead*) are seen on the P-face and corresponding pits (*arrow*) on the E-face. (See Chapter 14, "Freeze Fracture Replication," for definition of fracture faces.)

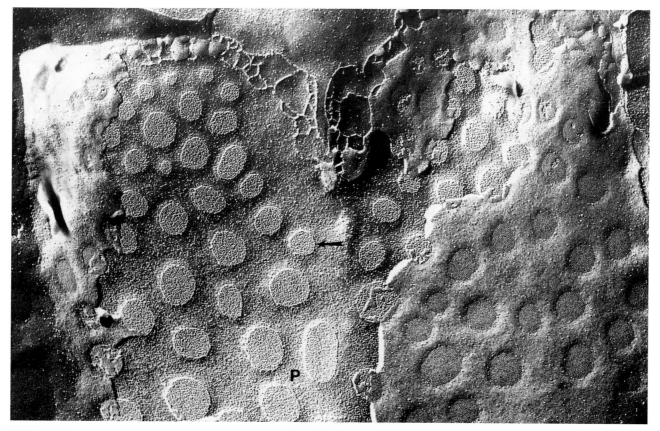

FIGURE 20.19 Gap junctional plaques (*arrow*) on an epithelial cell of a tunicate. The plaques are composed of tightly packed intramembranous particles on the P-face and corresponding depressions on the E-face. (*Micrograph courtesy of R. Dallai.*)

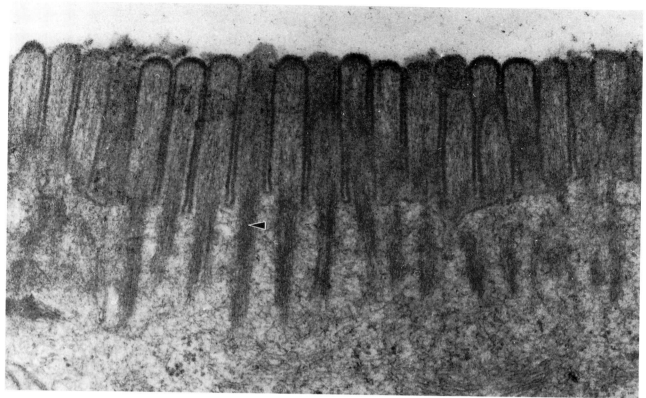

FIGURE 20.24   In these microvilli, actin filaments (*arrowhead*) are seen entering microvillus processes from within the cytoplasm.
(Micrograph from the mouse intestinal epithelium courtesy of W. Dougherty.)

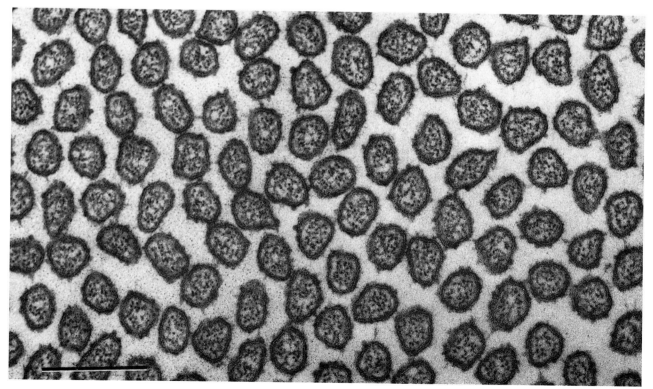

FIGURE 20.25   Microvilli in cross section. Actin filaments are seen within the cores of the filaments. Bar = 0.25 μm.

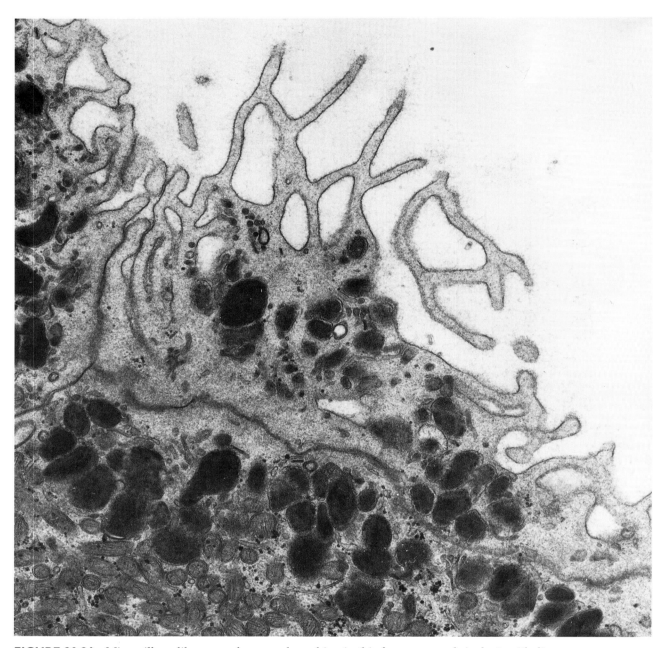

FIGURE 20.26 Microvillous-like appendages are branching in this frog mesonephric duct epithelium. *(Micrograph courtesy of R. Sprando.)*

FIGURE 20.27　Stereocilia projecting from the apical surface of an epithelial cell. (Stereocilia from the efferent duct of the rat.)

Microspikes and filipodia are long needlelike specializations seen mainly in tissue culture cells that are adjusting to their new environment. They contain an actin core and may extend from 5 to 10 μm (microspikes; Figure 20.29) or from 10 to 50 μm (filipodia; Figure 20.22). These cell processes, like lamellipodia described above, are best viewed by scanning electron microscopy. Their actin core allows them to form and to retract rapidly. Both structures seem to be devices for exploring the environment around the cell.

Flagella have basic similarities to cilia, although they are a more highly modified cell surface specialization. Flagella are longer than cilia and there are generally fewer flagella per cell (Figure 20.30; see also "Cilia and Flagella"). Flagella are apparatuses for cell motility.

Micropinocytotic vesicles form as small, rounded pits that pinch off into the cell to form vesicles (Figure 20.31). These vesicles are nonselective transporters of materials, primarily transporting to sites within the cell or across the cell mem-

FIGURE 20.28 Free lamellipodia are seen at one region (*top left*) of this tissue culture cell. Much of the remainder of the cell shows smaller extensions of various types.
(ATCC CCL-13 Chang liver cells courtesy of W. Kournikakis.)

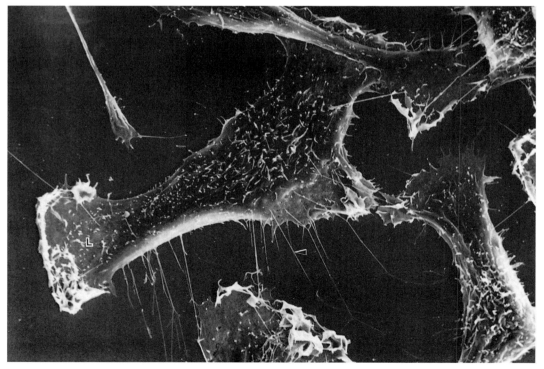

FIGURE 20.29 Flattened areas of two cultured cells form lamellipodia (L). Numerous microspikes form projections from an otherwise smooth cell surface.
(Micrograph courtesy of W. Kournikakis.)

500

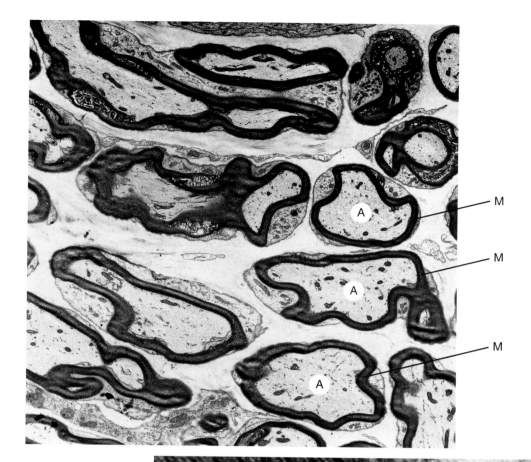

A

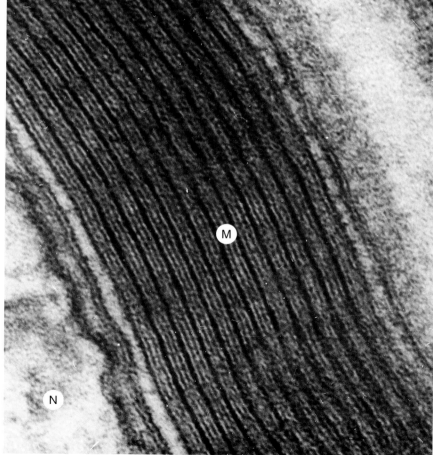

B

FIGURE 20.38 Numerous myelinated axons of a nerve bundle are shown in 20.38A. The light areas are axons (A) and the dense areas are myelin (M). In Figure 20.38B, a high-magnification micrograph of the myelin sheath of one axon and extensive wrapping of the neuron (N) by the plasma membrane of one Schwann cell is evident.

The Cytoskeleton

The cytoskeleton provides the architectural framework of the cell. It is also important in both maintenance and alteration of cell shape, in cell locomotion, and in cell organization and organelle movements. Composed primarily of filamentous elements, the cytoskeleton is a complex and dynamic system within the cell. The filaments have been classified according to size: *microtubules* are the largest; *microfilaments,* or actin filaments, are the smallest; and an intermediate-size class of filaments, not surprisingly, is named *intermediate filaments.* Other cytoskeletal elements have been suggested as present, but their validity as true cytoskeletal elements is not widely accepted at this time. A huge variety of proteins associated with the cytoskeleton may change their assembly/disassembly properties depending on the manner in which they are linked to the same or other cytoskeleton proteins or structures.

Microtubules

Microtubules are long, cylindrical elements having a diameter of about 25 nm and composed of protein subunits (usually 13) called *tubulin.* Tubulin occurs in microtubules as a heterodimer of alpha and beta subunits (Figure 20.39). If sectioned longitudinally, microtubules appear to take a straight or slightly curved course, giving the viewer the impression that they are relatively inflexible structures. In electron micrographs, the central region of microtubules appears less dense than the margins, a feature that, in addition to size, can be used to distinguish them in longitudinal section from other cytoskeletal elements (Figure 20.40). In cross section, the microtubules have a distinct translucent core (Figure 20.41). Microtubules may share subunits to form singlets, doublets, or triplets, the latter two components of cilia and centrioles, respectively (Figure 20.42).

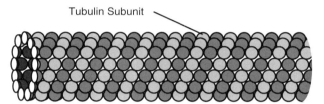

FIGURE 20.39   A small portion of a microtubule reveals its hollow core. Two types of tubulin subunits surround the core in a packed spiral array.

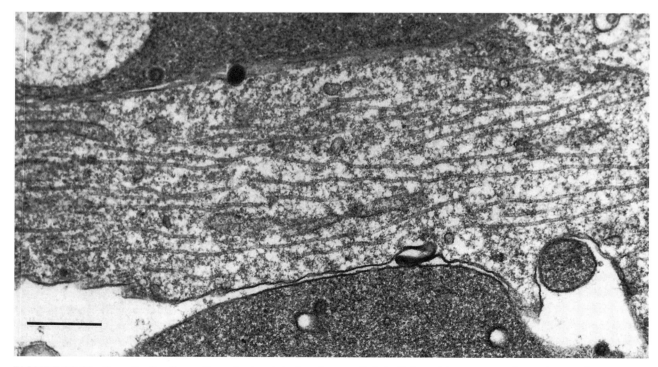

FIGURE 20.40   Longitudinally sectioned microtubules show a slightly lighter (electron-translucent) central region. (Sertoli cell from the rat testis.) Bar = 0.5 μm.

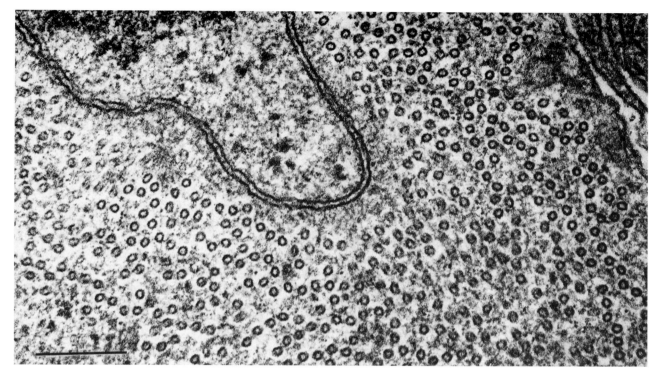

FIGURE 20.41 In cross sections, the microtubular profiles are rounded. The core of the microtubule is distinctly translucent. (Manchette microtubules from the rat testis.) Bar = 0.25 μm.

FIGURE 20.42 Sharing of tubulin subunits in doublet and triplet microtubules.

Microtubules are best known for their role in chromosome movement during mitosis. Here they attach to structures known as *kinetochores* (Figures 20.43 and 20.44) as well as to microtubules from the opposite pole of the cell and, in a way currently under investigation, they are thought to act in concert with other cytoskeletal elements to affect chromosome position and movement (Figure 20.45).

Most cells contain at least a few scattered microtubules and many contain regions of organized bundles of microtubules. The region responsible for organizing microtubules is known as the *microtubule organizing center*. The microtubule organizing center is usually located at the *centrosome,* a region where the centrioles and associated dense material reside. A certain subclass of microtubules is involved in laying down and orienting the cellulose of the plant cell wall. Microtubules function in the

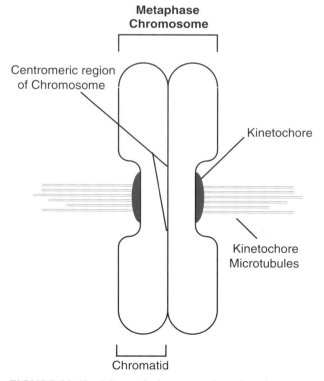

FIGURE 20.43 Microtubules are anchored to the chromosome by the kinetochore, a structure located at the centromere.

development and maintenance of an asymmetrical cell shape. They are thought to be responsible for organelle movements and function in secretory processes. Although they appear to be static structures in electron micrographs, they are usually in a state of flux and are capable of forming and dissociating rapidly.

The ordering of tubulin molecules gives functional polarity to the microtubule. Its plus (+) end shows the ability to polymerize more rapidly than the minus (−) end. Thus growth and dissociation of tubulin units is one means by which microtubules behave dynamically. Another way microtubules change cells is by binding with *microtubule associated proteins* (MAPs). These proteins interact with other proteins that anchor microtubules or alternatively bundle microtubules. Yet another property of microtubules is that they may associate with *micro-*

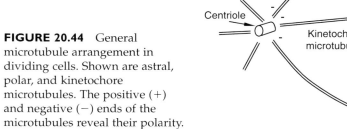

FIGURE 20.44 General microtubule arrangement in dividing cells. Shown are astral, polar, and kinetochore microtubules. The positive (+) and negative (−) ends of the microtubules reveal their polarity.

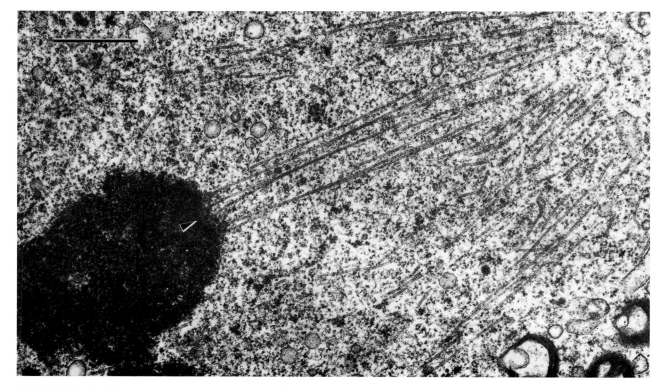

FIGURE 20.45 Microtubules of a metaphase cell are seen attaching to the kinetochore (*arrowhead*) of the chromosome—a structure that appears slightly more dense than the chromatin.

tubular motors such as dynein or kinesin. Protein-microtubule interaction of this type promote movement of one microtubule relative to another (flagellar beating or ciliary movement) or transport of substances along microtubules (Figure 20.46).

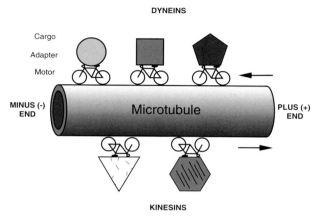

FIGURE 20.46 Dyneins and kinesins in association with microtubules promote movements such as flagellar beating, ciliary motion, and transport of molecules along the microtubule surface.

Microfilaments

The smallest of the recognized cytoskeletal filamentous elements, microfilaments, are composed of polymerized *actin* (Figure 20.47). Microfilaments are notoriously difficult to preserve in standard fixation protocols and, as a consequence, may not appear as distinct filaments (Maupin-Szamier and Pollard, 1978), but more as a fuzzy material (Figure 20.48). Tannic acid is a useful fixative additive for preservation of actin (Maupin and Pollard, 1983). If actin filaments are organized into geometric bundles, one's ability to preserve them is somewhat improved. Actin filaments measure about 5 to 7 nm in diameter and may be seen scattered throughout

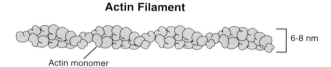

FIGURE 20.47 An actin filament is formed by wound strands of actin protein monomers.

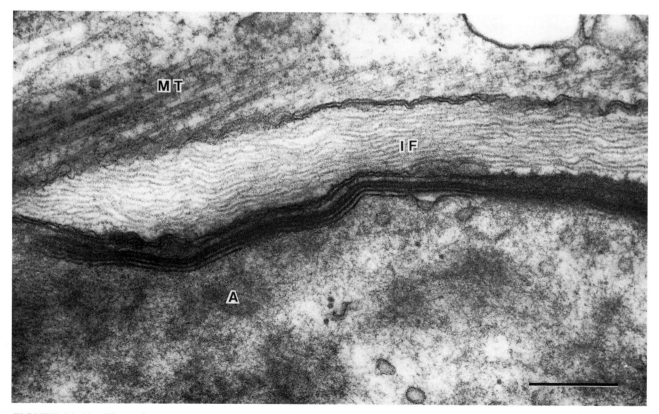

FIGURE 20.48 The indistinct appearance of actin filaments (A) in typical preparations is contrasted with that of intermediate filaments (IF) and longitudinally sectioned microtubules (MT). Intermediate filaments appear solid whereas microtubules show a hollow core. (Opossum spermatid and adjoining Sertoli cell.) Bar = 0.25 μm.

the cell or in highly organized groupings (Figure 20.49). Actin filament groupings may take several forms within the cell. Actin may appear bundled in tissue culture cells, and as such are termed *stress fibers*. Commonly, actin is seen at the cortical region of cells. One of the most dramatic examples of cortical actin is the bundles of actin that form the cores of microvilli (Figures 20.24 and 20.25).

The ability of groups of actin filaments to assume so many forms in cells is due to their association with proteins known as *actin-binding proteins*. Actin-binding proteins also regulate the degree to which actin filaments polymerize/depolymerize and associate with one another or with membranes. The myosin of skeletal or smooth muscle or other contractile systems may be regarded as an actin-binding protein.

Given the morphological diversity of groups of actin filaments and their plasticity within cells, it is not surprising that they perform a variety of functions within cells. Actin filaments may participate in contractile processes in concert with myosin, as in skeletal or smooth muscle contraction (Figure 20.50) or in the contractile ring of dividing cells that separates daughter cells during telophase. They are important in determining the shape of the cell surface, examples being the actin associated with microvilli, stereocilia, and a variety of other local surface specializations. Actin filaments are involved in motile processes and cell locomotion.

Intermediate Filaments

These filaments, also known as *10 nm filaments,* actually represent molecular forms of filaments that are of slightly variable composition and size (about 10 to 12 nm in diameter; Figure 20.51). Tonofilaments of desmosomes (containing keratin), neurofilaments, vimentin-containing filaments (mesenchymal cells), desmin-containing filaments (muscle), and glial filaments (glial cells) are the major intermediate filament types.

Intermediate filaments are not as difficult to preserve for ultrastructural examination as are

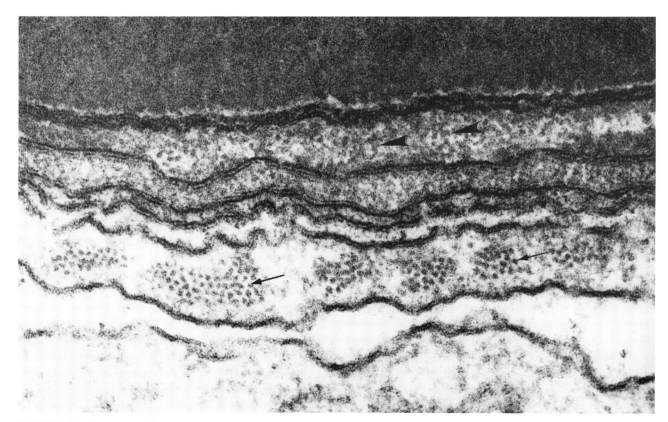

FIGURE 20.49 Actin filaments (*arrows*) are well-preserved bundles along the cell surface of adjoining Sertoli cells. Here this cytoskeletal complex participates in maintaining tight junctions. Actin filaments at another site (*arrowheads*) appear less discrete, which is more typical of the manner of preservation of actin filaments. (×225,000)

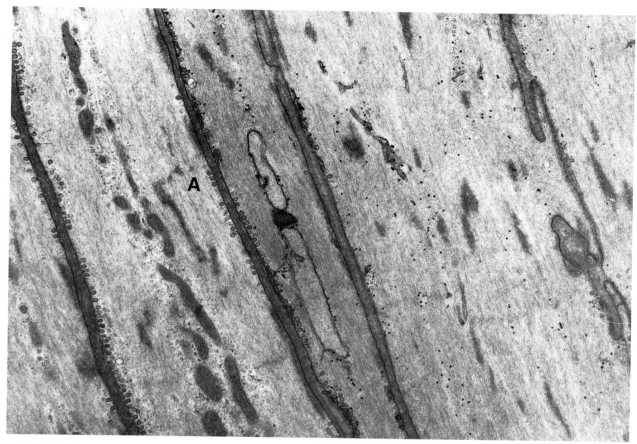

FIGURE 20.50 Actin filaments (A) located just under the cytoplasmic membrane. Such filaments are seen throughout smooth muscle cells such as those shown here. They show specialized contractility functions.

INTERMEDIATE FILAMENT

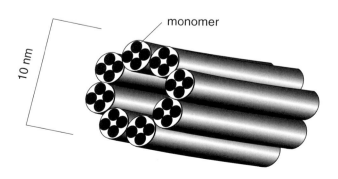

FIGURE 20.51 The 10 nm intermediate filament is composed of eight major cords. Each cord, in turn, is composed of four much smaller cords made from monomeric protein.

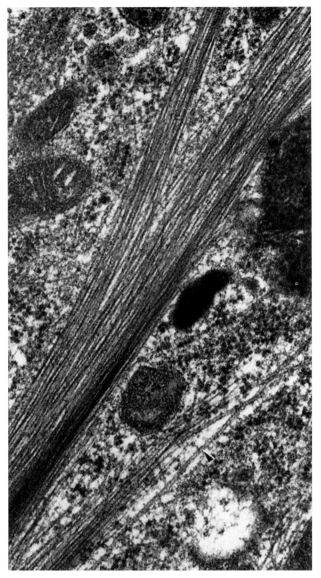

FIGURE 20.52   These intermediate filaments are present in a cultured cell. They appear to course throughout the cytoplasm in bundles (the arrowhead points to a microtubule). *(Micrograph courtesy of A. W. Vogl.)*

actin filaments (Figure 20.48). They are from 30 to 50% larger than actin filaments and generally take a straight or slightly curved (sometimes wavy) course within the cytoplasm (Figure 20.52). Other than occasional bundles of intermediate filaments, there are fewer examples of organized arrays of intermediate filaments than there are of actin filaments. Unlike actin filaments, which prefer a cortical location, intermediate filaments often appear near the nucleus. In spite of these differences, the two filament types are easily confused by microscopists.

The precise function of intermediate filaments is not known. Their positioning at sites within the cell is subject to considerable stress, suggesting that they resist tension within cells and serve a structural role. They may stabilize the nucleus in position.

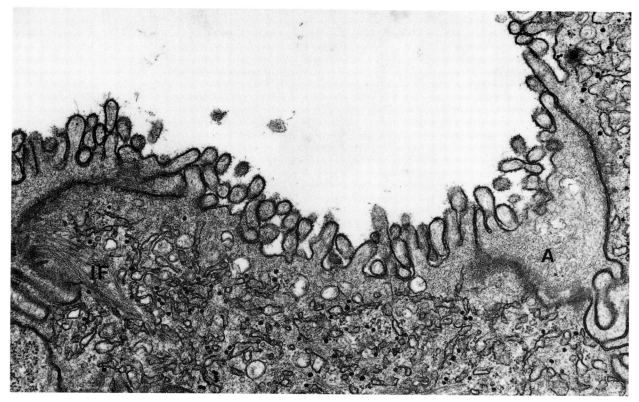

FIGURE 20.53 Cytoskeletal elements of this epithelial cell show specific locations within the cell. Actin (A) forms a network under the microvilli and extends into the microvilli. Intermediate filaments (IF) extend within the cell from the region of the desmosome. (Epithelium of the urethral gland duct of a mouse.)

Regionalization of Cytoskeletal Elements

Cells rarely show a random distribution of cytoskeletal elements. Invariably, their distribution is regional (Figure 20.53), reflecting their functional properties in a particular portion of the cell. For example, epithelial cells show actin filaments along and protruding into cell specializations (terminal web). Also, intermediate filaments extend as a bundle into the cell from desmosomes to join other desmosomes, thereby linking the tension at desmosomes from cell to cell. Another excellent example of the continually changing location of cytoskeletal elements (especially microtubules and actin filaments) is during mitosis.

The Nucleus

The nucleus is the most easily recognizable feature in eukaryotic cells because of its large size and relative constancy of structural features. Genetic information is contained within the nucleus. The main function of the nucleus is to duplicate the genetic information, as well as transcribe information necessary for synthetic processes. The size and shape of nuclei are highly variable. Many nuclei are spherical (see Figure 20.60), others are ellipsoidal, flattened, lobed, or highly infolded (Figure 20.54). Nuclear features include the nuclear envelope, the nuclear lamina, and the nucleolus.

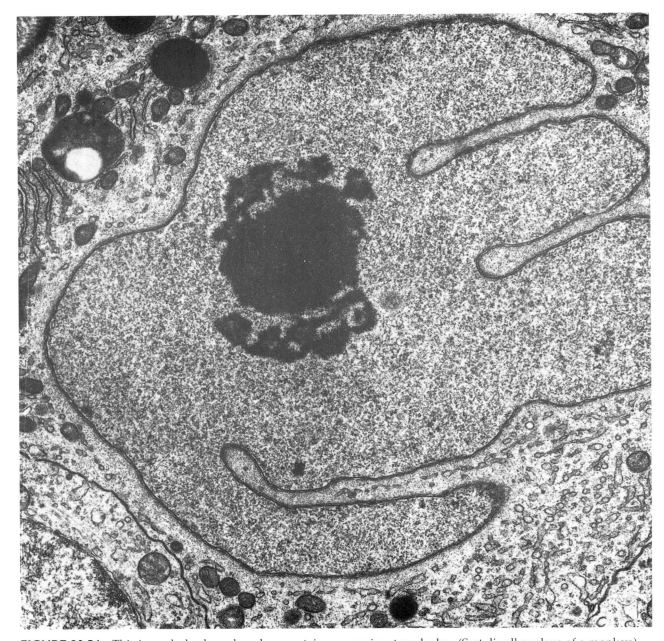

FIGURE 20.54 This irregularly shaped nucleus contains a prominent nucleolus. (Sertoli cell nucleus of a monkey.)

The Nuclear Envelope and Nuclear Lamina

The nuclear envelope, present in eukaryotic cells but not in prokaryotic cells (bacteria), is a double membrane consisting of *outer* and *inner membranes* that encompass the *nucleoplasm.* The space between the two membranes is called the *perinuclear space.* Rough endoplasmic reticulum is often in direct con-

tinuity with the nuclear envelope. At intervals, the two membranes of the nuclear envelope participate in forming *nuclear pores.* These structures are made from over 100 proteins assembled in octagonal symmetry (Figure 20.55). The arrangement of nuclear pore complexes are seen to advantage in grazing sections of the nucleus and in freeze fracture preparations (Figures 20.56 and 20.57). Nuclear pores allow passive entry of substances with low molecu-

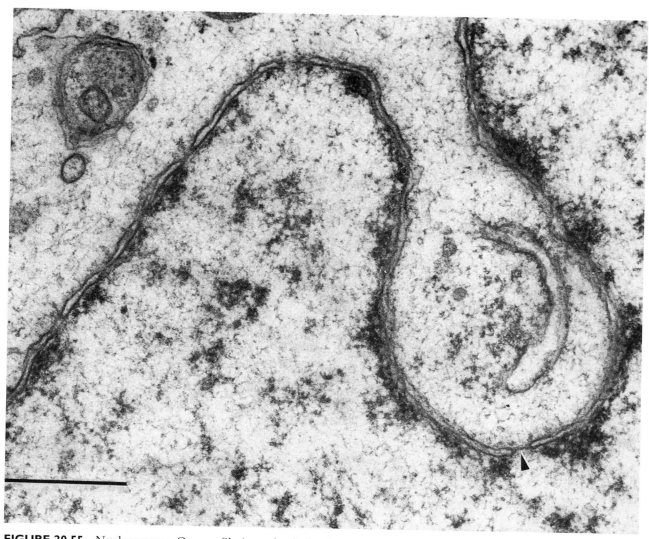

FIGURE 20.55 Nuclear pores. One profile (*arrowhead*) clearly shows the diaphragm bridging the pore. Bar = 0.25 µm.

lar weight and may participate in active transport of substances. Molecules greater than 60,000 MW do not readily enter the nucleus.

Along the inner aspect of the nuclear envelope a uniform layer of dense material may be found termed the *nuclear lamina* (Figure 20.58). The nuclear lamina is composed of proteins resembling intermediate filaments and is involved with maintaining the structural integrity of the nucleus. The nuclear envelope separates the DNA and RNA synthetic processes taking place in the nucleus from that of protein synthesis, which occurs in the cytoplasm. Nuclear pores function as selective gates for entry and exit of molecules to and from the nucleus thus regulating the environment of the nucleus.

Chromatin

In structural terms, the condensed or visible form of chromatin is known as *heterochromatin*. The relatively clear areas of the *nuclear matrix* are assumed to be occupied by an uncondensed form of chromatin known as *euchromatin*. From a functional standpoint the heterochromatin is very stable (inactive), tightly packed (Figure 20.59), and not undergoing transcription. The so-called euchromatin is very loosely packed, but only about 10% of it is available for transcription. Thus, cells with large proportions of euchromatin are thought to be active in synthesizing substances. Upon close examination, heterochromatin is usually made up of fine 15 to 20 nm granules, but at low magnification the granules are not resolved (Figure 20.60).

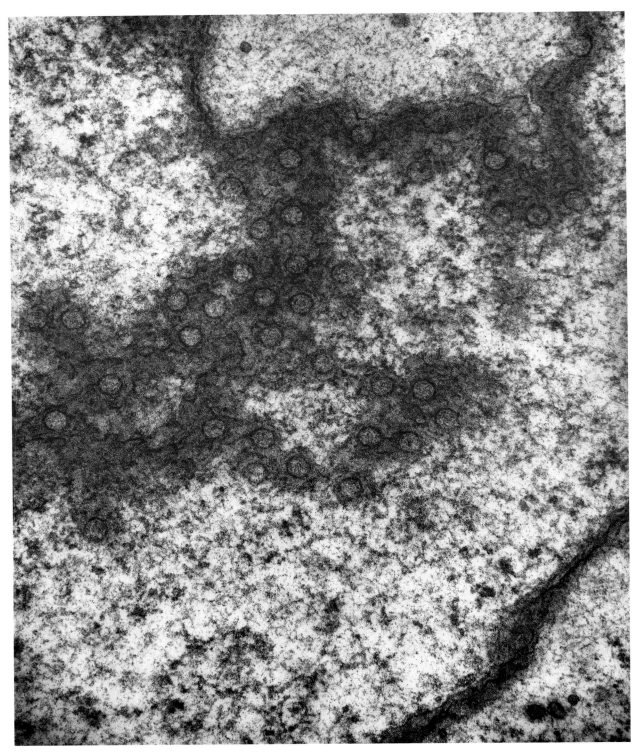

FIGURE 20.56   Nuclear pores. The indistinct appearance of the membranes of the nuclear envelope indicates that this nucleus was grazed by the sectioning knife. Several nuclear pore complexes are seen *en face* and appear spherical.

FIGURE 20.57 Freeze fracture of a nuclear envelope shows that the small rounded nuclear pores are not necessarily randomly distributed.

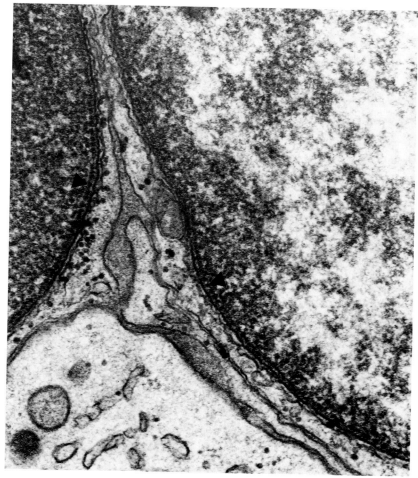

FIGURE 20.58 The nuclear lamina is the material just inside the nuclear envelope (*arrowheads*).

517

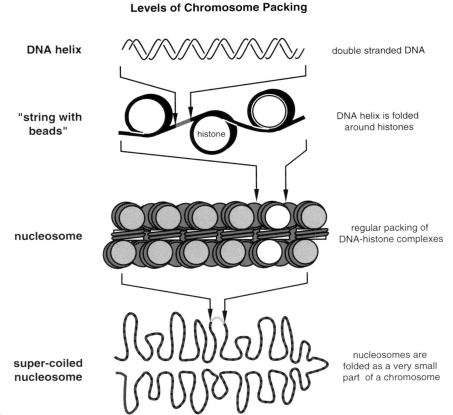

FIGURE 20.59 Chromosome packing is very complex. There are at least four major levels of packing.

Heterochromatin is frequently seen lining the nuclear envelope, but absent from regions of the nuclear envelope containing pores.

In eukaryotes, heterochromatin is assumed to be complexed with histones, which package the extremely long strands of DNA. DNA complexed with histones is called *chromatin*.

The Nucleolus

Of variable size and number, the nucleoli occupy only a small portion of the nucleus and are thus seen in only some thin sections of the nucleus. Although the organization of the components of the nucleolus differ widely from one cell type to another, there are three major recognizable structural components (Figures 20.61, 20.62, and 20.63). The *pars fibrosa* or *nucleolonema* is represented as strands of dense material that form a spongy network. Associated with the pars fibrosa are the *fib-rillar centers,* which are usually finely granular, rounded masses. These areas are portions of the nucleus containing DNA. However, from a functional standpoint the fibrillar centers are areas where new RNA is being synthesized. The third component, the *pars granulosa*, is usually near the pars fibrosa and appears as an aggregation of finely granular material. The pars granulosa represents newly developed ribosomes. RNA synthesis takes place in the nucleolus.

Dividing Animal Cells

During cell division, the nuclear envelope is resorbed, the nucleolus dissipates, and the cell goes through the classic phases of cell division: *prophase* (Figure 20.64), *metaphase* (Figure 20.65), *anaphase* (Figure 20.66) and *telophase* (Figures 20.67 and 20.68).

In germ cells of the testis, the cleavage furrow never pinches off in telophase, but remains as an

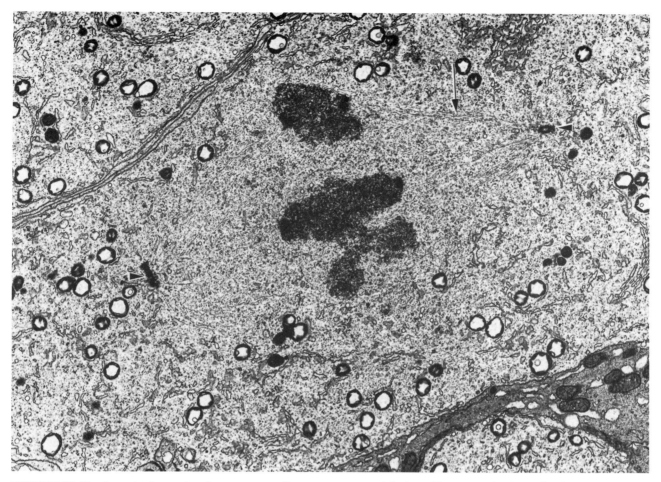

FIGURE 20.65 In metaphase, the chromosomes align on an equatorial plate. The centrioles (*arrowhead*) are apparent and the spindle apparatus (*arrow*) is visible. Bar = 1.0 μm.

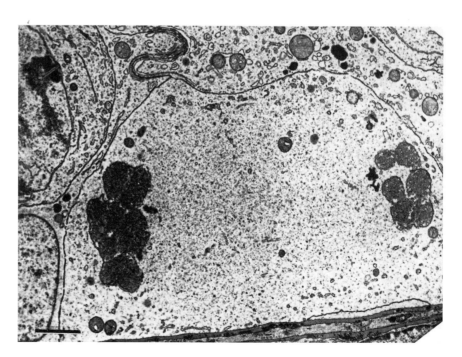

FIGURE 20.66 In late anaphase, the chromosomes have just reached the poles of the cell. Bar = 2.0 μm.

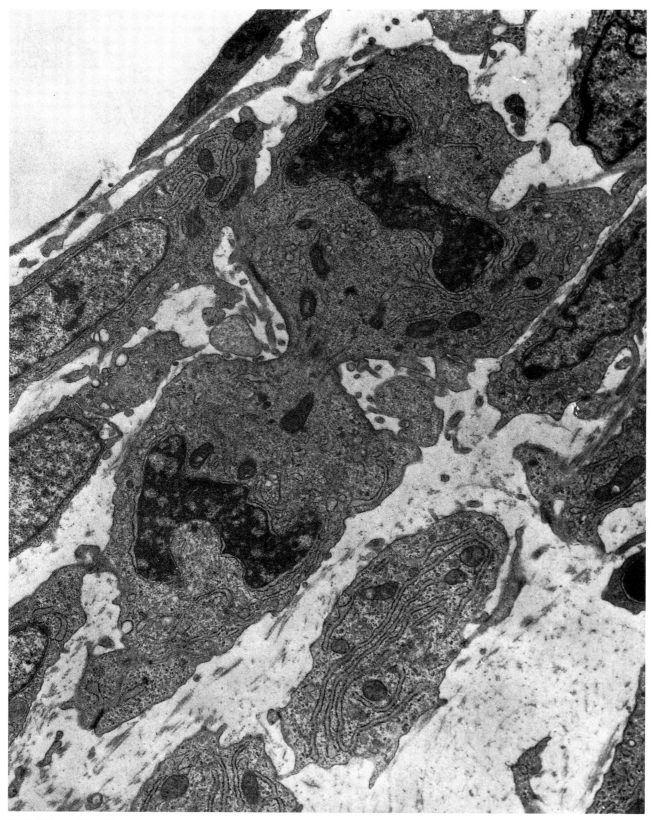

FIGURE 20.67   Telophase cell. A cleavage furrow has begun to divide the two daughter cells and the nuclear envelope has reformed.
(Micrograph courtesy of W. Dougherty.)

in sections, their connection to the inner membrane as it begins to fold inward is not always observed. The cristae may be lamellar (sheetlike) or tubular or may take other geometrical forms. Branching and anastomosing cristae have also been observed. The space between the outer and inner mitochondrial membranes is the *intracristal space* and extends inward with the cristae. The *mitochondrial matrix* lies internal to the inner mitochondrial membrane and appears homogeneous and finely granular, except for the occasional presence of *matrix granules*. These very dense bodies are calcium salts.

The structural features of mitochondria as well as variations in size, shape, and internal structure are illustrated in Figures 20.73 through 20.83.

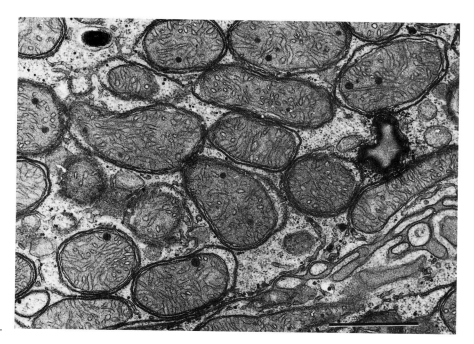

FIGURE 20.74 Mitochondria with tubular cristae. Bar = 1.0 μm.

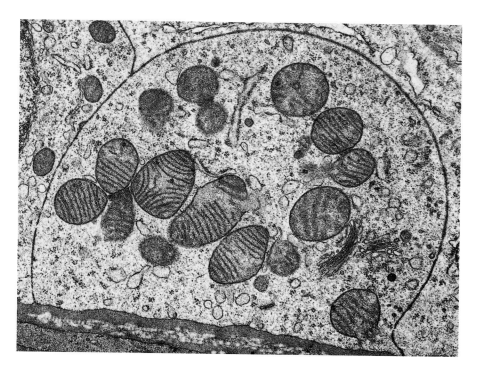

FIGURE 20.75 Mitochondria with exclusively lamellar cristae.

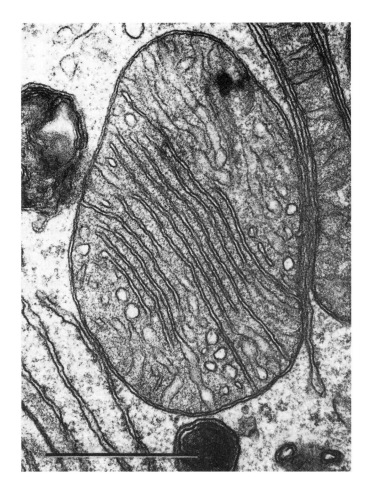

FIGURE 20.76   Mitochondria showing both lamellar cristae and tubular cristae. Bar = 0.25 μm.

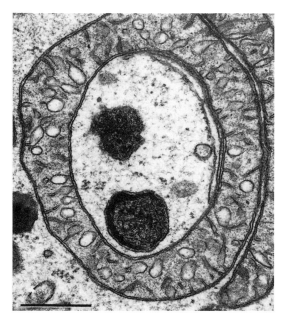

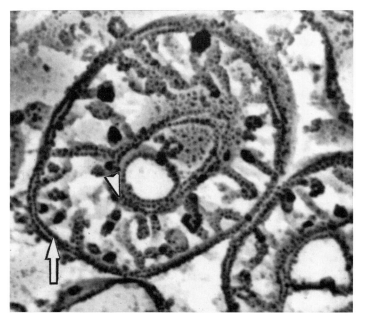

FIGURES 20.77 and 20.78   The mitochondrion in Figure 20.77 (*left*) is cup-shaped, but here appears ring-shaped because only the lip of the cup has been sectioned. Its cristae are distinctly tubular and expanded at their ends. In Figure 20.78 (*right*) the same mitochondrial type is visualized by scanning microscopy after it has been broken open. The inner mitochondria membrane is indicated by both the arrow and the arrowhead and cristae are visualized. Contrast has been reversed for comparison to Figure 20.77.
(Scanning micrograph courtesy of I. Fritz.)

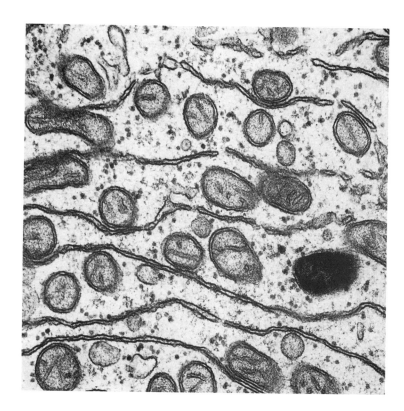

FIGURE 20.79 These small mitochondria contain only a few cristae.

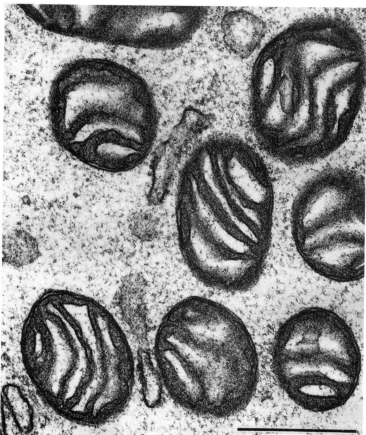

FIGURE 20.80 The intracristal space is greatly dilated in cristae of these mitochondria. Although they superficially resemble poorly fixed mitochondria described in Chapter 19, the modifications of these mitochondria are seen in well fixed tissues. Bar = 0.5 μm.

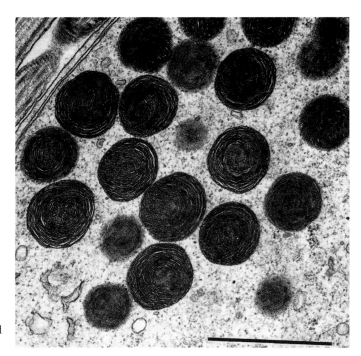

FIGURE 20.81 The cristae of these mitochondria are difficult to resolve because they are extensively spiraled within the mitochondrial matrix. Bar = 1.0 μm.

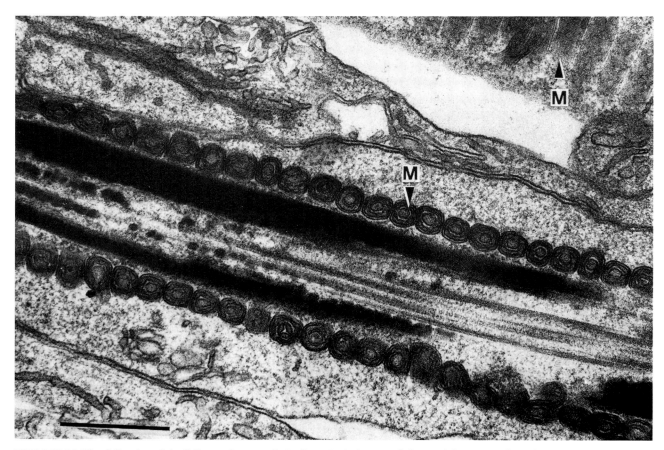

FIGURE 20.82 Mitochondria (M) are characteristically spiraled around the middle piece of the flagellum of mammalian spermatids and sperm. A transverse section (*extending from upper left to below*) shows their internal structure and a grazing section (*upper right*) suggests their spiral arrangement around the flagellum. The intracristal space is the less dense area of these mitochondria. Bar = 0.5 μm.

as seen in thin section, are considered to be interconnected and part of one totally enclosed membrane system, although the plane of section does not often reveal the continuities between sectioned cisternae. The *lumen* of the rough endoplasmic reticulum often appears dense or contains flocculent material due to the presence of secreted protein. Occasionally, the reticulum is seen to be in continuity with the two membranes forming the nuclear envelope (Figure 20.89). Because many of the membranes of the cell are continuous they are said to be part of the endomembrane system. The rough and smooth endoplasmic reticula (Figure 20.90) are more commonly seen to be in direct continuity. In epithelial cells, the rough endoplasmic reticulum is usually concentrated in parallel cisternae at the base of the cell in a juxtanuclear position. Although rough and smooth endoplasmic reticula are frequently in continuity, the rough

reticulum contains at least 20% more protein than does the smooth variety. Rough endoplasmic reticulum may be dilated or may contain dense material to reflect the storage of proteinaceous material in its lumen (Figures 20.91 and 20.92).

The rough endoplasmic reticulum segregates secreted material from the cytosol and is the site of synthesis of secretory proteins destined for export from the cell. The bound ribosomes congregate with both messenger RNA and specific transfer RNAs to produce a polypeptide chain that is extruded into the lumen of the rough endoplasmic reticulum and that eventually is cleaved off into its lumen.

Smooth Endoplasmic Reticulum

There is a dangerous tendency to refer to all of the nonribosome bound membranous vesicles within the cytoplasm as smooth endoplasmic reticulum. In

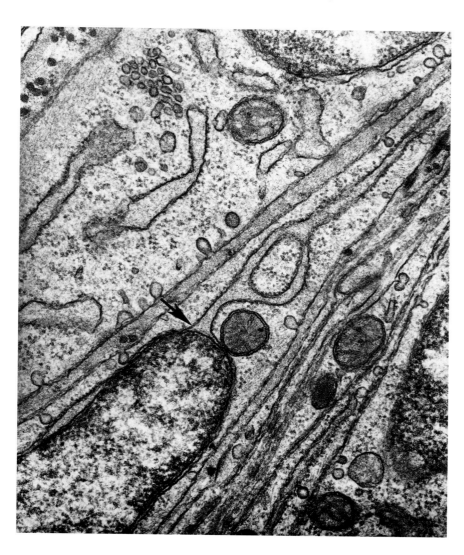

FIGURE 20.89 Continuity of the rough endoplasmic reticulum and the nuclear envelope is seen at the arrow.

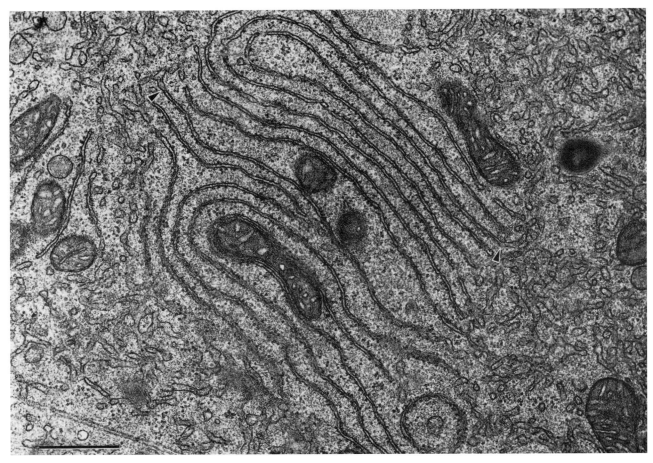

FIGURE 20.90 Ribosomes sparsely populate the external surface of the stacked cisternae of rough endoplasmic reticulum. The cisternae show a relatively collapsed lumen indicating that they contain relatively little secretory material. Communication of the rough endoplasmic reticulum with the smooth endoplasmic reticulum is indicated by the arrowheads. Bar = 0.5 μm.

a strict sense, the smooth endoplasmic reticulum is defined not only by the absence of ribosomes from its membranes, but also by the enzymes that are associated with its membranes. There are other vesicular elements within the cytoplasm, such as transport vesicles, which do not contain ribosomes and cannot be classified as smooth endoplasmic reticulum.

In most instances, the smooth reticulum displays characteristic morphological features and, for practical purposes, can be distinguished as such. Because it is usually composed of anastomosing tubules of irregular shape and diameter, it appears in micrographs cut in various planes as a "contorted" network of smooth membranous elements (Figure 20.94; also see "Lipid "section). Tubules of about 100 nm in diameter form a continuous membrane system that weaves in and out

of the plane of section. Occasionally, in some cells, the smooth reticulum forms flattened cisternae in concentric swirls. These may surround lipid droplets (Figure 20.95).

Smooth endoplasmic reticulum performs a variety of functions. It contains enzymes necessary for lipoprotein, cholesterol, and steroid synthesis. This structure is involved in lipid metabolism and metabolism of toxic substances as well as the breakdown of glycogen.

The Golgi Apparatus

The Golgi apparatus appears as a group of stacked membrane cisternae (*Golgi stack* or *dictyosome*) that together form a slightly curved or U-shaped structure. Associated with the Golgi stacks are numerous small *Golgi vesicles,* some of which are smooth

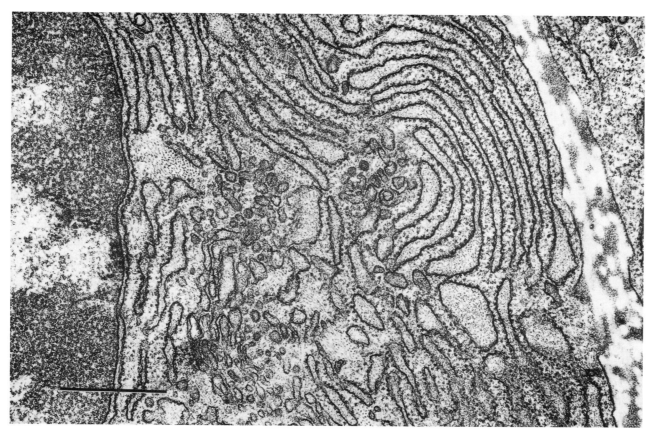

FIGURE 20.91 In contrast to that shown in Figure 20.90, this rough endoplasmic reticulum is heavily populated with ribosomes and its lumen is swollen. Bar = 0.5 μm.

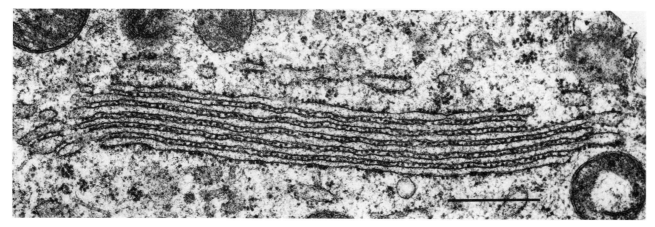

FIGURE 20.92 Rough endoplasmic reticulum forming a stack of parallel cisternae. The lightly granular material within the cisternae represents synthesized protein. Bar = 0.5 μm.

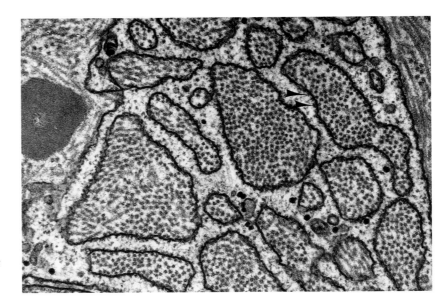

FIGURE 20.93 The rough endoplasmic reticulum of this cell is swollen due to accumulation of a product of unknown nature within the cisternae. Ribosomes are indicated by the arrowheads.

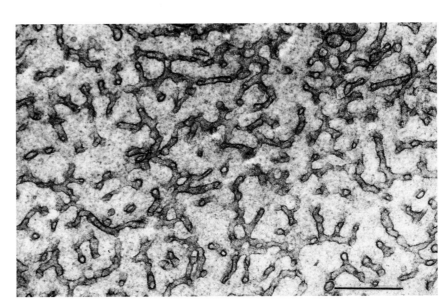

FIGURE 20.94 High-magnification micrograph of smooth endoplasmic reticulum in a steroid secreting cell depicting the characteristic anastomotic nature constituting the reticulum. The continuity of the reticulum can be traced for some distance in this planar micrograph. (Rat Leydig cell.) Bar = 0.5 µm.

surfaced and others of which display a bristle or fuzzy coat.

The Golgi complex is generally located near the nucleus and, in epithelial cells, is positioned between the nucleus and the apex of the cell (i.e., *supranuclear location*). All synthetic cells have one or many Golgi stacks that are generally interconnected. Individual stacks contain a variable number of cisternae, but generally less than ten. The Golgi stack is polarized. The convex face is known as the *forming* or *cis face* and the concave face is known as

the *maturing* or *trans face*. Near the trans face of many secretory cells are seen vacuoles containing dense material. They are slightly larger than the majority of those associated with the Golgi apparatus. These *condensing vacuoles* are considered as immature secretory granules. A variety of Golgi types are illustrated (Figures 20.96 through 20.100).

The Golgi is involved in modifying, packaging, and sorting proteins newly synthesized in the rough endoplasmic reticulum. Sugar residues are added and also modified in the Golgi. Golgi cister-

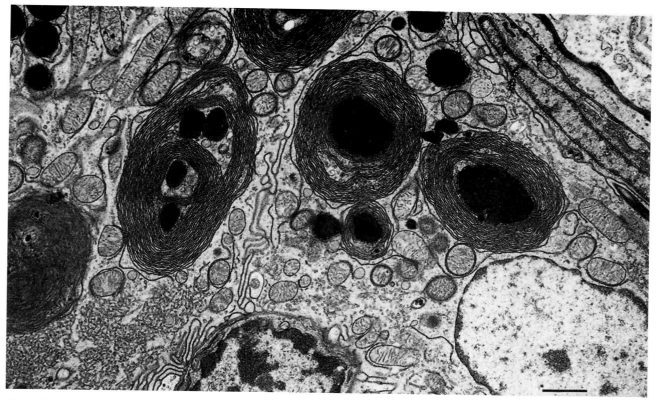

FIGURE 20.95 Swirls of smooth endoplasmic reticulum organized around lipid droplets predominate in this tissue, although major aggregations of the tubular reticulum are also found. (Mouse Leydig cell.) Bar = 1.0 μm.

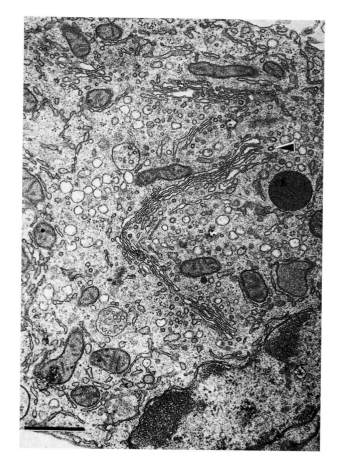

FIGURE 20.96 The typical Golgi complex is U- or V-shaped with a convex (cis) and concave (trans) face. The latter is frequently directed toward the nucleus (*lower right*). Numerous Golgi associated vesicles are seen. Some of these contain a bristle coat (*arrowhead*) making the vesicle wall appear thicker. Bar = 1.0 μm.

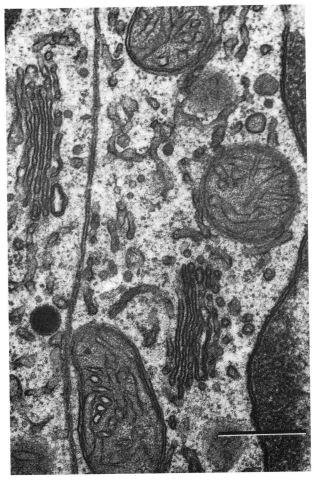

FIGURE 20.97 Simple stacks of the Golgi apparatus as seen in two adjoining cells. Bar = 0.5 μm.

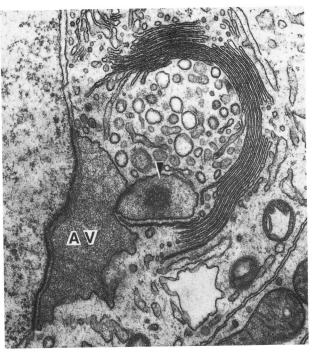

FIGURE 20.98 This Golgi apparatus from a spermatid shows a condensing vacuole with secretory product (*arrowhead*). The condensing vacuole is in the process of joining the acrosomal vesicle (AV), a structure which was produced previously by the Golgi. The secreted material will eventually produce the acrosome of the sperm.

nae within a particular stack are not similar biochemically and, although they may have a similar appearance in the electron microscope, they have different functional properties. There is considerable experimental evidence that the working of the various compartments (saccules) of the Golgi differ because each saccule has specific enzymatic properties relating to different functions.

In virtually every view of the Golgi, numerous vesicular elements are associated with this complex organelle (Figure 20.96). These are transport vesicles of one type or another that have surface messages telling them which organelle will receive them. The Golgi is probably the best example of an organelle that both sends and receives numerous transport vesicles as depicted in Figure 20.101.

Secretory Products

Secretory granules are membrane-bound structures, almost always with dense appearing interiors. Unequivocal proof that a membrane-bound structure is a secretory granule, and not, for example, a lysosome rests with the biochemical identification of its contents as a secretory product. The kinds of material being secreted are varied. In a practical sense, secretory products are abundant within cells and generally clustered, whereas lysosomes are rarely clustered. In epithelial cells,

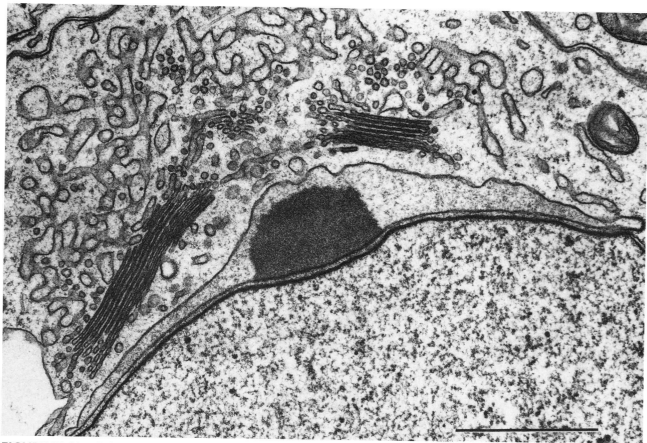

FIGURE 20.99 Golgi apparatus from a spermatid showing secretory material that has been deposited from the Golgi. Note the anastomotic smooth endoplasmic reticulum at the cis face of the Golgi. Bar = 1.0 μm.

secretory granules are usually positioned between the Golgi apparatus and the apical surface of epithelial secretory cells. The content of secretory granules is most often homogeneous (Figures 20.102 through 20.105). Secretory products are a characteristic of many cell types, but their appearance in electron micrographs is as variable as the types of products secreted. Unlike lysosomes, the contents of secretory products are generally homogeneous. They are commonly electron dense (Figures 20.102 and 20.103) but may appear clear (Figures 20.104 and 20.105) depending on their chemical nature.

Secretory granules are the storage site for secretory material produced in the Golgi apparatus. Under the appropriate stimulus, they exit the cell by a process known as *exocytosis*. Upon stimulation, secretory granules will commonly fuse with each other (Figure 20.105). Subsequently, fusion of the limiting membrane of the granule with the plasma membrane allows the contents of the vesicle to escape (Figure 20.106), whereas the bounding membrane of the secretory granule becomes part of the plasmalemma and is recycled.

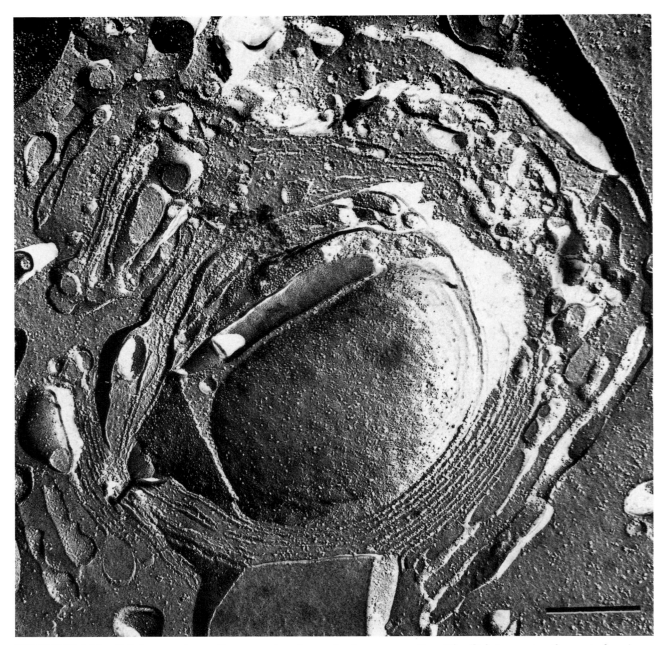

FIGURE 20.100 Golgi apparatus as it appears in a freeze fracture preparation. The Golgi occupies the central region of the figure. Most Golgi saccules are cross-fractured but one is fractured extensively in the plane of the membrane. Bar = 1.0 μm.

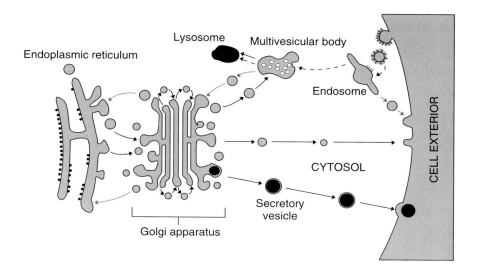

FIGURE 20.101   Some examples of cell trafficking.

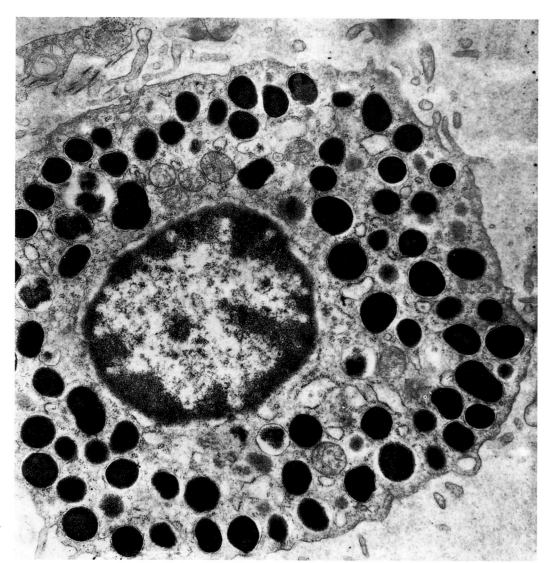

FIGURE 20.102
A mast cell containing numerous packets of dense secretory material (primarily histamine), each packet individually bound by a membrane.

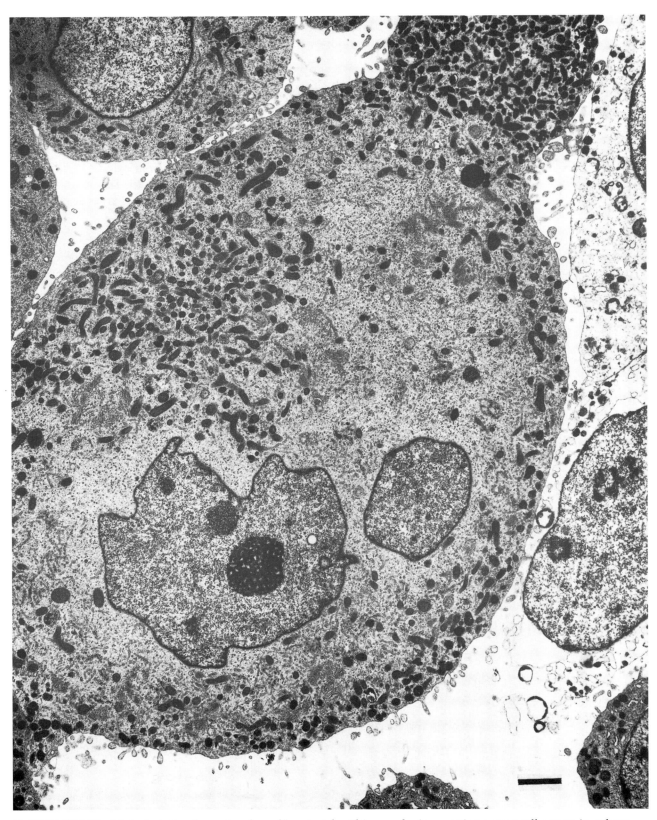

FIGURE 20.103 Melanin is the dense "packaged" material within a melanin-secreting tumor cell grown in culture. Bar = 1.0 μm.

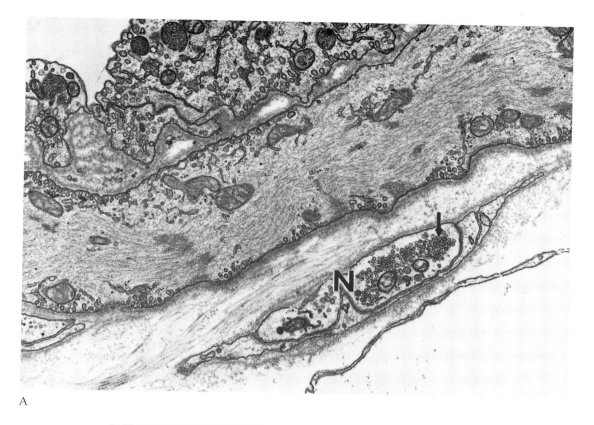

A

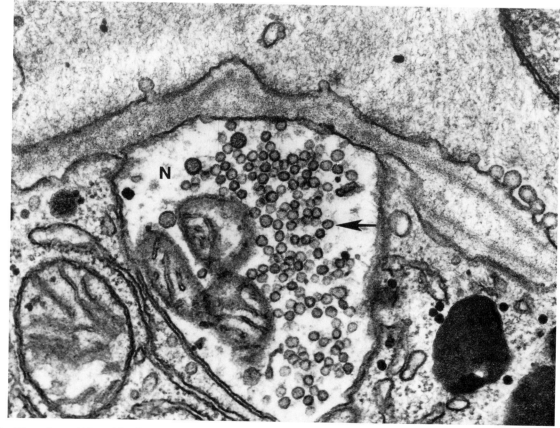

B

FIGURE 20.104 These low- (A) and high-magnification (B) micrographs of nerve terminals (N) reveal their membrane-bound secretory vesicles (*arrow*). Most neurosecretory vesicles appear clear but some are larger and have a dense material inside suggesting that there are two types of neurotransmitters present. In (A), smooth muscle is adjacent to the nerve terminal suggesting its intervention by the nerve terminal.

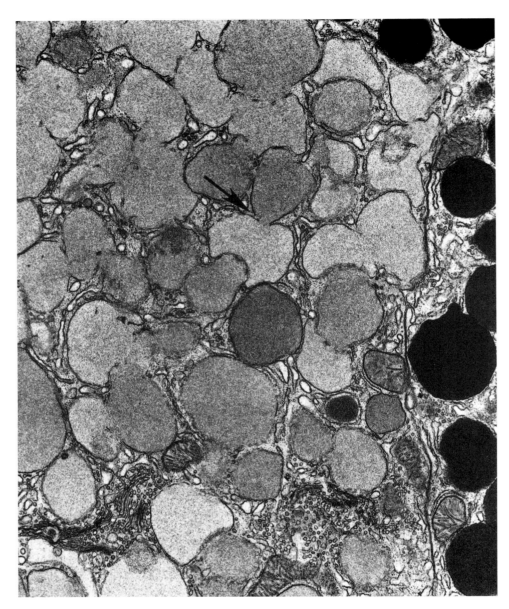

FIGURE 20.105   Fusion of secretory granules with each other (*arrow*) often precedes their exit from the cell. Many of the granules that were originally very dense like those shown at the bottom of the micrograph appear to have fused. (Urethral gland cells of a mouse.)

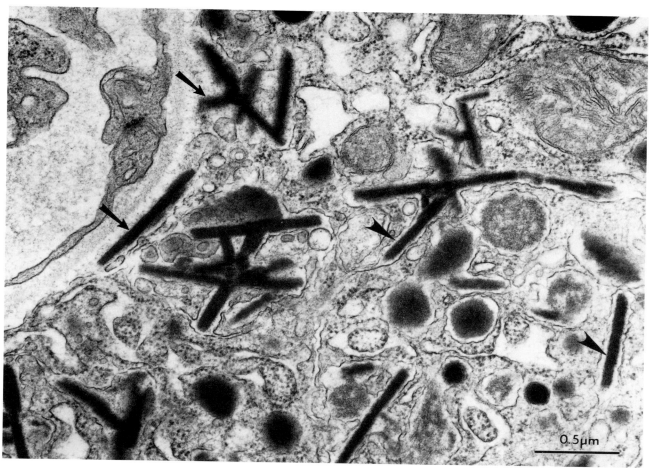

FIGURE 20.106 Exocytosis (extrusion) of growth hormone containing secretory material from the pituitary gland of the musk shrew. Upon stimulus some of the rod-shaped secretory products have exited the cell (*arrows*), whereas other similar rod-shaped materials remain in the secretory vesicles (*arrowhead*). Note that some of the spherical secretory granules are not exocytosed.
(Micrograph courtesy of M. Shiino.)

Centrioles

Centrioles are short (up to 0.5 μm long), cylindrical (0.15 μm) structures that are frequently seen in pairs called *diplosomes*. The centrioles of the diplosome frequently lie at right angles to one another. Centrioles are usually positioned near the nucleus close by the Golgi apparatus (Figure 20.107). The wall of the centriole cylinder is best seen in a cross section of the centriole. It is composed of microtubular elements, or *sub-fibrils*, in fused triplets. Nine sets of triplets are arranged in a pinwheel fashion (Figure 20.108). The designation of individual sub-fibrils of a triplet is by letters; the innermost circular subfibril *A*, the middle *B*, and the outermost *C*. Dense material is often seen radiating outward from the triplets (*pericentriolar satellites* or *matrix*). In longitudinal section, the short, parallel, microtubular walls readily identify the structure as a centriole.

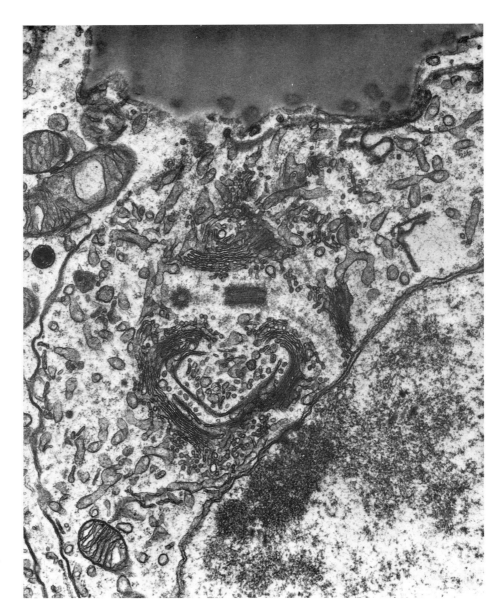

FIGURE 20.107   This pair of centrioles, or diplosome, is situated near the nucleus and the Golgi apparatus. This general area is known as the centrosome. The long axis of the centrioles lie at approximate right angles to one another and in neither centriole is its internal structure distinct, a result of how these centrioles were sectioned. Bar = 1.0 μm.

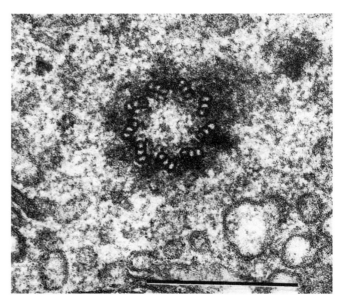

FIGURE 20.108   A centriole is sectioned transversely and displays the nine triplets, each containing three sub-fibrils. The dense material, or pericentriolar satellites, lying external to the centriole is closely associated with each triplet. Bar = 0.5 μm.

Centrioles function in organizing microtubules as, for example, in the development of the mitotic spindle or in the development of cilia or the flagellum. One centriole of a pair is distributed to each daughter cell during cell division. The centrioles are paired again during the S-phase where another centriole is synthesized. Procaryotes and plant cells lack centrioles, but do have comparable centrosomal areas at their poles for regulation of cell division.

Centrioles are located in a specific region of the cell known as the *centrosome*. The centriole pair (or diplosome) and associated dense material (Figure 20.108) comprise the centrosome. Microtubule growth and anchorage frequently occurs at this site.

Cilia and Flagella

Cilia and flagella are structures that arise from centrioles and are specialized for the purpose of motility. Generally, there are fewer flagella (usually one) on a per cell basis than there are cilia. Flagella are almost always longer than cilia (Figure 20.30), but because they contain more accessory structures, they may be considered a more elaborate form of a cilium.

Cilia

About 0.2 μm across and 10 to 15 μm long (also see the description in the section "The Cell Surface") cilia are projections from the cell surface usually into an air- or fluid-filled cavity (Figure 20.109). The detailed structure of cilia is best examined in cross-sectioned profiles of its shaft. Cilia are often referred to as 9 + 2 structures, meaning that they are composed of 9 fused peripheral doublets (*microtubule sub-fibrils*) and a central unfused pair of microtubules. The 9 + 2 arrangement is called an *axoneme* (Figure 20.109). Sub-fibril *A* of the cilium appears to be a complete microtubule, whereas sub-fibril *B* joins sub-fibril *A* and may be described as C-shaped. Cells bearing cilia are common in the respiratory epithelium; however, cilia may develop from many cell types in an organism not normally known to possess cilia. There is considerable substructure to the cilium that is not apparent on routine examination. Cilia show characteristic intramembranous particle distribution in freeze fracture micrographs (Figure 20.110).

At the base of the cilium is a *basal body*, which resembles a centriole that has been modified in some regions (Figure 20.111). There is considerable species variability in the structure of the basal body. In some species, the centriole undergoes little modification,

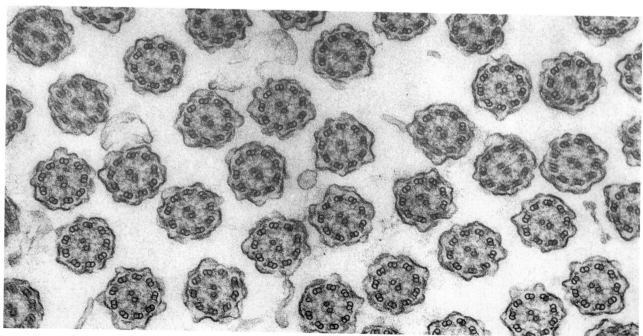

FIGURE 20.109 These ciliary axonemes from respiratory epithelium are sectioned transversely showing their 9 + 2 pattern of microtubules.
(Micrograph courtesy of W. Dougherty.)

FIGURE 20.110   Cilia (large protusions) and microvilli (small protusions) are shown in this freeze fracture micrograph. The cilia, but not the microvilli, show regional specializations at three separate sites (1, 2, and 3) as evidenced by the pattern of intramembranous particles. The bottom specialization (*arrow*) is often referred to as the *ciliary necklace.*
(Micrograph of a protozoan corona courtesy of R. Dallai.)

whereas in others there are appendages (e.g., striated rootlets) that extend more deeply into the cytoplasm and act as anchors for the cilium. Cilia beat in a wavelike motion. They move fluids or other structures, such as an ovum, over the surface of the ovaductal epithelium.

Flagella

Flagella are seen in many phyla of the animal kingdom. They are commonly found in bacteria and also are seen in mammals. A prime example of a flagellum in mammals forms the tail of the spermatozoon. The core of the sperm flagellum resembles the 9 + 2 microtubule organization (*axoneme*) of the cilium (see Figure 20.109).

Techniques that employ tannic acid in the fixative reveal the substructure (subunits) of the flagellar microtubules or microtubules in general. These subunits or *protofilaments* are generally 13 in number (Figure 20.112) but may be more or less depending on the species being examined. Species variations in the appearance of the flagellum are large in invertebrates. A typical model of flagellar structure is shown in Figure 20.113. Accessory structures are seen that characterize the flagellum regionally and divide it into various parts, the *middle, principal,* and *end pieces* (Figures 20.114 through 20.116). The wave motion of the flagellum is distinctly different from that of the cilium. Flagella usually propel the entire cell through a fluid medium.

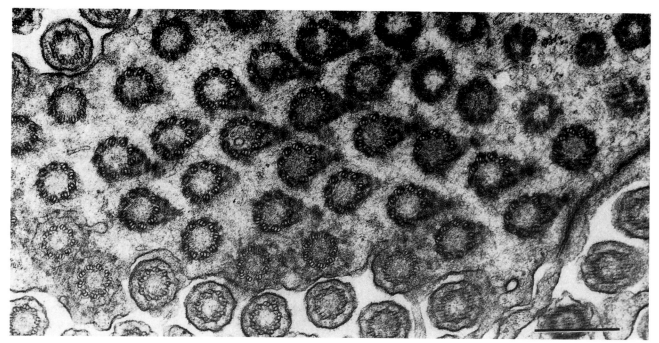

FIGURE 20.111 Basal bodies within the cytoplasm are cross-sectioned. Each basal body resembles a centriole, but gives rise to a cilium. Bar = 0.5 μm.
(Micrograph of respiratory epithelium courtesy of W. Dougherty.)

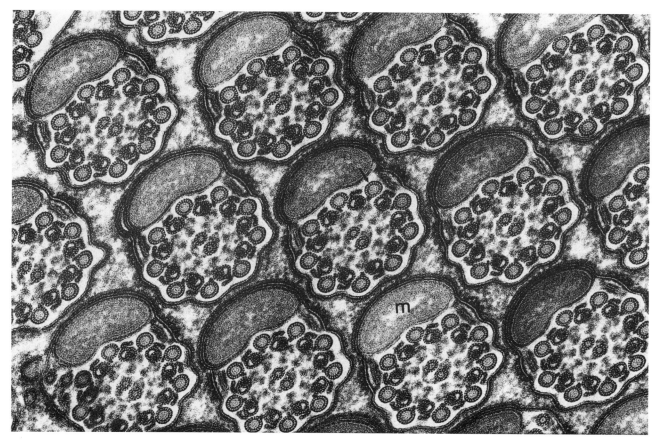

FIGURE 20.112 Packed flagella of insect sperm (*Ondontocerum albicore*) each showing mitochondria (m), two central microtubules, nine peripheral microtubules, and nine central microtubules. The tubulin protofilaments (*arrows*) of individual microtubules are visible after tannic acid addition to the primary glutaraldehyde fixation.
(Micrograph courtesy of R. Dallai.)

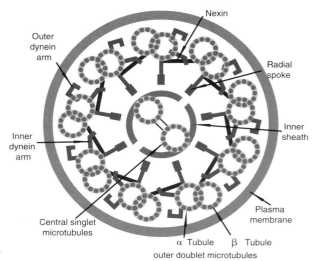

FIGURE 20.113   Flagellar substructure.

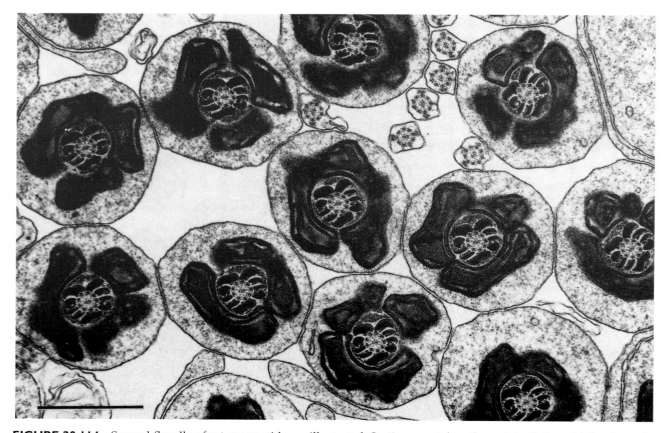

FIGURE 20.114   Several flagella of rat spermatids are illustrated. Sections are taken through both the middle piece of almost fully formed flagella and from the axonemal core of newly forming flagella. Bar = 1.0 μm.
(From Russell et al., 1990. Histological and histopathological evaluation of the testis. *Cache River Press, Clearwater, FL.)*

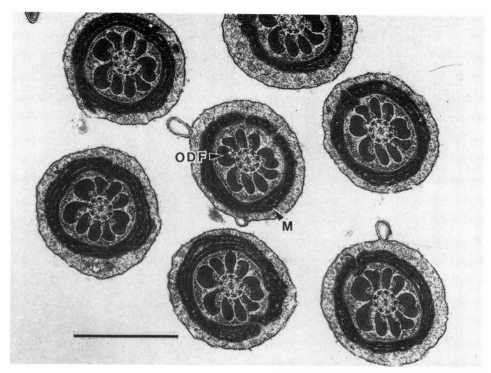

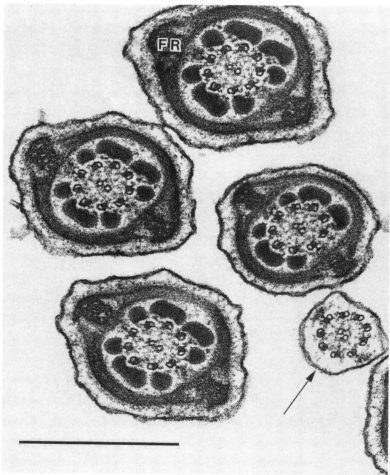

FIGURES 20.115 and 20.116 Flagella from the different regions of the rat sperm tail are shown. In the middle piece (*top*), mitochondria (M) spiral around the axoneme. Dense bodies termed outer dense fibers (ODF) are also seen. Mitochondria are no longer seen in the principal piece (*bottom*), but, instead, a structure known as the fibrous ring (FR) is seen. In the end piece (*arrow*), the axoneme is about the only internal structural feature noted. Bars = 1.0 μm.

The Lysosomal System

This assorted grouping of structures is related to the internalization and degradation of internalized materials.

Lysosomes

Bounded by a single membrane, lysosomes are strictly defined by the presence of hydrolytic enzymes rather than by their structural characteristics. Technically, a lysosome and a secretion granule cannot be distinguished by structure alone. Nevertheless, lysosomes can usually be identified within the cytoplasm as distinct entities. Lysosomes are frequently sparsely scattered throughout the cell, whereas secretory granules lie near the Golgi or in an epithelial cell clustered in a supranuclear position (Figure 20.117). Usually, lysosomes contain digested material (Figures 20.118 through 20.120) that provide a clue that may

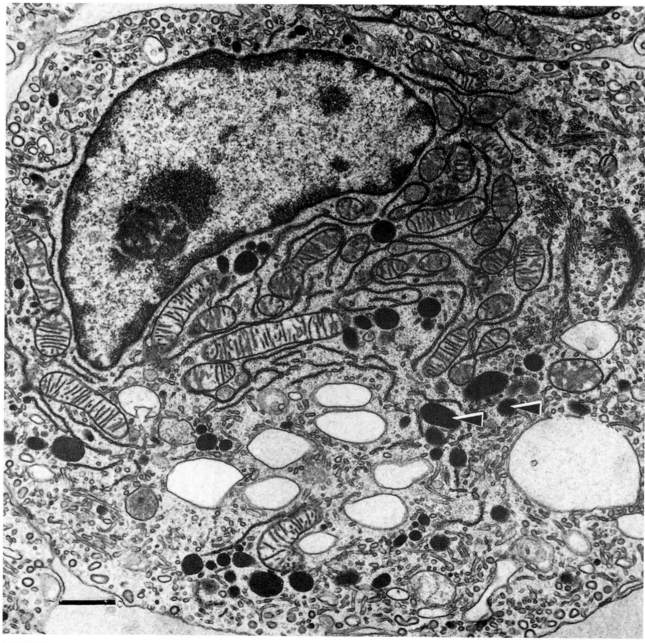

FIGURE 20.117 Lysosomes (*arrowheads*) within a macrophage are homogeneous appearing structures and are scattered throughout this cell. Their bounding membrane is not resolved in this low magnification micrograph. Bar = 1.0 μm.

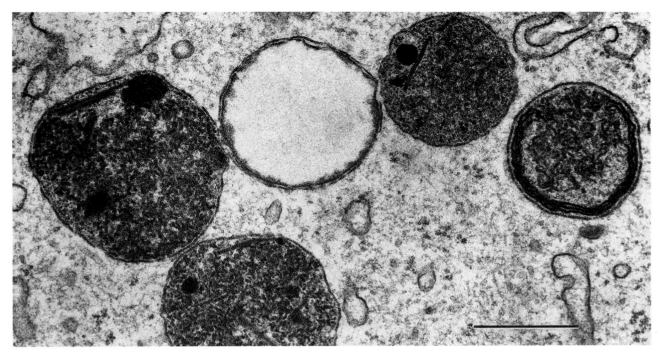

FIGURE 20.118 At high magnification, these secondary lysosomes contain a coarse granular and/or filamentous matrix in addition to densely stained material and membranous elements. A membrane bounds each lysosome. Bar = 0.5 μm.

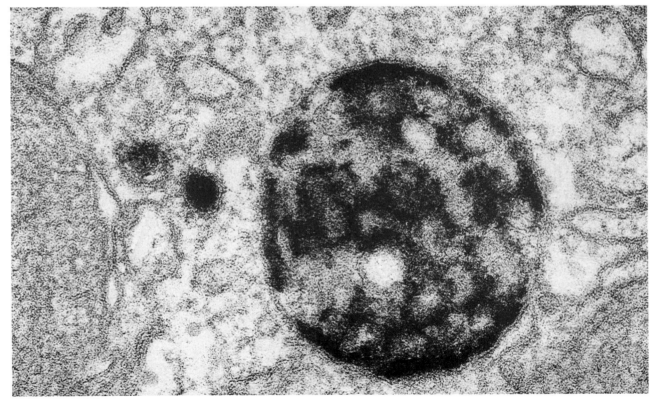

FIGURE 20.119 Two distinct lysosomal types are shown in this acid phosphatase preparation. The smaller is homogeneous without evidence of digested material. The larger is heterogeneous with areas not showing acid phosphatase but probably containing engulfed material.
(Micrograph from Alberts et al. 1989, courtesy of D. Friend and used with permission of Garland Pub. Co.)

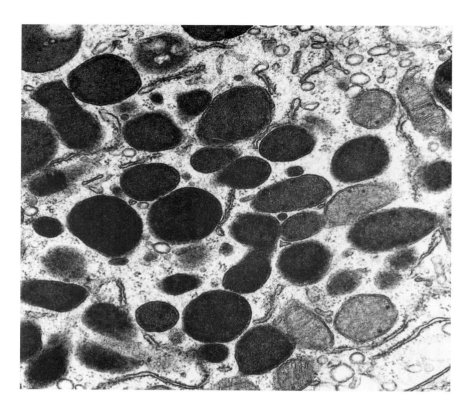

FIGURE 20.120 Numerous aggregated lysosomes are seen, some of which are relatively small and homogeneous and others that appear small and heterogeneous.

be used to distinguish lysosomes from secretory granules (the latter generally being more homogeneous as a class).

Lysosomes function to break down substances within a cell (autophagy) or substances that have been taken in from outside the cell (heterophagy). There are at least 40 types of hydrolytic enzymes within the bounds of lysosomes. These include phosphatases, nucleases, proteases, lipases, and sulfatases. Most enzymes in the lysosome operate at a pH of about 5.0. Usually, lysosomes do not receive material directly but receive membrane-bound sacs termed *endosomes* that have engulfed material from the cell exterior. They receive *phagosomes* that represent the intake of large bodies such as another cell or a bacteria. Finally, they receive *autophagosomes*, which are membrane-enclosed organelles from the same cell. Lysosomes are important in maintenance of body defenses against outside organisms. In addition, lysosomes may cause death of the cell (autolysis) by emptying their contents into the cytoplasm.

Multivesicular Bodies

The *multivesicular body* is bounded by a single membrane and contains numerous smaller membrane-bounded vesicles that are generally rounded and of relatively uniform size (Figures 20.121 through 20.123).

Digestive vacuoles are large structures that contain a large volume of ingested material. Generally, the material within the vacuole is partially digested and extremely heterogeneous (Figure 20.124).

Autophagic vacuoles form as the result of self-ingestion of constituents of the cell (Figure 20.125).

Heterophagic vacuoles contain other cells or material from other cells (Figure 20.126).

Residual bodies contain the remnants of undigested material. A form of residual body is the *lipofuscin granule*, which contains pigmented material (*lipochrome pigment*). Usually, a portion of the lipofuscin granule has material that appears like lipid (Figure 20.127).

Lipofuscin is regarded as undigested material left over from lysosomal activity and is seen in tissues as the result of the aging process.

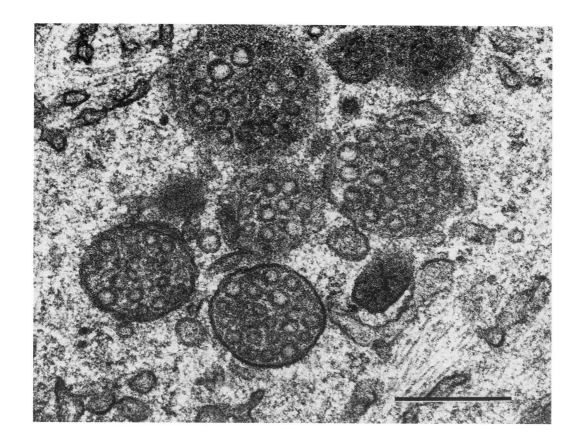

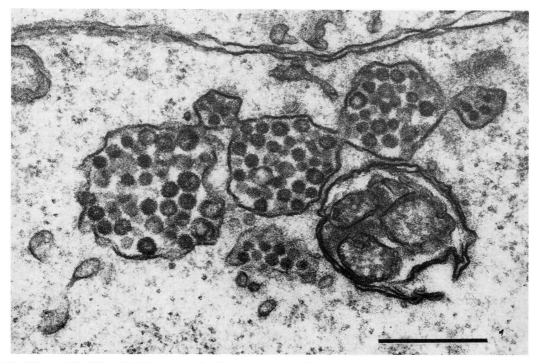

FIGURES 20.121 and 20.122 Multivesicular bodies are identified by the multiple small vesicles enclosed by a single membrane. Bars = 0.5 μm.

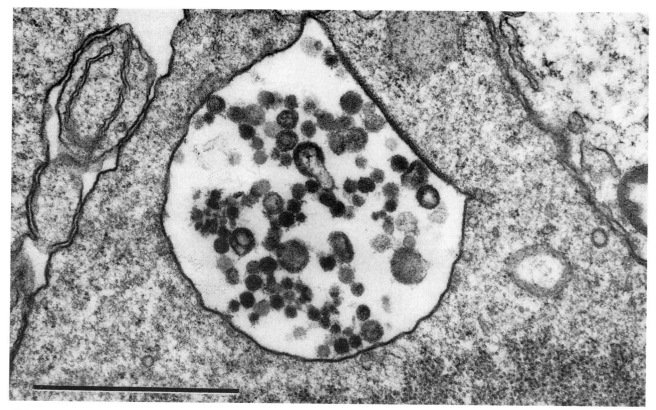

FIGURE 20.123 This large multivesicular body is bounded by a single membrane, part of which shows a fuzzy coat. Numerous small vesicles are seen within its interior. Bar = 0.5 μm.

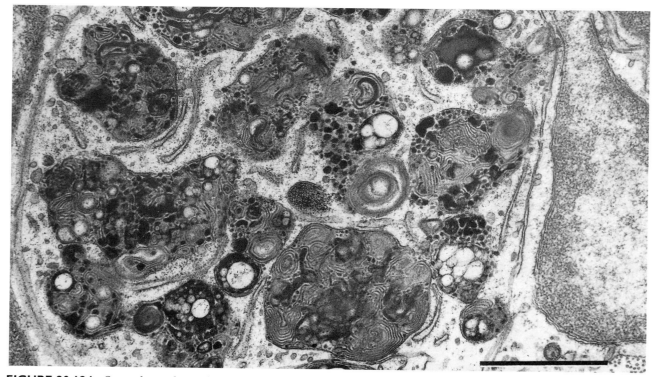

FIGURE 20.124 Several membrane bounded digestive vacuoles are seen within the cell. These comprise a large volume of this cell. Bar = 1.0 μm.

(Micrograph of an interstitial cell of the rat ovary courtesy of A. Mayerhofer.)

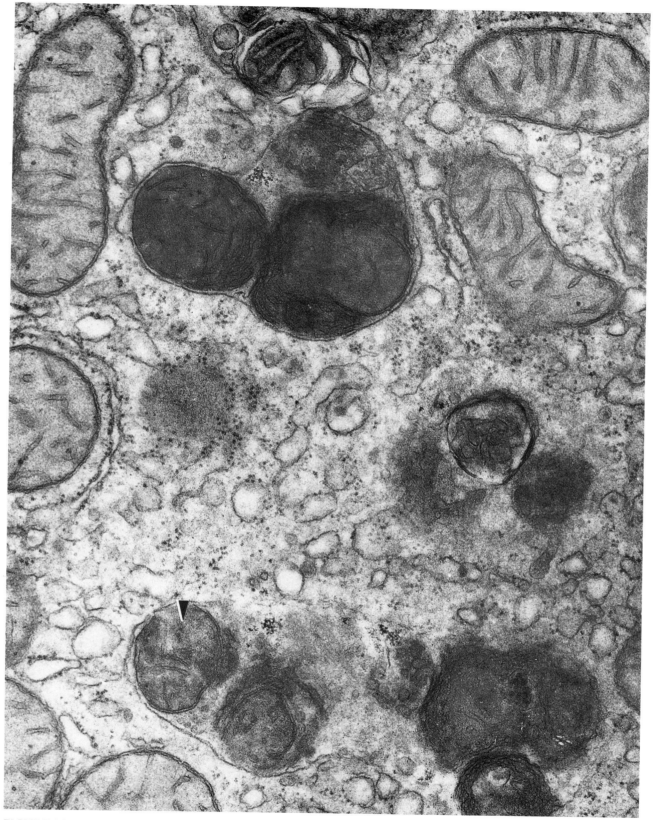

FIGURE 20.125 Autophagic vacuole containing partially degraded mitochondria (*arrowhead*) from the same cell. (*Micrograph courtesy of W. Dougherty.*)

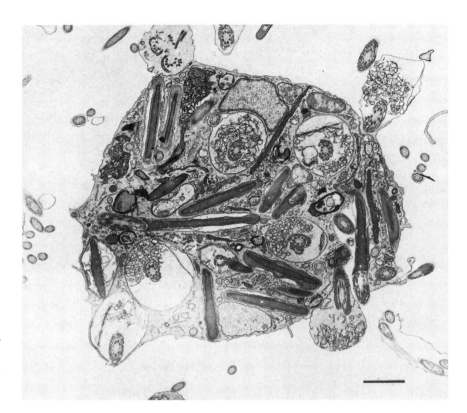

FIGURE 20.126 This macrophage in the reproductive tract of a rabbit has engulfed and walled off several sperm within heterophagic vacuoles. Bar = 2.0 μm.

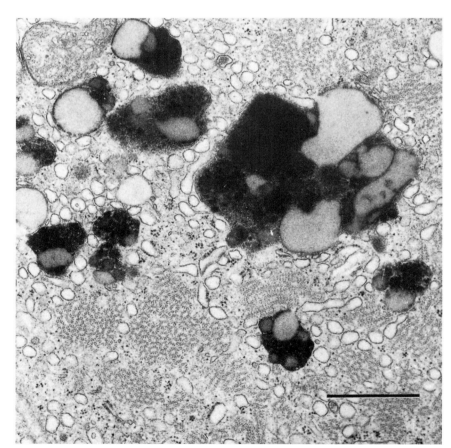

FIGURE 20.127 Lipofuscin granules, like these shown, are irregularly shaped and bounded by a membrane. Their contents are extremely heterogeneous, usually containing dense lipochrome pigment and sometimes less dense lipid droplets. (Human Leydig cell.)

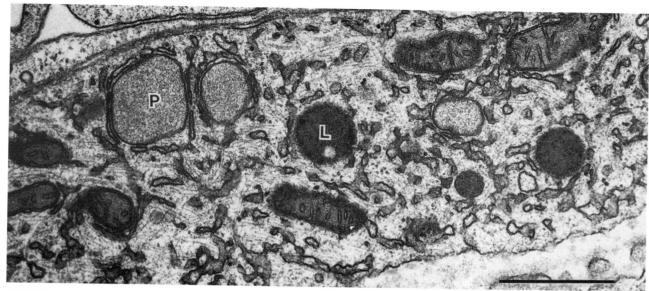

FIGURE 20.128 Peroxisomes and lysosomes are illustrated. The peroxisome (P) displays a finely granular matrix whereas the lysosome (L) is more densely granular and contains ingested material. (Rat Leydig cell.) Bar = 1.0 μm.

Peroxisomes or Microbodies

Peroxisomes are one form of microbody. Like the lysosome, the peroxisome is defined by its constituent enzymes (catalase and oxidases) and is positively identified only by the localization of these enzymes. These membrane-bounded structures resemble lysosomes, but usually possess a less dense matrix than do lysosomes (Figure 20.128).

In some cells, peroxisomes contain crystalloids, called *nucleoids*, which identify them as peroxisomes (Figure 20.129).

The function of peroxisomes in animal cells is not clear. It is known that they can both produce and break down hydrogen peroxide (H_2O_2). They use oxygen and may participate as detoxifying organelles. Their role is clearer in plant tissues where they are involved in photorespiration. They contain enzymes that are involved in oxidation of the side products of cellular respiration. *Glyoxysomes* appear similar to peroxisomes and are another form of microbody in plants. They are seen almost exclusively in germinating seeds and have a structure resembling the peroxisome. Glyoxysomes function in the conversion of fatty acids into sugars.

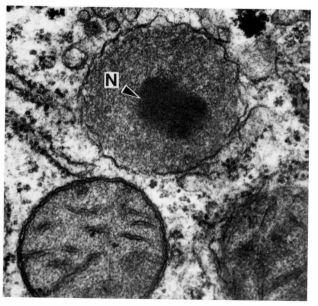

FIGURE 20.129 A nucleoid (N) is located centrally within this peroxisome.
(Micrograph courtesy of W. Dougherty.)

Annulate Lamellae

This infrequently found membranous complex appears as stacked cisternae with pore complexes that appear much like pores of the nuclear envelope. The cisternae of annulate lamellae may be continuous with the smooth and/or rough endoplasmic reticulum (Figure 20.130).

The function of annulate lamellae is not known. Annulate lamellae are frequently seen in rapidly dividing cells and germ cells. It is not known if this membranous complex arises from the nuclear envelope. Antigens found on the nuclear envelope are not found on annulate lamellae.

Glycogen/Glycosomes

Glycogen is commonly observed by electron microscopy as round 30 nm granules (β-particles) typical for that seen within muscles or in leuko-

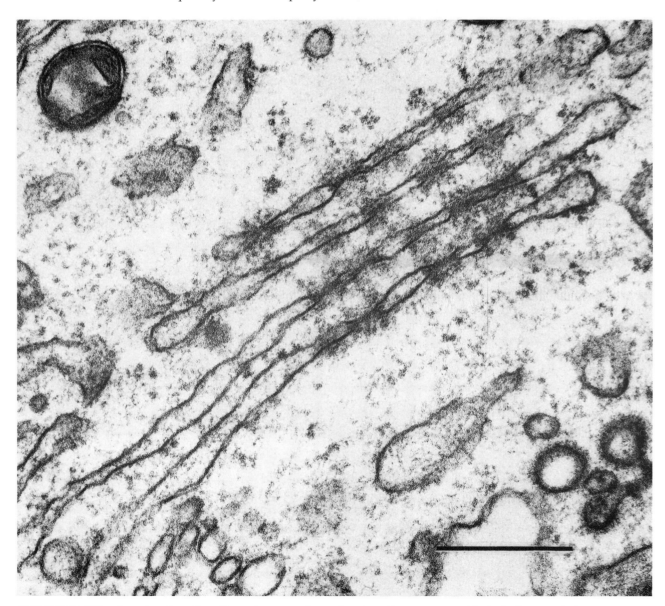

FIGURE 20.130 Annulate lamellae. Four cisternae of annulate lamellae are seen with their characteristic pore complexes. At one end they are continuous with the smooth endoplasmic reticulum and at the other end the rough reticulum. (From a germ cell in the rat testis.)

cytes (Figures 20.131 and 20.132), or is encountered as rosettes of granules (α-particles) typically found in liver tissue (Figure 20.133). Enzymes involved in glycogen synthesis and degradation are associated with glycogen (Figure 20.134). Because glycogen with the associated enzymes undergoes constant metabolic turnover, these structures recognized as dynamic cellular organelles, have been called *glycosomes.*

The granules stained with uranium and lead under the electron microscope (Figures 20.131 through 20.133) represent, in fact, the protein component of glycosomes, and not the storage form of glucose (glycogen) as was erroneously interpreted. Glycogen itself is not stainable by heavy metals, but it can be demonstrated by a cytochemical reaction for the electron microscope similar to the PA-Schiff technique used in light microscopy. Cytochemically stained glycogen appears as 3 nm particles that form aggregates corresponding in size to the protein component in glycosomes (Figure 20.135). When tissue is treated with uranyl acetate before dehydration (*en bloc* staining), protein is separated from glycogen and it solubilizes, whereas the remaining glycogen particles aggregate into large clumps. Because glycogen is not stained by the ordinary techniques in EM, these clumps remain in plastic as unstained spots and can mislead an inexperienced student that the plastic did not polymerize homogeneously (Figure 20.136). Glycogen, degraded by associated enzymes, provides a source of energy for the cell. In plants starch serves this purpose.

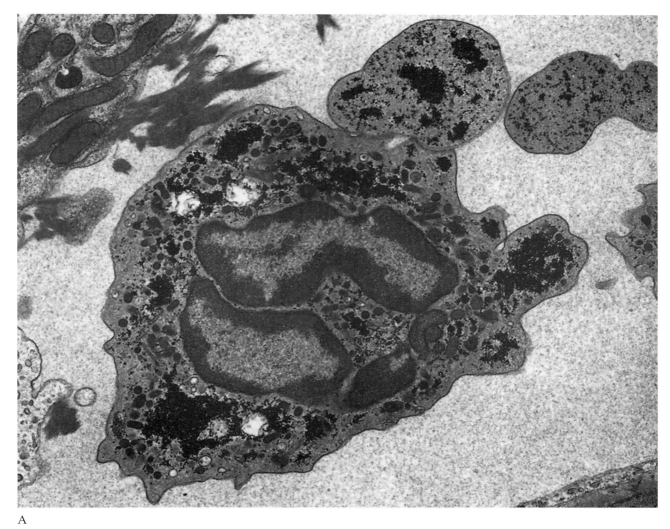

A

FIGURE 20.131A This leucocyte shows glycogen deposits throughout the cell; however, dense regional accumulations of particles are seen. Figure 20.131B is a higher magnification of one of the cell processes and shows the accumulations of glycogen. Bar in lower figure = 0.5 μm.

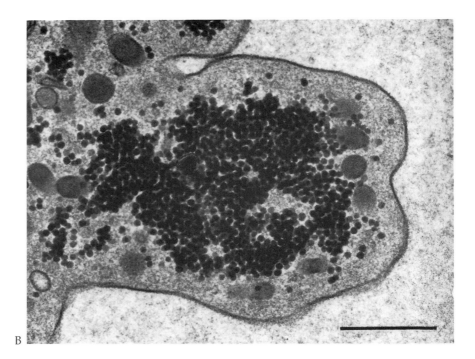

FIGURE 20.131B

B

FIGURE 20.132 Glycosomes among mouse skeletal muscle fibers. Section stained by uranyl acetate and lead citrate shows the protein component in glycosomes sharply demarcated as 30 nm spheres.

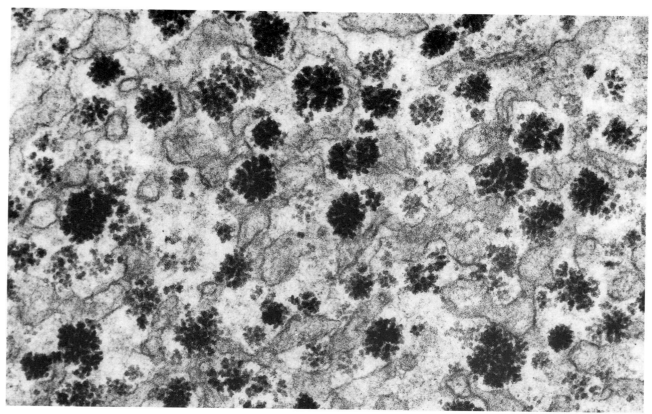

FIGURE 20.133 Rosettes of glycogen (alpha particles). *(Micrograph courtesy of W. Dougherty.)*

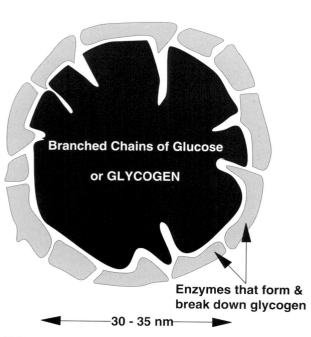

Branched Chains of Glucose

or GLYCOGEN

Enzymes that form & break down glycogen

←—— 30 - 35 nm ——→

FIGURE 20.134 The composition of glycogen.

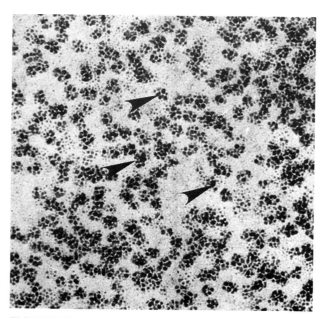

FIGURE 20.135 Glycosomes in dog cardiac conduction fibers. Sections stained cytochemically by Thiery's (1967) technique show glycogen as 3 nm particles forming 20 to 30 nm aggregates (*arrowheads*). *(Micrograph courtesy of K. Rybicka.)*

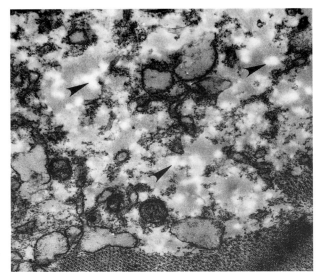

FIGURE 20.136 Tissue treated with uranyl acetate prior to dehydration and stained with uranium and lead salts. The protein component of glycosomes has solubilized and the clumps of glycogen remain unstained (*light spot at arrowheads*). (*Micrograph of cardiac conduction fibers courtesy of K. Rybicka.*)

Cell Inclusions

Lipid

The storage form of fat is called *lipid droplets* and is mainly composed of triglycerides. Lipids show great morphological heterogeneity within cells. The size of individual droplets is highly variable, generally less than 1 μm in diameter, but in some instances (fat) exceeding 100 μm. Lipid droplets have no bounding membrane and appear to be held together by their hydrophobic interaction with the aqueous environment. Their lack of membrane is a distinguishing feature, but it is sometimes difficult to discern whether or not a membrane is present at low magnification, especially if the lipid droplet has a constituent that stains more densely at its periphery and, thus, mimics a membrane (Figure 20.137).

Lipid droplets are usually rounded, but may appear in a variety of configurations. Lipid droplets may be leached out by the dehydration process, especially if osmium penetration is poor. Moreover, the kind of lipid determines its staining properties with osmium. Thus, lipid may appear

very electron dense or may stain hardly at all. Variations in lipid droplet form and configuration are illustrated in Figures 20.138 and 20.139.

The content of lipid appears less granular or has a more homogeneous texture than other dense staining structures within the cytoplasm, such as lysosomes or secretion granules. If sectioning artifacts such as chatter or knife marks are present, they are very prominently displayed against the background of the homogeneous lipid droplets, whereas they may not be noticed elsewhere in the section. These features may sometimes be used to identify lipid from other organelles (Figure 20.140).

Crystalloids

A variety of crystalline features are seen within cells, but few cells contain crystalloids. They usually display an irregularly geometric outline and are unbounded by membrane (Figure 20.141). Crystalloids are amenable to X-ray diffraction studies (see Chapter 15), whereby the structure of the crystal can be deduced. The fine structure of crystalloids reveals a repeating pattern within the crystal (Figure 20.142). Although they are mostly proteinaceous in nature, rarely has their function in cells, if any, been ascertained.

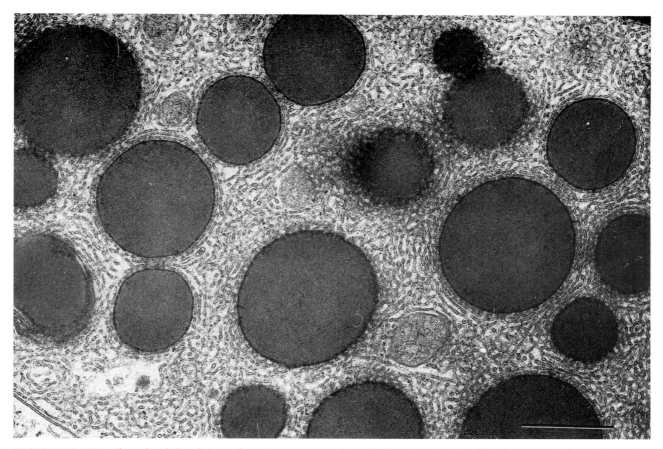

FIGURE 20.137 These lipid droplets are found among smooth endoplasmic reticulum. The interiors of the lipid droplets are finely granular, giving them a very homogeneous appearance. At its periphery, the lipid droplet has a denser staining band that might be easily mistaken for a membrane, except no trilaminar structure is apparent. Bar = 1.0 μm.

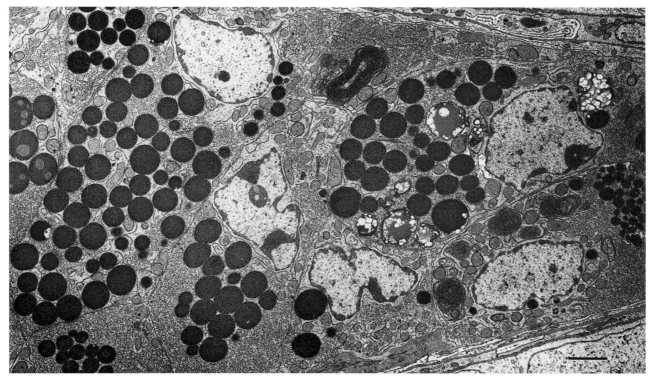

FIGURE 20.138 In this group of cells, some lipid droplets are densely stained, whereas other lipid is less densely stained. In addition, some lipid shows a small vacuolated area or areas in which the internal component is less well stained than the majority of the droplet. Bar = 2.0 μm.

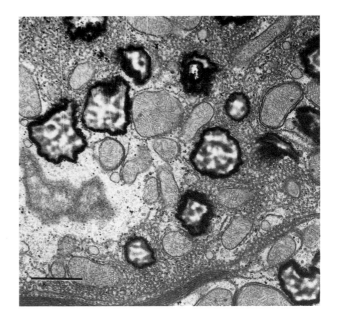

FIGURE 20.139 Irregularly shaped lipid with a mottled electron-translucent interior. Bar = 1.0 μm.

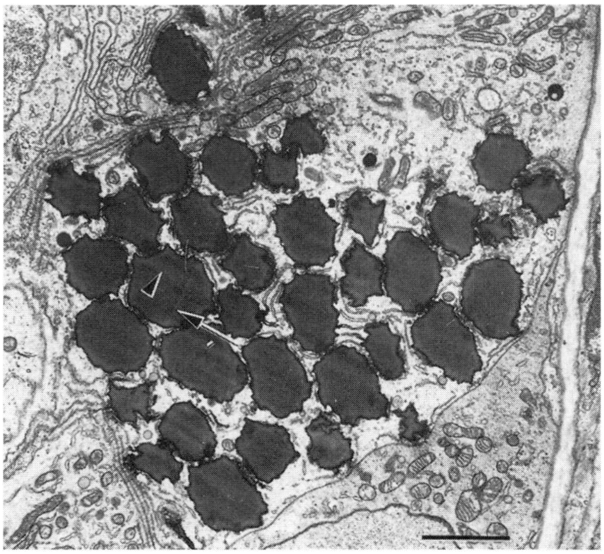

FIGURE 20.140 Chatter (*arrowhead*) and a knife mark (*arrow*) are evidenced in lipid droplets, but not elsewhere on the micrograph. Bar = 1.0 μm.

FIGURE 20.141 Crystalline structures (*arrowheads*), such as these usually possess angular profiles and are not bounded by a membrane. (Reinke crystal in a Leydig cell of the human testis.) Bar = 10.0 μm.

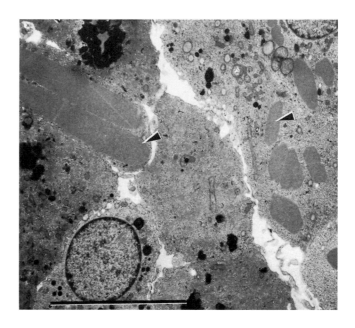

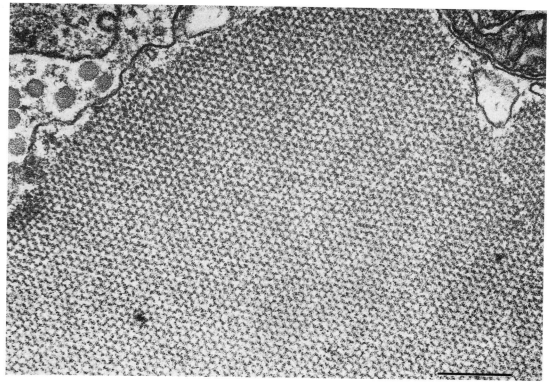

FIGURE 20.142 At high magnification, the crystalloid in Figure 20.141 shows a crystal lattice of repeating units. Bar = 0.5 μm.

Extracellular Material

Collagen

Collagen is the most abundant protein in the body. Originally, only a few types of collagen were known to exist; now more than 15 types have been described. More than one type of collagen may be present in the same fibril. The most familiar form of collagen is the banded form with a periodic repeat of the major band at about 67 nm (Figures 20.143 and 20.144).

Many collagen fibrils form a bundle, visible in the light microscope as a collagen *fiber*. Collagen is also organized into sheets of fibrils that lie over other collagen sheets, that are approximately perpendicular to the first layer (Figure 20.145). The collagens are a family of proteins secreted by fibroblasts that add tensile strength to tissues and organs and hold them together. For example, cow hide or leather is rich in collagen.

Elastic fibers show properties of extensibility and recoil and are also found extracellularly (Figures 20.146 and 20.147). Collagens and elastins are best visualized in preparations where cationic (positively charged) stains such as ruthenium have been employed.

FIGURE 20.143 Longitudinally and obliquely sectioned collagen fibrils are shown from rabbit tendon. These fibrils contain both Type I and Type III collagen. The major repeat band is about 45 nm from its nearest neighbor and falls short of the 67 nm measured for fresh tissue. The other bands between the major band are composed of dermatin sulfate proteoglycan. The dense material outside the collagen fibrils is ruthenium.
(Micrograph stained with ruthenium courtesy of Doug Bray.)

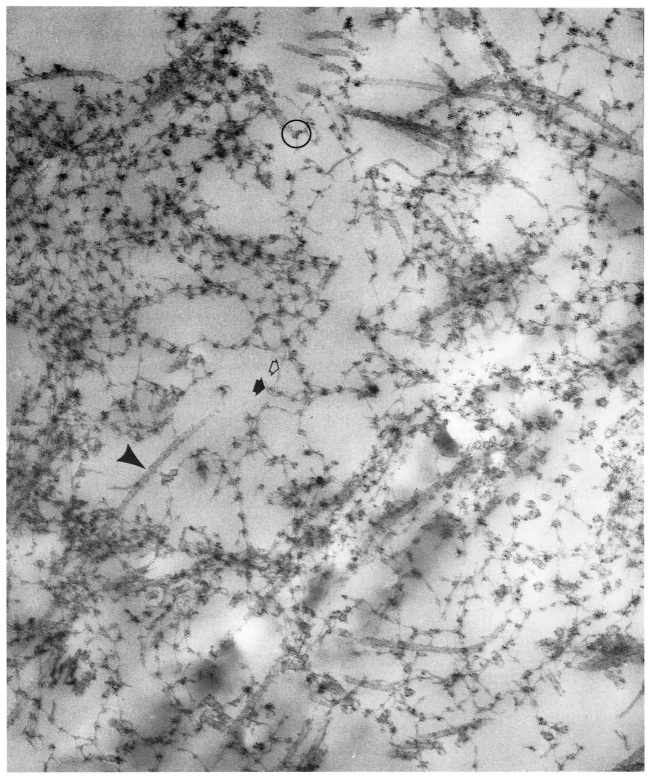

FIGURE 20.144 Section of Type VI collagen among small banded collagen fibrils (*arrowhead*). An antigenic component of the collagen has been localized with colloidal gold particles (*encircled*). Type VI collagen has thin (*open arrow*) and thick regions (*closed arrow*) and the antigenic molecule is only observed on the thick regions.
(Micrograph of rabbit tendon tissue treated with hyaluronidase and stained with ruthenium is courtesy of Doug Bray.)

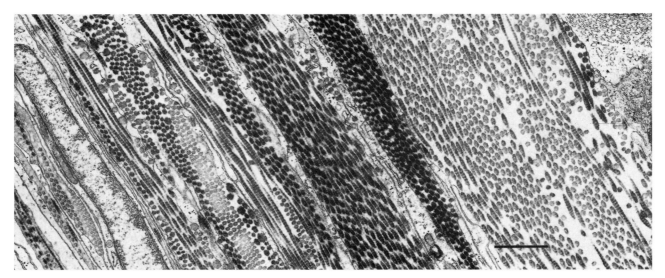

FIGURE 20.145   Sheets of collagen are shown in which the orientation of the fibrils is at approximate right angles to the adjacent sheet. Flattened cellular elements are occasionally interposed. (Capsule of the rat testis.) Bar = 1.0 μm.

ELASTIC FIBER

RELAXED STATE

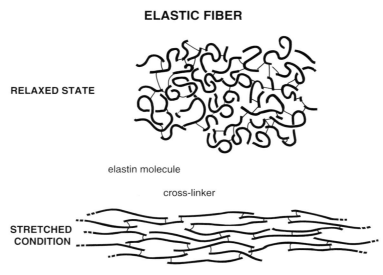

elastin molecule

cross-linker

STRETCHED CONDITION

FIGURE 20.146   Elastic fibers in the stretched and relaxed states.

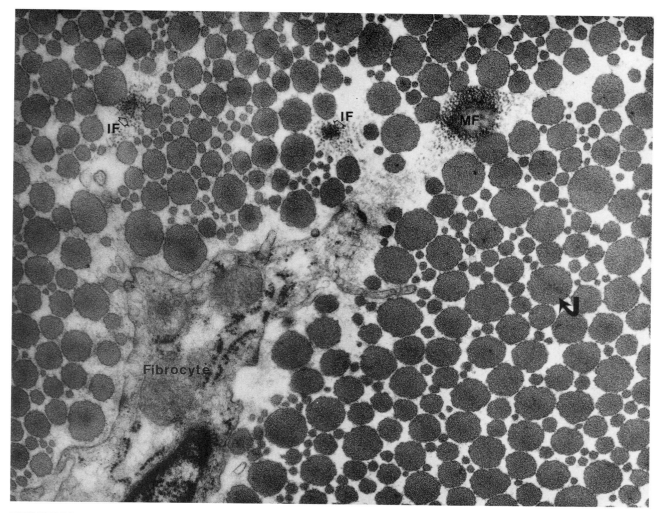

FIGURE 20.147 Banded collagen (*curved arrow*) measure from 20 to 300 nm in diameter and because they are viewed in cross section they appear spherical and without the banding patterns characteristic of longitudinally sectioned collagen. The collagen-secreting cell, the fibrocyte, is also indicated. Also shown is a cross section of one mature elastic fiber (MF) and two immature fibers (IF). The mature fibril contains an amorphous center and peripheral microfilaments.

(Micrograph of rabbit tendon tissue prepared conventionally courtesy of Doug Bray.)

Basal Lamina

At the base of epithelial cells, the basal lamina is a continuous sheet (about 150 nm thick) of extracellular material that is formed primarily by the epithelial cells and separates these cells from underlying connective tissue. The basal lamina generally follows the undulating contours of the epithelial cell. At low and medium magnifications, it has a homogeneous appearance, but at high magnification, a fine fibrillar appearance is evident (Figure 20.148).

The function of the basal lamina is primarily structural, forming a framework on which epithelial cells rest. The basal lamina may act as a partial filter or participate in tissue restructuring. Its composition and associations with other connective tissue elements is complex.

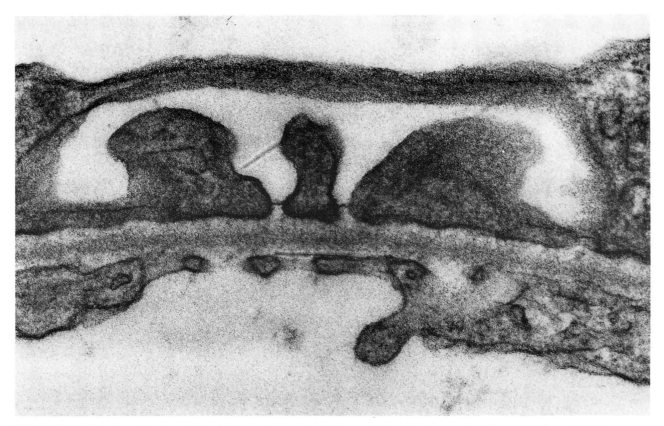

FIGURE 20.148 Basal lamina of the kidney glomerulus. This acellular structure extending centrally within the micrograph as a dome from left to right separates the endothelium of the capillary from the podocytes of the glomerulus.
(Micrograph courtesy of W. Dougherty.)

Matrix of Bone and Cartilage

Osteocytes (bone cells) and chondrocytes (cartilage cells) are present in a great volume of extracellular matrix. Bone is very hard and its matrix must be decalcified before it can be thin sectioned. The matrix of both cartilage and decalcified bone is complex, but may be recognized by the numerous collagenous (nonstriated in cartilage) fibrils running in various planes. In bone the fibrils are collagenous, being organized in layers or *lamellae* that run at right angles to each other. In cartilage the fibrils are less well organized than in bone (Figures 20.149 and 20.150) and of a different type (Figure 20.151).

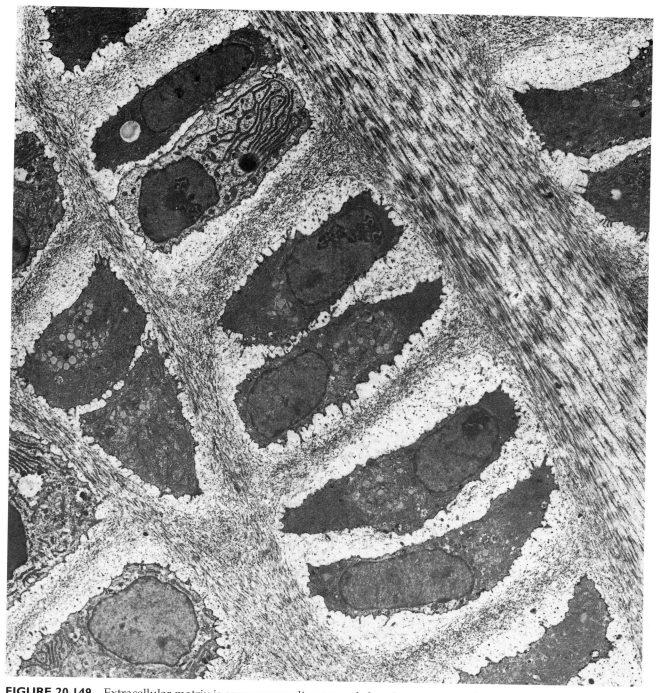

FIGURE 20.149 Extracellular matrix is seen surrounding several chondrocytes in cartilage.

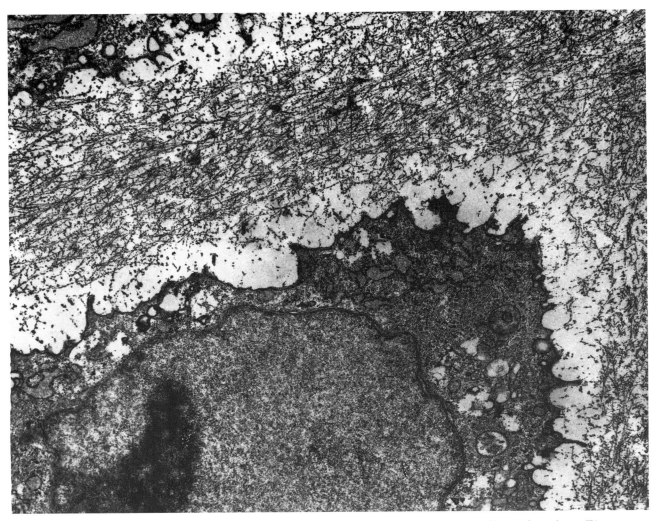

FIGURE 20.150   The relationship of the chondrocyte to the matrix is shown. Collagenous fibers of cartilage. (Pig cartilage.)

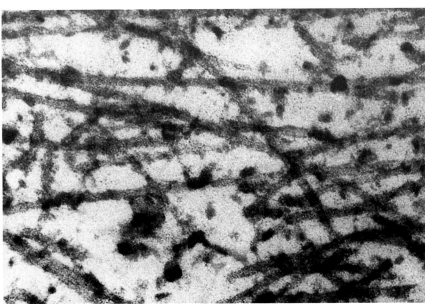

FIGURE 20.151   The collagenous fibrils of cartilage are a type that are not typically striated.

Special Features of Plant Tissues

Plant cells are similar, in many ways, to animal cells (Figure 20.152), but several structural variations are worthy of mention.

Chloroplasts

The chloroplast is bounded by a double membrane. The majority of the internal aspect of the chloroplast is termed the *stroma*. Within the stroma are closed sacs called *thylakoids*, which do not connect to the bounding membrane of the chloroplast, but which do interconnect to form a space called the *thylakoid space*. Thylakoids are usually arranged in parallel stacks called *grana* (Figures 20.153 through 20.156). Chloroplasts divide by fission or splitting (Figure 20.157).

The chloroplast is an essential energy-producing component of plants that functions in the process of *photosynthesis*. Besides generation of ATP, choloroplasts use this energy source to convert six CO_2 molecules into one glucose molecule. Often, the glucose is polymerized and stored as starch grains in the chloroplast. Mitochondria are also present in plant cells and may sometimes be mistaken for chloroplasts. However, mitochondria are smaller and their cristae are connected to the inner mitochondrial membrane. It is thought from an evolutionary standpoint that chloroplasts are derived from organisms that resemble bacteria.

Vacuoles

The plant vacuole is bounded by a single membrane called a *tonoplast*. Vacuoles are generally large structures, over half of the cell volume, and in many plant cells they are so large as to cause the cytoplasm to be confined to a narrow area at the periphery of the cell just along the cell wall (Figure 20.158).

Vacuoles are storage elements largely for water, but also for waste. As the plant cell grows, the vacuole is the primary structure that is enlarged. Vacuoles are related to lysosomes because they contain hydrolytic enzymes.

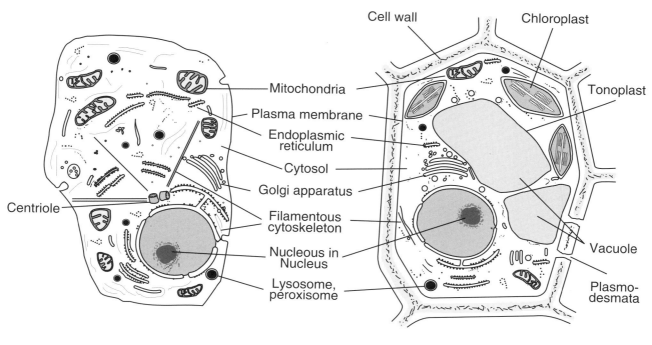

ANIMAL CELL   PLANT CELL

FIGURE 20.152   A comparison of plant and animal cells.

outer chloroplast membrane

inner chloroplast membrane

DNA

ribosomes

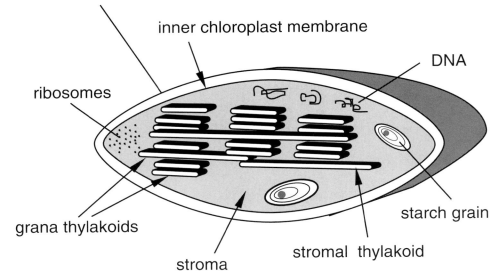

grana thylakoids

stroma

starch grain

stromal thylakoid

FIGURE 20.153   Drawing showing structure of chloroplast present in plant cells.

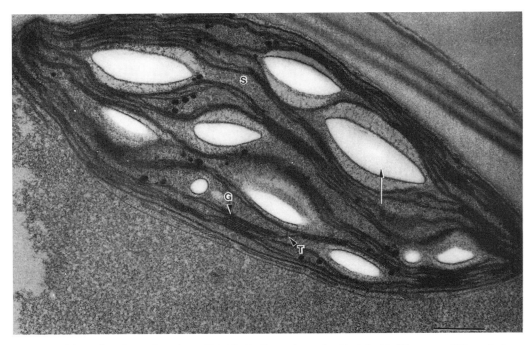

FIGURE 20.154   A chloroplast from the plant, *Takakia.* Indicated are the thylakoid (T), grana (G), and stroma (S). Accumulations of starch (*arrow*) are evident. Bar = 0.5 μm. *(Micrograph courtesy of B. Crandall-Stotler.)*

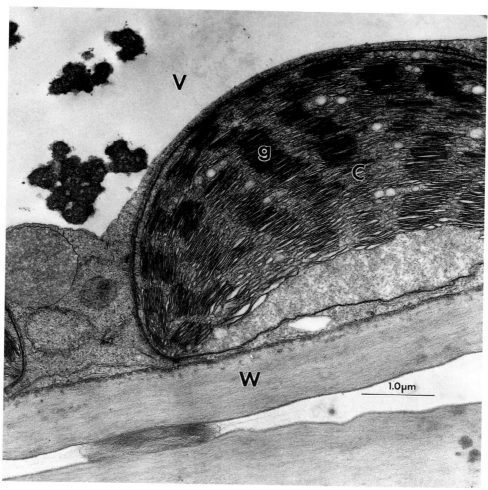

FIGURE 20.155 Portion of cell of the Pacific yew (*Taxus brevifolia*) needle in the mesophyll layer of the leaf. The portion of the chloroplast (C) shows typical grana (g) stack and clear inclusions. V, small portion of vacuole; W, cell wall. Specimen was fixed and embedded in 3 hours using microwave technology described in Chapter 2. (*Micrograph courtesy of R. S. Demaree and R. T. Giberson.*)

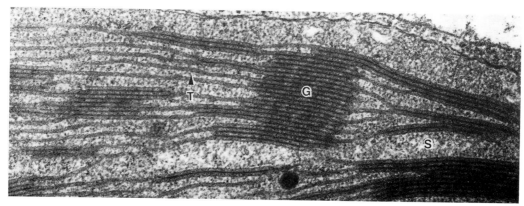

FIGURE 20.156 Higher magnification of a chloroplast revealing the detail of the stroma (S), thylakoids (T), and the thylakoids forming grana (G).
(*Micrograph courtesy of M. Gillott.*)

FIGURE 20.157 Splitting of a chloroplast by the process of fission. *(Micrograph courtesy of K. Renzaglia.)*

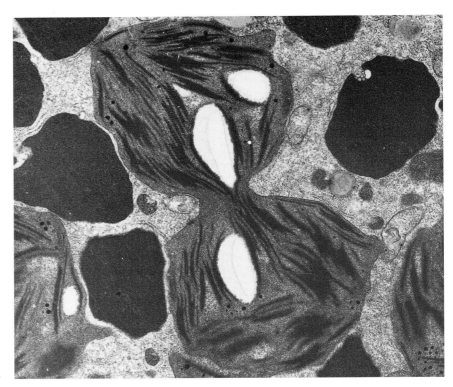

FIGURE 20.158 A plant cell containing a large vacuole (V) and a thick cell wall (CW). Chloroplasts are also evident. Bar = 1.0 μm. *(Micrograph of Takakia courtesy of B. Crandall-Stotler.)*

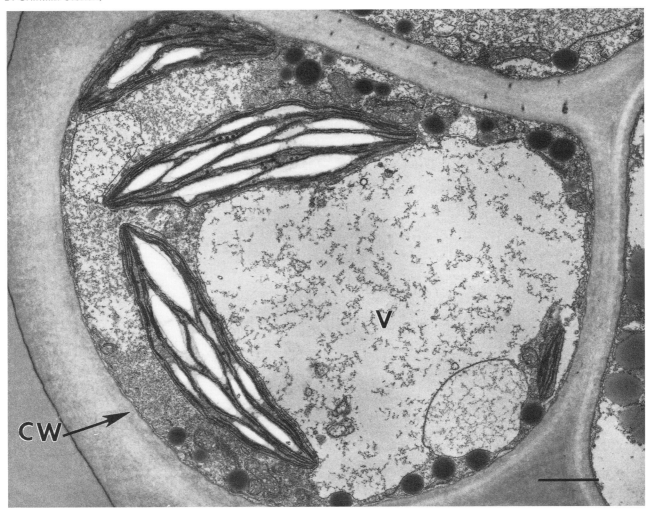

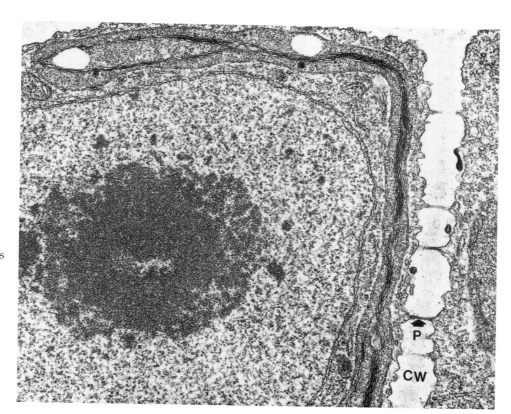

The Cell Wall

Each plant cell has an external surrounding wall that appears amorphous in structure (Figure 20.159). The thickness of the wall may vary from a fraction of a micron to many microns. Small cell-to-cell channels called *plasmodesmata* pierce the cell wall between adjoining cells to provide areas of cell-to-cell communication between many adjoining plant cells (Figure 20.159).

Cell walls are rigid structures that do not allow cell shape changes, but do act as restraining structures as cells osmotically swell up against them (*tur-*

gor). Being semipermeable structures, cell walls passively regulate many of the substances that reach plant cells and also may form functional fluid channels within plants. They are composed primarily of cellulose synthesized by the cell, which is exported to the outside of the cytoplasmic membrane. During the process of cell division, the separation of the nucleus and the cell proper (by means of a cross wall) may occur so closely in time as to appear to be one event. However, mitosis ends just before the final dividing cross wall is laid down by a specialized spindle structure, the *phragmoplast* (Figure 20.160).

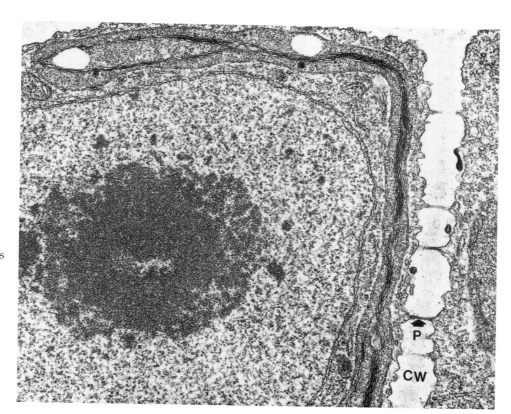

FIGURE 20.159 Portions of two plant cells are depicted. The cell wall (CW) is indicated and plasmodesmata (P) are seen penetrating the cell wall to join cells. Bar = 0.5 μm. (*Micrograph of the plant* Tortula *courtesy of K. Renzaglia.*)

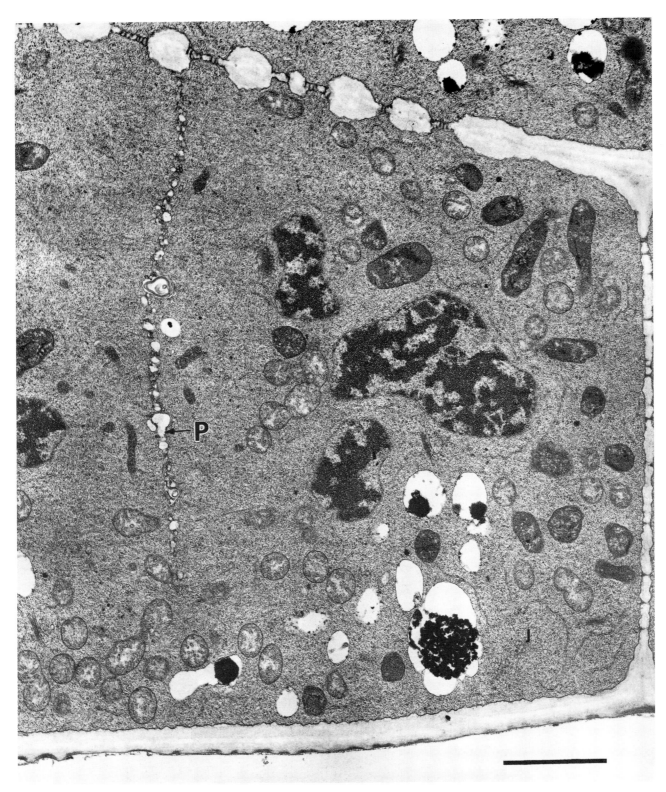

FIGURE 20.160   *Allium* cell completing mitosis. The phragmoplast (P) appears as a series of vesicles that will eventually coalesce to separate the two daughter cells. Bar = 2.0 μm.
(Micrograph courtesy of D. Molsen and L. Hanzely.)

Bacteria

Bacterial cells are composed of *genetic material (DNA), ribosomes,* a *cell membrane,* and a *cell wall* (except *mycoplasmas,* which are similar to bacteria but do not have a cell wall). A diagrammatic representation of a bacterium and its various components is shown in Figure 20.161. A typical thin section view is also shown (Figure 20.162).

Bacteria lack a nuclear envelope and for this reason are considered procaryotes in contrast to eucaryotes, which have their nuclear material packaged by a nuclear envelope. Transcription of DNA to RNA and translation of RNA to protein all take place in one compartment whereas in eucaryotes these processes are separately compartmentalized (Figure 20.163). Some bacteria have a flagellum and thus are motile. The typical bacterium ranges from a fraction of a μm to a few μm long (range 0.1 to 5 μm). The shape (spherical, rod, or

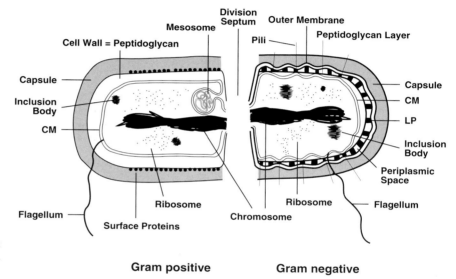

FIGURE 20.161 A diagrammatic representation of a bacterium and its various components.
(Modified from Joklik et al., from Zinsser Microbiology, used with permission of Appleton-Century Crofts.)

Gram positive   **Gram negative**

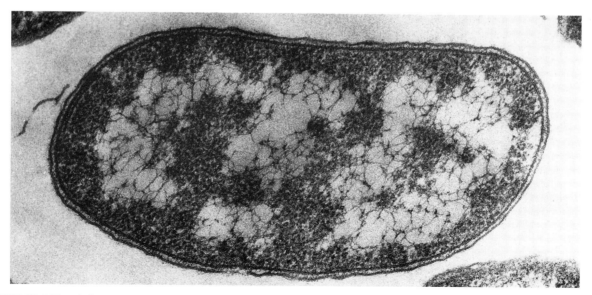

FIGURE 20.162 A thin section through a gram negative bacterium.
(Micrograph courtesy of G. Brewer.)

spiral) and organization (sheets, chains, or clusters) of groups of bacteria is diverse.

Bacteria may be divided into two major types: gram positive or gram negative (indicated in Figure 20.164). This classification is based on a staining reaction wherein gram positive cells retain the dye, crystal violet, while gram negative cells lose the dye upon decolorization with ethanol. Retention of the dye is based on a thicker gram positive cell wall that becomes impervious upon treatment with an iodine reagent used just before the decolorization step. A gram positive bacterial cell is shown in Figure 20.165. Some bacteria secrete extracellular material which may serve as protective capsules (Figure 20.166).

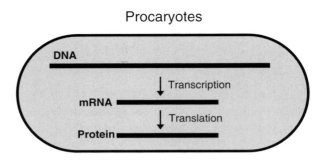

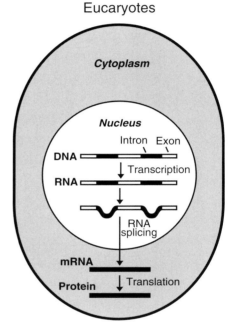

FIGURE 20.163 In procaryotes, transcription and translation take place in close proximity to each other while in eucaryotes they are separate in time and space.

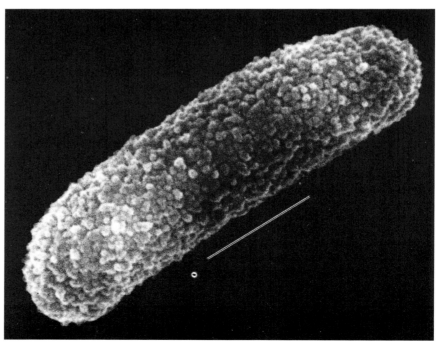

FIGURE 20.164 Scanning electron micrograph showing a gram negative bacillus in the rhizobial family of nitrogen-fixing organisms. Bar = 1.0 μm. *(Micrograph courtesy of N. Navarrete-Tindall.)*

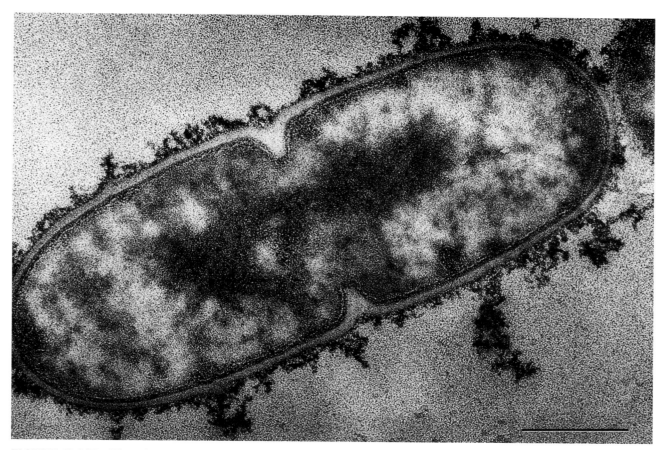

FIGURE 20.165 Ultra thin section of the gram positive bacterial cell, *Streptococcus mutans,* responsible for causing tooth decay. Bar = 0.25 μm.

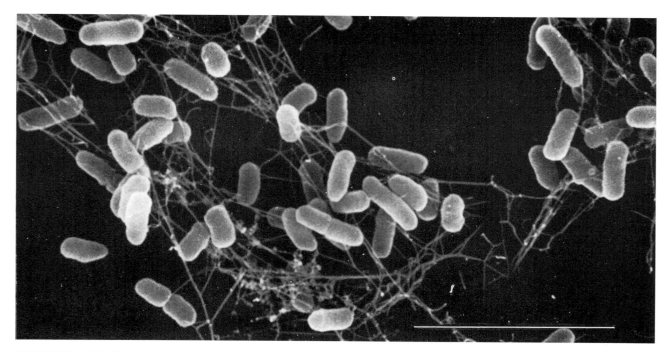

FIGURE 20.166 Scanning electron micrograph of pathogenic bacteria of the genus *Xanthomonas.* These bacteria secrete a stringy appearing extracellular substance or slime. Bar = 5.0 μm.

Algae, Fungi, Yeast, and Protozoa

Algae contain chloroplasts (see "The Special Features of Plant Tissues"; Figure 20.167). Algae may be unicellular or arranged in aggregates or in long branched or unbranched chains. Some divisions of algae (kelps and red algae) can become quite large with differentiated tissues approaching the complexity of higher plants.

Fungi are sometimes thought of as nongreen plants that lack chloroplasts and may consist of a plant body composed of threadlike structures called *hyphae*. In some classification schemes, fungi are viewed as a separate Kingdom. In fungi, food is absorbed from the external environment. The cell walls usually contain the polysaccharide chitin (also present in insect and crustacean exoskeletons), but usually not cellulose. Fungi are ubiquitous and may be found in extreme environments. Some are pathogenic for animals or plants but most are saprophytic organisms involved in recycling of biomass in nature (Figures 20.168 through 20.170).

Yeasts are one of the families of fungi that are widely distributed in substrates that are moist or contain sugar (Figures 20.171 through 20.173).

Protozoa are motile, unicellular organisms such as amoeba or paramecia that may be found in stagnant waters. They lack cell walls and chlorophyll. Some are quite simple cells, while other cells have rudimentary digestive apparati (mouth pores, gullets) that ingest food particles. Digestion takes place in cytoplasmic vacuoles and digested residue is excreted through an anal pore in the cell membrane. Most are actively motile by means of cilia, flagella, or ameboid movement using pseudopodia.

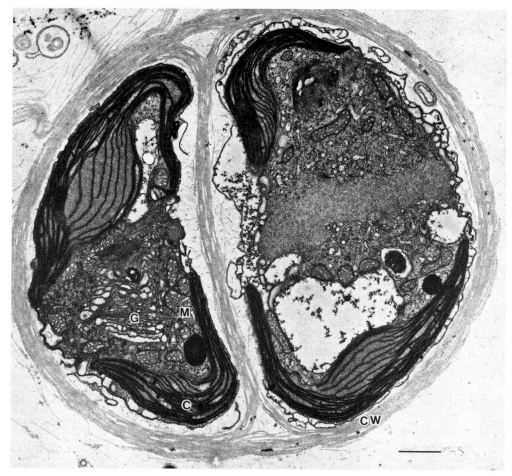

FIGURE 20.167 Algae. These algal cells show a cell wall (CW) of cellulosic scales as well as chloroplasts (C), mitochondria (M), and a Golgi apparatus (G). It is interesting to note that the scales originate from the cisternae of the Golgi apparatus and are deposited on the cell surface by exocytosis. Bar = 1.0 μm.
(Micrograph of Pleurochrysis scherfelii *courtesy of M. Brown, Jr.)*

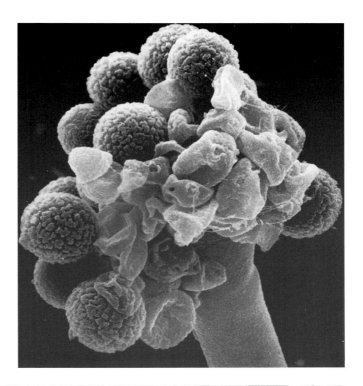

FIGURE 20.168 Scanning electron micrograph of the spore producing apparatus of an unidentified saprophytic fungus.

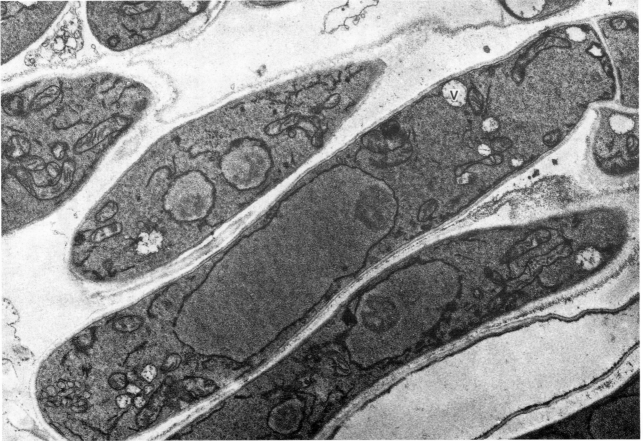

FIGURE 20.169 A longitudinal section through a diploid fungal basidium (*Schizophyllum commune*) shows the nucleus, nucleolus, vacuoles (V), mitochondria, and other organelles. *(Micrograph courtesy of W. J. Sundberg.)*

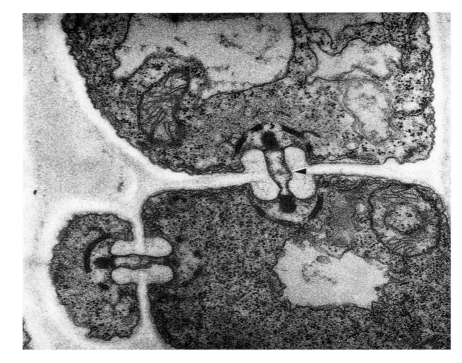

FIGURE 20.170 Transmission electron micrograph showing three fungal cells (*Schizophyllum commune*) connected by septal pores (*arrowhead*). Septal pores are larger and more structurally complex than the plasmodesmata described for plants and are generally seen in higher fungi. *(Micrograph courtesy of W. J. Sundberg.)*

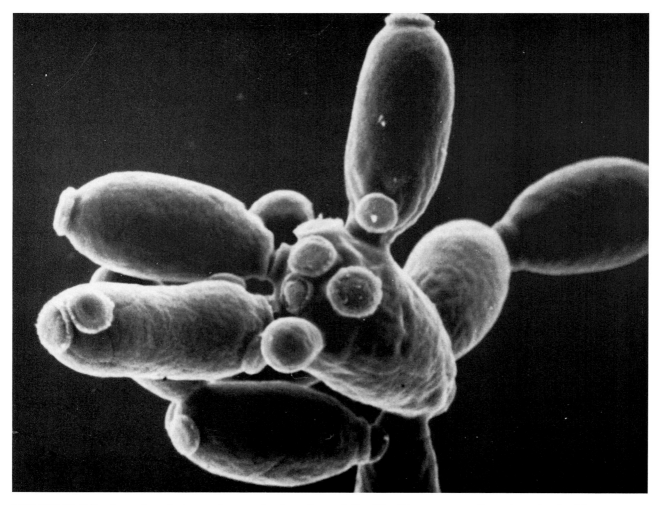

FIGURE 20.171 Scanning electron micrograph of a colony of *Candida albicans* yeast cells undergoing budding.

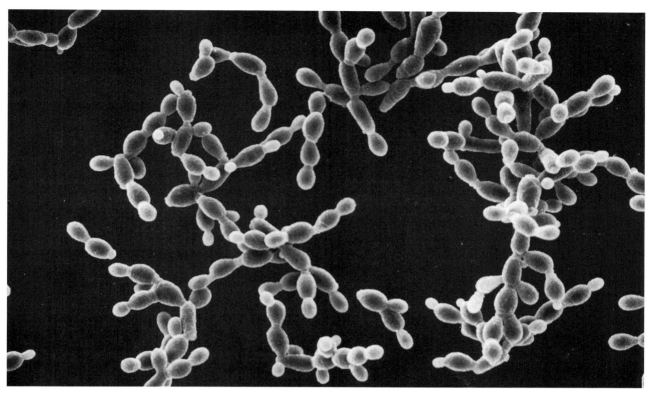

FIGURE 20.172   Scanning electron micrograph of common pathogenic yeast, *Candida albicans.*

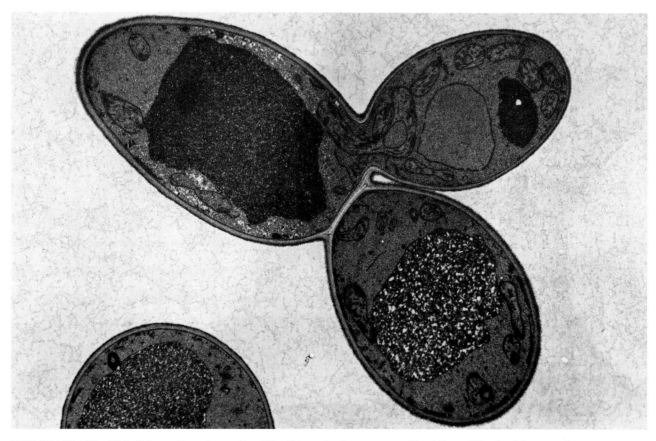

FIGURE 20.173   This thin section shows *Candida albicans* in the process of budding. (Fixed with potassium permanganate.)

Viruses

Electron microscopy is a requirement to visualize most viruses, since their diameter ranges from about 0.02 μm to 0.3 μm. The basic structure of viruses is relatively simple. They are composed of genetic material (DNA or RNA) covered by a protective coat. The variability of virus shape within this general plan is great. Often viruses take rounded or geometrical shapes (Figure 20.174 and Table 20.3).

The coat material of the virus, the *capsid*, is largely protein with some lipid. The form the coat may take is diverse. Some viruses may have a fuzzy appearing coat, whereas in others a lipid bilayer is visualized.

Electron microscopy is a powerful aid in the classification of viruses since it is possible to correctly identify the family of the virus based on a number of morphological features (size, symmetry of capsid, presence or absence of an envelope or unit membrane, surface projections, etc.). When combined with specific immune sera, the technique can give rapid and specific identification of the viruses to the species level. Very often, electron microscopy is the only technique available for the identification of newly discovered viruses such as the retroviruses causing AIDS and the virus responsible for hepatitis B.

Viruses gain entry into other cells by binding to highly specific cell *surface receptors.* After entrance and uncoating of the capsid, they multiply within the cell by using the cell's own anabolic metabolism. The *host cell* may be a bacterium, animal, or plant cell. Viruses may cause the death of the host cell when they are released from the cell.

The usual method to see detail within isolated viruses is by employing a negative staining technique (see Chapter 5).

We have included electron micrographs of some of the major groups of viruses in Figures 20.175 through 20.184. A short description of the major characteristics of each virus is included in the figure legend.

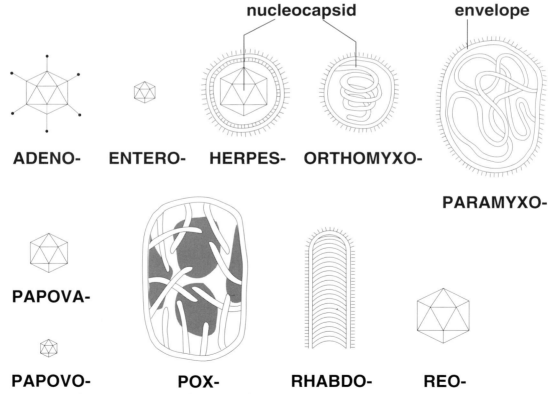

FIGURE 20.174 Variability in shapes and forms of different types of viruses.

TABLE 20.3 Viruses may be classified into families on the basis of physical and chemical properties. The following chart shows the major virus families

Nucleic Acid Core	Capsid Symmetry	Virion: Enveloped or Naked	Ether Sensitivity	No. of Capsomeres	Virus Particle Size (nm)*	Molecular Weight of Nucleic Acid in Virion ($\times 10^6$)	Physical Type of Nucleic Acid	No. of Genes (Approx.)	Virus Family
DNA	Icosahedral	Naked	Resistant	32	18–26	1.5–2.2	SS	7	Parvoviridae
				72	45–55	3–5	DS circular	10	Papovaviridae
				252	70–90	20–30	DS	50	Adenoviridae
		Enveloped	Sensitive	162	100†	92–102	DS	180	Herpetoviridae
	Complex	Complex coats	Resistant‡		230 × 300	130–240	DS	400	Poxviridae
RNA	Icosahedral	Naked	Resistant	32	20–30	2–2.8	SS	12	Picornaviridae
				?§	60–80	12–19	DS segmented	40	Reoviridae
	Unknown or complex	Enveloped	Sensitive	32?	40–70	4	SS	15	Togaviridae
		Enveloped	Sensitive		50–300	3–5	SS segmented	15	Arenaviridae
					80–130	9	SS	30	Coronaviridae
					~100	7–10	SS segmented	50	Retroviridae
	Helical	Enveloped	Sensitive		90–100	6–7	SS segmented	15	Bunyaviridae
					80–120	4	SS segmented	15	Orthomyxoviridae
					150–300	5–8	SS	30	Paramyxoviridae
					70 × 175	3–4	SS	20	Rhabdoviridae

*Diameter, or diameter × length.

†The naked virus (i.e., the nucleocapsid) is 100 nm in diameter; however, the enveloped virion varies up to 200 nm.

‡The genus *Orthopoxvirus*, which includes the better studied poxviruses (e.g., vaccinia, variola, cowpox, ectromelia, rabbitpox, monkeypox) is ether-sensitive. Some of the poxviruses belonging to other genera are ether-resistant.

§Reoviruses contain an outer and an inner capsid. The inner capsid appears to contain 32 capsomeres, but the number on the outer capsid has not been definitely established. A total of 92 capsomeres has been suggested.

Source: Brooks, George F., Janet S. Butel, L. Nicholas Ornston, Ernest Jawetz, Joseph L. Melnick, and Edward A. Adelberg. *Jawetz, Melnick, & Adelberg's Medical Microbiology: 20th Edition.* © Appleton & Lange, 1995. Reprinted by permission.

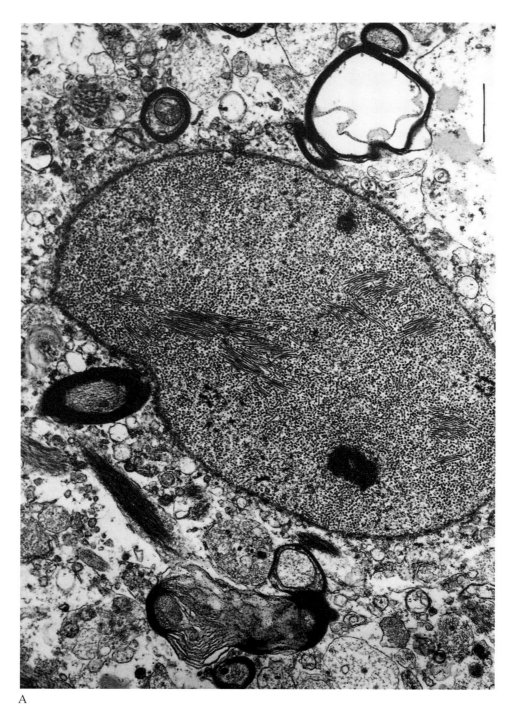

A

FIGURE 20.175 Papovaviruses are double-stranded DNA viruses with icosahedral nucleocapsids. They are divided into two main subgroups: papillomaviruses (causing warts in humans) that average 50 to 55 nm in diameter and the polyomaviruses that are 40 to 45 nm in diameter and associated with such terminal diseases as progressive multifocal leukoencephalopathy (JC virus) as well as in immunosuppressed patients (BK virus). (A) Nucleus of oligodendrocyte from brain biopsy of patient with progressive multifocal leukoencephalopathy. Nuclear inclusion contains fibrils and nucleocapsids of JC virus. Bar = 1.0 μm. (B) Isolated view of nucleus described in (A). Bar = 1.0 μm. (C) Higher magnification of polyomavirus seen in different thin section. Bar = 100 nm. (D) Polyomavirus (probably BK virus) seen by negative staining. Bar = 100 nm. (E) Papovavirus seen in negatively stained preparation. Bar = 100 nm. *(A, B and E courtesy of W. L. Steffens and M. B. Ard; C and D courtesy of S. E. Miller.)*

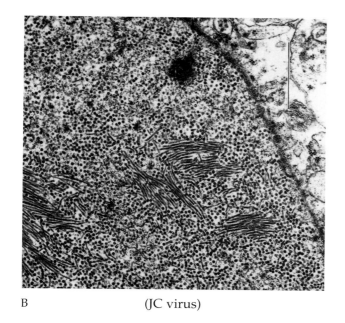

B   (JC virus)

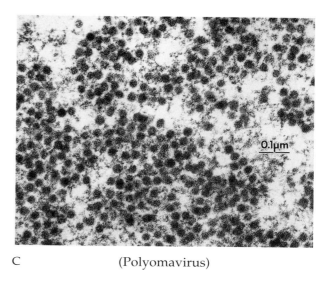

C   (Polyomavirus)

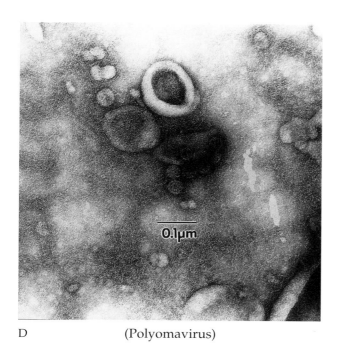

D   (Polyomavirus)

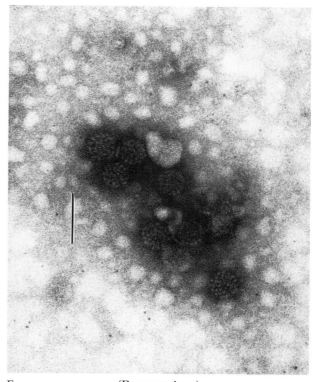

E   (Papovavirus)

FIGURE 20.175 Continued

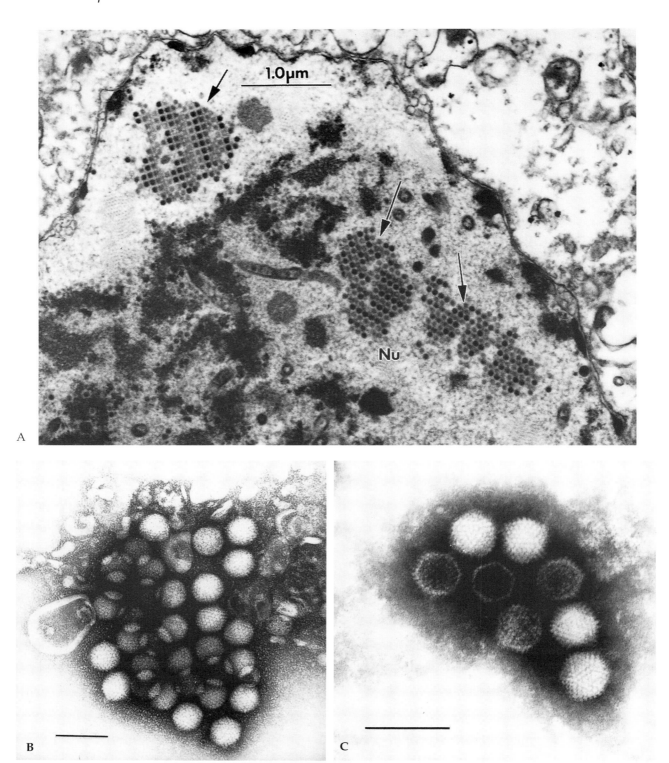

FIGURE 20.176 Adenoviruses are so named because they were first isolated from human adenoid tissues. These viruses cause a variety of diseases (respiratory illness, sore throat, eye infections, fever, intestinal complications) ranging from mild to life-threatening in immunocompromised patients. (A) Transmission electron micrograph (TEM) of tissue showing adenoviral inclusions (*arrows*) in the nucleus (Nu) of an infected cell. Bar = 1.0 μm. (B and C) TEM of negatively stained tissue extracts showing adenovirus particles. Bars = 100 nm.
(A and B courtesy of S. E. Miller; C courtesy of W. L. Steffens and M. B. Ard.)

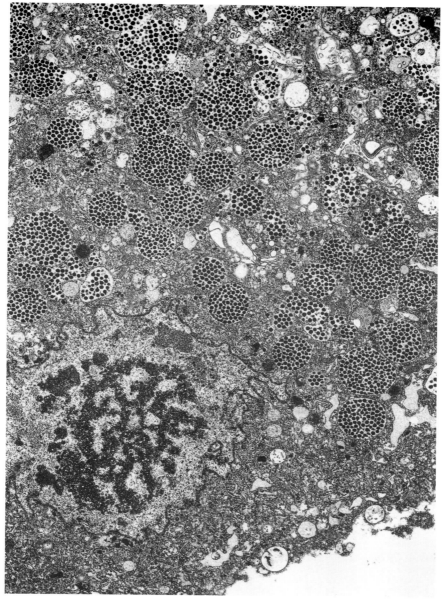

A

FIGURE 20.177   Herpesviruses comprise a large family of DNA viruses most of which are pathogenic for animals, including humans. The viruses are surrounded with two envelopes (although only one is seen in negative staining) enclosing a nucleocapsid about 150 nm. In this family are the *Herpes simplex* (HSV) types 1 and 2 causing fever blisters and genital herpes, respectively; varicella-zoster (VZV), which causes chickenpox and shingles; cytomegalovirus (CMV), which causes respiratory, urological and neurological diseases; Epstein-Barr virus (EBV), which causes Burkitt's lymphoma and probably infectious mononucleosis; and *Herpesvirus simiae,* which causes the deadly monkey B-virus infection when transmitted from monkey to man. (A) CMV particles in a thin section of human lung cell (pneumocyte). The numerous particles are contained within vesicles of the cytoplasm. (B) CMV particle (*arrow*) negatively stained with phosphotungstic acid. Specimen was obtained from the urine of a congenitally infected infant. Note the single prominent membranous envelope surrounding the nucleocapsid (*arrow*). (C) Avian laryngotracheitis herpesvirus grown in tissue culture and negatively stained with phosphotungstic acid. (D) VZV particles obtained from the vesicle fluid of a patient suffering from shingles. Negatively stained with phosphotungstic acid. (E) Section through cell infected with *Herpesvirus.* Nucleocapsids are found in the nucleus (*small arrows*) and an enveloped virion is seen in the cytoplasm (*large arrow*). Bar = 1.0 μm.
(A courtesy of J. Harb; C courtesy of W. L. Steffens and M. B. Ard; E courtesy of S. Miller.)

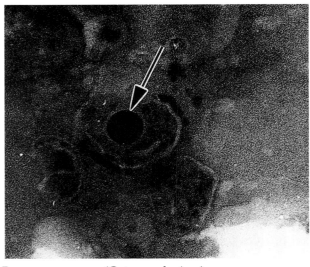

B   (Cytomegalovirus)

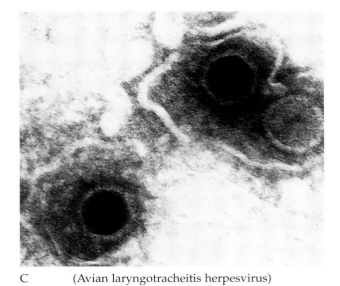

C   (Avian laryngotracheitis herpesvirus)

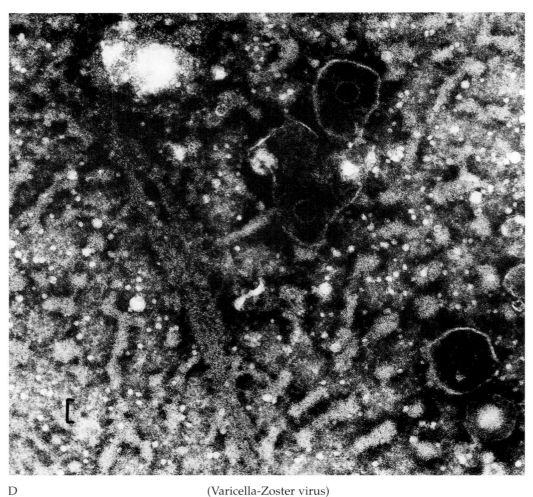

D   (Varicella-Zoster virus)

FIGURE 20.177 Continued

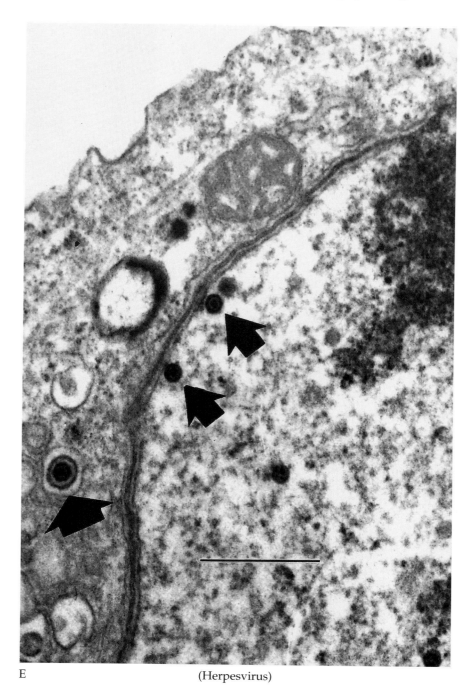

E (Herpesvirus)

FIGURE 20.177 Continued

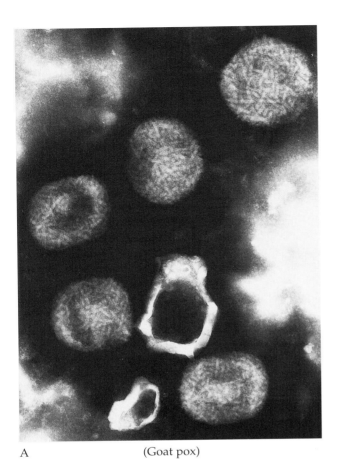

A   (Goat pox)

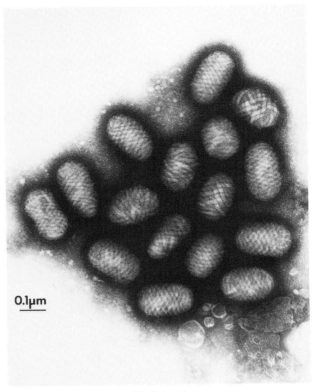

B   (Orf virus)

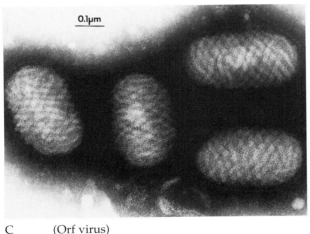

C   (Orf virus)

FIGURE 20.178 The poxvirus family contains several members, including variola major (smallpox, with mortality rates of 25%, which hopefully has been eradicated worldwide), variola minor (alastrim, with mortality rates of 1%), and vaccinia virus (a less worrisome pox disease of humans and animals). In addition, the family includes poxviruses that cause cowpox and monkeypox, molluscum contagiosum (an uncommon human skin affliction), milker's nodule, and orf virus (which causes contagious pustular dermatitis). Poxviruses have double stranded DNA. They are large viruses, 230 × 300 nm, quite similar in appearance and brick-shaped. (A) Goat pox particles taken from lesion scraping. Negatively stained preparation. (B and C) Orf virus particles contrasted using negative staining. This pox virus is readily recognized due to the cross-hatched pattern of the complex viral envelope. Compare this structure with the more random pattern seen in (A). (D) Poxvirus particles seen in a portion of a cell of thin sectioned tissue. Note the large, dense particles (*arrows*) present in the cytoplasm and the obvious endoplasmic reticulum (ER). Nucleus (Nu) is shown in the bottom half of the micrograph. (E–F) Pox particles revealed in thin section, electron micrographs of mammalian tissue culture cells. Virus appears as dense, brick-shaped particles (*arrows*) in the cytoplasm.
(A courtesy of W. L. Steffens and M. B. Ard; B, C, E, and F courtesy of J. F. Putterill; D courtesy of S. E. Miller.)

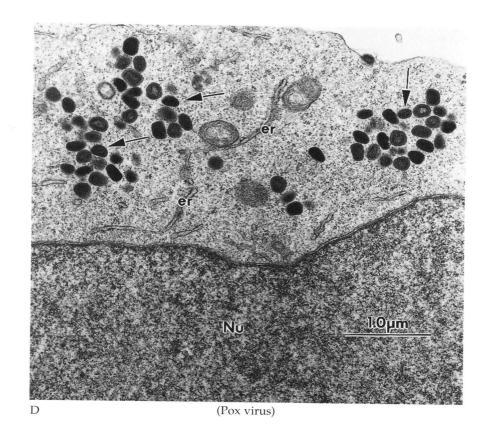

D   (Pox virus)

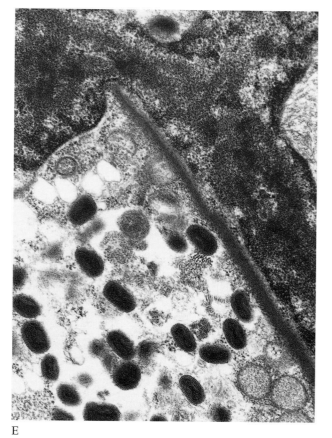

E

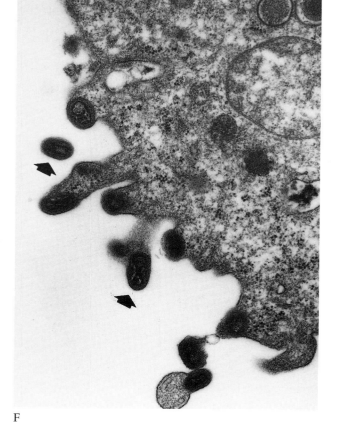

F

FIGURE 20.178   Continued

FIGURE 20.179 Enteroviruses are members of the picornaviridae (*pico* = small, single stranded RNA virus) family. These viruses are quite small, 25 to 30 nm, lack envelopes and are consequently termed "naked," consisting only of nucleocapsid. Members include poliovirus, coxsackie virus (isolated originally in Cocksackie, NY), echoviruses (enteric cytopathogenic human orphan). Most are found in the alimentary tract and may cause such diseases as poliomyelitis, meningitis, and serious infections of the heart. Transmission electron micrograph of negatively stained tissue culture extract showing avian enterovirus in feces. *(Courtesy of W. L. Steffens and M. B. Ard.)*

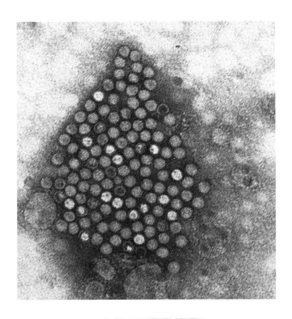

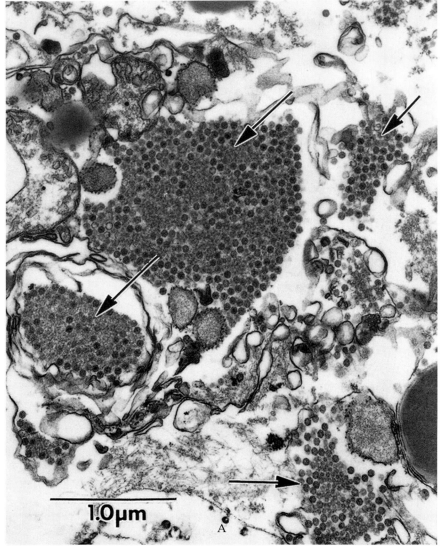

FIGURE 20.180 Reoviruses can be found in many animal species and can be isolated from individuals with respiratory and enteric diseases as well as from healthy children. These icosahedral viruses exhibit a double stranded RNA genome and average 70 nm in diameter. Like the enteroviruses, they can be readily found in the intestinal tract. (A) Reovirus particles forming large aggregations (*arrows*) in cytoplasm of infected mammalian cell. Tissue has been thin-sectioned and stained for transmission electron microscopy. (B and C) Whole reovirus particles, negatively stained. (D) Rotaviruses (*rota* = wheel) cause acute diarrhea in a wide range of animals, including humans. The rotavirus averages 70 to 75 nm in diameter and has what appear to be two capsid layers in some preparations. Negatively stained virus particles examined in the TEM. *(A, B, and D courtesy of S. E. Miller; C courtesy of W. L. Steffens and M. B. Ard.)*

1.0μm

A   (Reovirus)

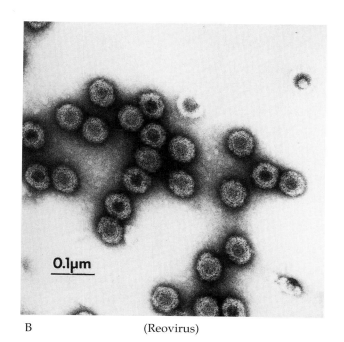

B   (Reovirus)

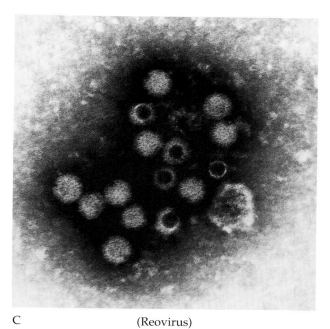

C   (Reovirus)

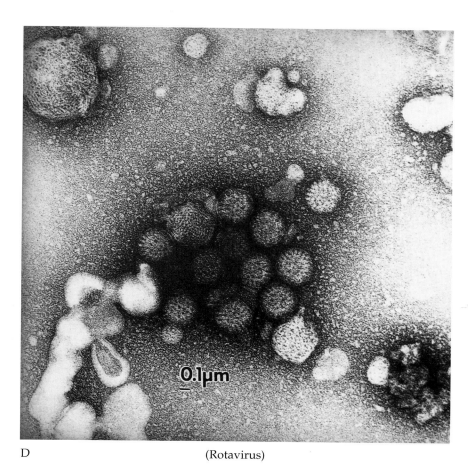

D   (Rotavirus)

FIGURE 20.180 Continued

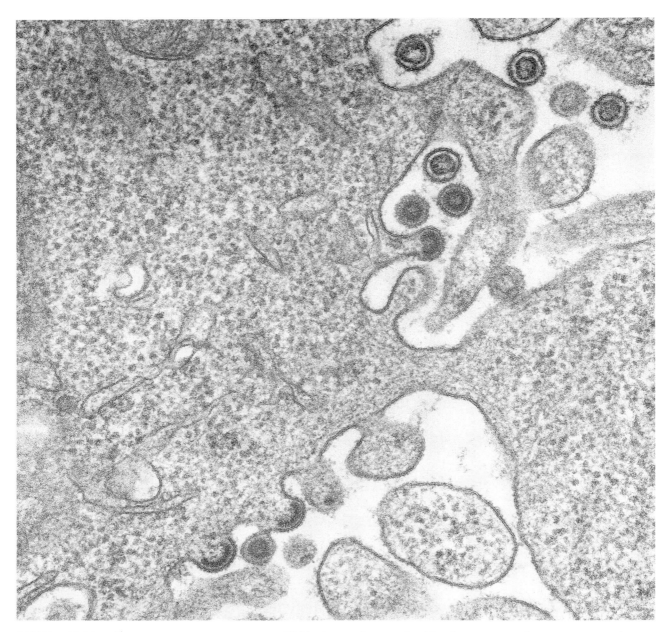

FIGURE 20.181 Retroviruses are tumor causing RNA viruses characterized by the presence of a reverse transcriptase enzyme in the virion. They may be classified according to their appearance in thin sections: B particles have an eccentric core, C particles have a centric core, while D particles are intermediate. The accompanying thin section electron micrograph shows mature and budding C-type particles in a leukemia cell. *(Courtesy of D. Friend.)*

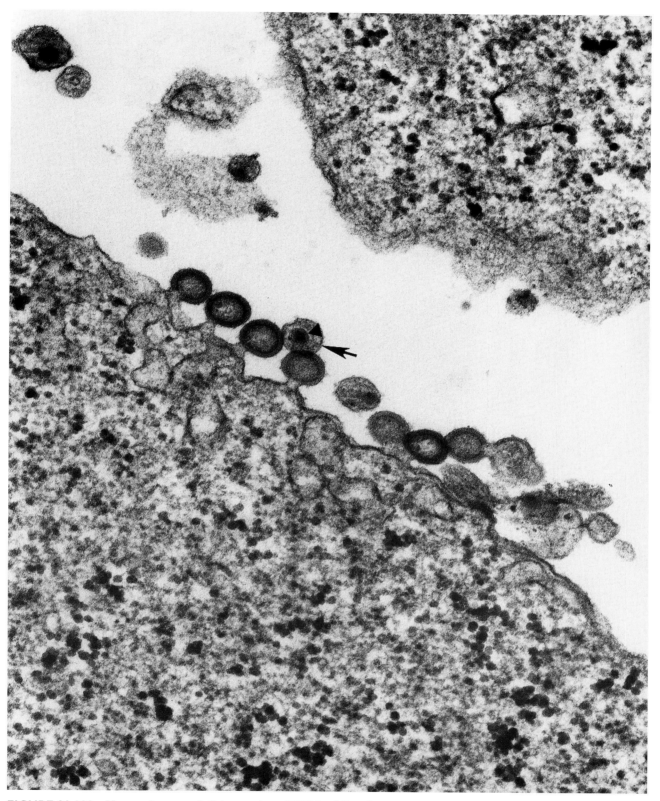

FIGURE 20.182 Human immunodeficiency virus (HIV) budding from a tissue culture cell. Among the particles shown is a mature virus containing a core (*arrowhead*) and a capsid (*arrow*).
(Courtesy of J. Pudney.)

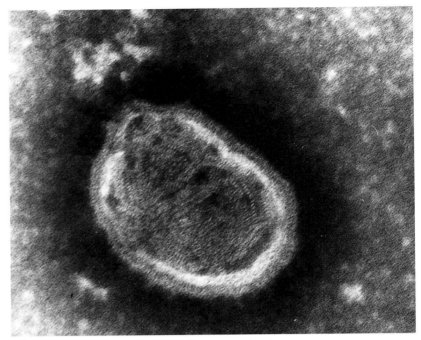

A   (Canine distemper virus)

FIGURE 20.183 Myxoviruses are single stranded RNA viruses with an envelope surrounding a helically shaped nucleoprotein core or nucleocapsid. Viruses range in size from approximately 100 nm for influenza to 175 to 300 nm for parainfluenza. Man is affected by the influenza, parainfluenza, and mumps viruses whereas the Newcastle disease virus (NDV) affects birds and the canine distemper virus (CDV) attacks dogs. Both of the latter viruses cause devastating diseases in nonimmune animals. (A) This image is of CDV taken from cerebrospinal fluid. (B) Measles virus budded from cytoplasmic membrane. Bar = 0.1 μm. (C) Measles virus, negatively stained whole virus particle. Bar = 0.1 μm. (D) Measles virus minus envelope reveals the helical nucleocapsid. Bar = 0.1 μm. *(A courtesy of W. L. Steffens and M. B. Ard; B–D courtesy of S. E. Miller.)*

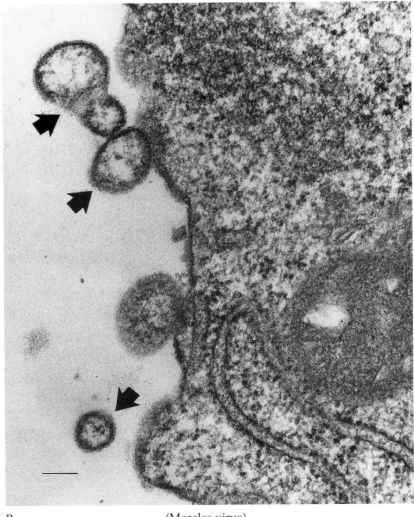

B   (Measles virus)

Rough Endoplasmic Reticulum

Blöbel, G., and B. Doberstein. 1975. Transfer of proteins across membranes. I. Presence of proteolytically processed and unprocessed nascent immunoglobulin light chains on membrane-bound ribosomes of murine myeloma. *J Cell Biol* 67:835–51.

De Pierre, J. W., and G. Dallner. 1975. Structural aspects of the membrane of the endoplasmic reticulum. *Biochem Biophys Acta* 415:411–72.

Hortsch, M., D. Avossa, and D. I. Meyer. 1986. Characterization of secretory protein translocation: ribosome-membrane interaction in endoplasmic reticulum. *J Cell Biol* 103:241–53.

Smooth Endoplasmic Reticulum

Emans, J. B., and A. L. Jones. 1968. Hypertrophy of liver cell smooth endoplasmic reticulum following progesterone administration. *J Histochem Cytochem* 16:561–71.

Higgins, J. A., and R. J. Barrnett. 1972. Studies on the biogenesis of smooth endoplasmic reticulum membranes in livers of phenobarbital treated rats. *J Cell Biol* 55:282–98.

Mori, H., and A. K. Christensen. 1980. Morphometric analysis of Leydig cells in the normal rat testis. *J Cell Biol* 84:340–54. (Smooth endoplasmic reticulum is related to steroid synthesis in this article.)

The Golgi Apparatus

Farquhar, M., and G. Palade. 1981. The Golgi apparatus (complex)—(1954–1981) from artifact to center stage. *J Cell Biol* 91:77s–103s.

Mollenhauer, H. H., and D. J. Morré. 1991. Perspectives on Golgi apparatus form and function. *J Elect Micros Techn* 17:2–14.

Pavelka, M. 1987. Functional morphology of the Golgi apparatus. *Adv Anat Embryol Cell Biol* 106:1–94.

Rambourg, A., and Y. Clermont. 1990. Three-dimensional electron microscopy: structure of the Golgi apparatus. *European J Cell Biol* 51:189–200.

Rothman, J. E. 1985. The compartmental organization of the Golgi apparatus. *Sci Am* 253:74–89.

Whaley, W. G. 1975. *The Golgi apparatus.* Heidelberg: Springer-Verlag.

Secretory Products

Burgess, T. L., and R. B. Kelly. 1987. Constitutive and regulated secretion of proteins. *Annu Rev Cell Biol* 3:343–93.

Jamieson, J., and G. E. Palade. 1971. Synthesis, intracellular transport and discharge of secretory proteins in stimulated pancreatic exocrine cells. *J Cell Biol* 50:135–58.

Centrioles

Bartholdi, M. F. 1991. Nuclear distribution of centromeres during the cell cycle of human diploid fibroblasts. *J Cell Sci* 99:255–63.

Brinkley, B. R. 1990. Toward a structural and molecular definition of the kinetochore. *Cell Motil and Cytoskel* 16:104.

De Harven, E. 1968. The centriole and the mitotic spindle. In *The nucleolus,* New York: Academic Press Inc., pp. 197–227.

Fulton, C. 1971. In *Centrioles. Origin and continuity of cell organelles,* vol. 2. J. Reinert and H. Ursprung, eds., Springer-Verlag, pp. 170–221.

Karsenti, E., and B. Maro. 1986. Centrosomes and the spatial distribution of microtubules in animal cells. *Trends Biochem Sci* 11:460–63.

Rattner, J. B., and D. P. Bazett-Jones. 1988. Electron spectroscopic imaging of the centrosome in cells of the Indian Muntjak. *J Cell Sci* 91:5–11.

Stubblefield, E., and B. R. Brinkley. 1967. Architecture and function of the mammalian centriole. *Symp Int Soc Cell Biol* 6:175–218.

Vorobjev, I. A., and Y. S. Chentsov. 1982. Centrioles in the cell cycle. I. Epithelial cells. *J Cell Biol* 93:938–49.

Cilia and Flagella

Cilia

Gibbons, I. R. 1981. Cilia and flagella of eukaryotes. *J Cell Biol* 91:107s–24s.

Steinman, R. M. 1968. An electron microscopic study of ciliogenesis in developing epidermis and trachea in embryos of Xenopus. *Am J Anat* 122:19–55.

Wolfe, J. 1972. Basal body fine structure and chemistry. *Adv Cell Molec Biol* 2:151–92.

Flagella

Fawcett, D. W. 1975. The mammalian spermatozoön. A review. *Dev Biol* 44:394–436.

The Lysosomal System

Multivesicular Bodies

Bainton, D. 1981. The discovery of lysosomes. *J Cell Biol* 91:66s–76s.

De Duve, C. 1963. The lysosome. *Sci Am* 208:64–72.

Essner, E., and A. B. Novikoff. 1961. Localization of acid phosphatase activity in hepatic lysosomes by means of electron microscopy. *J Biophys Biochem Cytol* 9:773–84.

Helenius, A., I. Mellman, D. Wall, and A. Hubbard. 1983. Endosomes. *Trends Biochem Sci* 8:245–50.

Hottzman, E. 1976. *Lysosomes, a survey.* New York: Springer-Verlag.

Microbodies

De Duve, C. 1983. Microbodies in the living cell. *Sci Am* 248:74–84.

De Duve, C., and P. Baudhuin. 1966. Peroxisomes (microbodies and related particles). *Physiol Rev* 46:323–57.

Fahimi, H. D., and H. Sies. 1987. *Peroxisomes in biology and medicine.* Heidelberg: Springer-Verlag.

Frederick, S. E., P. J. Gruber, and E. H. Newcomb. 1975. Plant microbodies. *Protoplasma* 84:1–29.

Lazarow, P. B., and Y. Fujiki. 1985. Biogenesis of peroxisomes. *Annu Rev Cell Biol* 1:498–530.

Tolbert, N. E., and E. Essner. 1981. Microbodies: peroxisomes and glyoxysomes. *J Cell Biol* 91:271s–83s.

Annulate Lamellae

Chen, T-Y, and E. M. Merisko. 1988. Annulate lamellae: Comparison of antigenic epitopes of annulate lamellae membranes with the nuclear envelope. *J Cell Biol* 107:1299–1306.

Kessel, R. G. 1985. Annulate lamellae: A last frontier in cellular organelles. *Int Rev Cytol* 133:43–120.

Merisko, E. M. 1989. Annulate lamellae: An organelle in search of a function. *Tissue Cell* 21:343–54.

Glycogen/Glycosomes

Revel, J. P. 1964. Electron microscopy of glycogen. *J Histochem Cytochem* 12:104–14.

Rybicka, K. K. 1979. Glycosomes (protein-glycogen complex) in the canine heart. Ultrastructure, histochemistry and changes induced by acidic treatment. *Virchows Arch B Cell Path* 30:335–47.

Rybicka, K. K. 1996. Glycosomes: The organelles of glycogen metabolism. *Tissue and Cell* 28:253–66.

Scott, R. B., and W. S. J. Still. 1968. Glycogen in human peripheral blood leucocytes. II. The macromolecular state of leucocyte glycogen. *J Clin Invest* 47:353–59.

Cell Inclusions

Lipid

Ashworth, C. T., J. S. Leonard, E. H. Eigenbrodt, and F. J. Wrightsman. 1966. Hepatic intracellular osmiophilic droplets. Effect of lipid solvents during tissue preparation. *J Cell Biol* 31:301–18.

Suter, E. 1969. The fine structure of brown adipose tissue. *Lab Invest* 21:246–58.

Crystalloids

Glusker, J. P., and K. N. Trueblood. 1985. *Crystal structure analysis: A primer.* Oxford: Oxford University Press.

Nagano, T., and I. Ohtsuki. 1971. Reinvestigation of the fine structure of Reinke's crystal in the human testicular interstitial cell. *J Cell Biol* 51:148–61.

Extracellular Material

Collagen

Martin, G. R., and R. Timpl. 1987. Laminin and other basement membrane components. *Annu Rev Cell Biol* 3:57–85.

Merker, H. J. 1994. Morphology of the basement membrane. *Micros Res Tech* 28:95–124.

Watt, F. M. 1986. The extracellular matrix and cell shape. *Trends Biochem Sci* 11:482–85.

Matrix of Bone and Cartilage

Cormick, D. 1987. *Ham's histology,* 9th ed. Philadelphia: Lippincott, pp. 264–323.

Jamde, S. S. 1971. Fine structural study of osteocytes and their surrounding bone matrix with respect to their age in young chicks. *J Ultrastruct Res* 37:279–300.

Special Features of Plant Tissues

Cutter, E. G. 1978. *Plant anatomy,* 2d ed. Part 1, cells and tissues. London: Arnolds.

Esau, K. 1965. *Plant anatomy.* New York: Wiley.

Esau, K. 1977. *Anatomy of seed plants.* New York: Wiley.

Fahn, A., 1974. *Plant anatomy.* Oxford, New York: Pergamon Press.

Merva, G. E. 1995. *Physical principles of the plant biosystem.* ASAE Textbook No. 9, American Society of Agricultural Engineers, St. Joseph, Michigan.

Troughton, J., and L. A. Donaldson. 1972. *Probing plant structure: A scanning electron microscope study of some anatomical features in plants and the relationship of these structures to physiological processes.* New York: McGraw-Hill.

Chloroplasts

Alberts et al., 1989. Special features of plant cells. In *Molecular biology of the cell,* 2d ed. New York: Garland, pp. 1137–86.

Anderson, J. M. 1986. Photoregulation of the composition, function, and structure of thylakoid membranes. *Annu Rev Plant Physiol* 37:93–136.

Basic, A., P. J. Harris, and B. A. Stone. 1988. Structure and function of plant cell walls. In *Biochemistry of plants: A comprehensive treatise,* J. Preiss, ed., vol. 14, Carbohydrates. San Diego, CA: Academic Press, pp. 298–371.

Bogorad, L. 1981. Chloroplasts. *J Cell Biol* 91:256s–70s.

Hoober, J. K. 1984. *Chloroplasts.* New York: Plenum Press.

Miller, K. R. 1979. The photosynthetic membrane. *Sci Am* 241:102–13.

The Cell Wall

Gunning, B. E. S., and R. L. Overall. 1983. *Plasmodesmata.* New York: Springer-Verlag.

McNeil, M., A. G. Darvill, S. C. Fry, and P. Alfersheim. 1984. Structure and function of the primary cell walls of plants. *Annu Rev Biochem* 53:625–63.

Vacuoles

Matile, P. 1978. Biochemistry and function of plant vacuoles. *Annu Rev Plant Physiol* 29:193–213.

Bacteria

Fuller, R., and D. W. Lovelock. 1976. *Microbiol ultrastructure: The use of the electron microscope.* New York: Academic Press. 331 pp.

Algae, Fungi, Yeast, and Protozoa

Bold, H. C., and M. J. Wynne. 1978. *Introduction to the algae: Structure and reproduction.* Englewood Cliffs, NJ: Prentice-Hall.

Levandowsky, M., and S. H. Hutner. 1980. *Biochemistry and physiology protozoa.* New York: Academic Press.

Prescott, D. M. 1988. *Cells.* Boston: Jones and Bartlett Publishers.

Viruses

Ackermann, H-W. And L. Berthiaume (eds.). 1995. *Atlas of virus diagrams.* ISBN: 0–8493–3457–2. Boca Raton, FL: CRC Press.

Casjens, S., ed. 1985. *Virus structure and assembly.* Jones and Bartlett, Boston: Publishers.

Dalton, A. J., and F. Haguenae. 1973. *Ultrastructure of animal viruses and bacteriophages, an atlas.* New York: Academic Press. 413 pp.

Field, A. M. 1982. Diagnostic virology using electron microscopic techniques. *Advanced Virus Res,* 27:1–69.

Fraenkel-Conrat, H. 1985. *The viruses. Catalogue, characterization, and classification.* New York: Plenum Publishing Co.

Hsiung, G. D., and C. K. V. Fong. 1982. Diagnostic virology illustrated by light and electron microscopy. *J Electron Microsc Tech* 4:265–301.

Hsiung, G. D., C. K. V. Fong, and M. J. August. 1979. The use of electron microscopy for diagnosis of virus infections: an overview. *Prog Med Virol* 25:133–59.

Miller, S. E. 1986. Detection and identification of viruses by electron microscopy. *J Electron Microsc Tech* 4:265–301.

Miller, S. E. 1995. Diagnosis of viral infection by electron microscopy. In E. H. Lennette, D. A. Lennette, and E. T. Lennette (eds.), *Diagnostic procedures for viral, rickettsial and chlamydial infections,* American Public Health Assoc., Washington, DC, pp. 35–76.

Palmer, E. L., and M. L. Martin. 1988. *Electron microscopy in viral diagnosis.* Boca Raton, FL: CRC Press. 208 pp.

Simmons, K., H. Garoff, and A. Helenius. 1982. How an animal virus gets into and out of its host cell. *Sci Am* 246:58–66.

Chapter 21

Safety in the Electron Microscope Laboratory

IT IS IMPORTANT FOR RESEARCHERS not only to be trained in the proper use of all equipment and reagents in the electron microscope laboratory, but also to be aware of potential hazards (fire, chemical, electrical, physical) associated with these items. Microscopists must know the proper procedures to follow in all hazardous situations and the proper method of disposing of dangerous wastes. The environmental impact of using various chemicals (and even equipment) must be discussed so that microscopists can make informed decisions about pursuing particular lines of research. Although electron microscopists and students often become engrossed in learning new techniques and making scientific discoveries, we must not lose track of the importance of safety in conducting our work. We must be aware of the consequences of our actions and be prepared to correct our mistakes in a manner that is consistent with the preservation of the quality of life and the environment.

Personal Safety in the Laboratory

All electron microscopists should be aware of basic operating conditions or guidelines to protect the safety of co-workers, equipment, and themselves. Even though each electron microscope laboratory will undoubtedly develop its own specific sets of rules for safe operation, certain basic safety principles and practices are covered here for the benefit of those who may be considering establishing or refining a set of safe operational procedures in their laboratory.

Safety Apparatus and Safe Practices

1. **Fume hoods** (Figure 21.1) are needed in all electron microscope laboratories because most of the chemicals used in electron microscopy are toxic and some are known carcinogens. A quick way to check if a fume hood is functioning properly is to place several drops of an odorous chemical such as beta mercaptoethanol on a filter paper in a petri dish inside of the operating hood for about 30 minutes. If one detects the odor outside of the confines of the cabinet, then the hood should be checked for proper operation. In one such check, an electron microscopist determined that

the main duct of a fume hood was leaking into the overhead ceiling space of an adjacent office. Fume hoods should be evaluated regularly for proper flow rate by trained personnel to ascertain that they are preventing the escape of fumes into the work environment.

Chemicals and apparatus should not be stored permanently in the working fume hood since they might interfere with the efficiency of the ventilation system. If one chooses to store dangerous chemicals in a fume hood, a special hood or vented cabinet should be designated for storage only and checked regularly for proper operation. Compatibilities of the chemicals stored in the hood must be checked to avoid placing unsafe combinations in the same location.

A surgeon's mask is not a substitute for a fume hood. Masks will not protect the wearer from noxious fumes or finely powdered chemicals (such as lead citrate). Surgical masks are designed only to prevent the liberation of aerosols of infected droplets from the wearer.

2. **Gloves** should be worn whenever handling potentially toxic chemicals. Perhaps the most useful gloves are made of nitrile, polyvinyl chloride, or 4H™ (Monsanto). Such gloves are

FIGURE 21.1   Fume hoods are essential pieces of safety equipment for working with chemicals used in electron microscopy. Hoods should be evaluated on a regular basis to verify that they are operating properly. Gloves are still necessary to protect the hands against contact with the chemicals.

resistant to most chemicals used in electron microscopy and do not interfere with the dexterity needed for most operations. For safety, however, one must verify that the glove is appropriate for the situation prior to using the chemical.

Care should be taken not to contaminate objects with soiled gloves. A common mistake is to touch books, papers, reagent bottles, telephones, or doorknobs with soiled gloves, thereby contaminating them. One researcher working late at night ran out of one of the components of an epoxy embedding medium. In an attempt to locate more of the reagent, the researcher entered several rooms while wearing gloves contaminated with resin monomer. The following day, another researcher entering one of the rooms noticed a sticky substance on the doorknob. Fortunately, they were able to deduce the nature of the substance and wash their hands and the doorknob using hot water and a strong detergent.

Gloves should be removed carefully so as not to contaminate the hands. One way of doing this is to remove the glove from one hand so that the glove is turned inside out. The inverted glove may then be used to take hold of and similarly invert the glove on the other hand. One glove may then be placed inside of the other and tied into a knot to seal the contamination inside of the gloves. Be aware that most latex gloves are powdered with talc, a substance to which some individuals may be sensitive. Unpowdered (or minimally powdered with cornstarch) gloves may be substituted. If a dangerous chemical is being used, most savvy researchers use doubly gloved hands and change the outer glove as it becomes contaminated.

Selection of Gloves

Gloves must be selected carefully to adequately protect the individual from contact with dangerous chemicals. It is dangerous to assume that the common latex glove will work in all instances. One should always consult the glove manufacturer's chemical resistance guide to determine the *degradation properties* (a measure of the glove's tendency to swell, discolor or change due to chemical contact) as well as the *permeation properties* (a measure of how readily the chemical will pass through the glove material). It is important to realize that all gloves are permeable,

depending on the chemical handled, composition and thickness of the glove, condition of the glove, and the length of time in contact with the chemical. There is no such thing as an "ideal" glove that will work in all instances. Some flexible laminate gloves (Silver Shield™ or 4H™) offer excellent protection over a wide range but they limit dexterity and are difficult to use when wet. Perhaps the best glove consists of two gloves worn together. For instance, wearing a reusable thicker glove (made of nitrile, latex, neoprene, butyl, or Viton™) over a flexible laminate will give the best protection with reasonable dexterity. If reusable gloves are employed, the condition and cleanliness of the gloves must be monitored. Inspect the gloves before use and replace them if they cannot be decontaminated or show signs of cracking or wear. Disposable gloves of latex or nitrile provide protection when working with aqueous materials, biologicals, and radioisotopes but should not be used with hazardous or aggressive chemicals (strong acids, organic solvents). Disposable gloves are thin (4 mil) compared with reusable gloves (18 to 28 mil).

As a guide to help determine which glove to use, we provide some basic information on chemical resistance. Consult the institutional safety officer or the glove company if there is any uncertainty in the selection process.

Following is a list of glove materials and their resistance:

Viton™	Excellent resistance to chlorinated and aromatic solvents. Expensive.
Butyl	Good choice for aldehydes, ketones, and esters. Expensive.
Neoprene	Wide range of resistance to solvents, acids, caustics and alcohols.
Nitrile	Same as neoprene and with good puncture and abrasion resistance.
Latex	Resists acids and caustics. Best if combined with other gloves.
PVC	Resists acids but not petroleum solvents.

Reference: Dartmouth College, Environmental Health and Safety brochure.

3. **Eye protection** may be necessary when working with cryogenic agents such as liquid nitrogen and when handling most chemicals. Goggles or shields should be used to cover glasses or contact lenses since they do not provide adequate eye protection. In fact, regular contact lenses may actually trap material between the lens and eye and interfere with the

expeditious removal of materials that have splashed into the eye. Soft contact lenses may absorb toxic fumes and transfer them onto the eye surface. If possible, substitute readily removable eyeglasses for contact lenses when handling chemicals such as osmium tetroxide, formaldehyde, acrolein, and other volatiles.

One should know how to operate the safety eyewash stations. In an emergency, it may be necessary to locate and operate the station with one or both eyes closed, so practice with both eyes closed. If the eyes come into contact with chemicals, they must be held open and exposed to the tempered stream of water for 15 minutes!

4. **Suitable clothing** should be worn. Laboratory coats, aprons, and arm protectors may be needed when handling certain chemicals or cryogens. Snap, rather than button, closures are desirable in laboratory clothing to permit its rapid removal if contaminated or on fire. Always check with the laboratory supervisor if there are any doubts about proper safety attire. If clothing becomes contaminated, remove it immediately and dispose of or launder it in the proper manner. Open-toed shoes or sandals are unsafe and should not be allowed in the laboratory. Do not wear expensive or hard to clean clothing when handling laboratory chemicals. It is a good idea to have a change of clothing nearby. Remove jewelry such as watches, rings, bracelets, and necklaces to prevent their contact with chemicals and electrical sources.

5. **Verify new or untried operational procedures or protocols** with laboratory supervisors before using them. This may save time and reagents as well as ensure the researcher's safety. Researchers should never work alone on untried protocols unless they are totally aware of all the dangers and potential problems likely to be encountered.

6. **Eating, drinking, and smoking are forbidden** in the electron microscopy laboratory. This is not only for the operator's protection, but also to protect equipment from contamination. Do not store food or drinks in a laboratory refrigerator or use the laboratory microwave for heating food. Do not use laboratory glassware for food or beverages. Special rooms and refrigerators should be set aside for the consumption and storage of food and drinks.

7. **Mouth pipetting is not permitted.** Use bulbs or other suitable equipment for such procedures.

8. **Work areas must be kept clean, uncluttered, and free of physical obstruction.** Put away unneeded reagents or apparatus when the procedure is finished and clean the work area thoroughly to remove any traces of chemicals. Dispose of contaminated materials in a safe and ecologically sound manner. Do not toss dangerous chemicals or broken objects into conventional trash bins since cleaning and maintenance people may be harmed.

9. **Be familiar with the location and operation of all safety equipment** or reagents such as safety showers, eyewash stations, spill control units, fire extinguishers, and first aid kits.

10. **Post important telephone numbers near all phones.** Such numbers should include fire department, ambulance, physician or nurse, poison control center, pollution control, power plant, security office, and the home and office numbers of the laboratory supervisor.

11. **Label all reagents** with complete chemical names of contents, date prepared, and with the initials or name of the preparer. Indicate any special precautions, such as need for refrigeration or freezing, avoidance of light and mechanical shocks, electrical hazards, and so on.

12. **Post material safety data sheets** (MSDS) (Figure 21.2), normally supplied by the manufacturer close to where the chemicals will be used, and require all users to read the sheets. MSDSs detail the characteristics of the chemicals and the precautions to be followed when handling the reagents.

13. **Arrange for regular safety inspections and drills.** This not only helps identify and eliminate potential hazards, but also gives practice in the proper actions to take if an accident occurs. *A plan of action* is essential for all electron microscope laboratories. Everyone must be familiar with proper safety procedures regarding exits, chemicals and spills, fires, electricity, and equipment.

14. **Dispose of all chemicals in accordance with local and federal regulations.** If in doubt, contact proper pollution control offices.

15. **Wash hands and arms before leaving the laboratory,** especially after handling chemicals.

16. **Verify the security of the laboratory before leaving:** (a) extinguish all flames or sources of ignition; (b) turn off all unnecessary gases, water, vacuum, and electricity; (c) lower the sash to the fume hood; (d) turn off all nonessential lights; and (e) lock all doors.

MATERIAL SAFETY DATA SHEET
(Adapted from USDL Form LSD-005-4)

SECTION I. IDENTIFICATION OF PRODUCT	
CHEMICAL NAME Lead Acetate	FORMULA $Pb(C_2H_3O_2) \cdot 3H_2O$

SYNONYM OR CROSS REFERENCE
none available

SECTION II. HAZARDOUS INGREDIENTS	
MATERIAL Lead Acetate	NATURE OF HAZARD Irritant, poisonous

SECTION III. PHYSICAL DATA

BOILING POINT
    decomposes at 100 C
	MELTING POINT
	    75 C

VAPOR PRESSURE
    not available
	SPECIFIC GRAVITY
	    2.55

VAPOR DENSITY
    not available
	% VOLATILE BY VOLUME
	    not available

WATER SOLUBILITY
    60 gm in 100 gm water
	EVAPORATION RATE
	    not available

APPEARANCE
    white crystalline granules
	ODOR
	    slightly acetic

SECTION IV. FIRE AND EXPLOSION HAZARD DATA

FLASH POINT
    not flammable
FIRE EXTINGUISHING MEDIA
    as appropriate for surrounding fire
SPECIAL FIRE-FIGHTING PROCEDURES
    protective clothing, self contained breathing apparatus with full
    faceplate; may decompose to acetic acid, carbon monoxide and toxic
    fumes of lead oxide
UNUSUAL FIRE AND EXPLOSION HAZARD
    not considered to be explosive; may combine with azides to form
    explosive compounds (e.g., lead azide)
FLAMMABLE LIMITS
    not applicable

FIGURE 21.2 Material safety data sheet (first page of two) for lead acetate stain. It is advisable to post these sheets on a bulletin board in the work area where the chemicals are to be used.

Pathogens and Radioisotopes

Most electron microscope laboratories are not prepared to handle materials contaminated by microbes or radioisotopes. Normally, such specimens are best handled outside of the electron microscope facility with the fixed specimens then taken to the electron microscope facility where they may be further processed. If pathogens or radioisotopes are to be used in an electron microscope facility, prior clearance with the laboratory supervisor must be obtained. Clear guidelines for the safe handling of pathogens and radioisotopes have been established by professional organizations and have been mandated by federal law. For example, one must be certified by a radiological control officer prior to handling radioisotopes, and an infection control officer may become involved in the handling of human pathogens. Federal approval is required before any of these materials can be transported either privately or commercially.

Human fluids and tissues must be handled with particular care because they may contain infectious agents. Staff must be aware of the subtle precautions to take with contaminated materials that enter the electron microscope laboratory. For example, negatively stained clinical specimens are still potentially infectious, and forceps as well as specimen cartridges from the electron microscope may become contaminated by contact with them. Even radioisotopes contained in specimens that have been embedded in plastic are dangerous since it is possible to inhale fragments of the specimen blocks generated during the trimming steps for ultramicrotomy.

Chemical Safety

Nearly all chemicals used in electron microscopy are toxic to humans and dangerous to the environment unless handled and disposed of properly. All electron microscopists must be acquainted with the potential dangers associated with the chemicals they are about to use, dispose of, or store.

Handling Chemicals in a Safe Manner

The first step in the safe handling of chemicals is to read the *material safety data sheet* (MSDS) for the chemical provided by the supplier. Most EM supply houses have these sheets, which give information on the dangerous properties of the chemical. Typical information in the MSDS may include name (chemical, synonyms, formula), physical data (boiling and freezing points, appearance and odor, etc.), fire and explosion hazard data (as well as fire fighting information), health hazard data (effects of exposure, emergency and first aid procedures), reactivity data (chemical compatibilities), ingredients (if other than 100%), spill or leak procedures, special protection information (respirators, gloves, eye protection, etc.), special precautions (handling and storage), and regulatory information. It is advisable to post such sheets on a bulletin board in the area where the chemicals are likely to be used and to insist that all first-time users read and understand the information on the sheet. Some discussion of the sheet is probably necessary to ensure compliance.

Many chemical manufacturers also label reagent bottles using the "704 Diamond" system recommended by the National Fire Protection Association. This diamond is subdivided into quadrants that classify the chemical in terms of its flammability (top quadrant, usually red background), reactivity (right quadrant, usually yellow background), health hazard (left quadrant, usually blue background), and specific hazards (bottom quadrant, usually white background). Higher numbers (range = 0 to 4) within the quadrants indicate greater risks. In Figure 21.3, for example, we see the safety diamond for acetone. In some institutions, safety or fire officers may place such a diamond on the door of each room to indicate the presence of various types of chemicals in each location.

With few exceptions, all operations involving chemicals should be carried out in a properly oper-

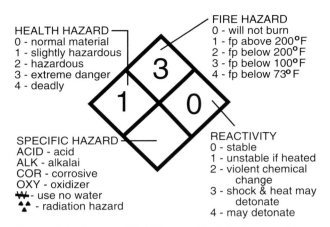

FIGURE 21.3   The "704 Diamond" of the National Fire Protection Association. This system is used to identify the various physical and health hazards of different chemicals. This is the type of symbol one would expect to encounter for acetone.

ating fume hood and while wearing gloves. This recommendation is frequently ignored during the weighing out of chemicals, because the fume hood may set up a draft that will affect the weighing and may actually spread the chemicals being measured. Nonetheless, it is usually possible to adjust the sash of the fume hood to minimize the drafts or to place a wind screen around the scale during the weighing process. Another possible approach is to place the scale inside of a glove bag and to carry out the operation inside of the plastic shroud. After weighing, the plastic or glassine disposable weighing dish should be rinsed into an appropriate waste container and the dish disposed of in the trash. Do not discard contaminated weighing dishes in the regular trash but place them in the containers designated for contaminated solid wastes.

Be aware that a number of chemicals may be readily absorbed through the skin (acrylamide, benzene, benzidine, carbon tetrachloride, dimethyl sulfoxide, dinitro-benzene and -toluene, xylene, dioxane, mercury, nicotine, nitrobenzene, phenol, picric acid, sodium cacodylate, o-toluidine, trichloroethane—to name a few) as well as by breathing the vapors they give off.

A newly-arrived chemical should be marked with the date received and initialed by the person who received it. Before using any chemical, read the labels twice to verify that it is the proper chemical. Some chemicals have similar names (sodium chlorite versus sodium chloride), but their effects are quite different. Do not use chemicals that are ambiguously labeled or whose labels have deteriorated. Check that the reagent is not too old to use. If no expiration dates are posted on the label, check with the laboratory supervisor to determine if it is still usable. If a stopper or lid is stuck, attempt to open it very carefully and seek assistance after several attempts. Stuck lids (especially on plastic embedding monomers) are normally caused by not wiping off the threads or cap of the reagent bottle after use.

When removing chemicals from their container, use a clean spatula for dry chemicals or a clean pipette for light liquids. Pour more viscous liquids, such as plastic monomers, slowly into the working container and take care not to generate airlocks in the bottle that could lead to splashes during pouring. If possible, wipe off the outside of the stock container using either a dry paper towel (for liquids) or a towel moistened in water (for dry reagents). But take care not to contaminate the contents or to damage the label during the wiping. The towels must be treated as hazardous waste and disposed of properly. Remove only the minimal amount needed and do not return any unused materials back to the stock bottle; dispose of any remaining chemical after determining that no one else needs it. Materials that have been stored in the refrigerator or freezer usually should be allowed to come to room temperature before opening to prevent moisture condensation on the contents.

Always pour contents slowly into water or into less concentrated solutions while stirring the solution. Remember that concentrated acids are added very slowly to the larger volume of water as the water is being stirred. Never look down into an open vessel unless it is empty; otherwise one risks splashing material into the face or coming in contact with dangerous fumes. Neither taste the contents of an unknown solution nor sniff it by placing the nose directly over the opening. Any sniffing should be done by gently wafting a tiny amount toward the nose using the cupped hand. During the sniffing, the container should be on a table and not held in the hand in case there is an involuntary response to a noxious substance.

Return the stock bottle to its proper location in the laboratory and inform the supervisor if any irregularities were noted (cracked or chipped container, wrong lid, discoloration, absorbed water, decomposition, nearly exhausted, etc.). If one has prepared a reagent, it should be completely labeled to indicate contents, date prepared, date expired,

and name of individual who prepared it. After using the chemicals, wash hands (and face, if necessary) and leave the work area clean and in proper order.

Storage of Chemicals

For convenience, most laboratories store chemicals in alphabetical order on shelves. Although this may be adequate in some instances, it is not the safest method of storage since it is possible to store incompatible chemicals (sodium nitrite and sodium thiosulfate, for example) in close proximity to each other. As a guide to the placement of chemicals, refer to Table 21.1.

Rayburn (1990) developed a set of recommendations for storing chemicals. A modified version of his recommendations is included in the panel that follows.

Storage Recommendations (Rayburn, 1990)

- Do not store incompatible chemicals in close proximity to each other.
- Store acids in a special cabinet designed for corrosives with nitric acid isolated from the others. The containers should rest on resistant trays capable of retaining the acids in case of leakage.
- Store flammable chemicals in a specially designed cabinet that is vented.
- Do not store chemicals in a fume hood that is used for routine work.
- Use shelving that is anchored to the wall rather than free-standing shelves.
- Do not store chemicals on top shelves or above eye level.
- Provide restraints at the edges of shelving to keep chemicals from falling off.
- Do not store chemicals on the floor. An exception is large storage drums that are kept in storage rooms.
- In general, store solids and liquids separately.
- Store very toxic chemicals or controlled substances in a locked area.
- Keep chemicals away from sunlight, heat, moisture, and flames.
- Minimize the amounts of chemicals stored in the laboratory. Use a specially designated, restricted storage area for large volumes.

TABLE 21.1   Some Incompatible Chemicals

Chemical	Incompatible with
Acetic acid	Chromic acid, nitric acid, hydroxyl compounds, ethylene glycol, perchloric acid, peroxides, permanganates
Acetone	Concentrated nitric and sulfuric acid mixtures
Anhydrous ammonia	Mercury, chlorine, calcium, hypochlorite, iodine, bromine, hydrofluoric acid
Ammonium nitrate	Acids, powdered metals, flammable liquids, chlorates, nitrates, sulfur, finely ground organics or combustible compounds
Aniline	Nitric acid, hydrogen peroxide
Arsenic compounds	Any reducing agent, certain acids
Azides	Acids, lead compounds such as lead nitrate or citrate
Calcium oxide	Water
Carbon (activated)	Hypochlorites, all oxidizers
Chlorates	Ammonium salts, all acids, powdered metals, sulfur, finely ground organics or combustibles
Cyanides	Acids
Flammable liquids	Ammonium nitrate, chromic acid, all peroxides, nitric acid, halogens (chlorine, bromine, etc.)
Hydrocarbons (butane, propane, benzene, toluene)	Chromic acid, peroxides, halogens
Hydrogen peroxide	Alcohols, acetone, organics, aniline, copper, chromium, iron, most metals or their salts, all combustibles
Hypochlorites	Acids, activated carbon
Iodine	Ammonia, hydrogen, acetylene
Mercury	Ammonia, acetylene, fulminic acid
Nitrates	Sulfuric acid
Nitrites	Thiosulfates
Nitric acid (conc)	Acetic acid, aniline, chromic acid, hydrogen sulfide, flammable liquids and gases, copper, brass, heavy metals
Oxalic acid	Silver, mercury
Oxygen	Oils, grease, hydrogen, flammable liquids, solids or gases
Perchloric acid	Acetic anhydride, bismuth and alloys, alcohol, paper, wood, grease, oils
Peroxides	Acids, all combustibles (avoid friction, store cold)
Phosphorous (white)	Air, oxygen, alkalis, reducing agents
Potassium permanganate	Glycerol, ethylene glycol, benzaldehyde, sulfuric acid
Sodium nitrate	Ammonium compounds
Sulfides	All acids
Sulfuric acid	Chlorates, permanganates

Source: National Academy of Sciences, National Research Council. 1980. Prudent practices for handling hazardous chemicals in laboratories. Committee on Hazardous Substances in the Laboratory. Washington: National Academy Press.

Some Chemicals Commonly Used in Electron Microscopy

In electron microscopy, there are some chemicals that are particularly dangerous for one reason or another. A common protocol for specimen preparation might include fixation in a formaldehyde/glutaraldehyde mixture buffered in cacodylate, postfixation in osmium tetroxide, dehydration in acetone or ethanol, exchanged with propylene oxide, embedding in an epoxy resin (possibly Spurr's), followed by staining of the sections in uranyl and lead salts. Since these are commonly used chemicals, they will be discussed in some detail.

Cacodylate salts are approximately 50% arsenic by weight and should be used only under very special circumstances, since they are very toxic, probably carcinogens, and will produce an allergic sensitization. They are readily absorbed through the skin (entry into the bloodstream is evidenced by a metallic or garlicky taste in the mouth). If cacodylate solutions evaporate to dryness (as along the sides of the bottle) the powders may become airborne. When possible, *substitute other salts* (phosphates, for example) for cacodylate in the production of buffers.

Formaldehyde is known to cause nasal cavity squamous cell carcinomas upon inhalation, as well as other types of skin cancers upon continued contact. In addition, formaldehyde (and probably glutaraldehyde) may sensitize one to the aldehyde itself or to molecules that have reacted with the aldehyde, leading to various allergies. Sensitization to formaldehyde may lead to an allergic reaction to an immunochemically related substance such as Urotropin, Mandelamine, Urised, or Methanamine (Fisher, 1976). One should minimize contact with the gas liberated from formaldehyde solutions by working in a fume hood and by wearing gloves.

Osmium tetroxide solutions are toxic, volatile, and highly irritating to the mucous membranes. Contact with skin and membrane surfaces must be avoided by using gloves and working in a fume hood. Wash contacted areas immediately with soap and water for 10 to 15 minutes. Spilled solutions may be reduced to metallic osmium using either sodium ascorbate powder or corn oil followed by cat litter to facilitate removal. The adsorbed solids must then be disposed of in an approved manner since they are still hazardous wastes.

Acetone and ethanol are both used as solvents, dehydrants, and general cleaning agents in all electron microscope laboratories. Both are flammable, toxic upon inhalation, and capable of facilitating the penetration through the skin of chemicals that have been dissolved in them.

Propylene oxide is highly flammable and has been shown to cause cancers of the nasal cavity as well as other types of cancers. If possible, less toxic solvents (ethanol, acetone) should be substituted in its place.

Picric acid should always be stored in aqueous solutions because the dried salts will explode if exposed to mechanical shock.

Most of the *embedding resins* are allergenic and some are probably carcinogens (epoxy components). The amines present in the epoxy resins are immunochemically related to aminophylline and ethylenediamine antihistamine so that allergic cross-reactions may result when these medications are used.

The *heavy metal salts* (lead, uranium) used commonly to enhance the contrast of sectioned materials are highly toxic, and uranium salts are alpha and beta emitters of radiation. Lead acetate and lead phosphate are suspected carcinogens (and uranium is probably also carcinogenic due to small traces of radioactivity likely to be present). These salts must be weighed in an enclosed environment, used in a manner that minimizes contact with the dissolved salts (especially if dimethyl sulfoxide has been added), and disposed of in accordance with local and federal guidelines.

Most of the chemicals used in *cytochemical reactions* are toxic or carcinogenic (lead salts, diaminobenzidine, toluene diisocyanate, etc.).

Disposal of Spent Chemicals

After use, most chemicals must be disposed of in accordance with specified local and federal guidelines. It is important to discriminate between spent or useless chemicals and those for which one has no further need. Most institutions have clearinghouses for donating unwanted chemicals to other researchers in the same or even in other institutions, if permitted. On the other hand, spent or otherwise unusable chemicals must be disposed of following certain procedures. Use the *minimum amount* of reagent feasible. This will not only conserve expensive chemicals, but also generate less waste to dispose of later. Use *less toxic substitutes* whenever possible, for example, phosphate instead of cacodylate buffer, LR White instead of Spurr's embedding resin.

Whenever possible keep the wastes in separate specially designated containers rather than mixing them together. This will not only prevent the combination of incompatible chemicals, but also greatly facilitates possible recycling of the chemicals.

Solid chemical wastes must be placed into individual, sealed containers, labeled as to content and amount, and sent to appropriate disposal agencies. Small amounts of nontoxic substances, such as sodium chloride, may be disposed of in the trash providing that this does not violate any local ordinances. Large volumes of even nontoxic substances should not be discarded in such a manner since they may be put into landfills and eventually contaminate the water supply.

Liquid wastes should be kept in separate, chemically resistant containers rather than mixed together into a single container. Leakproof caps must be provided, and the containers must be completely labeled. It may be possible to pour small amounts (less than 100 ml) of water soluble nontoxic liquids (certain buffers, ethanol, etc.) down the drain if followed by flushing with running water for several minutes. If the liquids have been contaminated with toxins (osmium, for example), then this is not permitted. Some communities do not permit pouring any flammable materials (including ethanol) down the drain. It may be possible for certain recyclers to purify or reuse certain liquids such as acetone, ethanol, or methanol; however, this is only possible if they are kept separate and notes are kept as to the presence of other components in the liquid.

Solutions containing heavy metals should be kept in separate containers and precipitated whenever practical. Uranyl and lead salts may be precipitated with phosphate buffers. Osmium tetroxide solutions may be mixed with corn oil until the oil blackens and the osmium is reduced. Carcinogens and very toxic materials should be placed in tightly sealed bottles or containers and then placed in a large, heavy duty outer container of absorbent materials adequate to absorb the contents of the inner container in case of leakage. Mark the container with the name of the contents, date, and any precautionary notes needed ("Danger: Carcinogen" or "Danger: Highly Toxic").

Used rotary pump oils should be stored in their original containers, labeled as used, and given to an oil recycler. Be careful, however, that the oils have not been contaminated with toxic materials such as epoxy resin components or heavy metals. Diffusion pump oils should be kept separate from rotary pump oils in their original containers, which are subsequently sealed, labeled, and taken to a disposal agency.

Embedding resins should be mixed and polymerized in an oven rather than disposed of as liquid wastes. A good way of dealing with epoxy resins (especially the toxic components in Spurr's resin) is to prepare the resin mixture in a large container (500 ml glass bottle, for example) without adding the catalyst. Dispense small amounts of the incomplete resin mixture into preweighed glass scintillation vials, cap the vials, and place them in a freezer. As needed, thaw a vial, add the appropriate amount of catalyst, and mix together in the scintillation vial. Any unused resin should be polymerized by placing the loosely capped vial into an oven. Collect used resin in a glass container and polymerize the resin after each experiment. When the container is filled with the polymerized layers, seal and label it, and dispose of the polymer in accordance with institutional policy.

Mercury should be picked up using special agents that chemically combine with elemental mercury and complex it. The complexed mercury is then sealed inside of a glass container, labeled, and taken to an approved disposal agency.

Safety Monitoring

A number of devices are available for monitoring exposure to formaldehyde, organic vapors, and mercury vapors. Although they are expensive and "after the fact" devices, they may be useful to determine if unsuspected exposures are taking place. Detector/alarms are also available for smoke, heat, radioactivity, natural gas, and even water (floods in the lab). These devices may be obtained from various safety supply houses and occasionally from general laboratory supply catalogs.

Cleaning Up Hazardous Spills

If materials that are only slightly hazardous have been spilled (weak acids or bases, oils, developers, etc.), one should absorb the spill using paper towels, newspapers, or spill-control pillows. The absorbed wastes can then be transferred into a plastic bag and placed into a container for solid wastes. The floor should be thoroughly cleaned with soap and water and the area dried to prevent slips and falls by other laboratory workers.

If elemental mercury has been spilled, try to contain the spill in a small area and warn coworkers not to walk over the spill. Use gloves and remove any sources of heat that might volatilize the mercury. Do not use ordinary vacuum cleaners; they will only aerosolize and spread the mercury all over the laboratory. If the mercury has gotten into porous surfaces (floors, lab benches, etc.), then special chemical inactivators must be used. One decontaminant, called "Hg-x" (available from most laboratory supply houses), is mixed with water and the contaminated area wiped with the mixture. Alternatively, zinc powder moistened with 5 to 10% sulfuric acid to form a paste is spread over the contaminated area and allowed to dry. Both chemicals complex the mercury to facilitate its removal. After the materials have been removed, the area should be checked for the presence of mercury using an appropriate monitor or "mercury sniffer." Several cleanings may be necessary to remove all of the mercury.

If acids or alkalies have been spilled, it is best to contain the spill using spill-control pillows designed for acids and bases. Small volumes of dilute acids/bases may be neutralized using commercially available spill-control products. It is not a good idea to try to neutralize large volumes of acids/bases (especially concentrated ones) since this may only liberate heat and lead to the splattering of concentrated acids/bases. One should wear heavy gloves and take care not to step on any of the spilled materials since they will be very slippery. After the spill has been absorbed with the appropriate agents, the residues should be placed into a plastic bag, labeled, and transferred into the chemical disposal bin. The area should be cleaned with soap and water and dried thoroughly.

Flammable or volatile chemicals pose a special problem. Immediately extinguish all open flames and shut down all sources of ignition (sparking equipment, incubators, etc.). Alert other workers present in the lab of the situation and seek assistance if necessary. Open any windows and the fume hood sash to help dissipate the fumes. Absorb the spill using spill-control pillows or other absorbent materials designed for volatile agents. Place the saturated pillow into a plastic bag and move the bag into a fume hood or outside of the main laboratory if safely possible. Return to the site and, after the remaining material has evaporated, clean up the contaminated area using soap and water if possible. If cleanup procedures are in doubt, mark the contaminated area with a warning sign (also place a warning on the door to the room) and check with the laboratory supervisor.

A *spill control cart* should be placed in a convenient location so that necessary supplies can be moved quickly to the contaminated area. Such a cart can be made using a standard laboratory cart outfitted with a number of provisions such as heavy gauge neoprene gloves, disposable plastic shoe covers or booties, rubber or vinyl aprons, disposable coveralls, safety eye goggles and face shield, disposable respirators (acid/gas, dust/mist), commercial acid and base neutralizer powders, mercury absorbers, spill-control pillows for acids/bases and volatiles, several plastic dustpans, whisk broom, squeegee with rubber blade, sponge mops, detergent, several plastic buckets, plastic bags, flashlight, paper towels, and warning signs.

Exposure to Chemicals

Exposure to a chemical may occur by direct contact with the hands, inhalation of fumes or powders, splashing into eyes or mucous membranes, and ingestion. If one contacts the chemical with the unprotected hand, then the immediate response should be to wash the hands using warm water and a strong soap such as Lava or a dishwashing detergent. Do not use alcohol, acetone, or other solvents since they may only facilitate absorption of the chemical through the skin.

Splashes into the eye, nose, or lips must be removed immediately using an eyewash station or by placing the face under a stream of warm running water. A mild soap may be used for the lips and nose, but only running water is to be used for the eye. It is important that the eye be held open in the stream of water and that contact lenses be removed to facilitate cleansing of the eye. Roll the eye during the rinsing, which should continue for 15 minutes. Seek medical attention immediately.

If a chemical has been spilled onto the clothing or other parts of the body, remove any contaminated clothing (modesty aside) or jewelry immediately and wash the body part with warm water and soap for at least 10 minutes. Contact with phenol, DMSO-containing solutions, or cacodylate acid is especially worrisome since these compounds pass through the skin and into the bloodstream rapidly. Dispose of the contaminated clothing in the same manner that chemical wastes are treated. Seek medical attention.

Inhalation of a toxic substance is serious since the substance is absorbed quite rapidly into the bloodstream. Those personally involved should leave the contaminated area and seek medical attention. Attempt to cough up as much bodily secretions as possible and rinse the mouth with water. Remove stunned or unconscious victims from the contaminated area taking care not to inhale any materials in the process yourself. It may be necessary to administer artificial respiration, but take care not to inhale any residual toxins that may be expelled from the victim. Leave warning signs on the door to the laboratory and indicate the contaminated area within the laboratory.

If materials have been ingested, rinse the mouth with warm water repeatedly. If corrosives were ingested, do not attempt to neutralize the chemical because this will only generate heat that may further damage sensitive tissues. Most physicians recommend diluting the corrosive materials using large quantities of water or milk. Do not attempt to induce vomiting if corrosives have been swallowed since this may rupture the esophagus.

If other toxic substances have been swallowed, drink large amounts of water or milk to dilute the toxin. Vomiting may be induced only if corrosives or hydrocarbons (petroleum distillates) have *not* been swallowed. Seek medical attention and be prepared to inform the medical team about the nature and approximate amount of chemical that was ingested.

The OSHA Standard

In 1990, the Occupational Safety and Health Administration (OSHA) established a set of rules in the Federal Register for dealing with dangerous chemicals in the workplace. Entitled *Occupational Exposures to Hazardous Chemicals in Laboratories*, this standard applies to all academic, clinical, and industrial laboratories. The purpose of the standard is not to issue new exposure limits, but to define a so-called *chemical hygiene plan*, or CHP, that establishes practices and procedures for the protection of the workers in a laboratory. An important part of the CHP is the requirement that laboratories develop a set of *standard operating procedures* (SOPs) that cover general safety precautions while using a chemical, exposure control measures (fume hoods, respirators, gloves, etc.), spill control measures, accident responses, and disposal methods to follow upon completion of the procedure. A chemical hygiene officer must be designated, and the employer is required to verify that fume hoods and other protective equipment are in proper operating condition. In essence, the employer is required to provide training and information on

- the chemical hygiene plan,
- protective measures the employees should take,
- specific hazards (health, physical, fire, etc.) associated with the chemicals,
- procedures or equipment used to detect the presence of the chemicals in the workplace.

OSHA has estimated employers' costs for compliance to the standard to be approximately $15 million, but that such compliance will result in at least a 10% reduction in illness and injuries caused by the misuse of chemicals in the workplace.

Fire Safety

Three elements are necessary to start a fire: *fuel, oxygen,* and an *ignition source.* Once a fire has started, heat generated by the burning process further decomposes the fuel into highly reactive free radicals that rapidly combine with available oxygen to generate even more heat. This results in an unrestrained *chain reaction* that will continue to increase in intensity as long as the three elements are available. A fire may be prevented or stopped by removing any of the three necessary elements and interfering with the chain reaction.

Preventing Fires

One may *restrict the fuel* by using only minimal amounts of flammable materials. For instance, instead of using a quart or pint container of propylene oxide or acetone during the embedding process, use 100 ml working containers and keep the larger stock containers in an approved safety cabinet. Limit access to the stock solvents to authorized personnel trained in the proper dispensing of the solvents.

Limit access to oxygen by keeping all containers capped when not in use. Especially with volatile solvents, make sure that the containers are closed immediately after dispensing the necessary amounts. When embedding, after transferring the solvent from the working container into the specimen vials, cap both systems and remove the working container from the work area after it is no longer needed. Limit storage of combustible materials in the immediate work vicinity to prevent the development of a fuel stockpile.

Keep ignition sources away from available fuels. Although it should be obvious that an ignition source never should be near combustible materials, researchers may become so engrossed in the procedure being followed that the ignition sources may be overlooked. For example, it may not be uncommon to use solvents in close proximity to electrically powered equipment such as stirrers, ultrasonic baths, or even ovens. Since solvent vapors may travel some distance, a spark or heating coil may ignite the vapor phase, which may then serve as a conduit back to the fluid phase. For example, it is a common but dangerous practice to degrease microscope parts in an ultrasonic bath in containers of volatile solvents such as acetone, chloroform, or ether. One should never store solvents in a standard refrigerator because a spark from the thermostat, compressor, or other relays may cause an explosion. Instead, a specially designed *laboratory-safe refrigerator* may be used to store volatile chemicals. Self-defrosting refrigerators cannot be safely modified to store volatile materials.

Even experienced individuals may occasionally make blunders that result in damage to the laboratory and trauma to the person. The author recalls an incident that occurred in his own laboratory during the routine maintenance of an electron microscope. A coworker was using a bunsen burner to flame platinum apertures some six feet away from a 300 ml open beaker of acetone. The vapors from the acetone drifted over to the flame, igniting the vapor trail back to the fluid phase. Literally in a flash, the acetone in the beaker burst into flames. In an attempt to cover the beaker to limit access to oxygen, it was tipped over and burning acetone was spilled along the countertop and down onto the floor. Not only was the countertop, floor, and a valuable camera damaged by the burning acetone, but the coworker sustained burns along his forearms and shin as he attempted to cover and stamp out the fire. In this instance, the fire was limited by the lack of additional fuel.

Stopping Fires

Once a fire has been started, proper measures must be taken rapidly to put it out by interfering with the chain reaction. Since it may not always be possible to remove the fuel from an established fire, fire fighting is done most often by limiting the source of oxygen and lowering the heat that sustains the reaction. This may be accomplished by using the appropriate fire extinguisher (Figure 21.4), depending on the class of fire in progress.

Classes of Fires and Proper Extinguishers

Class A Fires Wood, cloth, paper, or other common combustibles are involved. This type of fire is best extinguished with a pressurized tank usually containing either plain water or water in combination with wetting agents and aqueous film-forming foam (AFFF). This type of extinguisher is often found in corridors and in offices. A special fire blanket may also be used.

FIGURE 21.4 Fire extinguishers used for various types of fires. (*left*) Class A: aqueous solution, (*left*) Class B/C: carbon dioxide, (*middle*); Class B/C: dry powder (*right*). Not shown is the Class D type of extinguisher used for certain types of metals such as magnesium or sodium.

Class B Fires Flammable liquids are involved. These fires are extinguished by using pressurized carbon dioxide, halogenated compounds such as Halon (bromotrifluoromethane, bromochlorodifluoromethane), dry chemicals (ammonium phosphate, sodium/potassium bicarbonates, potassium chloride), or possibly by means of AFFF.

Class C Fires Live electrical circuits are involved either as the cause of the fire or as simply being near the fire. Use carbon dioxide, Halons, or dry chemical extinguishers. Aqueous-based or electrically conductive reagents must not be used to fight such fires.

Class D Fires Combustible metals such as magnesium, sodium, or potassium are involved. Extinguishers contain dry chemicals such as sodium chloride containing a thermoplastic binder that forms a solid suffocating crust over the fire.

When selecting a fire extinguisher, consider not only the class of fire but also whether or not valuable equipment may be damaged by the extinguisher. For instance, if a fire were to occur in the electron microscope room (or darkroom) where live electrical circuits are involved, select a Class C fire extinguisher. Unfortunately, dry chemicals may damage delicate electronic components by corroding and short circuiting them, while carbon dioxide may damage the components with the extreme cold generated. In this case, Halon would be the best choice.

To use the fire extinguisher, direct the spray at the *base of the fire*, not at the flames. Remember, the object is to suffocate and cool the chain reaction that is occurring at the level of the fuel. Slowly sweep the extinguishing agent over the base of the fire until all flames are extinguished. Even after the flames have subsided, continue to apply the agent to prevent a flare up of the fire.

Should a fire break out in a laboratory, prompt action must be taken to prevent the spread. Therefore, one must *have a plan* of procedures to follow.

Plan of Action for Fire Fighting

1. Evacuate all persons to a safe location. Leave the building, if necessary. Everyone should gather at a prearranged assembly point. This is important to determine if someone is still behind in the building. To avoid further injury, do not move injured persons unless the fire appears to be out of control.

2. If it is possible to extinguish the fire yourself, do so immediately. After determining the type of fire, select the proper fire extinguisher and position yourself between the fire and the room exit. Do not permit the fire to block your exit. After the fire is safely out, notify the fire department so that they may evaluate the need for further measures.

3. If it is not possible to quickly extinguish the fire, close off the room and activate the fire alarm to alert others in the building. Telephone the fire department, giving the exact location of the fire (street address, building name, floor, and room number). Indicate the nature of the fire (electrical, flammable liquids, etc.) and if dangerous chemicals are involved. Proceed to the assembly point and be prepared to meet the fire department with further directions as needed.

Electrical Safety

The electron microscope laboratory has numerous pieces of equipment, besides the microscope itself, that use electrical power. Any of these smaller pieces of equipment may produce a potentially damaging or lethal shock or serve to ignite flammable materials. Although such shocks and fires may result from defective equipment, most often they result from the unsafe practices of the user.

Darkrooms

The darkroom is a particularly dangerous location since electrical equipment is used in close proximity to water. It is important to keep such equipment well away from the processing sinks so they do not tumble into the sinks. Cords should be checked regularly for fraying, and they must not be allowed to dangle over the sinks. Safelights and timers must be secured so that they will not fall into the sink. Battery or spring-driven timers are much safer to use in the darkroom.

Darkroom users must be cautioned not to handle electrical equipment with wet hands and to keep the area around the sink dry. It is dangerous to contact the enlarger and the sink at the same time since the body may be providing an electrical bridge between a defective enlarger/timer and the grounded sink. If an energized item should drop into the sink, avoid the instinct to retrieve it. Instead, unplug it first or, better yet, cut off the circuit breaker and then unplug it. The retrieved equipment should be labeled as potentially dangerous, and it should not be used until it has been evaluated by a trained individual.

Vacuum Evaporators and Sputter Coaters

These systems use large amounts of current to thermally evaporate noble metals or high voltages to sputter metal targets. Although most contain safety interlocks, caution must be exercised during their use and maintenance. If electrodes are not properly attached and safe grounding procedures are not followed, a short circuit to the equipment case or operator may be established. Always follow the manufacturer's directions when working with such equipment. Always unplug the equipment when cleaning or servicing it.

Proper Grounding of Equipment

All electrical circuits in laboratories that use electrical equipment near water should have *ground fault interrupt devices* that will discontinue the electrical connection when an individual completes the circuit from the equipment ground to the earth ground (e.g., contacts a defective piece of electrical equipment and a sink at the same time). In addition, all electrical equipment should use a three-pronged or other properly grounded connector. One should not use two-to three-pronged adapters without ascertaining that a proper ground has been established (i.e., check with an electrician). Users of equipment must be shown the location and proper use of circuit breakers or fuse boards to cut off the power to equipment in an emergency.

Servicing of Electron Microscopes and Small Equipment

Except for individuals with the proper training, users should not attempt to service electrical components. If a problem is suspected, disconnect the equipment, mark it as potentially defective, and notify the laboratory supervisor. Although most microscopists do not service the electrical boards inside of the electron microscopes, be aware that electrical dangers exist even while doing routine maintenance on the microscope.

There is a story about a service engineer who was performing routine maintenance on a TEM. In a hurry to replace the filament, he attempted to remove the cathode without first discharging the high voltage built up in the gun. Two other observers who were being instructed by the first engineer witnessed a bluish white discharge arc from the electron gun over to the instructor who had grounded himself on the console of the microscope. Although visibly shaken, the instructor was able to continue routine maintenance of the microscope after a short discussion on electrical safety. In another similar incident, a service engineer claims to have had the soles of his shoes welded to the floor as he worked on an undischarged high voltage tank. In this instance, the engineer was so shaken by the incident that he transferred to an administrative position where the potential hazards were of a different type.

First Aid for Shock Victims

If contact with a live circuit has taken place, the first step should be to cut off the current at the breaker box. If this is not possible, separate the victim from the electrical source using a nonconductive material such as a lab coat, wooden chair, or rubber hose. Make sure that you do not become involved in the circuit (i.e., avoid water spills or making contact with grounded objects such as water faucets or other electrical equipment). If victims are unconscious, attempt to arouse them by gentle shaking. If

breathing has stopped or a pulse cannot be felt, apply cardiopulmonary resuscitation (CPR) following approved methods. At least one person in the laboratory, in addition to the instructor, should be versed in CPR. Send someone to call for medical help immediately.

Physical and Mechanical Hazards

The safe and proper handling of sharp objects and other potentially hazardous equipment or laboratory facilities are covered in some detail in the following section.

Sharp Objects

A number of objects used in electron microscopy may inflict cuts or lacerations if improperly handled. Used glass knives and pipettes should be kept in specially designated "sharps" containers or boxes. When the box is filled, contact the appropriate disposal officer.

Used razor blades and syringe tips also should be collected in a special "sharps" container and disposed of in accordance with local regulations. Contaminated syringes should not be placed in the box until they are sterilized. Housekeeping staff should be warned about the dangers associated with these disposal areas in the laboratory.

Critical Point Dryers (CPDs)

Since dangerous pressures build up inside of these units, it is important that they be shielded. One method is to place a Lexan sheet between the pressure chamber and the operator so that it will deflect material away from the face if the vessel explodes. Never look directly into the viewing window of such chambers, but use a metal mirror located between the Lexan sheet and the chamber. The minimum precaution that should be taken with CPDs is to wear a full face shield made of Lexan or polycarbonate.

The CPD should be checked for damage or bends to the high pressure lines from the tank to the chamber, and the seals should be checked and replaced as necessary. No lubricants should be used near the door or on any gaskets associated with the door seal.

Avoid inhalation of the fumes exhausted from such units (CO_2, amyl acetate, Freons) because they may asphyxiate or sicken the operator. The hoses used to vent gas from such units should be constructed of copper, stainless steel, or other cryogen-safe materials. The hose must be firmly anchored and vented into a fume hood.

Vacuum Pumps

Rotary and diffusion pumps may present several hazards. The rapidly spinning pulley and belts of older types of rotary pumps should be shielded to prevent entanglement of operators' clothing or pinching of the fingers. Both types of pumps become quite hot during operation, and it is very easy to burn oneself when working with them. The used oils from such pumps must be considered hazardous waste (potential carcinogens) and disposed of in a manner consistent with local pollution control guidelines.

Rotary pumps should not be permitted to exhaust into the room. Even mist traps placed on the outlet of rotary pumps are not 100% efficient all of the time. A better method is to exhaust the pumps into a fume hood or outside the building (Figure 21.5). A few diffusion pumps still use mercury that must be very carefully disposed of after consultation with experts.

Vacuum Evaporators

The major risks associated with such units are implosion of the vacuum chamber bell jar, electrical shocks, burns to the hand or damage to the eye caused by heated filaments, and trauma caused by contact with the belt or pulley of the spinning mechanical pump. Since most evaporators use a glass bell jar to enclose the chamber, it is essential to check the glass before each use for nicks, hairline fractures, or scratches. If any flaws are detected, do not use the jar and check with the laboratory supervisor. The bell jar should be surrounded by a plastic shroud or wire shield to contain glass fragments in case of implosion. Eye protection should be worn as an additional safeguard against flying glass. To avoid eye damage, never stare directly into the extremely bright, heated filament. Instead, use welder's goggles or a fogged photographic film to view the evaporation process. Ordinary sunglasses usually are not dark enough to protect the eyes. To avoid burns, always allow enough time for the electrodes to cool before reusing them.

Sputter Coaters

The risks associated with sputter coaters are less from implosion of the vacuum chamber than from electrical shocks from the high voltage sources used

FIGURE 21.5 Rotary pumps give off a carcinogenic oil mist that must be efficiently trapped by special filters placed over the pump outlet as shown on the two pumps on the left and right. A safer approach is to vent the fumes to the outside by means of pipes (*middle pump*). Absorbent paper with an impermeable plastic backing is placed under the pumps to contain any oil that may leak from the pumps.

to generate the plasma. Consequently, safety interlocks should be checked periodically to insure that one cannot activate the high voltage when the chamber is open. Inspect the glass chamber enclosure (bell jar) to determine if it has developed scratches or nicks along the sealing edge or if it is cracked and in need of replacement.

Ovens

Several problem areas exist when laboratory ovens are used. For example, a fire may start if volatile reagents such as acetone, propylene oxide, or ether are placed into most ovens. Most ovens vent directly into the working area rather than into a fume hood, releasing fumes and chemical vapors in the work area. Remember that some resin components (Spurr's epoxy resin, for example) are potential carcinogens and that most are allergenic. Such ovens should vent into exhausted fume hoods or directly outside the building. It is dangerous to use a vacuum oven to outgas resins since the more volatile components of the resin will be distilled over into the vacuum pump oil and possibly liberated into the room if the pump is not vented to the outside.

Cryogenic Gases and Vacuum Dewars

One of the obvious dangers of working with liquefied cryogenic agents is the potential for frostbite upon contact with cold liquid, gases, or chilled surfaces. Liquid nitrogen, helium, and carbon dioxide may all cause serious burns on contact with living tissues. Whenever dispensing such cryogens, properly insulated gloves should be used and a face shield (rather than goggles) should be worn. Sandals or open-toed footwear should be avoided due to the possibility of spilling cryogens onto the feet. Non-absorbent laboratory clothing should be worn since the liquids will soak into and wet most fabrics. Consequently, if liquid cryogens are spilled onto clothing, the clothing must be removed as rapidly as possible since the cryogen may soak the fabric, freezing it onto the skin.

An insidious situation exists when using liquid helium and storage vessels containing large amounts of liquid nitrogen. Such cryogens are capable of condensing oxygen and ozone into the containers. If the contaminated cryogens are then used in conjunction with propane in a jet-freezing procedure, then a very explosive mixture may be generated. Such procedures should be conducted in a properly functioning fume hood to prevent the accumulation of propane fumes in the cryogen, and the storage vessel should not be allowed to stand for long periods of time to minimize ozone/oxygen condensation.

Liquid carbon dioxide is frequently used as the transitional fluid in the critical-point drying procedure, as well as to cool specimens during freeze substitution. Since this liquid tends to solidify upon release from its container, one must take care that the dry ice does not block the outlets and gas lines.

All of these cryogens eventually will be converted into the gas phase with a tremendous increase in volume. Therefore, care must be taken not to use them in confined spaces or the air may be displaced. Since there are few warning symptoms of asphyxiation, one may be rendered unconscious and suffocate unless someone else is nearby. A fume hood would prevent this from happening.

The vacuum Dewars used to contain the various cryogens are normally made of glass and are subject to breakage, especially when chilled with the cryogen. Consequently, the Dewars should be totally encased in plastic or metal (or at least wrapped in two layers of plastic electrician's tape) to prevent injury by glass fragments if one were to implode. Never use vacuum vessels purchased in variety stores in place of true Dewar flasks since the thinner, less tempered glasses will break upon repeated freezing and thawing. The Dewars should be filled slowly with the cryogens and capped lightly when not in use to prevent the condensation of water inside of the flask. If possible, store the vessels in an inverted position to permit condensed water to run out. Thermal and mechanical shocks and trapped water are the two major causes of failure of Dewar flasks. When filling or dispensing liquid cryogens, eye protection must be worn since the cryogens always boil vigorously when poured into warmer vessels.

Compressed Gas Safety

There is a classic story, reported originally in the May 1969 Chemical Section Newsletter of the National Safety Council, that dramatizes the power that compressed gas cylinders may unleash. In Humphreys' 1977 rendition of the incident:

> Six 220 cubic foot cylinders, part of a fire extinguishment system, had been moved away from their wall supports to allow painters to complete the painting of the area. While moving them back into position, it was noticed that one cylinder was leaking. The painter had the cylinder leaning against his shoulder, and was attempting to scoot it across the floor. At this time, the valve separated from the cylinder and projected backward, hitting the side of a stainless steel cabinet.
>
> The man suddenly found himself with a jet-propelled 215 pound piece of steel. He wrestled it to the floor, but was unable to hold it. The cylinder scooted across the floor, hitting another cylinder, knocking it over and bending its valve. The cylinder then turned 90 degrees to the right and traveled 20 feet, where it struck a painter's scaffold, causing a painter to fall 7 feet to the floor. After spinning around several times, it traveled back to its approximate starting point, where it struck a wall.
>
> At this point, the cylinder turned 90 degrees to the left and took off lengthwise across the room chasing an electrician in front of it. It crashed into the end wall 40 feet away, breaking

loose four concrete blocks. It turned again 90 degrees to the right, scooting through a door opening, still chasing the electrician. The electrician ducked into the next door opening, but the cylinder continued its travel in a straight line for another 60 feet, where it fell into a truck well, striking the truck well door. The balance of the cylinder pressure was released as the cylinder spun harmlessly around in the truck well area. The painter who fell from the scaffold received multiple fractures of his leg.

The above cylinder contained pressure of about 900 psi. One wonders what the result would have been if the cylinder had been pressurized to 2,200 psi.

A photograph identifying the various parts of a gas cylinder and regulator valve is shown in Figure 21.6. The cylinder may have a series of numbers stamped into the metal. For instance, the designation "DOT 3AA 2300" refers to the Department of Transportation specification of the tank for 2,300 psi at 70°F. Directly under this may be a serial number such as "Z48976" to identify the specific tank. Other identifiers, such as "10 PB 89+," indicate the most recent inspection date or date of manufacture. In this case, an inspector with the initials PB inspected the tank in October of 1989. The plus sign reveals that this tank may be filled to 10% over the 2,300 psi service pressure. In addition, the tank must be clearly marked as to the contents and whether or not the gas is hazardous.

All pressurized tanks are equipped with a *safety device* (S) that permits gas to escape from the cylinder in the event that excessive pressures develop inside the tank. A *hand wheel* (H) is used to open or close the valve, but not to adjust the pressure of the escaping gas. An *outlet connection* (O) is provided for attaching a pressure regulator valve to the main cylinder valve. The threads of these connections are designed to accommodate special regulators or attachments and to prevent the connection of improper accessories to the tank (e.g., the wrong type of regulator). A two-stage *pressure regulator valve* (PR) is shown in Figure 21.6. The gauge nearest the cylinder indicates the pressure inside of the main tank, while the second gauge indicates the discharge pressure of the gas. The discharge pressure is regulated by the handle on the large portion of the regulator. In most of these valves, the discharge pressure may be increased by turning the valve (see arrow) clockwise, rather than counterclockwise as one might expect. A *final delivery valve* (not present in Figure 21.6) may be present to turn the flow of gas from the regulator on or off. It has no effect on the dispensing pressure, but only on the rate of discharge. All tanks have a *cylinder cap* (not shown) that is screwed over the cylinder valve to protect it when the tank is not in use or during shipping. This cap must be securely screwed in place on all tanks that do not have a regulator or on tanks that are not in use.

Cylinders must be handled properly to prevent disastrous and potentially fatal accidents. When a cylinder is received, it must be checked for damage and for proper contents. An in-house "necklace" label should be placed over the cap of the tank to indicate its contents, that it is full, the date received, and the initials of the person receiving the tank. Do not attempt to open the main tank valve until the tank has been attached to a pressure regulator or other piece of equipment. Otherwise, the tank may give off a high pressure discharge of gas that cuts like a knife. Transport the cylinder in a specially designed gas tank transport cart with the cap in place and with the tank secured to the cart with chains or ties. Never attempt to move the cylinder by rolling it or sliding it along the floor as it may drop to the floor and damage the main cylinder valve.

Gas cylinders must be stored in a dry, well-ventilated area that does not get excessively hot or cold (−29 to 52°C). Do not store incompatible gases (oxygen and hydrogen, for example) in close proximity. The tanks must be securely chained or held in place to prevent their falling over. When removing

FIGURE 21.6 Gas cylinder with cylinder valve and pressure regulator valve (PR) attached. A description of the various parts is found in the text. H, hand wheel for opening tank; O, threaded inlet for attaching PR valve; S, safety pressure relief valve; arrow, handle for adjusting pressure of gas from tank.

tanks, use the special cart to transport them to the location and immediately chain or secure the tank into position rather than leaving it on the cart.

When attaching the tank to the apparatus or pressure regulator, use the proper tool (wrench) rather than a makeshift device such as a pliers. It is best to have such a tool chained near the area where the tank is used. The apparatus should be securely attached to the main valve so that it will not shift position when lateral pressure is applied to the device. After attaching the device to the main cylinder valve and before opening it, verify that the attached device or pressure regulator is closed off. Now, partially open the main cylinder valve and check for any leaks between the main valve and the attached device. If any leaks are found, close the main valve immediately and seek assistance from the laboratory supervisor. If no leaks are found, check the tank pressure (it should be neither too high nor too low for the particular gas) and verify that the discharge pressure valve is closed or that the final dispensing valve is closed.

Adjust the flow of gas using the pressure adjustment valve (if present) and open the final valve. If any problems are encountered during the use of the compressed gas, turn off the main tank valve first before attempting to resolve the problem. Wear appropriate eye, hand, and body protection when working with the tanks and use no lubricants on the tank or the pressure regulators. It is best to discontinue using the tank before it is completely empty to prevent contaminants from entering the tank during the refilling step at the suppliers. Remove the original necklace label and replace it with another label indicating that the tank is empty. Have the empty tank removed as soon as possible to prevent others from using it and to remove clutter from the laboratory. The tank must be transported back to storage on the tank cart with the cylinder cap in place.

Microwave Ovens

Electron microscopists are making increasing use of microwave ovens for fixation, embedding, and staining purposes as well as to warm up various reagents. Just as in the home, such ovens should be regularly checked to make sure the door seal is intact using inexpensive monitoring devices. The oven should not be used to heat foods or drinks, since the inside may be contaminated with toxic chemicals. The ovens should be kept clean to prevent the carryover of reagents from one experiment to another. Solutions containing azides, picric acid, or other potentially explosive chemicals should not be used inside of the ovens. For example, if the oven had been contaminated during lead citrate staining and another researcher were to use the oven to warm up a solution containing the preservative sodium azide, it may be possible to generate lead azide, the explosive chemical used in blasting caps. In order to contain dangerous chemicals, it is advisable to place them inside of a microwave-safe receptacle inside of a sealable plastic bag. Laboratory microwaves should be exhausted into fume hoods or outside the building.

CAUTION: Verify that the bag will not expand and explode due to the evolution of vapors. The bag and container should be opened in a fume hood.

Radiation

Most modern electron microscopes are extremely well shielded against X-ray leakage. Nonetheless, an instrument should be inspected for leakage with a beta-gamma counter if the instrument has been recently installed or moved, after additional detectors have been installed, or after major modifications to the instrument. Side-entry specimen stages in the TEM may leak X rays during specimen exchange if the filament is left on. Although older instruments are more likely to show leakage, newer instruments should be suspected until proven otherwise. Annual monitoring of an instrument is recommended even if no major changes to the instrument have taken place during the year. This may be achieved by operating the instrument at the highest accelerating voltage with the apertures slightly out of alignment and the beam expanded. Check the gun, column, specimen chamber, and viewing screen especially in the areas of gaskets. Essentially, zero counts above background should be expected. Leakage greater than 0.5 mR/hr measured at 5 cm from the microscope is considered significantly hazardous by the Electron Microscopy Society of America Radiation Committee of 1973. A vacuum evaporator equipped with an electron beam gun type of electrode may generate significant levels of X rays and should be evaluated before extensive use.

All uranium salts are radioactive to various degrees since several radioisotopes may be present (^{238}U, ^{235}U, ^{234}U). Even though most uranyl salts are said to be radioactivity "depleted" since the last two isotopes have been removed, one should still consider the salt to be radioactive until checked with an appropriate counter (alpha and beta particles are emitted). Most workers probably will be alarmed by the high count rates.

The toxic effects of the heavy metal uranium are as much of a concern as the radioactive emanations. Soluble uranium salts (uranyl acetate, nitrate, etc.) as used in electron microscopy are less worrisome than the insoluble compounds (UO_2 and U_3O_8) since soluble salts are more rapidly cleared from the body. Most ingested uranium is excreted in the urine within 24 hours (Hursh and Spoor, 1973) with only 2% to 10% becoming incorporated into bone. Therefore, use caution when preparing stains so that the powders are not inhaled; prepare the stains in a fume hood and wear rubber gloves. All spills must be cleaned up immediately using a wet paper towel. One should check with regional authorities on the proper ways of disposing of uranyl salts since some localities consider it to be radioactive while others classify it only as a toxic waste. To reduce volume, one may precipitate the aqueous uranyl solutions using a phosphate buffer and collect the sediment, uranyl phosphate, for disposal.

Centrifuges

The major problems associated with using centrifuges have to do with the improper balancing of centrifuge tubes, which results in broken tubes, contamination of the work environment with the tube contents, and potential damage to the equipment. If proper operating procedures are followed, such accidents may be avoided.

Before using the centrifuge, check if it is in proper operating condition by asking other users and by noting the log book. Make certain that the proper rotor is being used. Different brands of rotors may resemble each other, and it may even be possible to install the wrong rotor on the centrifuge. In addition, make sure that the rotor is rated for the speeds that you desire. Verify that the centrifuge tubes are proper for the rotor and that they can withstand the g forces that will be generated and the chemicals that will be placed in them. If a swinging bucket rotor is used, make certain that the buckets are properly attached, balanced, and free to swing out during the centrifugation. The tubes should not be so long as to interfere with the free swing of the buckets. Use only centrifuge tubes, not test tubes, that are safety rated for the g forces used and nonreactive with the chemicals being centrifuged.

Stand clear of the rotor as it is accelerating or slowing down. Take care to keep long hair or jewelry away from the spinning rotor. It is very poor practice to slow down the rotor by using one's hand or other object pressed against the rotor. Clean up any spills inside of the centrifuge and especially in the rotor. If contaminated, follow proper procedures to clean the rotor and store it in the proper manner. Note any unusual noises or operational problems in a log book (or attach a sign to the equipment to alert other users).

Pipetting

The use of mouth pipetting is strictly forbidden in all laboratories. The danger of ingesting toxic liquids or fumes from the liquids is simply too great to risk such a procedure. There are numerous bulbs and other devices for pipetting that are always available in the laboratory (Figure 21.7). If such devices cannot be found, ask for assistance and wait until one is found before proceeding. The author recalls an incident in which a senior researcher was mouth pipetting some 2% glutaraldehyde when the phone rang, startling the researcher. Even though the small volume of fixative was quickly expelled from the mouth, followed by rinsing with water, the researcher still sustained temporary tissue damage to the soft palate.

Volatile liquids have a tendency to expand when taken into a pipette for the first time. This may result in the expulsion of liquid from the tip of the pipette. One should be prepared for this phenomenon and take care not to point the pipette in the direction of another researcher.

Glass pipettes should be disposed of in a proper manner when broken or after use. Unless contaminated with toxic chemicals, they are best placed in a "sharps" container with other broken glass and sent to a recycler.

NOTE: Recyclers may not accept Pyrex types of glass, only softer varieties. Check with the recycler first

---

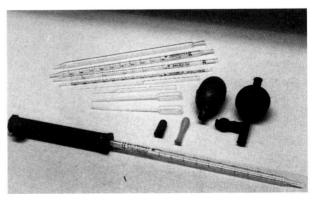

FIGURE 21.7 Various types of pipetting aids that may be used in the electron microscope laboratory.

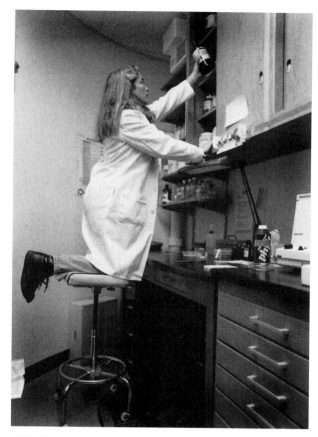

FIGURE 21.8 Reach for objects placed high on shelves using the proper safety equipment. Do not use regular stools or chairs for such purposes.

Falls

Although less dramatic than other accidents, falls are probably second only to cuts and burns as major types of accidents in the laboratory. Cluttered laboratory aisles and liquids spilled on floors are common reasons for falls in laboratories. Standing on chairs or other unsafe surfaces also results in many accidents each year (Figure 21.8). Surprisingly, nearly all laboratories in which these incidents occurred had ladders or stools that should have been used for such procedures.

Training and Orientation Programs

Most electron microscopists recognize the importance of proper training in the safe manner to work in the laboratory environment. Most have an established set of rules that define the procedures associated with working in a particular laboratory. Unfortunately, simply posting the rules in the laboratory will not always ensure that everyone is aware of, let alone willing to comply with them. Some formal training is needed.

In the electron microscopy courses given at the author's institution, the first period is spent instructing the students in such basic topics as:

- location of exits and routes to follow in various situations
- location and proper use of fire extinguishers and first aid kit
- proper methods of handling and disposing of chemicals
- safe operation of all equipment
- electrical safety in the darkroom
- the use of fire alarms and telephones to alert others
- importance of proper laboratory attire
- the operational rules for the laboratory

It is important to incorporate an awareness of safety consistently throughout training in electron microscopy. In addition to having a presentation on overall safety, it is important to reinforce safety concepts each time a new area is covered. As modified from the safety book by Rayburn (1990), several questions to ask each time *before* embarking on a procedure are:

1. What are the potential hazards associated with this procedure and the materials being used in it?
2. What can one do to minimize the risks in this procedure?
3. What are the *worst* events that could occur during the procedure?
4. If an incident does occur, what is the plan of action?
5. What safety equipment is needed? Is it nearby, operational, and does everyone know how to operate it?

A regular safety inspection is necessary to ensure compliance with established rules and to spot any neglected dangerous situations before an accident occurs. A checklist, such as the one shown in Figure 21.9, is useful for such inspections because it prevents one from overlooking some areas.

Safety Inspection Check Sheet

LOCATION _____ INSPECTOR _____ DATE _____

A check mark indicates problem areas that were located during the recent inspection of your electron microscope laboratory. Please correct the areas indicated and notify the Safety Office when you are ready to be reinspected.

Chemical Safety
____ Incompatible chemicals stored together
____ Outdated chemicals in use
____ Improperly labeled reagents
____ Excessive amounts of flammables in lab
____ Acid cleaning solutions improperly used
____ Liquids stored too high on shelves
____ Chemicals stored in fume hoods
____ Chemical wastes not disposed of properly
____ Storage shelves crowded or dangerous
____ Chemicals not properly inventoried
____ Containers defective or not intact

Equipment Safety
____ Fume hood not operating properly
____ Refrigerator not explosion safe for the reagents present
____ Improper electrical grounding
____ Refrigerator not labeled to indicate presence of hazardous materials
____ Equipment dirty and not in good working condition
____ Vacuum evaporators, desiccators, or Dewars not properly shielded
____ Belt guards not present on rotary pumps or other equipment
____ High voltage areas of electron microcopes and other equipment not properly marked as such
____ Gas, vacuum, air, and water hoses not securely clamped
____ Microscopes not certified safe for X rays

Fire Safety
____ Fire extinguishers improper for site; not operational; not readily accessible
____ Electrical appliances unsafe
____ Wiring problem: frayed, improperly used extension cords
____ Open flames left unattended
____ Smoking permitted in dangerous area
____ Exits not clearly demarcated
____ Obstructions in exit route
____ Electrical circuits overloaded

Physical/Mechanical
____ Fume hoods cluttered and used for storing chemicals
____ Food and drink stored in lab refrigerators

____ Gas tanks not secured properly
____ Precautionary warning signs not posted near equipment (critical point dryers, evaporators, microscopes, centrifuges)
____ Refrigerators cluttered, need cleaning, defrosting
____ Improper disposal of sharp objects
____ Cleaning needed around benches, sinks, hoods, storage shelves
____ Floors: need cleaning, unsafe
____ Equipment not provided for safe reaching onto high shelves
____ Radioactive or biohazardous materials used inappropriately
____ Dangerous areas, activities not so marked or explained

General
____ Emergency shower, eyewash not marked or readily accessible
____ Protective equipment (eyes, gloves, respirator, whole body) not available or in use when needed
____ Emergency telephone numbers not posted near each phone
____ Safety equipment (spill pillows, radioactivity monitors) not available
____ Smoking, eating, drinking in lab
____ Improper disposal of needles and syringes
____ Mouth pipetting
____ Gloves not removed before leaving contaminated area
____ Lab coats not worn when needed
____ Chemicals mixed in open areas
____ Spilled materials not cleaned up
____ Reagent containers not properly labeled
____ Shelves overloaded or dangerously stocked (heavy, dangerous materials on top shelves)
____ Books, notes used in areas near dangerous reagents
____ Long hair, jewelry worn during experimental procedures
____ Electron microscopes not checked for radiation leakage
____ No training provided for new individuals
____ Material safety data sheets not posted or readily available

Other Hazards _____
_____

FIGURE 21.9 A safety inspection sheet useful for evaluating your laboratory work environment. Areas needing improvement may be marked on the sheet. Microscopists may wish to xerox the page for use in an inspection.

13. Discuss how two condenser lenses are used to vary the amount of illumination striking a specimen. Specify the functions of each lens.

14. What is depth of field and depth of focus? Why are they important in electron microscopy?

15. What are anticontaminators? Describe the function of anticontaminators in two very different locations in the electron microscope.

16. What is alignment? Why must one align an electron microscope?

17. Outline the general steps that must be done in the alignment process.

18. Discuss the following terms: translational lens movement, lens tilt, optical axis, current and voltage centration.

19. How may one correct astigmatism in an electromagnetic lens? Discuss the design of stigmators.

20. Discuss the four major operational modes of the transmission electron microscope. List the conditions necessary to achieve each of the modes.

21. Why is it necessary to calibrate the magnification and determine the resolving power of an electron microscope? Discuss at least one way to determine each of these capabilities.

22. Describe one method to simultaneously determine the magnification and resolving power of an electron microscope.

23. Make a photocopy and fill in the table dealing with image defects, symptoms, causes, and cures.

QUESTIONS FOR CHAPTER SIX TABLE

Imaging Defect	How Recognized	Possible Causes	Cure
Astigmatism			
Barrel Distortion			
Chromatic Aberration			
Chromatic Change in Magnification			
Illumination Shifting			
Image Drift			
Low Contrast			
Pincushion Distortion			
Poor Resolution			
Specimen Contamination			
Spherical Aberration			
Questionable Magnification			
Vignetting of Image in EM			

Chapter 7

1. Compare the SEM to the TEM and light microscopes in terms of: types of images obtained, resolution capabilities, types of specimens examined, specimen preparation procedures, use of vacuums inside the instrument, lenses, mode of viewing and recording.

2. Discuss the concept that the lenses of the SEM are not image-forming lenses.

3. How is magnification varied in the SEM? What effect does this have on spot size?

4. Discuss considerations for operating the SEM in the high resolution mode.

5. Is it possible to obtain both high resolution and great depth of field? Discuss your answer.

6. Compare the types of images obtained in the secondary versus the backscattered imaging mode of the SEM.

7. Consult Chapters 2 and 3 and list the similarities and differences in the processing of specimens for TEM and SEM.

8. What factors will influence resolution in the SEM?

9. Of what value are stereomicrographs?

10. Discuss factors that contribute to the three dimensionality of the SEM image.

11. Discuss why astigmatism, chromatic and spherical aberration degrade resolution of the SEM.

12. What is "noise" and how can it be minimized in the SEM?

Chapter 8

1. In the darkroom, produce a contrast series and exposure series similar to those shown in this chapter. How much variability in contrast and exposure is acceptable in your eyes? Obtain other opinions.

2. Review the previous chapters dealing with the transmission and scanning electron microscopes and generate a list of all possible ways that one may increase contrast in an electron micrograph (i.e., instrument adjustments and darkroom procedures). How can one decrease contrast?

3. How does the exposure of a photographic emulsion to photons differ from the exposure of the emulsion to electrons? Consider the terms speed, resolution, contrast, and efficiency in this discussion.

4. What factors will affect the speed and resolution of an emulsion? Are high speeds and high resolution mutually exclusive when referring to electron micrographic emulsions?

5. Why is an underexposed electron micrograph often mistakenly said to be "grainy"?

6. Why is it better to err on the side of overexposure versus underexposure of the negative?

7. Suppose you examine one of your TEM negatives and determine that the margins of the film (which are normally clear) have a gray cast to them. What might have caused this grayness?

8. How can one determine how much a negative was enlarged on a print? How does one determine magnification of the final print?

Chapter 9

1. When should immunocytochemistry be used in place of other localization methods?

2. Divide into two groups. One group will give a technical problem with one aspect of immunocytochemistry, and the other group will respond on how the problem may be overcome or how another approach will help. Reverse roles.

3. From the list of current localization reports given above, select two to describe the purpose, methods, results, and impact of the experiment.

4. How do preembedding labeling and postembedding labeling differ?

5. Obtain the current issue of *Journal of Histochemistry and Cytochemistry* and review the approaches to current immunocytochemical localization experiments.

6. Describe how antibodies are obtained for immunocytochemistry.

7. Why is protein-A sometimes called a pseudo-immunocytochemical tag?

8. What is the major difference between the direct and indirect methods of immunocytochemical labeling?

9. How do the various immunocytochemical tags appear under the electron microscope?

Chapter 10

1. How is a sound knowledge of enzyme histochemistry an important preparation for understanding enzyme cytochemistry?

2. Review two recent reports using enzyme cytochemistry. Examine the methods section to determine the basis for the localization, and outline the procedure.

3. In the protocol described in Chapter 10 (under "A Typical Protocol") for demonstration of acid phosphatase activity, why was phosphate buffer not utilized in the incubation medium? How would one determine which buffers would be suitable?

4. Are methods, other than enzyme cytochemistry, available to localize enzymes? Discuss the reason for your answer.

5. Could you envision the use of an enzyme to localize a substrate at the electron microscope level? If so, outline a method by which this could be accomplished.

Chapter 11

1. What is the difference between an autoradiograph and an ordinary photograph taken with a handheld camera?

2. What kinds of information does autoradiography provide that make it a suitable technique to answer biological problems?

3. What are the basic steps in performing autoradiography?

4. Autoradiography, at the electron microscope level, is not in widespread use. What factors limit the technique's usefulness and popularity?

5. Obtain one or two of the recent reports (see Chapter 11 for a partial list), and describe the details of the purpose of the experiment, the methods, the results, and the significance. Discuss whether some other technique would have answered the question.

Chapter 12

1. Consider how the inability to use colored dyes at the electron microscope level has limited the number of techniques available for electron microscopy.

2. Obtain a recent volume of the *Journal of Histochemistry and Cytochemistry* and categorize the basis for each localization technique used (e.g., immunologic, enzymatic, autoradiographic, etc.).

Chapter 13

1. What is the difference between obtaining relative and absolute stereological data? If one wishes to obtain absolute data, then how does one perform stereology as compared with obtaining relative data?

2. What are the different parameters measured in stereology, and what test system is used for each? What formula is used for each parameter?

3. Choose a body tissue other than the one illustrated, and compose a chart for the various compartments through the organelle level.

4. Two key, but difficult, procedures in stereology are the determination of nuclear diameter, especially in nonspherical cells, and the determination of section thickness. Summarize how to perform these procedures.

5. Design a *detailed* protocol to determine mitochondrial volume (V) in a cell type of your choice. Consider what must be done at each level of the experiment.

6. How would unbiased sampling be obtained in question 5?

7. How does one determine if sufficient data has been gathered from a test system?

8. Take an $8\frac{1}{2}$ " $\times$ 11" sheet of paper and draw about thirty irregularly shaped enclosed structures that will occupy about $\frac{1}{3}$ to $\frac{1}{2}$ of the surface area. Photocopy the paper twice and use one sheet for point counting, another for digitizing (if a digitizer is available), and another for cutting and weighing to determine volume density. Compare the results from the three methods.

9. Obtain one or more of the recent references cited previously and be able to show how the final results were obtained. Pay particular attention to the methodology section of the reference.

10. Figure A.1 is a computer drawn illustration of a hypothetical sectioned cell. Using the formulas provided in Chapter 13 determine the volume density of the square organelles, the triangular organelles, and the oval structures. To do this you must prepare an appropriate test system.

11. Using the information obtained from question 10, apply the formula to obtain the volume of the organelles and/or structures assuming that the cell volume is 2,000 μm^3. Do your values match closely with those obtained by your classmates/coworkers? What can be done with the test system and the sample size to make them match even more closely?

12. Figure A.1 shows a hypothetically sectioned cell. Using the formulas provided in Chapter 13 determine the surface density (S_v) and the surface (S) area of the cell membrane and the membrane-bound internal organelles shown as triangles. Assume the volume of the cell has been predetermined to be 2,000 μm^3. To do this task you must first make an appropriate grid test system utilizing the information provided in Chapter 13.

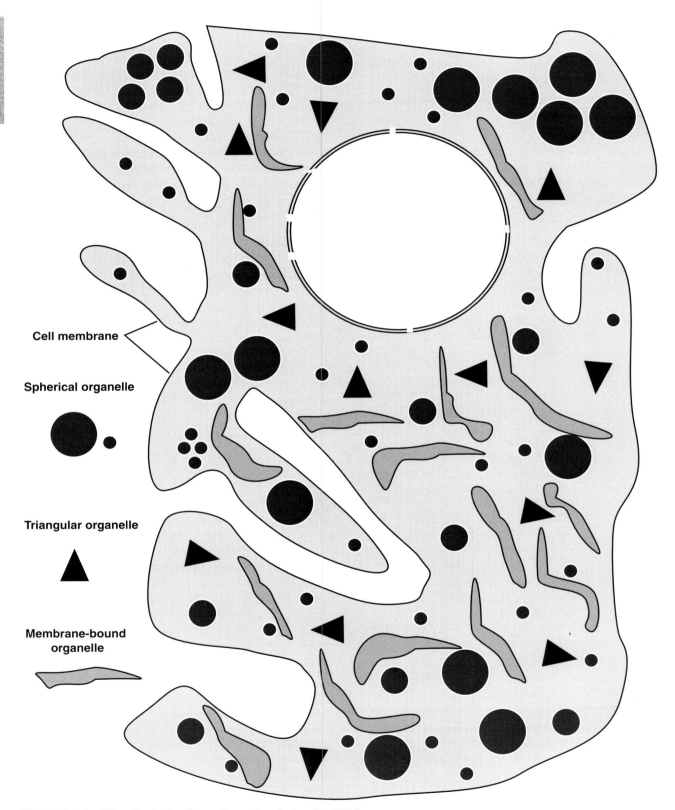

Cell membrane

Spherical organelle

Triangular organelle

Membrane-bound
organelle

FIGURE A.I   Hypothetical cell seen in sectional view by TEM.

Chapter 14

1. Rotate freeze fracture micrographs in Figures 14.15 and 14.16 and describe the three-dimensional features. Do they appear different when rotated from their original orientation? Why are arrows sometimes placed on micrographs to indicate the direction of shadowing?

2. What is the difference between freeze fracturing and freeze etching? Would it not be better to say that etching is an additional step in the freezer fracturing process?

3. List the steps necessary to produce a replica.

4. What rules must be followed to determine membrane faces?

5. What are complementary replicas?

6. Select two of the recent publications provided in this chapter and summarize them.

7. Have someone else cover over the labeling and legends to Figures 14.15 and 14.16. Use the instructions provided in the chapter to determine areas of the cytoplasm and the membrane faces.

8. In Figure A.2 (a) determine the orientation of the micrograph, (b) identify the membranes that have been numbered, and (c) provide membrane face designations for each membrane.

Chapter 15

1. Could a standard transmission electron microscope be considered an analytical instrument? Explain your answer.

2. Review the designs of the various (secondary, backscattered, and transmitted electron, cathodoluminescent) detectors used in SEM and TEM instruments. What types of information may be obtained from each detector type?

3. Compare the designs of the standard SEM and TEM to the STEM instrument. What are the capabilities of each instrument in terms of: resolving power, accelerating voltage, magnification, ability to be fitted with analytical detectors, and types of specimens that may be examined?

4. Outline the events that take place (on the atomic level) to yield an X ray for analytical purposes.

5. Compare characteristic X rays to continuum X rays in terms of: mode of generation, energy levels, information obtainable, and mode of detection.

6. Diagram and explain the basic features of an EDX unit versus a WDX unit.

7. What are the strong points and weak points of each of the two types of X-ray detection systems (EDX, WDX)?

8. Why is it not possible to obtain both excellent structural preservation and accurate retention of intracellular ions?

9. Describe a procedure that will allow adequate ultrastructural preservation and permit accurate localization of intracellular ions using conventional procedures.

10. Discuss why quantitative microanalysis is so difficult in bulk versus thin sectioned specimens.

11. Discuss the basic principles involved in electron energy loss spectroscopy.

12. Compare the use of X-ray microanalysis and EELS for the detection of light elements (atomic number under 11).

13. Explain the basic phenomenon involved in electron diffraction.

14. Tell how you would prepare a suspension of cells that contained water insoluble crystals for examination by electron diffraction. It is important that the cell ultrastructure be as well preserved as possible to show where in the cell the crystals are located.

15. Why must one check the camera length of a diffraction camera on a regular basis?

16. Calculate the camera constant for the electron microscope system that obtained the data in Table 15.2 (the gold diffraction standard). Make your measurements from the gold diffraction pattern shown in Figure 15.28.

17. Suppose you are given unknown biological crystals taken from the spleens and brains of individuals suffering from Alzheimer's disease. Tell how you would use analytical techniques to (a) determine if the two types of crystals were the same, (b) determine the chemical identity of the crystals, (c) determine if aluminum is present in the crystal.

Chapter 16

1. What are the advantages of using a high voltage microscope? How has the high voltage microscope contributed to our knowledge of biological structure?

2. Read the papers published by Porter in 1945 and by Wolosewick and Porter published over 30 years later (1976). Compare them and discuss how the

FIGURE A.2 Micrographs for self-interpretation. Identify each numbered surface by giving the name of the membrane and its faces from the following list: E-face of the inner leaflet of the nuclear membrane, P-face of the outer leaflet of the nuclear membrane, P-face of the plasma membrane, organelle membranes, nuclear pores, and cytoplasm.

high voltage microscope contributed to the latter paper.

3. Read the journal article by Porter and Tucker (1981) and discuss how the investigators tested their hypothesis of the existence of a microtrabecular lattice.

Chapter 17

1. Define a tracer by giving three of its applications.

2. Describe the different types of procedures, in this and other chapters, that have employed horseradish peroxidase.

3. What junctional types (tight, gap, and desmosome) would allow a tracer to pass through the junction and which would not? What junctional types would have their features highlighted by the tracer?

Chapter 18

1. Discuss the advantages and disadvantages of digital images versus conventionally produced photographic images.

2. What is a pixel? Discuss pixels per inch versus dots per inch.

3. Why would the output from a dye sublimation printer appear to be more photographic in appearance than a conventional laser printer because they both have the same degree of resolution (300 ppi)?

4. Discuss the difference between image processing and image analysis.

5. Suppose you have an SEM image containing electronic noise that occurs as a parallel series of lines spaced 1 mm apart. How could you remove or minimize the lines from an electronic image?

6. If one wishes to count a group of particles, some of which are touching each other or the edge of the image frame, how may one count only the individual, intact particles and exclude any of the touching particles?

Chapter 19

1. Go to the literature of electron microscopy, especially the earlier literature, and find examples of four of the artifacts described in this chapter.

2. In the section entitled "Survey of Biological Ultrastructure" (Chapter 20), attempt to approximate the magnifications of the figures based on the general description of cell size and membrane

appearance provided in this chapter. In the low-magnification range, your estimate should be equal or better than 50 to 70% of the actual magnification. At medium and higher magnification ranges, try to guess within 50% of the actual magnification.

3. In published micrographs, measure structures as described above, and determine how close your calculated magnification comes to the published magnification. You are doing well if you are within 50 to 100% of the actual magnification given for the published figure (assuming the published magnification is accurate).

4. Without referring to the text, list (1) as many specific artifacts that one might find on a micrograph as possible, (2) the problem each causes, and (3) the solution(s) to the problem.

5. The micrograph shown in Figure A.3 was produced by a student who had celebrated the evening before sectioning and staining the tissue. How many artifacts can be found, and what is the specific cause of each?

6. The scanning electron micrograph in Figure A.4 and the transmission electron micrograph in Figure A.5 are both the same organism (*Equisetum* or horsetail). Describe how the information obtained differs from the SEM to the TEM. What plane of section is the transmission electron micrograph relative to the scanning electron micrograph? (Courtesy of K. Renzaglia.)

Chapter 20

1. Fully interpret the micrograph in Figure A.5 using Chapters 19 and 20 as your guide.

Chapter 21

1. Perform a safety inspection on your research environment and indicate any potential problems that were discovered. Discuss how the problems will be remedied.

2. Outline which fire extinguishers are to be used for the various fires. Check the rooms in the electron microscope laboratory for the appropriateness of the fire extinguishers.

3. What are the advantages and disadvantages of the various fire extinguishers?

4. On an established fire, what steps should be taken to stop the fire? Arrange for a demonstration (or film to be shown) by the local fire department or institutional safety officers.

FIGURE A.3 Student micrograph.

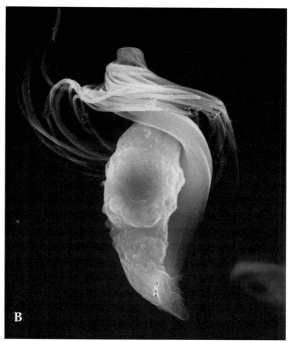

FIGURE A.4 SEM of *Equisetum* sperm. *(Courtesy of K. Renzaglia.)*

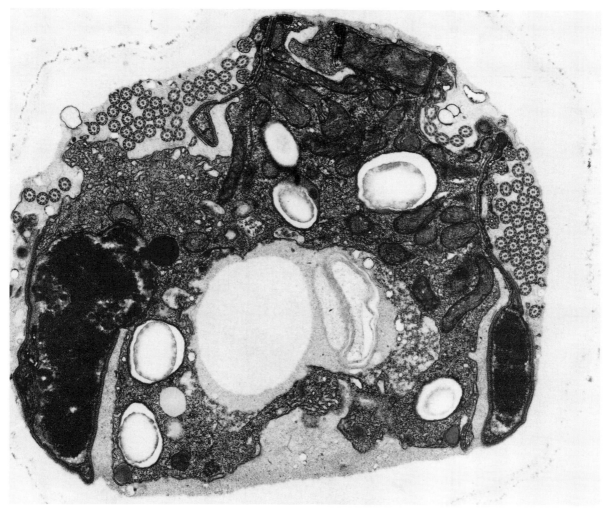

FIGURE A.5 Sectioned TEM image of *Equisetum* sperm. *(Courtesy of K. Renzaglia.)*

5. Study the material safety data sheets for the chemicals in use in the electron microscope laboratory. Are there any chemicals for which data sheets are not available? If so, obtain and post them in the laboratory.

6. Develop an evacuation plan in the event of a fire or accident involving chemicals used in the EM laboratory.

7. How does one deal with spilled osmium (salt and liquids), cacodylate buffer, propylene oxide, or epoxy resins? Develop a plan of action with the laboratory supervisor.

8. Make a list of chemicals that would be classified as fire hazards, toxins, or carcinogens.

9. Discuss the proper way of disposing of specific chemicals used in the EM laboratory.

10. Evaluate the fume hoods for proper functioning and the presence of clutter.

11. Evaluate several types of gloves for resistance to the commonly used solvents in the EM laboratory.

12. Arrange to meet with recyclers to discuss the reutilization of various laboratory reagents or apparatuses.

Index

Knives, ultramicrotome. *See* Ultramicrotomes
Knoll, Max, 7

Label-fracture technique, 357
Labels/labeling
    antibodies, 266–67
    blocking nonspecific, 276
    embedding, 39, 41
    gold, localization of cell surface receptors using, 229
    multiple, 277
    preembedding or postembedding, 275–76
    prints, 257–58
Lactoperoxidase, 411
Ladd, W., 5
Lamellipodia, 493
    definition of, 478
Lanthanum, 409, 410
Lanthanum hexaboride filaments, 170
Laplacian filter, 430
Latent image, 251, 296
Latent-image speck, 244
Lattice test, 199
Lead citrate (lead stains), 123, 127–29
    CO_2-free water for, 128
Lectins, localization techniques using, 313
Leica Knifemaker, 83, 84
Leica Trimmer, 78, 79
Length
    digital image analysis and, 439–40
Lenses
    enlarging, 252
    of light microscopes, 163
    of scanning electron microscopes, 205, 208–13
        final condenser lens, 208–13
        first and second condenser lenses, 208
    of transmission electron microscopes, 155–64
        condenser lenses, 171–73, 189–90
        defects in, 158–61
        design of electromagnetic lenses, 157–58
        focal length, 156–58
        general design, 156–57
        increasing strength of, 158
        intermediate (diffraction) lens, 178–79
        objective lens. *See* Objective lens
        projector lens, 179
        resolving power, 155–56
Leydig cells, 338–39
Light microscope, 4
    autoradiography using, 299
    localization experiments with, 274
    transmission electron microscope compared to, 163

Line scan analysis mode, 379
Lipid, 568
    definition of, 478
Lipid bilayer of plasmalemma, 480–82
Lipid droplets, 568
Lipochrome pigment, 558
Lipofuscin granule, 558
LKB (Leica) ultramicrotome, 98
    cryoultramicrotomy kit for, 110
LKB Pyramitome, 78
Localization techniques. *See also*
        Autoradiography/radioautography;
        Enzyme cytochemistry;
        Immunocytochemistry
    for actin, 312
    for carbohydrates and oligosaccharides, 313–14
    for glycogen and membranes, 315
    for Golgi complex/multivesicular body, 315
    for ions, 315
    for nucleic acids (DNA/RNA), 317
    for protein, 317
    for sterols, 317
Low angle diffraction, 394
Lowicryl acrylic resin, 31
Low pass filter, 425
LR White embedding resin, 36–37, 38
    in immunocytochemical experiments, 277
    rapid method for processing tissues
        embedded with, 44
Lumen, 537
Lysosomal apparatus, definition of, 478
Lysosomes, 556–58
    definition of, 478

Macula adherens, 483, 486
    definition of, 478
Magnesium chloride, 23
Magnification
    estimation of, 474–75
    interpretation of micrographs and, 446–47
    of micrographs, 260
    with scanning electron microscopes, 209
        calibration, 236–37
    with transmission electron microscopes
        calibration, 197–98
        checking, 197
        chromatic change, 159, 160
        concept of, 150–51
        definition, 162
        focal length, 157–58
        useful, 162